Mechanical Properties and Engineering Applications of Special Soils

Mechanical Properties and Engineering Applications of Special Soils

Guest Editors

Xianwei Zhang
Xinyu Liu
Ran An

MDPI

Basel • Beijing • Wuhan • Barcelona • Belgrade • Novi Sad • Cluj • Manchester

Guest Editors
Xianwei Zhang
State Key Laboratory of Geomechanics and Geotechnical Engineering
Chinese Academy of Sciences
Wuhan
China

Xinyu Liu
School of Civil and Hydraulic Engineering
Huazhong University of Science and Technology
Wuhan
China

Ran An
School of Civil Engineering
Hefei University of Technology
Hefei
China

Editorial Office
MDPI AG
Grosspeteranlage 5
4052 Basel, Switzerland

This is a reprint of the Special Issue, published open access by the journal *Applied Sciences* (ISSN 2076-3417), freely accessible at: https://www.mdpi.com/journal/applsci/special_issues/CWJ6TMM41N.

For citation purposes, cite each article independently as indicated on the article page online and as indicated below:

Lastname, A.A.; Lastname, B.B. Article Title. *Journal Name* **Year**, *Volume Number*, Page Range.

ISBN 978-3-7258-5353-3 (Hbk)
ISBN 978-3-7258-5354-0 (PDF)
https://doi.org/10.3390/books978-3-7258-5354-0

Contents

About the Editors . . . . **vii**

Ran An, Xinyu Liu and Xianwei Zhang
Editorial: Mechanical Properties and Engineering Applications of Special Soils
Reprinted from: *Appl. Sci.* **2024**, *14*, 10447, https://doi.org/10.3390/app142210447 . . . . **1**

Armen Z. Ter-Martirosyan, Lyubov Yu. Ermoshina and George O. Anzhelo
Viscosity of Clayey Soils: Experimental Studies
Reprinted from: *Appl. Sci.* **2024**, *14*, 5974, https://doi.org/10.3390/app14145974 . . . . **7**

Zhongnian Yang, Rongchang Wang, Wei Shi, Zhenxing Sun and Xianzhang Ling
Investigation of the Shear and Pore Structure Characteristics of Rubber Fiber-Reinforced Expansive Soil
Reprinted from: *Appl. Sci.* **2024**, *14*, 5794, https://doi.org/10.3390/app14135794 . . . . **22**

Tiecheng Yan, Xingyuan Zhang, Sutong Cai, Zefeng Zhou, Ran An and Xianwei Zhang
Mechanical Characteristics and Damage Constitutive Model of Fiber-Reinforced Cement-Stabilized Soft Clay
Reprinted from: *Appl. Sci.* **2024**, *14*, 1378, https://doi.org/10.3390/app14041378 . . . . **39**

Muawia Dafalla
Predicting Swell in Clay-Sand Mixtures Used in Liners
Reprinted from: *Appl. Sci.* **2023**, *13*, 11161, https://doi.org/10.3390/app132011161 . . . . **52**

José Manuel Moreno-Maroto, Óscar Merlo, Joel Torres-Serra, Jacinto Alonso-Azcárate, Mark Tyrer and Vicente Navarro
Determining the Macrostructural Stability of Compacted Wyoming Bentonites by a Disaggregation Method
Reprinted from: *Appl. Sci.* **2023**, *13*, 11159, https://doi.org/10.3390/app132011159 . . . . **66**

Lei Zhu, Ying Guo, Gan Cheng and Xiangyang Liu
Research on the Accumulated Plastic Strain of Expansive Soil under Subway Loading
Reprinted from: *Appl. Sci.* **2023**, *13*, 9994, https://doi.org/10.3390/app13189994 . . . . **78**

Bilge Aksu Alcan and Semet Çelik
The Effect of Different Fiber Reinforcement on Bearing Capacity under Strip Foundation on the Sand Soil: An Experimental Investigation
Reprinted from: *Appl. Sci.* **2023**, *13*, 9769, https://doi.org/10.3390/app13179769 . . . . **94**

Bo Liu, Danguang Pan, Wanying Zhang, Yankai Lu and Guowei Ni
Mechanical Behaviors and Numerical Simulation Analysis of a New Isolation System
Reprinted from: *Appl. Sci.* **2023**, *13*, 7667, https://doi.org/10.3390/app13137667 . . . . **123**

Jinjin Liu, Chuanqin Yao, Wenbo Su and Yizhe Zhao
Research Progress on the Influence of Thermo-Chemical Effects on the Swelling Pressure of Bentonite
Reprinted from: *Appl. Sci.* **2023**, *13*, 5580, https://doi.org/10.3390/app13095580 . . . . **137**

Yifeng Zhou, Xing Xiao, Zhenglong Zhou and Qi Wu
Experimental Study on Maximum Dynamic Shear Modulus of Yangtze River Overconsolidated Floodplain Soft Soils
Reprinted from: *Appl. Sci.* **2023**, *13*, 4733, https://doi.org/10.3390/app13084733 . . . . **155**

Pravin Badarayani, Bogdan Cazacliu, Erdin Ibraim, Riccardo Artoni and Patrick Richard
Sand Rubber Mixtures under Oedometric Loading: Sand-like vs. Rubber-like Behavior
Reprinted from: *Appl. Sci.* **2023**, *13*, 3867, https://doi.org/10.3390/app13063867 . . . . **166**

Yiqing Xu, Xianwei Zhang, Gang Wang, Xinyu Liu and Aiwu Yang
Role of Diatom Microstructure in Determining the Atterberg Limits of Fine-Grained Diatomaceous Soil
Reprinted from: *Appl. Sci.* **2023**, *13*, 2287, https://doi.org/10.3390/app13042287 . . . . **179**

Xuejun Chen, Xiaotong Gao, Hui Li, Mingming Xue, Xiaohui Gan and Yu Song
Model Test Analysis of Groundwater Level Fluctuations on Karst Cover Deformation Taking the Monolithic Structure of Guilin as an Example
Reprinted from: *Appl. Sci.* **2023**, *13*, 1747, https://doi.org/10.3390/app13031747 . . . . **197**

Mehmet Ugur Yilmazoglu and Askin Ozocak
Bearing Capacity of Shallow Foundations on Unsaturated Silty Soils
Reprinted from: *Appl. Sci.* **2023**, *13*, 1308, https://doi.org/10.3390/app13031308 . . . . **214**

Daniel Zuluaga-Astudillo, Juan Carlos Ruge, Javier Camacho-Tauta, Oscar Reyes-Ortiz and Bernardo Caicedo-Hormaza
Diatomaceous Soils and Advances in Geotechnical Engineering—Part I
Reprinted from: *Appl. Sci.* **2023**, *13*, 549, https://doi.org/10.3390/app13010549 . . . . **236**

Qian Wang, Yan Wang, Wenguo Ma and Dongwang Tao
Dynamic Characteristics of Post-Cyclic Saturated Loess
Reprinted from: *Appl. Sci.* **2023**, *13*, 306, https://doi.org/10.3390/app13010306 . . . . **271**

Xinxin Dong, Xiaohua Bao, Hongzhi Cui, Changjie Xu and Xiangsheng Chen
Effects of Cement Treatment on Mechanical Properties and Microstructure of a Granite Residual Soil
Reprinted from: *Appl. Sci.* **2022**, *12*, 12549, https://doi.org/10.3390/app122412549 . . . . **292**

Xinlei Zhang, Jun Guo, Yumin Chen, Yi Han, Ruibo Yi, Hongmei Gao, et al.
Mechanical Properties and Engineering Applications of Special Soils—Dynamic Shear Modulus and Damping of MICP-Treated Calcareous Sand at Low Strains
Reprinted from: *Appl. Sci.* **2022**, *12*, 12175, https://doi.org/10.3390/app122312175 . . . . **313**

Honghu Liu, Jing Liu, Xianwei Zhang and Xinyu Liu
The Shrink–Swell Process of the Granite Residual Soil with Different Weathering Degree in a Gully System in Southern China
Reprinted from: *Appl. Sci.* **2022**, *12*, 11200, https://doi.org/10.3390/app122111200 . . . . **331**

Jianqiang Wang, Yu Song, Shuaishuai Dong, Song Ding, Yukun Geng and Xiaotong Gao
Triaxial Experimental Study of Zinc Contaminated Red Clay under Different Temperature Conditions
Reprinted from: *Appl. Sci.* **2022**, *12*, 10742, https://doi.org/10.3390/app122110742 . . . . **342**

Yu Song, Song Ding, Yukun Geng, Shuaishuai Dong and Hongbin Chen
Study on the Mechanical Strength of Calcium Oxide-Calcium Phosphate Cured Heavy Metal Zinc Contaminated Red Clay Soil
Reprinted from: *Appl. Sci.* **2022**, *12*, 10041, https://doi.org/10.3390/app121910041 . . . . **356**

Yu Song, Shuaishuai Dong, Song Ding, Yukun Geng, Yaoyi Huang and Hui Li
Effect of Lead Ion Contamination on the Microstructure of Guilin Red Clay
Reprinted from: *Appl. Sci.* **2022**, *12*, 9888, https://doi.org/10.3390/app12199888 . . . . **368**

About the Editors

Xianwei Zhang

Xianwei Zhang is a Professor at the State Key Laboratory of Geomechanics and Geotechnical Engineering Safety, Institute of Rock and Soil Mechanics, Chinese Academy of Sciences. He obtained both his BSc and PhD degrees from Jinlin University, China. He has extensive experience in research within the fields of geotechnical engineering and engineering geology. Dr. Zhang has now been in charge of more than twenty research projects, including five that have been supported by the National Natural Science Foundation of China. He has published over 100 research papers as the first or corresponding author. He is an editorial board member for three international journals.

Xinyu Liu

Xinyu Liu is currently a research fellow at the School of Civil and Environmental Engineering, Nanyang Technological University, Singapore. His research works include advanced experimental investigations of the geotechnical behavior of various types of problematic soils. As the first or corresponding author, he has published more than 40 journal papers, including those in Géotechnique, the Journal of Geotechnical and Geoenvironmental Engineering, and the Canadian Geotechnical Journal. Dr. Liu has been funded by the National Natural Science Foundation of China, the China Postdoctoral Foundation, and the State Key Laboratories of China.

Ran An

Ran An obtained a bachelor's degree in civil engineering from Hefei University of Technology in 2015 and a doctorate in geotechnical engineering in 2020. Ran An is an associate professor in the Department of Civil Engineering at Hefei University of Technology and is also a member of the International Society for Soil Mechanics and Geotechnical Engineering (ISSMGE). Ran An mainly focuses on the prevention and control of special soil mechanics and geological disasters. Ran An has participated in over ten scientific research projects, among which he served as the project leader for seven. Up to now, Ran An has published 68 scientific research papers, of which 32 have been indexed by SCI/EI. Also, Ran An is a reviewer for more than 30 international journals of geotechnical engineering.

applied sciences

MDPI

Editorial

Editorial: Mechanical Properties and Engineering Applications of Special Soils

Ran An [1], Xinyu Liu [2] and Xianwei Zhang [3,4,*]

1 School of Civil and Hydraulic Engineering, Hefei University of Technology, Hefei 230009, China; anran@wust.edu.cn
2 School of Civil and Hydraulic Engineering, Huazhong University of Science and Technology, Wuhan 430074, China; liuxinyu17@mails.ucas.ac.cn
3 State Key Laboratory of Geomechanics and Geotechnical Engineering, Institute of Rock and Soil Mechanics, Chinese Academy of Sciences, Wuhan 430071, China
4 School of Engineering Science, University of Chinese Academy of Sciences, Beijing 100049, China
* Correspondence: xwzhang@whrsm.ac.cn

1. Introduction

Special soils refer to the soils with unique physical, mechanical, and engineering properties that are developed under specific geological environments and weathering conditions [1]. Generally, the characteristics of special soils, such as soft soil, loess, residual soil, expansive soil, frozen soil, moraines, and contaminated soil, present significant challenges in geotechnical design and construction on a global scale. In comparison to common sedimentary clay and sand, the mechanical properties of special soil exhibit significant differences, thereby introducing considerable complexities in engineering applications [2]. A comprehensive understanding of the characteristics of special soils is essential to ensuring safe and cost-effective engineering practices. Figure 1 provides typical examples of engineering geological disasters associated with different types of special soils [3–8]. It is evident that the soil properties pose a significant risk to the safety of engineering structures and construction. Based on a multitude of engineering case studies, engineering accidents associated with special soils differ from those in homogeneous soils primarily due to the inherent anisotropy and inhomogeneity. In particular, the mechanical properties of special soils play a crucial role in shaping the progression of geological disasters and engineering accidents.

To ensure the stability and safety of construction, special attention should be devoted to the adverse engineering characteristics of special soils. Moreover, appropriate engineering measures must be implemented for soil treatment and reinforcement. For engineering applications, it is essential to implement corresponding foundation treatment methods, such as replacement, compaction, grouting, and additional measures, in order to enhance the bearing capacity and stability of the foundation [9]. During the process of building design and construction, it is imperative to adequately consider the heterogeneity and take appropriate structural measures to mitigate the impact of heterogeneous settlement on the building [10]. Furthermore, as a result of advancements in materials science, an increasing number of novel materials are being utilized for the treatment and enhancement of special soils. For instance, nanomaterials and fiber-reinforced materials have demonstrated promising potential in enhancing soil's strength and durability [11,12].

The Special Issue "Mechanical Properties and Engineering Applications of Special Soils" provided a significant opportunity for pioneering advancements, thereby fostering the progression of soil mechanics research into a new stage. Global collaboration among researchers is essential for exploring the intricacies of specialized soil and driving the progress of soil mechanics and sustainable development in engineering geology. Throughout this process, it is essential to not only focus on the leading trends in soil mechanics theory but also emphasize the translation of research findings into practical engineering applications to contribute technical guidance towards modern construction.

Figure 1. Geological hazards of different types of special soil: (**a**) foundation settlement of diatomite in Colombia; (**b**) gully erosion of residual soil in South Africa; (**c**) cracking of frozen soil foundation in the Tibet Plateau; (**d**) cracking of expansive soil in China; (**e**) foundation pit collapse of soft soil in China; (**f**) earthquake liquefaction of sand in Japan.

2. An Overview of the Published Articles

The first published paper, from Ter-Martirosyan et al. (contribution 1), aimed to investigate the effect of shear rate on the viscosity coefficient and strength characteristics of clayey soil. The findings held potential for enhancing the accuracy of settlement calculations for structures over time.

The study proposed by Yang et al. (contribution 2) examined the impact of rubber fibers on the shear strength and pore structure characteristics of expansive soil. The research findings, including optimal amounts of rubber fibers, soil porosity, and maximum strength, hold significant promise for applications of expansive soil.

From the third paper, Yan et al. (contribution 3) improved the shear strength of cemented soft clay by fiber reinforcement and revealed its mechanism. This study contributed to the advancement of high-performance soft soil, which is essential for ensuring the stability of slopes and foundations.

The study by Dafalla et al. (contribution 4) presents a novel approach for predicting the swelling properties of clay–sand mixtures. It offered insights into a method of predicting the expansion of clay–sand mixtures commonly used in liners.

Moreno-Maroto et al.'s research (contribution 5) focused on examining the overall structural stability of compacted bentonite. The findings from the experiment have important implications for future studies involving various materials and testing conditions such as compaction and moisture levels.

The sixth paper submitted by Zhu et al. (contribution 6) investigated the impact of different loading methods on the dynamic strain of saturated expansive soil. The findings hold significant implications for the design and construction of subway and underground structures in Hefei City.

The research carried out by Alcan et al. (contribution 7) sought to examine how various fiber reinforcements affect the load-bearing capacity of sand in the settlement of strip foundations. The results demonstrated that the reinforcement effectively redistributes stress generated by the special soil over a wider area.

Liu et al.'s study (contribution 8) discussed the mechanical properties of sliding and rolling friction systems and their application in vibration isolation frame structures. This study held great significance for the investigation of novel vibration isolation mechanisms.

The focus of Zhou et al.'s study (contribution 9) was the undisturbed soft soil in the Yangtze River floodplain. The findings served as valuable raw data for informing engineering construction requirements within the floodplain.

The paper submitted by Badarayani et al. (contribution 10) was focused on the one-dimensional loading/unloading behavior of sand–rubber mixtures. A new approach has been proposed for the reutilization of extensive amounts of discarded tires.

The primary focus of the study conducted by Xu et al. (contribution 11) was to investigate the Atterberg limit of diatomaceous soil. The findings contributed to a deeper understanding of the Atterberg limit of diatomaceous soil and offer insights for its classification.

Chen et al.'s research (contribution 12) examined the engineering practice and practical case of a karst site in Guilin with complex geological conditions, especially regarding groundwater. The findings offered references for the prevention and management of karst collapses.

Yilmazoglu and colleagues (contribution 13) focused on studying the maximum load-bearing capacity of shallow foundations on sandy soil. This paper represented a significant advancement in the characterization of the ultimate bearing capacity of shallow foundations situated on unsaturated sandy soil.

Wang et al. (contribution 14) discuss the nonlinear dynamic characteristics of loess after earthquakes. Results of the study were applied to the engineering design, which reduced the damage to soil structure caused by strong earthquakes.

Dong et al.'s work (contribution 15) was devoted to conducting macroscopic and microscopic evaluations of granite residual soil treated by cement. This study has provided valuable insights for the proper solidification and management of residual soil in geotechnical engineering practice.

The study by Zhang et al. (contribution 16) delved into the dynamic characteristics of calcareous sand treated with MICP technique. The findings of this investigation offered a quantitative analysis of the dynamic parameters of cemented sand.

The study conducted by Liu et al. (contribution 17) examined the contraction and expansion characteristics of weathered granite soil in southern China, as well as its influence on gully formation. The findings offered new perspectives on the process of gully erosion.

Wang et al.'s study (contribution 18) investigates the influence of temperature and zinc pollution concentration on the mechanical properties of red clay. These findings provided a solid foundation for the management of heavy metal-contaminated soil in Guilin, a city in southern China.

In the research conducted by Song et al. (contribution 19), the focus was on analyzing the shear strength and electrical resistivity of red clay using Zn^{2+} as the subject of investigation. The results of this study provided valuable insights into soil remediation approaches for sites contaminated with heavy metals.

The paper submitted by Song et al. (contribution 20) investigated the impact of lead ion contamination on the microstructural features of red clay. The research findings had significant implications for addressing heavy metal pollution in red clay.

Liu et al.'s study (contribution 21) offered a comprehensive overview of the research findings on the variations in swelling pressure of bentonite during thermochemical reactions, emphasizing discrepancies in swelling pressure values obtained through different testing methods.

The paper written by Zuluaga-Astudillo et al. (contribution 22) analyzed multiple sets of results regarding the physico-mechanical index of diatomaceous soil. The findings of this study highlight the complex geotechnical properties and significant potential of diatomaceous soil.

The 22 papers selected for publication in this Special Issue hve greatly advanced research on the mechanical properties of special soils and their engineering applications.

3. Conclusions

The mechanical properties of special soil are intricate and variable, exerting a significant influence on the stability and safety of engineering projects. Through in-depth investigation into its mechanical properties, implementation of effective engineering measures, and utilization of novel materials, the potential advantages of special soil in engineering applications can be completely realized, thereby enhancing engineering quality and safety. As science and technology continue to advance, there will be broader and deeper exploration into the mechanical properties and engineering applications of special soil. To further facilitate global scholarly exchange in this field, we are pleased to announce the upcoming release of a second issue dedicated to this topic.

We would like to extend our sincere appreciation to all the authors who have made significant contributions to this Special Issue, as well as to the editors of *Applied Sciences* for their diligent efforts in promoting this Special Issue.

Author Contributions: Conceptualization, X.Z. and R.A.; validation, X.L. and X.Z.; formal analysis, R.A.; writing—review and editing, X.Z. and X.L.; funding acquisition, X.Z. and R.A. All authors have read and agreed to the published version of the manuscript.

Funding: This research received no external funding.

Acknowledgments: We sincerely thank all the authors who contributed valuable manuscripts to this Special Issue.

Conflicts of Interest: The authors declare no conflicts of interest.

List of Contributions

1. Ter-Martirosyan, A.Z.; Ermoshina, L.Y.; Anzhelo, G.O. Viscosity of Clayey Soils: Experimental Studies. *Appl. Sci.* **2024**, *14*, 5974. https://doi.org/10.3390/app14145974
2. Yang, Z.; Wang, R.; Shi, W.; Sun, Z.; Ling, X. Investigation of the Shear and Pore Structure Characteristics of Rubber Fiber-Reinforced Expansive Soil. *Appl. Sci.* **2024**, *14*, 5794. https://doi.org/10.3390/app14135794
3. Yan, T.; Zhang, X.; Cai, S.; Zhou, Z.; An, R.; Zhang, X. Mechanical Characteristics and Damage Constitutive Model of Fiber-Reinforced Cement-Stabilized Soft Clay. *Appl. Sci.* **2024**, *14*, 1378. https://doi.org/10.3390/app14041378
4. Dafalla, M. Predicting Swell in Clay-Sand Mixtures Used in Liners. *Appl. Sci.* **2023**, *13*, 11161. https://doi.org/10.3390/app132011161
5. Moreno-Maroto, J.M.; Merlo, Ó.; Torres-Serra, J.; Alonso-Azcárate, J.; Tyrer, M.; Navarro, V. Determining the Macrostructural Stability of Compacted Wyoming Bentonites by a Disaggregation Method. *Appl. Sci.* **2023**, *13*, 11159. https://doi.org/10.3390/app132011159
6. Zhu, L.; Guo, Y.; Cheng, G.; Liu, X. Research on the Accumulated Plastic Strain of Expansive Soil under Subway Loading. *Appl. Sci.* **2023**, *13*, 9994. https://doi.org/10.3390/app13189994
7. Aksu Alcan, B.; Çelik, S. The Effect of Different Fiber Reinforcement on Bearing Capacity under Strip Foundation on the Sand Soil: An Experimental Investigation. *Appl. Sci.* **2023**, *13*, 9769. https://doi.org/10.3390/app13179769
8. Liu, B.; Pan, D.; Zhang, W.; Lu, Y.; Ni, G. Mechanical Behaviors and Numerical Simulation Analysis of a New Isolation System. *Appl. Sci.* **2023**, *13*, 7667. https://doi.org/10.3390/app13137667
9. Zhou, Y.; Xiao, X.; Zhou, Z.; Wu, Q. Experimental Study on Maximum Dynamic Shear Modulus of Yangtze River Overconsolidated Floodplain Soft Soils. *Appl. Sci.* **2023**, *13*, 4733. https://doi.org/10.3390/app13084733
10. Badarayani, P.; Cazacliu, B.; Ibraim, E.; Artoni, R.; Richard, P. Sand Rubber Mixtures under Oedometric Loading: Sand-like vs. Rubber-like Behavior. *Appl. Sci.* **2023**, *13*, 3867. https://doi.org/10.3390/app13063867

11. Xu, Y.; Zhang, X.; Wang, G.; Liu, X.; Yang, A. Role of Diatom Microstructure in Determining the Atterberg Limits of Fine-Grained Diatomaceous Soil. *Appl. Sci.* **2023**, *13*, 2287. https://doi.org/10.3390/app13042287
12. Chen, X.; Gao, X.; Li, H.; Xue, M.; Gan, X.; Song, Y. Model Test Analysis of Groundwater Level Fluctuations on Karst Cover Deformation Taking the Monolithic Structure of Guilin as an Example. *Appl. Sci.* **2023**, *13*, 1747. https://doi.org/10.3390/app13031747
13. Yilmazoglu, M.U.; Ozocak, A. Bearing Capacity of Shallow Foundations on Unsaturated Silty Soils. *Appl. Sci.* **2023**, *13*, 1308. https://doi.org/10.3390/app13031308
14. Wang, Q.; Wang, Y.; Ma, W.; Tao, D. Dynamic Characteristics of Post-Cyclic Saturated Loess. *Appl. Sci.* **2023**, *13*, 306. https://doi.org/10.3390/app13010306
15. Dong, X.; Bao, X.; Cui, H.; Xu, C.; Chen, X. Effects of Cement Treatment on Mechanical Properties and Microstructure of a Granite Residual Soil. *Appl. Sci.* **2022**, *12*, 12549. https://doi.org/10.3390/app122412549
16. Zhang, X.; Guo, J.; Chen, Y.; Han, Y.; Yi, R.; Gao, H.; Liu, L.; Liu, H.; Shen, Z. Mechanical Properties and Engineering Applications of Special Soils—Dynamic Shear Modulus and Damping of MICP-Treated Calcareous Sand at Low Strains. *Appl. Sci.* **2022**, *12*, 12175. https://doi.org/10.3390/app122312175
17. Liu, H.; Liu, J.; Zhang, X.; Liu, X. The Shrink–Swell Process of the Granite Residual Soil with Different Weathering Degree in a Gully System in Southern China. *Appl. Sci.* **2022**, *12*, 11200. https://doi.org/10.3390/app122111200
18. Wang, J.; Song, Y.; Dong, S.; Ding, S.; Geng, Y.; Gao, X. Triaxial Experimental Study of Zinc Contaminated Red Clay under Different Temperature Conditions. *Appl. Sci.* **2022**, *12*, 10742. https://doi.org/10.3390/app122110742
19. Song, Y.; Ding, S.; Geng, Y.; Dong, S.; Chen, H. Study on the Mechanical Strength of Calcium Oxide-Calcium Phosphate Cured Heavy Metal Zinc Contaminated Red Clay Soil. *Appl. Sci.* **2022**, *12*, 10041. https://doi.org/10.3390/app121910041
20. Song, Y.; Dong, S.; Ding, S.; Geng, Y.; Huang, Y.; Li, H. Effect of Lead Ion Contamination on the Microstructure of Guilin Red Clay. *Appl. Sci.* **2022**, *12*, 9888. https://doi.org/10.3390/app12199888
21. Liu, J.; Yao, C.; Su, W.; Zhao, Y. Research Progress on the Influence of Thermo-Chemical Effects on the Swelling Pressure of Bentonite. *Appl. Sci.* **2023**, *13*, 5580. https://doi.org/10.3390/app13095580
22. Zuluaga-Astudillo, D.; Ruge, J.C.; Camacho-Tauta, J.; Reyes-Ortiz, O.; Caicedo-Hormaza, B. Diatomaceous Soils and Advances in Geotechnical Engineering—Part I. *Appl. Sci.* **2023**, *13*, 549. https://doi.org/10.3390/app13010549

References

1. Zhang, X.; Liu, X.; Xu, Y.; Wang, G.; Wang, F. Swelling–shrinkage mechanisms of diatomaceous soil and its geohazard implication. *Eng. Geol.* **2023**, *314*, 107009. [CrossRef]
2. Caicedo, B.; Mendoza, C.; López, F.; Lizcano, A. Behavior of diatomaceous soil in lacustrine deposits of Bogotá, Colombia. *J. Rock Mech. Geotech. Eng.* **2018**, *10*, 367–379. [CrossRef]
3. Olivier, G.; Van de Wiel, M.J.; De Clercq, W.P. Intersecting views of gully erosion in South Africa. *Earth Surf. Proc. Land* **2023**, *48*, 119–142. [CrossRef]
4. Luo, X.; Yu, Q.; Ma, Q.; Guo, L. Study on the heat and deformation characteristics of an expressway embankment with shady and sunny slopes in warm and ice-rich perma-frost regions. *Transp. Geotech.* **2020**, *24*, 100390. [CrossRef]
5. Gao, H.; An, R.; Zhang, X.; Wang, G.; Liu, X.; Xu, Y. Dynamic evolution of desiccation cracks and their relationship with the hydraulic properties of expansive soil. *Int. J. Geomech.* **2024**, *24*, 04023299. [CrossRef]
6. Guo, P.; Gong, X.; Wang, Y.; Lin, H.; Zhao, Y. Analysis of observed performance of a deep excavation straddled by shallowly buried pressurized pipelines and underneath traversed by planned tunnels. *Tunn. Undergr. Space Technol.* **2023**, *132*, 104946. [CrossRef]
7. Gao, W.; Ge, S.; Gao, Y.; Yuan, S. Prediction of utility tunnel performance in a soft foundation during an operation period based on deep learning. *Appl. Sci.* **2024**, *14*, 2334. [CrossRef]
8. An, R.; Wang, Y.; Zhang, X.; Chen, C.; Liu, X.; Cai, S. Quantitative characterization of drying-induced cracks and permeability of granite residual soil using micron-sized X-ray computed tomography. *Sci. Total Environ.* **2023**, *876*, 163213. [CrossRef] [PubMed]
9. Liu, Y.; Li, X.; Tu, Y.; Lu, Y. Mining Leachates Effect on the Hydraulic Performance of Geosynthetic Clay Liners under Different Temperatures. *Water* **2023**, *15*, 1132. [CrossRef]

10. Hu, Y.N.; Ji, J.; Sun, Z.B.; Dias, D. First order reliability-based design optimization of 3D pile-reinforced slopes with Pareto optimality. *Comput. Geotech.* **2023**, *162*, 105635. [CrossRef]
11. Zhao, Y.; Ling, X.; Gong, W.; Li, P.; Li, G.Y.; Wang, L.N. Mechanical Properties of Fiber-Reinforced Soil under Triaxial Compression and Parameter Determination Based on the Duncan-Chang Model. *Appl. Sci.* **2020**, *10*, 9043. [CrossRef]
12. Zhang, D.; Cao, Z.; Fan, L. Evaluation of the influence of salt concentration on cement stabilized clay by electrical resistivity measurement method. *Eng. Geol.* **2014**, *170*, 80–88. [CrossRef]

applied sciences

Article

Viscosity of Clayey Soils: Experimental Studies

Armen Z. Ter-Martirosyan *, Lyubov Yu. Ermoshina * and George O. Anzhelo

Department of Soil Mechanics and Geotechnical Engineering, National Research Moscow State Civil Engineering University, 26, Yaroslavskoye Shosse, 129337 Moscow, Russia; nocgeo@mail.ru
* Correspondence: gic-mgsu@mail.ru (A.Z.T.-M.); lyubov.ermoshina1801@mail.ru (L.Y.E.)

Abstract: Due to the high rate of the development of housing, transportation and hydraulic engineering construction in the last hundred years, the study of the phenomenon of creep of clay soils has become a subject of scientific research. In the study, experimental investigations of clay soil were conducted using a simple shear device in kinematic loading mode, aimed at examining the influence of shear rate on the viscosity coefficient of the clay soil and its strength characteristics. The tests were performed at four different shear rates and three different vertical load values. Based on the results of experimental and theoretical studies, the viscosity coefficients of clay soil were obtained, and a new rheological equation was proposed, which simultaneously takes into account the influence of Coulomb friction, structural cohesion, cohesion of water–colloidal bonds and viscous resistance of the soil. It has been shown that the shear rate has a significant impact on the viscosity coefficient of clay soil, and the viscosity coefficient itself is a variable quantity, depending both on the magnitude of the applied load and the duration of its application. The obtained results can be used for further improvement of methods for calculating the settlement of structures over time, as well as for predicting the time until the bearing capacity of foundation soils is exhausted.

Keywords: simple shear device; kinematic loading mode; shear rate; viscosity coefficient; viscous resistance; clay soil

1. Introduction

Over the last hundred years, the study of the phenomenon of the creep of clay soils has become a subject of scientific research in connection with the active development of housing, transportation and hydraulic engineering construction. The increased attention of scientists to the creep process occurring in clay soils was caused by the fact that in a number of cases, unacceptably large deformations of various buildings, structures and roads were observed, disturbing their normal operation, and in some cases, leading to destruction.

The process of the creep of clay soils in nature manifests itself in the form of long-term settlements, displacements, rolls, horizontal displacements of structures, as well as in the form of the slow sliding of clay masses along natural slopes and slopes of structures made of soil materials.

The historical development of soil rheology began with the study of the shear creep deformations of soils. Vyalov S.S. [1] mentioned in his book that the phenomenon of shear creep is caused by two mutually opposite processes: the process of hardening (the process of damped creep causes a decrease in shear strains) and the process of soil unstrengthening (the development of the process of non-damped creep causes an increase in shear strains).

Goldshtejn M.N. believed that in the process of shear both destruction and restoration of contacts between mineral particles in the skeleton of clayey soil occur [2].

Maslov N.N. proposed to take into account the change in the viscosity of the soil skeleton in time in order to take into account the hardening of clayey soil under shear [3,4].

Based on the results of experimental studies of clay in the ring shear device, Ter-Stepanyan G.I. suggested that shear strains change discontinuously during testing under constant load [5].

To describe the process of soil creep, Meschyan S.R. [6–8], Goldin A.L. [9] and Ter-Martirosyan Z.G. [10] applied the Boltzmann–Volterra theory of hereditary creep. This theory as applied to soils was first used by Florin V.A. [11,12] and was subsequently developed by Maslov N.N. and Arutyunyan N.Kh. [13] to describe the creep process of hardening media (in particular, concrete), which made it possible to take into account the hardening of soils in the process of creep.

A great contribution to the development of soil rheology was made by Tertsagi K., Puzyrevsky N.P., Gersevanov N.M., Vyalov S.S., Maslov N.N., Florin V.A., Cytovich N.A., Zareckij Y.K., Goldshtejn M.N., Goldin A.L., Meschyan S.R., Ter-Stepanyan G.I., Shukle L., Rzhanitsyn A.R., Arutyunyan N.H., Ter-Martirosyan Z.G. and many other researchers [1–18].

Currently, a large number of scientific works are devoted to the study of the rheological properties of clay, sandy, melted and frozen soils, in which the authors derive various equations of state describing the rheological properties of soil on the basis of theoretical, experimental and numerical studies.

Based on the performed experimental studies of clayey soil in the direct shear apparatus, Boley K. and Strokova L.A. [19] showed in their work that the time of action of compacting load before shear tests has a significant effect on the structure and properties of clays, which undoubtedly affects the strength and creep indices of clayey soil.

A hypoplastic model (a model in which there is no clear division into elastic and plastic deformation zone) for clayey soil, which includes eight parameters that can be determined in laboratory conditions on the basis of performed experiments, was proposed by Tafil and Triantafyllidis [20].

A nonlinear creep model using a fractional order derivative and a small number of parameters capable of describing the creep process of clay was proposed by Ren et al. [21]. The authors compared the proposed creep model with the Bürgers model describing the behavior of viscoelastic materials. In this work, experimental soil investigations were also conducted to study the nonlinear creep properties of clay.

A nonlinear creep model for frozen sand was proposed in [22]. The authors of this paper have validated this model based on the results of laboratory tests on frozen sand. Tests were carried out on sand samples with different particle size distribution and density in dry conditions and at different levels of shear stress and temperature.

A model reflecting the nonlinear rheological characteristics of swelling soils was proposed in [23]. The results of the performed studies showed that the proposed model well describes the viscoelastic-plastic creep deformation of the swelling soil, and the results of the analysis of this model were close to the results obtained from the performed laboratory tests of the soil in triaxial compression devices.

A viscoplastic model that is capable of describing creep and plastic deformation of rock during loading–unloading cycles was proposed by Haghighat et al. [24]. The authors of this paper performed a validation of this model based on the results obtained from triaxial tests.

A seven-component rheological model to describe the behavior of weak soil with organic matter content under different loading conditions was proposed in [25]. The model used parallel three-element components to describe the viscosity, plasticity, and elastic deformation of the soil over time.

The results of scientific studies of rheological properties of sandy soils are presented in the work of Ter-Martirosyan Z.G. et al. [26]. The authors performed experimental studies to determine the viscosity of sandy soil by performing triaxial tests in kinematic and dynamic loading modes. The paper also considers the problem of evaluating the effect of dynamic loads on the viscosity of sandy soils.

A three-dimensional elastic viscoplastic state model that is capable of describing the behavior of soils with isotachy viscosity in time was proposed in [27]. The authors of the paper gave a rather detailed description of their model and its basic equations, as well as performed the implementation of this model in a software package using the finite element method. The model was then used to simulate the results of laboratory tests, paying particular attention to the advantages and disadvantages of the model in reproducing the behavior of soft clays over time.

Studies of clay soils containing the clay mineral montmorillonite, which has pronounced creep and swelling properties, were presented by Yin and Tong [28]. Based on the data of experimental studies of clayey soil samples consisting of a mixture of bentonite and silica sand, the authors proposed a one-dimensional elastic viscoplastic model that takes into account both creep and swelling of clayey soil.

Based on the above, we can conclude that in order to describe the rheological properties of clayey soils, various methods are now widely used, which are based on the use of mechanical models, the theory of plastic flow and hereditary creep, etc. Based on these, various equations of state are derived, containing several parameters (depending on the model under consideration, the number of determined parameters will change), which can be determined during experimental studies of soil using various instruments. One such device is the simple shear device.

The deformation conditions of the soil specimen in simple shear tests are very similar to the deformation conditions occurring at the base of a number of structures (e.g., soil adjacent to the side of a pile, slope deformations of soil massifs, etc.). The study of rheological properties of soils in conditions of simple shear is currently quite relevant because it allows the consideration of the processes occurring in time in the soil mass, which is an important component in the design of various buildings and structures.

The test in a simple shear device can be characterized as follows: a process in which sliding (displacement) occurs along planes parallel to the plane of a stationary stamp (upper or lower, depending on the design of the device) at a value close to their distance from this plane. In simple shear, the cubic specimen takes the shape of an inclined parallelepiped at the end of the test.

Depending on the design and assembly method of the simple shear device, it is possible to determine not only the usual strength (specific cohesion (c), angle of internal friction (φ), angle of dilatancy (ψ)) and deformation (ground shear modulus (G), shear strain (γ)) characteristics of tested soil samples, but also to determine rheological parameters (ground viscosity (η)).

The concept of viscosity originates from Newton's discovery in 1687 of the basic law of viscous liquids, according to which the internal friction in the flow of a fluid, arising «due to insufficient slip of the fluid particles», is proportional to the shear rate at which these fluid particles move relative to each other. A fluid obeying this law is called an ideal viscous (Newtonian) fluid.

In a Newtonian fluid at any shear stress different from zero, a viscous flow develops in time with a constant shear strain rate ($\dot{\gamma} = const$), and this shear rate is directly proportional to the shear stress.

The viscosity coefficient of the soil can be determined from tests in which the relationship between the angular strain rate and the acting shear stress on the soil specimen can be determined. Thus, it can be concluded that the viscosity coefficient or, in other words, dynamic viscosity defines a measure of the resistance of the soil to its deformations. Soil viscosity is measured in Poise (1 Poise = 0.1 Pa·sec).

The viscosity of different media varies over a fairly wide range. For example, the viscosity of air is $1.8 \cdot 10^{-4}$ Poise, the viscosity of water is—$1.0 \cdot 10^{-2}$ Poise, the viscosity of various oils generally ranges from 0.5 to 10 Poise and the viscosity of the Earth's crust reaches about $5.0 \cdot 10^{22}$ Poise.

Based on the performed experimental studies, as well as the analysis and generalization of both his results and the results obtained by other researchers, Maslov N.N. [3] determined the following average values of the dynamic viscosity coefficient of melted clay depending on its state (consistency): soft plastic clay—$\eta = 10^{10}$–10^{11} Poise; tight plastic clay—$\eta = 10^{12}$–10^{13} Poise; semi-hard clay—$\eta = 10^{14}$–10^{15} Poise; and hard clay—$\eta = 10^{15}$–10^{17} Poise.

The methodology of determining the coefficient of viscosity of soil, as well as the great variety of properties of the soils themselves have a great influence on the obtained values of viscosity coefficients for different soils, which can explain the large variation in the obtained values.

In most cases, according to Maslov N.N., there is a misinterpretation of the value itself. Indeed, taking viscosity as a constant quantity ($\eta = const$) is valid only for a perfectly viscous (Newtonian) medium. In soils, the relationship between stress and flow rate is nonlinear, and as a consequence, the viscosity coefficient of the soil will be a variable value depending on both the magnitude of the applied load and the time of its action.

For example, at the beginning of the test, when the ground deformations are still small, the viscosity of the ground may be 10^{13} Poise and closer to the end of the deformation stage, the viscosity coefficient value may increase to 10^{14} Poise. This phenomenon was confirmed by E.S. Sobolev in his work [29] based on experimental studies on sandy soil samples in a triaxial compression device in the kinematic loading mode. Based on the results of the performed studies, the dependences of sandy soil viscosity on the angular strain rate were obtained. The presented work also described a methodology that allows the determination of the viscosity of air-dry and water-saturated sandy soil based on triaxial tests in kinematic loading mode. This methodology is based on the application of Maxwell's elastic-viscous rheological model [29].

Based on the above, we can conclude that the deformation of soils occurs at a variable rate, unlike a Newtonian medium, but at a certain point in time, the flow velocity in soils begins to approach a constant, which allows us to apply the basic law of viscous fluid at this stage, in which the process of viscous flow occurs at a constant rate.

Based on the above, the following key conclusions can be drawn:

- The deformability and strength of clay soils differ significantly from the deformability and strength of sandy, semi-rocky and rocky soils, especially when the time factor is taken into account;
- The rheological properties of clay soils are always manifested in the interaction of soils with structures and the surrounding geological environment, which equally applies to the processes of creep, relaxation and changes in soil strength over time. These processes significantly influence the nature of the formation and transformation of the stress–strain state of clay soil masses in space and time, ultimately determining the operating conditions and stability of structures;
- The historical development of soil rheology, particularly in relation to the calculations of foundation structures, initially focused on the study of creep deformations under shear conditions;
- Rheological tests of clay soils differ from conventional, standard tests in that they pay special attention to the study of processes occurring in time. These processes manifest depending on the testing regime, testing conditions, and the moisture content of the clay soils.

The purpose of this study is to investigate the influence of shear rate on the viscosity coefficient of clay soil (and its change during the testing process) as well as its strength characteristics, including specific cohesion and its components.

2. Materials and Methods

To investigate the rheological properties of clay soil in the present work, experimental studies were carried out in a simple shear device (sample diameter d = 71.4 mm, sample height h = 23 ± 1 mm) in kinematic loading mode. Figure 1 shows a general view of the simple shear device (a) as well as the main components that make up the device (b).

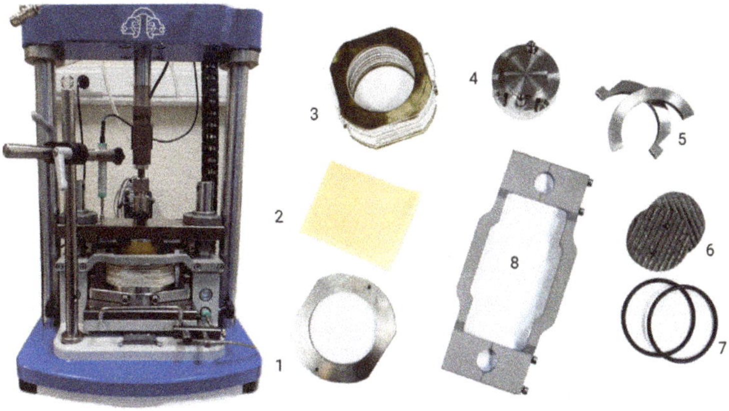

(**a**) (**b**)

Figure 1. Simple shear device: (**a**) general view; (**b**) main components: 1—steel ring with two threaded holes; 2—rubber sheath; 3—stack of brass and PTFE rings; 4—upper stamp; 5—metal casing; 6—porous disks with ribbed surface; 7—rubber rings; 8—half-clamps, tightened by four screws to ensure plane parallel shear.

One of the main advantages of this design of the simple shear device is that it allows the realization of a homogeneous stress–strain state inside the soil specimen during the test, since each individual elementary layer of the specimen moves by the same amount during shear (these conditions of deformation of the soil specimen are very similar to the deformation conditions occurring in the base of a number of structures, e.g., soil adjacent to the side surface of a pile, slope deformations of soil massifs, etc.).

It is important to note that the deformation conditions described above cannot be recreated in a direct shear device, because the contact area in this device between the upper and lower parts of the soil sample changes along the predetermined shear surface. During the test, the actual contact area between the two parts of the sample decreases, leading to an error in calculating the applied stresses (while the normal stress is assumed to be constant, its actual value increases). Additionally, during the failure of soil samples, a shear plane defined by a predetermined gap is initially prescribed. If a high-strength inclusion is encountered on the structurally prescribed shear plane, the slip surface will circumvent it. The result of such private experience is an overestimated private value of ultimate shear resistance, which complicates subsequent statistical processing: some tests have to be discarded, and the calculated values of shear resistance parameters decrease relative to the normative ones.

The procedure for preparing soil samples for testing, as well as conducting the tests themselves in a simple shear device, consists of sequentially performing the following stages:

1. Making twin samples of clay soil;
2. Assembly of the sample in the simple shear device;
3. Setting the test parameters.

Let us examine each of these stages in more detail.

2.1. Making Twin Samples of Clay Soil

Due to the lack of the necessary number of undisturbed samples, twin samples of clayey soil with the required density and moisture parameters were specially made.

In GOST 30416-2020 "Soils. Laboratory testing. General" [30] presents a methodology for making samples of dispersed soil of disturbed composition with specified parameters of density and moisture. However, the presented method does not describe exactly how the calculated amount of water should be added to ensure even distribution of moisture within the sample and to avoid the formation of lumps of dry soil.

Based on the above, the preparation of twin clay soil samples was carried out using a special method described below [31].

Part of the dry soil sample was evenly spread from a small height (to avoid dust formation) in a thin layer in a small metal mixing tray. The amount of water calculated according to the formula [30] was poured into the atomizer-pulverizer to provide the required moisture content. Next, the water was sprayed onto the surface of the scattered soil. After moistening the top layer, the soil was mixed and leveled. The next portion of the dry soil was spread with subsequent moistening. The steps were repeated until all of the dry soil and water in the atomizer-pulverizer were used up. The moistened soil was then placed in an exicator for at least two hours to allow for a more even distribution of moisture within the sample. Next, the soil was evenly placed in a metal compaction mold to achieve the required density of the soil sample. The characteristics of the tested soil samples are presented in Table 1.

Table 1. Characteristics of tested soil samples.

Type of Clay Soil	Plasticity Index	Index of Liquidity	Density, g/cm^3	Moisture, %	Cohesion, kPa	Internal Friction Angle, Deg.
	I_P	I_L	ρ	w	c	φ
Fluid-plastic loam	0.11	0.80	2.20	22	1.30–36.40 *	19.89–24.98 *

*—Values of strength parameters varies depending on the shear rate. A detailed description of the reasons for the variation in the scatter of these parameters is presented in Section 3.2. Effect of shear rate on ultimate shear resistance of clay soil.

2.2. Assembly of the Sample in the Simple Shear Device

A lower round stamp, corresponding to the diameter of the sample, with a metal porous disk was fixed on a square base. A soil sample encased in a rubber membrane was placed on top of the metal porous disk. In the lower part of the sample, the rubber membrane was fixed with a rubber ring, over which a metal casing was installed. Next, a stack of 25 individual rings (including 12 fluoroplastic and 13 brass rings) with special "ears" designed to ensure that all rings work together during the test was placed on top of the rubber membrane. The rings were alternated with each other to reduce friction between them. The rings were connected to each other via two vertical sliding guides that tilt during the test as shear strain accumulates. A steel ring with threaded holes was placed on top of the stack of rings, over which clamps were fastened to prevent the rings from fanning out during shear. Then, the mechanism of vertical load application was activated and the upper stamp (with a metal porous disk fixed on it) was aligned with the upper surface of the soil sample. Metal porous disks installed in the upper and lower parts of the sample are designed to realize drainage during compaction. The porous disks had a ribbed surface for better adhesion to the test soil sample. In GOST R 71042-2023 "Soils. Determination of strength characteristics by simple shear method" [32] presents a methodology describing the stage of preparation of soil samples for testing, as well as conducting the tests in a simple shear device with subsequent processing of the results.

2.3. Setting the Test Parameters

After assembling the soil sample in the simple shear device, the following parameters were set in the test program: the value of vertical load on the soil sample (kPa), shear displacement rate (mm/min) and relative shear strain (%).

3. Results

3.1. Experiment to Determine the Viscosity Coefficient of Clay Soil

In order to determine the viscosity coefficient of the fluid-plastic loam, experimental studies were carried out in a simple shear device in the kinematic loading mode ($\dot{\gamma} = const$) at four different shear displacement rates ($\dot{u}$ = 0.005 mm/min, $\dot{u}$ = 0.05 mm/min, $\dot{u}$ = 0.5 mm/min, $\dot{u}$ = 5 mm/min) and at three different values of vertical stress acting on the soil specimen during the test (σ_{n1} = 200 kPa, σ_{n2} = 400 kPa, σ_{n3} = 600 kPa). The tests were carried out until the ultimate relative shear strain ε = 18% was reached.

After processing the results of the laboratory tests, the graphs of the dependences of shear stresses on shear strains ($\tau - \gamma$) were obtained at various values of vertical stress σ_n (Figure 2).

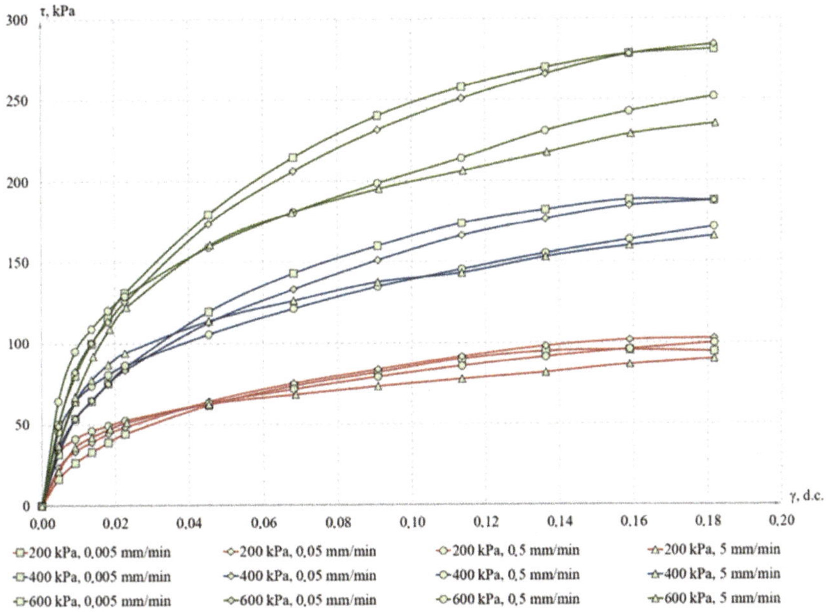

Figure 2. Graph of the dependence of shear stresses on shear strains ($\tau - \gamma$) at different values of σ_n.

Taking into account that the obtained graphs of dependence of shear stresses on shear strains ($\tau - \gamma$) have a pronounced bilinear character, the coefficients of soil viscosity (η) (Table 2) were determined for two characteristic parts of the graph (Figure 3).

Table 2. Viscosity coefficients of clayey soil.

Vertical Stress, kPa	Shear Displacement Rate, mm/min	Soil Viscosity at the First Section, kPa·min	Soil Viscosity at the Second Section, kPa·min
σ_n	$\dot{u}$	η_1	η_2
200	5	226	133
	0.5	2310	1707
	0.05	21,296	21,780
	0.005	194,920	221,320
400	5	414	259
	0.5	3802	3023
	0.05	36,740	40876
	0.005	374,000	426,800

Table 2. *Cont.*

Vertical Stress, kPa	Shear Displacement Rate, mm/min	Soil Viscosity at the First Section, kPa·min	Soil Viscosity at the Second Section, kPa·min
σ_n	$\dot{u}$	η_1	η_2
600	5	539	418
	0.5	5680	4466
	0.05	55,264	61,732
	0.005	576,840	610,720

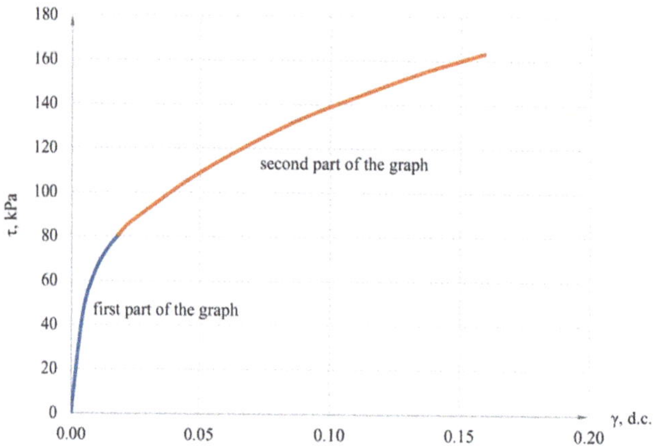

Figure 3. Characteristic graph of the dependence of shear stresses on shear strains ($\tau - \gamma$) with division into characteristic sections.

The viscosity coefficients (η) for two characteristic sections of the graph were determined based on the simple Newton's equation:

$$\eta = \frac{\tau}{\dot{\gamma}}, \tag{1}$$

where τ is the shear stress [kPa] and $\dot{\gamma}$ is the shear rate [1/s].

3.2. *Effect of Shear Rate on Ultimate Shear Resistance of Clay Soil*

According to the well-known concept of N.N. Maslov [3,4], the ultimate shear resistance of clay soil (τ^*) can be represented as the sum of Coulomb friction ($\sigma tg\varphi$), structural cohesion (c_c) and cohesion of water-colloid bonds (c_w), which depend on moisture content, i.e.,

$$\tau^* = \sigma tg\varphi + c_c + c_w, \tag{2}$$

Agreeing with this representation of the ultimate shear resistance of clay soil (τ^*), it should be noted that it does not take into account the effect of the shear rate ($\dot{\gamma}$). Indeed, under kinematic shear ($\dot{\gamma} = const$) in a soil medium possessing viscosity, viscous resistance may manifest ($c_{\dot{\gamma}}$).

Assuming that the viscous resistance is due to the viscosity of the soil as a whole, then under kinematic shear, viscous resistance will arise ($c_{\dot{\gamma}} = \dot{\gamma}\eta$). Based on the above, the total cohesion (c) in clay soil can be represented as the sum of structural cohesion (c_c), cohesion of water-colloid bonds (c_w) and viscous resistance ($c_{\dot{\gamma}}$), i.e.,

$$c = c_c + c_w + c_{\dot{\gamma}}, \tag{3}$$

It is clear that in order to account for the influence of shear rate ($\dot{\gamma}$) on the shear resistance of a clay soil, it is necessary to conduct at least three tests at different shear rates ($\dot{\gamma}_1 < \dot{\gamma}_2 < \dot{\gamma}_3$).

In the laboratory of the REC "Geotechnics" named after Z.G. Ter-Martirosyan (National Research Moscow State University of Civil Engineering) experimental studies were conducted on samples of fluid-plastic loam using a simple shear device in a kinematic loading mode ($\dot{\gamma} = const$) (at a constant shear displacement rate equal to $\dot{u} = \dot{\gamma}h$) at four different shear rates ($\dot{\gamma}_1 > \dot{\gamma}_2 > \dot{\gamma}_3 > \dot{\gamma}_4$) and under three different constant vertical loads ($\sigma_1 < \sigma_2 < \sigma_3$) applied to the soil sample during testing.

Figures 4–7 show the results of the tests of flow-plastic loam in the simple shear device in the kinematic loading mode ($\dot{\gamma} = const$).

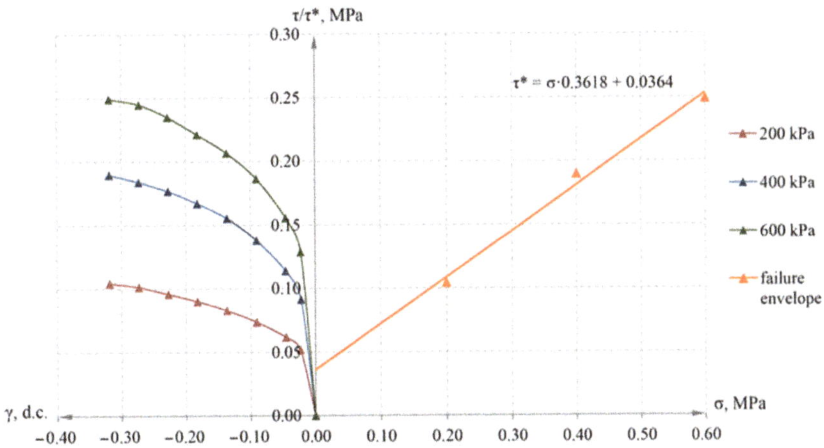

Figure 4. Graph of the dependence of acting shear stresses on shear strains ($\tau - \gamma$) and graph of the dependence of ultimate shear stresses on normal stresses ($\tau^* - \sigma$) during testing of clay soil in a simple shear device in the kinematic loading mode at a constant shear displacement rate of $\dot{u}$ = 5 mm/min and at different values of σ_n.

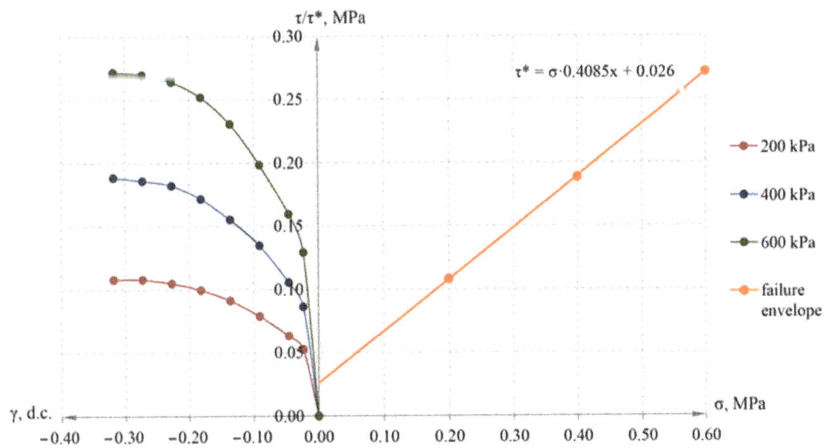

Figure 5. Graph of the dependence of acting shear stresses on shear strains ($\tau - \gamma$) and graph of the dependence of ultimate shear stresses on normal stresses ($\tau^* - \sigma$) during testing of clay soil in a simple shear device in the kinematic loading mode at a constant shear displacement rate of $\dot{u}$ = 0.5 mm/min and at different values of σ_n.

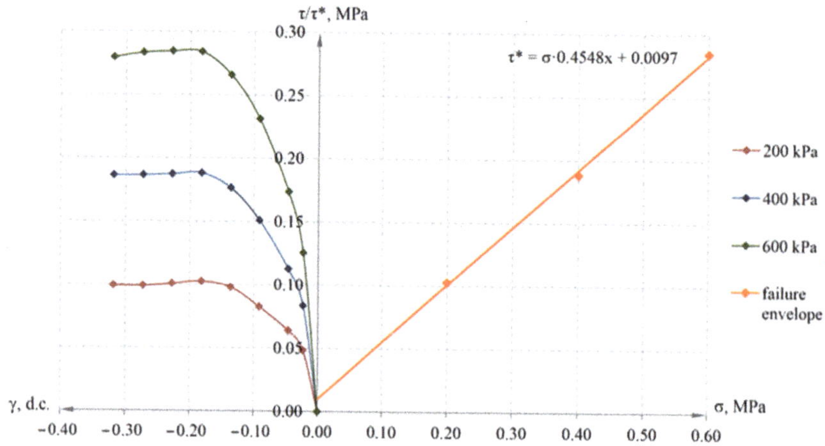

Figure 6. Graph of the dependence of acting shear stresses on shear strains ($\tau - \gamma$) and graph of the dependence of ultimate shear stresses on normal stresses ($\tau^* - \sigma$) during testing of clay soil in a simple shear device in the kinematic loading mode at a constant shear displacement rate of $\dot{u}$ = 0.05 mm/min and at different values of σ_n.

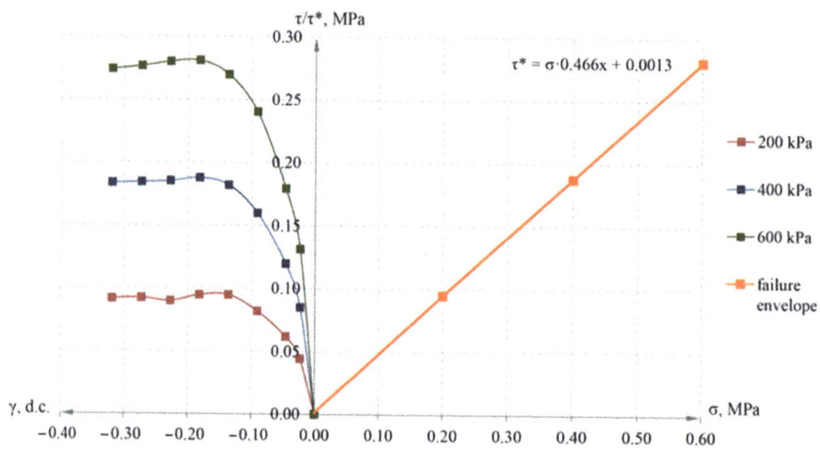

Figure 7. Graph of the dependence of acting shear stresses on shear strains ($\tau - \gamma$) and graph of the dependence of ultimate shear stresses on normal stresses ($\tau^* - \sigma$) during testing of clay soil in a simple shear device in the kinematic loading mode at a constant shear displacement rate of $\dot{u}$ = 0.005 mm/min and at different values of σ_n.

Analyzing the obtained results presented in Figures 4–7, one can conclude that there is a negligible change in the internal friction angle (when testing at different shear displacement rates), which can be explained by the limited amount of the conducted experimental studies. However, based on the obtained results, it can be concluded that the shear rate ($\dot{\gamma}$) has a significant effect on the cohesion of clay soil, which is confirmed graphically. The correlation coefficients $r_{\sigma\tau^*}$ for each shear displacement rate were as follows: at $\dot{u}$ = 5 mm/min $r_{\sigma\tau^*}$ = 0.9894; at $\dot{u}$ = 0.5 mm/min $r_{\sigma\tau^*}$ = 0.9999; at $\dot{u}$ = 0.05 mm/min $r_{\sigma\tau^*}$ = 0.9988; at $\dot{u}$ = 0.005 mm/min $r_{\sigma\tau^*}$ = 1, which is sufficient for the purpose of the study.

It should be noted that numerous experimental studies conducted on samples of Quaternary clay with disturbed and undisturbed structures in the kinematic loading mode and at various values of vertical stress σ_n (0.100; 0.200; 0.300 and 0.500 MPa), presented in the work of Kornoukhov M.B. [33], showed (Figure 8) that the ultimate lines at different shear rates ($\dot{\gamma}_1 > \dot{\gamma}_2 > \dot{\gamma}_3 > \dot{\gamma}_4$) have the same slope equal to $\varphi = const$, and this angle does not depend on the shear rate ($\dot{\gamma}$).

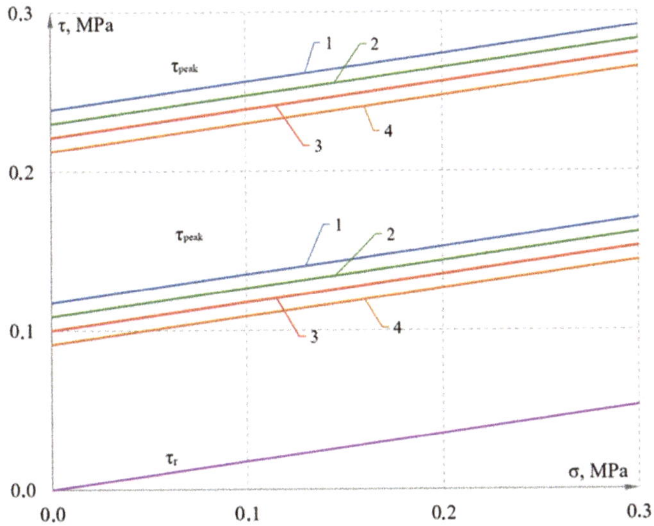

Figure 8. Graph of the dependence of the peak and residual strength of Quaternary clay with undisturbed structure and disturbed structure on the vertical stress (σ_{n_1} = 0.1 MPa; σ_{n_2} = 0.2 MPa and σ_{n_3} = 0.3 MPa), acting on the sample at different shear rates: 1—$\dot{\gamma}_1$ = 0.066 min^{-1}; 2—$\dot{\gamma}_2$ = 0.0041 min^{-1}; 3—$\dot{\gamma}_3$ = 0.00026 min^{-1}; 4—$\dot{\gamma}_4$ = 0.00002 min^{-1} [33].

The ultimate shear resistance of clay soil (τ^*) significantly depends on the internal friction angle (φ) and the cohesive strength (c). The total cohesion (c), in turn, significantly depends on the shear rate ($\dot{\gamma}$) and as the shear rate increases, the ultimate shear resistance of clay soil increases proportionally to ($ln\dot{\gamma}$) (Figure 9).

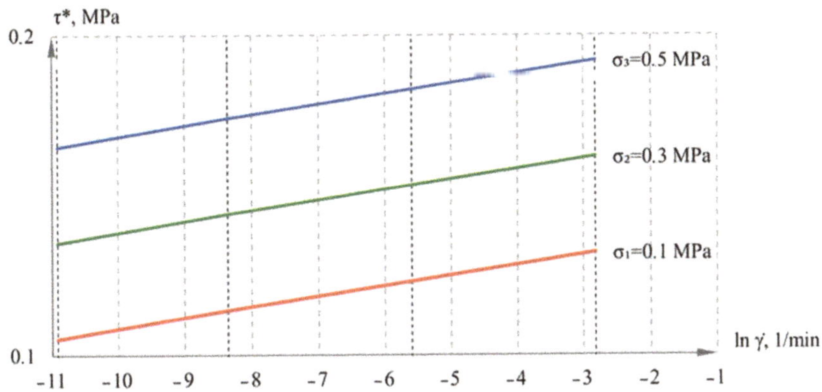

Figure 9. Graph of dependence of peak strength of disturbed structure Quaternary clay on the deformation rate [33].

Since, according to the experimental results, it was obtained that the structural cohesion (c_c) and the angle of internal friction (φ) are independent of the shear rate ($\dot{\gamma}$), we present the viscous resistance ($c_{\dot{\gamma}}$) as the following relationship:

$$c_{\dot{\gamma}} = c_{\dot{\gamma}0} + \alpha ln\left(\dot{\gamma}/\dot{\gamma}_1\right), \tag{4}$$

where α is the experimental parameter.

Based on the analysis of the experimental research results (performed on clay soil samples) and also on the rheological model of a Maxwell viscoelastic body, we present the rheological equation for kinematic shear as applied to a soil medium in the following form:

$$\dot{\gamma} = \frac{\tau - \tau^*}{\eta} + \frac{\dot{\tau}}{G}, \tag{5}$$

where $\tau^* = \sigma tg\varphi + c_0$ (with $c_0 = c_c + c_w$) [kPa]; η is the overall viscosity of the soil [kPa·s]; $\dot{\tau}$ is the rate of change in shear stress [kPa/s]; and G is the shear modulus of the soil [kPa].

By performing a certain grouping in Equation (5), we obtain a differential equation of the following form:

$$\dot{\tau} + \tau \cdot \frac{G}{\eta} = \frac{G}{\eta} \cdot \left(c_{\dot{\gamma}} + \tau^*\right), \tag{6}$$

The general solution of Equation (6) is known [34] and is as follows:

$$\tau = e^{-\int \frac{G}{\eta} dt} \cdot \left\{ \int \frac{G}{\eta} \cdot \left(c_{\dot{\gamma}} + \tau^*\right) \cdot e^{\int \frac{G}{\eta} dt} dt + C \right\}, \tag{7}$$

$$\tau(t) = e^{-\frac{G \cdot t}{\eta}} \cdot e^{\frac{G \cdot t}{\eta}} \cdot \left(c_{\dot{\gamma}} + \tau^*\right) + C \cdot e^{-\frac{G \cdot t}{\eta}}, \tag{8}$$

We will determine the constant of integration C from the initial Condition (8) at $t = 0$:

$$C = -\tau^* - c_{\dot{\gamma}}, \tag{9}$$

where $\tau^* = \sigma tg\varphi + c_0$ [kPa].

By substituting (9) into (8), we obtain the solution of the equation for constant shear rate ($\dot{\gamma} = const$):

$$\tau(t) = (\sigma tg\varphi + c_c + c_w + c_{\dot{\gamma}}) \cdot (1 - e^{-\frac{G \cdot t}{\eta}}), \tag{10}$$

when $t \to \infty$ $\tau(\infty) = \sigma tg\varphi + c_c + c_w + c_{\dot{\gamma}}$, with $c_{\dot{\gamma}} = \dot{\gamma}\eta$.

The obtained result corresponds to our understanding of the ultimate shear resistance (3). However, in this case, the viscous resistance ($c_{\dot{\gamma}}$) increases in direct proportion to the shear rate $(\dot{\gamma})$, which does not correspond to the experimental results (4). This dependence can be obtained if the following relationship is taken as a rheological model:

$$(\dot{\gamma}/\dot{\gamma}_1)^{\alpha/\eta} = e^{\left(\frac{\tau - \tau^*}{\eta} + \frac{\dot{\tau}}{G}\right)}, \tag{11}$$

where $\tau^* = \sigma tg\varphi + c_c + c_w$ [kPa].

Equation (5) can also be represented in the following form:

$$\frac{\alpha}{\eta} ln(\dot{\gamma}/\dot{\gamma}_1) = \frac{\tau - \tau^*}{\eta} + \frac{\dot{\tau}}{G}, \tag{12}$$

The solution of this equation is known [34] and is given by:

$$\tau(t) = (\sigma tg\varphi + c_c + c_w + \alpha ln(\dot{\gamma}/\dot{\gamma}_1)) \cdot (1 - e^{-\frac{G \cdot t}{\eta}}), \tag{13}$$

when $t \to \infty$ $\tau(\infty) = \sigma tg\varphi + c_c + c_w + \alpha ln(\dot{\gamma}/\dot{\gamma}_1)$.

It follows from Equation (13) that $c_{\dot{\gamma}} = \alpha ln(\dot{\gamma}/\dot{\gamma}_1)$.

The obtained result coincides with the result of experiment (4), and using Equation (13), it is possible to construct the $(\tau^* - ln\dot{\gamma})$ diagram (Figure 9), where the asymptotic lines at different shear rates are parallel and offset from each other according to the law $\alpha ln(\dot{\gamma}/\dot{\gamma}_1)$.

To obtain the dependence $(\tau - \gamma)$ it is necessary to replace the time t with the quantity $\gamma/\dot{\gamma}$, in Equation (13), then we obtain the following equation:

$$\tau(\gamma/\dot{\gamma}) = (\sigma tg\varphi + c_c + c_w + \alpha ln(\dot{\gamma}/\dot{\gamma}_1)) \cdot (1 - e^{-\frac{G}{\eta}\cdot(\gamma/\dot{\gamma})}), \tag{14}$$

4. Discussion

Due to the fact that the viscosity coefficient (η) of soil depends on many factors such as time, the stress–strain state of the soil mass, temperature, humidity, rearrangement of soil particles during its compaction and creep, determining this parameter during experimental studies and subsequent interpretation of the obtained results is challenging and leads to a wide dispersion of the values obtained.

Based on the experimental studies conducted using a simple shear device in a kinematic loading mode, it has been established that the shear rate ($\dot{\gamma}$) has a significant effect on the viscosity coefficient (η) of fluid-plastic loam, namely, an inverse correlation is observed (when the shear displacement rate decreases, the viscosity coefficient increases) (Table 2).

It is shown that the viscosity coefficient is a varying value depending on both the value of the applied load and the time of its action. Therefore, when using different rheological soil models containing the viscosity parameter, it is necessary to take into account the variation of this parameter over time, rather than taking it as a constant value, as this will allow a more accurate prediction of the final settlement over time of different buildings and structures.

It was determined that at low values of shear displacement rates ($\dot{u}$ = 0.005 mm/min and $\dot{u}$ = 0.05 mm/min), the viscosity coefficient on the first section of the graph is lower than on the second section. However, with an increase in the shear displacement rate ($\dot{u}$ = 0.5 mm/min and $\dot{u}$ = 5 mm/min), opposite results were obtained (Table 2).

It should also be noted that the results of experimental studies conducted using a simple shear device in kinematic loading mode ($\dot{\gamma} = const$) showed that the shear rate ($\dot{\gamma}$) has a significant impact on the cohesion (c) of clay soil.

It has been found that under kinematic shear ($\dot{\gamma} = const$) of clay soil, the ultimate shear resistance (τ^*) depends not only on internal friction (φ), structural cohesion (c_c) and cohesion of water colloid bonds (c_w), but also on the viscous resistance of the soil $\left(c_{\dot{\gamma}}\right)$.

Experiments show that with the increase in shear rate ($\dot{\gamma}$), the viscous shear resistance ($c_{\dot{\gamma}}$) increases proportionally to the logarithm of the shear rate $\left(c_{\dot{\gamma}} = c_{\dot{\gamma}_0} + \alpha ln(\dot{\gamma}/\dot{\gamma}_1)\right)$ and this affects the rate of increase in shear stress during shear testing under kinematic loading conditions $(\dot{\gamma} = const)$.

Comparing Equations (10) and (13), it should be noted that at the same shear rate $(\dot{\gamma} = const)$ the shear stresses increase more slowly when $c_{\dot{\gamma}} = \alpha ln(\dot{\gamma}/\dot{\gamma}_1)$ (13) than when $c_{\dot{\gamma}} = \dot{\gamma}\eta$ (10). On this basis, it follows that considering the effect of shear rate on the growth rate of shear stresses is necessary.

Previously, in the works of many authors [1–4,6,11,12,14,16], it was noted that the total cohesion can be divided into two components: structural cohesion (c_c) and cohesion of water–colloid bonds (c_w). However, based on the results of experimental studies presented in Figures 4–9, and the analysis of the obtained results, it is clearly shown that the total cohesion (c) also depends on the shear rate ($\dot{\gamma}$).

Thus, it can be considered that a new rheological Equation (12) has been obtained, which can describe the dependence $(\tau - t)$ according to (10) and the dependence $(\tau - \gamma)$ according to (14) during experimental studies on clay soil samples in the kinematic loading mode ($\dot{\gamma} = const$) and under various vertical loads ($\sigma_1 < \sigma_2 < \sigma_3$) acting on the soil sample during testing.

At present, there are quite a lot of rheological models (Newton's model, Kelvin–Voigt viscoelastic model, Maxwell's elastic-viscous model, Bingam–Shvedov elastic-viscoplastic model, etc.) describing the behavior of soils in time. However, none of the presented models show the effect of shear rate on the strength characteristics of clay soil. When analyzing the works of other researchers performed on similar topics, unfortunately, no work with a similar setup of experimental studies and analysis of the obtained results was found, except for the work of Kornoukhov M.B. [33]. In the comparative analysis, a high convergence with the experimental results obtained in the present work was observed.

5. Conclusions

Based on the above, it can be concluded that soil rheology is one of the most important parts of soil mechanics, as it considers the formation and change in time of the stress–strain state of soils (mainly clay soils), and also has a significant impact on the development of long-term settlement in the foundations of various buildings and structures.

This paper presents experimental studies showing the effect of shear rate on the viscosity of flow-plastic loam and proposes dependence (12) describing this phenomenon.

The performed experimental studies also show the influence of shear rate on the total cohesion of clay soil, which correlates with earlier studies presented in the work of Kornoukhov M.B. [33].

It is shown that the total cohesion can be decomposed not only into previously known components, such as structural cohesion and cohesion of water–colloid bonds, but also a third component, viscous resistance, can be identified.

The results obtained in this study can be used for further improvement of methods for calculating settlement of structures in time, as well as for predicting the time of exhaustion of the bearing capacity of foundation soils.

As part of further research, additional laboratory tests are planned to be conducted using a simple shear device in a kinematic loading mode on clay soil samples (with different plasticity index I_P and liquidity index I_L) at various shear rates and different values of vertical loads acting on the soil sample, in order to identify and determine correlation dependencies.

Author Contributions: Conceptualization, methodology A.Z.T.-M.; software, validation, L.Y.E. and G.O.A.; formal analysis, writing—review and editing, all authors; investigation, A.Z.T.-M., L.Y.E. and G.O.A.; Supervision, project administration, A.Z.T.-M.; Resources, A.Z.T.-M. and G.O.A.; visualization, L.Y.E. All authors have read and agreed to the published version of the manuscript.

Funding: The research was funded by the National Research Moscow State University of Civil Engineering (grant for fundamental and applied scientific research, project No. 02-392/130).

Institutional Review Board Statement: Not applicable.

Informed Consent Statement: Not applicable.

Data Availability Statement: The original contributions presented in this study are included in the article. Further inquiries can be directed to the corresponding author.

Conflicts of Interest: The authors declare no conflicts of interest.

References

1. Vyalov, S.S. *Rheological Foundations of Soil Mechanics: Textbook for Construction Universities*; Vysshaya Shkola: Moscow, Russia, 1978; p. 447.
2. Goldshtejn, M.N. *Mechanical Properties of Soils*; Stroyizdat: Moscow, Russia, 1977; p. 256.
3. Maslov, N.N. *Long-Term Stability and Deformation of Displacement Retaining Walls*; Energiya: Moscow, Russia, 1968; p. 467.
4. Maslov, N.N. *Fundamentals of Soil Mechanics and Engineering Geology*; Vysshaya Shkola: Moscow, Russia, 1968; p. 632.
5. Ter-Stepanyan, G.I. Phenomenon of sudden restructuring of soil structure under deformation. *Eng. Geol.* **1988**, 3, 11–26.

6. Meschyan, S.R. *Creep of Clay Soils*; Publishing House of the Academy of Sciences of the Armenian SSR: Yerevan, Armenia, 1967; p. 320.
7. Meschyan, S.R. *Rheological Processes in Clay Soils (Considering Special Factors)*; Ayastan: Yerevan, Armenia, 1992; p. 395.
8. Meschyan, S.R. *Experimental Rheology of Clay Soils*; Gitutyun: Yerevan, Armenia, 2005; p. 498.
9. Goldin, A.L. Study of Compaction Processes of Cohesive Soils Taking Creep into Account. Ph.D. Dissertation, All-Union Scientific Research Institute of Hydraulic Engineering named after. B.E. Vedeneeva, Leningrad, Russia, 1965.
10. Ter-Martirosyan, Z.G. *Rheological Parameters of Soils and Foundation Calculations*; Strojizdat: Moscow, Russia, 1990; p. 200.
11. Florin, V.A. *Fundamentals of Soil Mechanics. Volume 1*; Strojizdat: Leningrad, Russia, 1959; p. 356.
12. Florin, V.A. *Fundamentals of Soil Mechanics. Volume 2*; Gosstroyizdat: Leningrad, Russia, 1961; p. 543.
13. Arutyunyan, N.K. *Some Questions of the Theory of Creep*; State Publishing House of Technical and Theoretical Literature: Moscow, Russia, 1952; p. 324.
14. Tertsagi, K. *Theory of Soil Mechanics*; Gostroyizdat: Moscow, Russia, 1961; p. 507.
15. Gersevanov, N.M. *Theoretical Foundations of Soil Mechanics and Their Practical Applications*; Strojizdat: Moscow, Russia, 1948; p. 248.
16. Cytovich, N.A. *Soil Mechanics*; State Publishing House of Literature on Construction, Architecture, and Building Materials: Moscow, Russia, 1963; p. 437.
17. Zareckij, Y.K.; Chumichev, B.D.; Shchebolev, A.G. *Viscoplasticity of Ice and Frozen Ground*; Science: Novosibirsk, Russia, 1986; p. 182.
18. Ter-Martirosyan, Z.G.; Ter-Martirosyan, A.Z. *Soil Mechanics in High-Rise Construction with Developed Underground Part: Textbook*; ASV: Moscow, Russia, 2020; p. 946.
19. Boley, K.; Strokova, L.A. Determination of creep characteristics of clayey soils. *News Tomsk. Polytech. Univ.* **2007**, *2*, 42–44.
20. Tafili, M.; Triantafyllidis, T. A simple hypoplastic model with loading surface accounting for viscous and fabric effects of clays. *Int. J. Numer. Anal. Methods Geomech.* **2020**, *44*, 2189–2215. [CrossRef]
21. Ren, P.; Wang, P.; Zhang, H.; Tang, Y. Nonlinear behavior of clay creep and its fractional derivative creep model. *Eng. Mech.* **2020**, *37*, 153–160. [CrossRef]
22. Zhu, Z.-Y.; Luo, F.; Zhang, Y.-Z.; Zhang, D.-J.; He, J.-L. A creep model for frozen sand of Qinghai-Tibet based on Nishihara model. *Cold Reg. Sci. Technol.* **2019**, *167*, 102843. [CrossRef]
23. Ning, X.; Xiao, N.; Zhang, C.; He, B.; Xie, J. Study on the nonlinear creep model of expansive soil. *J. Nat. Disasters* **2017**, *26*, 149–155. [CrossRef]
24. Haghighat, E.; Rassouli, F.S.; Zoback, M.D.; Juanes, R. A viscoplastic model of creep in shale. *Geophysics* **2020**, *85*, 155–166. [CrossRef]
25. Gu, R.-G.; Zou, Y.; Fang, Y.-G.; Hu, Y.-G. Rheological model of soft soils using nonlinear instantaneous elastic modulus. *Rock Soil Mech.* **2018**, *39*, 237–241. [CrossRef]
26. Ter-Martirosyan, Z.G.; Sobolev, E.S.; Ter-Martirosyan, A.Z. Rheological models developed on the basis of triaxial tests. *Int. J. Appl. Eng. Res.* **2016**, *11*, 1719–1723.
27. Bodas Freitas, T.M.; Potts, D.M.; Zdravkovic, L. A time dependent constitutive model for soils with isotach viscosity. *Comput. Geotech.* **2011**, *38*, 809–820. [CrossRef]
28. Yin, J.-H.; Tong, F. Constitutive modeling of time-dependent stress-strain behaviour of saturated soils exhibiting both creep and swelling. *Can. Geotech. J.* **2011**, *48*, 1870–1885. [CrossRef]
29. Sobolev, E.S. Creep and Vibration Creep of Sandy Soils at the Bases of Buildings and Structures. Ph.D. Dissertation, Federal State Budgetary Educational Institution of Higher Professional Education "Moscow State University of Civil Engineering", Moscow, Russia, 2014.
30. *GOST 30416-2020*; Soils. Laboratory Testing. General. Izdatelstvo Russian Institute of Standardization: Moscow, Russia, 2021; p. 21.
31. Mirny, A.Y. *Triaxial Testing of Soils: Theory and Practice: Monograph*; "Academic Science"—GeoMarkketing LLC: Moscow, Russia, 2015; p. 144.
32. *GOST R 71042-2023*; Soils. Determination of Strength Characteristics by Simple Shear Method. Russian Institute of Standardization: Moscow, Russia, 2023; p. 18.
33. Kornoukhov, M.B. Shear Resistance of Clay Soils during the Development of Deformation. Ph.D. Thesis, Moscow Civil Engineering Institute, Moscow, Russia, 1990.
34. Bronshtein, I.N.; Semendyaev, K.A. *Handbook of Mathematics*; State Publishing House of Physical and Mathematical Literature: Moscow, Russia, 1962; p. 608.

applied sciences

MDPI

Article

Investigation of the Shear and Pore Structure Characteristics of Rubber Fiber-Reinforced Expansive Soil

Zhongnian Yang [1,*], Rongchang Wang [1], Wei Shi [1], Zhenxing Sun [1] and Xianzhang Ling [1,2]

1 School of Civil Engineering, Qingdao University of Technology, Qingdao 266520, China; wrcoan@163.com (R.W.); shiwei123@qut.edu.cn (W.S.); sun.zhenxing@outlook.com (Z.S.); lingxianzhang@qut.edu.cn (X.L.)

2 School of Civil Engineering, Harbin Institute of Technology, Harbin 150006, China

* Correspondence: yzhnqd@qut.edu.cn

Abstract: In recent years, many researchers have evaluated the sustainable use of waste tire rubber as an aggregate in soil. Its effectiveness has been widely acknowledged. The main objective of this work is to study the influence of rubber fibers on shear strength and pore structure characteristics in relation to expansive soil. In this context, we conducted a series of experiments that were carried out on reinforced expansive soil with rubber fiber contents of 0, 5, 10, 15, and 20%. The results show that the shear strength and maximum dilatation angle increase gradually with rubber fiber content. Due to the pore water pressure and creep effects, the deviator stress and effective cohesion of the samples under the consolidated drained conditions were higher than those under the undrained conditions. The converse was true for the internal angle. The addition of an appropriate amount of (5–10%) rubber fiber can effectively inhibit the development of soil cracks and reduce the porosity of the samples. The results obtained can highlight the beneficial effects of rubber fiber, which is highly desirable in many backfill applications.

Keywords: rubber fiber; expansive soil; shear strength; dilatation angle; pore evolution

1. Introduction

Expansive soil is rich in strong hydrophilic minerals such as montmorillonite and illite. It is generally considered a problem soil with a high expansion potential, with expansion contractility, multi-fracture, and over-consolidation [1]. The internal factors affecting soil expansion and contraction include mineral composition and microstructure. Meanwhile, the external factors include water and climatic conditions. Expansive soil is known as "catastrophic soil" because of the cyclic process of shrinkage and expansion: it characteristically expands when it absorbs water and contracts when it loses water [2]. This generally damages the superstructure, particularly the slight structure, e.g., cracking and destruction of foundations [3], roadbeds [4], dams [5], and channels [6]. According to statistics, the losses caused by expansive soil worldwide amount to billions of dollars. Within this, 10 million square meters of houses were destroyed due to expansive soil in China, and the economic losses exceeded USD 1 billion [7]. Therefore, it is necessary to improve expansive soil foundations to reduce the adverse effects of expansive soil on project services.

The currently proposed methods to improve the engineering properties of expansive soil mainly involve chemical, physical, and biological aspects. Chemical methods improve the soil conditions through the use of additives, such as fly ash, cement, and lime [8–12]; physical methods improve soil conditions via geosynthetics, soil substitution, and moisture control [13–17]; and biological methods improve the soil conditions through the catalytic action of biological enzymes [18–22]. The use of soil replacement methods may considerably increase construction costs. Although the use of additives such as cement effectively suppresses the expansion and contraction of expansive soil, the carbon dioxide emitted during the production of cement is uncontrollable. This has an adverse impact on the environment and a low long-term advantage. Therefore, it is necessary to study energy-saving and environmentally friendly methods to improve expansive soil.

The resource utilization of solid waste is an important part of China's green, low-carbon, and circular economic system. Additionally, it is a strategic measure to achieve "carbon peak and carbon neutrality". In February 2022, to promote the standardized utilization of renewable resources, eight departments, including the Ministry of Industry and Information Technology, issued the "Implementation Plan on Accelerating the Comprehensive Utilization of Industrial Resources". It promotes the comprehensive utilization of renewable resources such as scrap steel, waste non-ferrous metals, waste tires, waste paper, and waste mobile phones. Research has shown that solid waste recycling and reuse are key to achieving eco-friendly, eco-efficient, and sustainable infrastructure development on a global scale.

The growing number of scrap tires, one of the largest sources of solid waste, has generated severe disposal challenges. It is estimated that a billion tires are discarded globally annually. This number is projected to increase to 1.2 billion by 2030 [23]. In the past ten years, rubber generation and recycling volumes in China have increased, as shown in Figure 1. In 2022, the amount of waste tires generated was 13.5 million tons, and the recycling volume was 6.5 million tons. This corresponded to a recycling rate of waste tires of 48.15%. The rate is relatively low, and the recycling capacity is relatively insufficient. This implies that there are relatively abundant waste tires for recycling and advantageous reuse. With the increasing awareness of environmental protection and resource conservation, the reuse of waste resources has become a new trend in scholarly research. Studies have shown that adding rubber fibers to soil can mitigate swelling potential, improve ductility, and increase shear strength. Abbaspour et al. [24] found that rubber fibers can mitigate the swelling characteristics of expansive soils by up to 44% of the swelling deformation of expansive soils. Özkul and Baykal [25] evaluated the effect of rubber fibers on the shear properties of clays using triaxial and permeability tests and found that rubber fibers increased the shear strength of clays; however, the change in permeability was not significant.

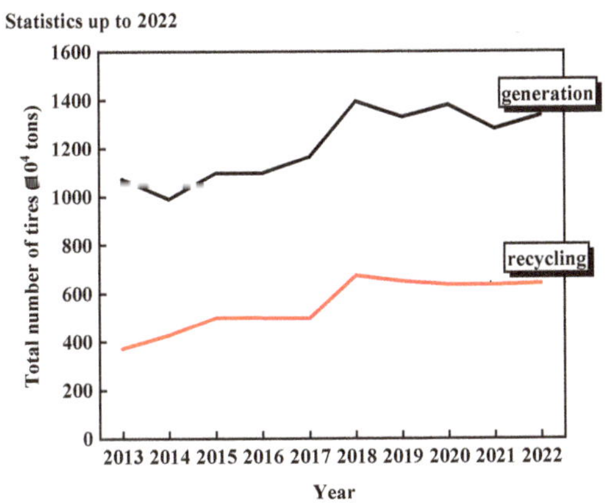

Figure 1. Rubber generation and recycling volumes in China.

The main problem of expansive soil is that its strength is generally low. It generally varies with alterations in the load and drainage conditions. Moreover, it may differ in strength and deformation behavior under drained and undrained loading conditions. For example, Li et al. [26] evaluated the effects of confining pressure, initial dry density, initial moisture content, and drainage conditions on the mechanical properties of expansive soil based on consolidated drained and undrained tests. They determined that a hyperbolic model could be used to fit the strain–hardening curves in the drained tests, and an exponential curve model could be used to fit the strain–softening curves in the undrained tests. Tajdini et al. [27] investigated the effects of rubber chips on the mechanical properties of kaolinite clay via undrained and drained triaxial tests. They determined that the shear strength, dilatation angle, and shear stiffness of reinforced soil were higher under drained conditions than under undrained conditions. Fiber reinforcement is an effective method to improve the shear strength and ductility of soil. It is necessary to study the drainage response of fiber-reinforced soil to evaluate its long-term stability. Furthermore, it is necessary to investigate the effect of drainage conditions on the mechanical properties of fiber-reinforced soil.

The purpose of this study was to explore the drainage conditions and the mechanical properties of rubber fibers to improve expansive soil. The effects of rubber fibers on the shear strength, maximum dilatation angle, and pore diameter of expansive soils were quantified. Additionally, the effect of rubber fibers on the pore distribution of expansive soil was analyzed quantitatively. The test results can help evaluate the improvement effect that rubber fibers had on expansive soil, realize the utilization of waste tires to improve the environment, and provide theoretical support for the use of waste rubber in expansive soil engineering.

2. Materials and Methods

2.1. Test Preparation

2.1.1. Test Materials

The expansive soil used in this study was the Henan remodeled expansive soil. Its gradation curve is shown in Figure 2. Physical property indices, such as optimum water content, maximum dry density, liquid–plastic limit, and other physical properties of the expansive soil, were tested according to the standard geotechnical test methods. The results are shown in Table 1. The expansive soil was dried and crushed through a 1 mm sieve to make soil powder for subsequent use.

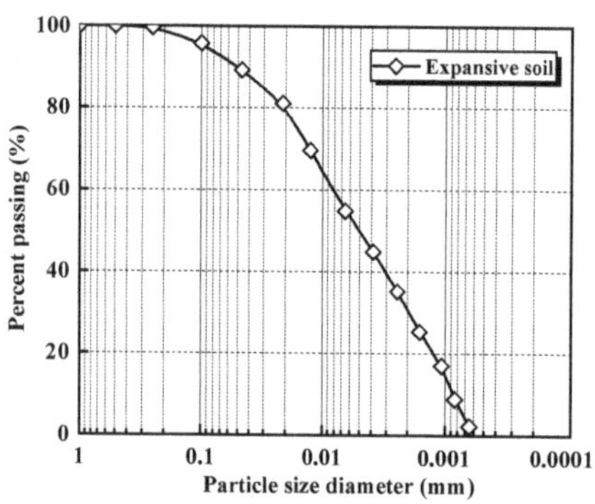

Figure 2. Particle grading curve of expansive soil.

Table 1. Physical properties of expansive soil.

Properties	Value
Natural moisture content	6.10%
Optimum moisture content	16.00%
Maximum dry density	1.853 $g \cdot cm^{-3}$
Specific gravity	2.73
Plastic limit	22.30%
Liquid limit	57.80%
Plasticity index	35.50
Free swell ratio	71.00%

Rubber is a composite material with high elasticity, durability, and friction resistance. The rubber fibers used in the tests were produced at the tire processing plant in Deyang, Sichuan, China. It was sieved to remove fine particles and coarse debris. Then, fibers with a length of approximately 15 mm and a diameter of approximately 1 mm were selected.

2.1.2. Sample Preparation and Saturation

The soil was initially prepared with an optimal moisture content. Then, it was mixed with fibers until the mixture was homogeneous. Subsequently, the sample was prepared. A mixture of rubber fibers and expansive soil was prepared via layered compaction to improve the random and uniform distribution of fibers in the sample. During the compaction process, the layers were scraped to ensure good bonding between adjacent layers of the mixture. The prepared samples were saturated using the pumping saturation method, placed in a vacuum saturation cylinder for 2 h, and then saturated by immersion in water for 48 h.

In this study, the rubber fiber content was defined as the ratio of the total mass of rubber to that of the rubber fiber-reinforced expansive soil. This is shown in Equation (1). Soltani et al. [28] found that when the rubber content exceeds 10%, the performance of the specimen changes greatly. Therefore, 0, 5,10, 15, and 20% rubber contents were selected for this study.

$$f = \frac{m_R}{m_s + m_\omega + m_R} \times 100\% \tag{1}$$

where m_R is the mass of rubber fibers, m_s is the mass of soil, and m_ω is the mass of water in the mixture.

2.2. *Triaxial Test*

The test instrument was a TSZ automatic triaxial instrument with computer control and automatic data acquisition systems. A pressure transducer was used to measure the pore water pressure inside the sample. A load cell and linear variable differential transducer (LVDT) measured the load and strain.

The displacement-controlled undrained shear tests were performed at a rate of 0.08 mm/min. The confining pressure p'_0 was set to 100, 200, and 300 kPa, respectively. The displacement-controlled drained shear tests were performed at a rate of 0.0096 mm/min with the above-mentioned confining pressure. The triaxial test scheme for the rubber fiber-reinforced expansive soil is presented in Table 2.

Table 2. Triaxial test scheme for the rubber fiber-reinforced expansive soil.

Sample	Rubber Fiber Content f (%)	Shearing Rate (mm min^{-1})	Drainage Condition
ESR-1	0		
ESR-2	5		
ESR-3	10	0.08	undrained triaxial test
ESR-4	15		
ESR-5	20		
ESR-6	0	0.0096	drained triaxial test
ESR-7	10		

2.3. *Nuclear Magnetic Resonance (NMR) Test*

The nuclear magnetic moment formed by the nuclear spin absorbs energy and transitions to a high-energy state. This releases energy in the form of electromagnetic waves to form a receptible radio signal known as NMR. It is a highly effective tool for studying water migration, pore size, and characteristic distribution in porous media. A PQ001-PM NMR analyzer (Suzhou Niumag Analytical Instrument Corporation, Suzhou, Jiangsu Province, China) was used. The triaxial test scheme for the rubber fiber-reinforced expansive soil is listed in Table 3.

Table 3. Nuclear magnetic resonance test scheme.

Sample	Rubber Fiber Content f (%)	Frequency (kHz)	Echo Interval (ms)
ESR-8	0		
ESR-9	5		
ESR-10	10	333	0.08
ESR-11	15		

3. Results and Discussion

3.1. Undrained Behavior of Expansive Soil-Rubber Fiber (ESR)

Figure 3 shows the deviator stress, effective stress path, and pore water pressure for the ESR with different rubber fiber contents at a confining pressure of 200 kPa. The test results show that the deviator stress of the ESR increased with the rubber fiber contents. The deviator stress of ESR-1 is the smallest and that of ESR-3 is the largest.

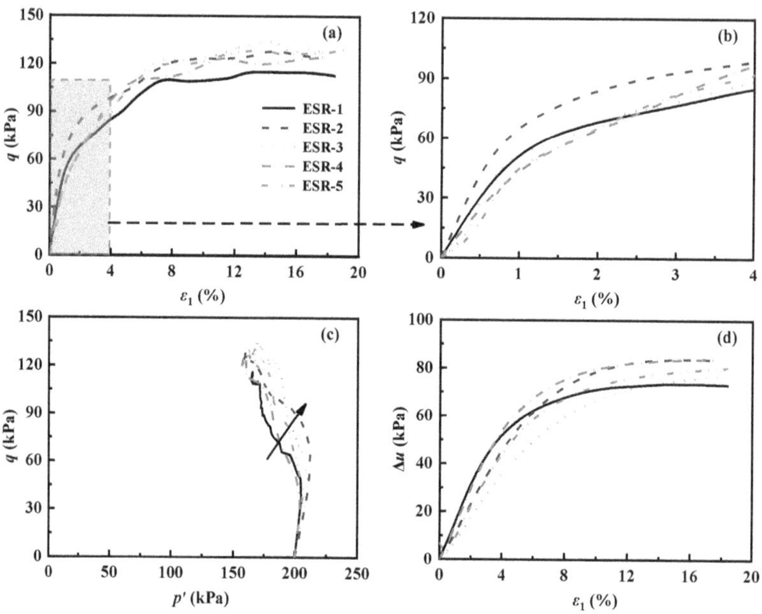

Figure 3. Results of undrained triaxial tests of ESR with different rubber fiber contents at 200 kPa confining pressure: (**a**,**b**) deviator stress–strain, (**c**) effective stress path strain behavior, (**d**) pore water pressure.

Compared with ESR-1, the stress–strain response of ESR varied with the addition of rubber fibers, as shown in Figure 3a. ESR-1 exhibits an initial strain hardening behavior in the nonlinear deviator stress–strain curve. It attains a peak deviator stress of 116.2 kPa under 14% axial strain. As the shear proceeds, the ESR tends to display strain softening behavior. This is manifested by a decrease in deviator stress. The peak deviator stress decreases from 116.2 kPa to 112.1 kPa. This agrees with the results of Alam et al. [29]. That study investigated the mechanical response of root-reinforced loess by triaxial tests. ESR-2, ESR-3, ESR-4, and ESR-1 showed similar deviator stress–strain trends. Moreover, significant deviator stress peaks were observed, namely, 128.6 kPa, 134.5 kPa, and 124.4 kPa, respectively. Conversely, no significant peak deviator stress was observed for ESR-5. Figure 3d shows the pore water pressure curves for the ESR. These exhibit a trend similar to that of the deviator stress. The maximum pore water pressure (80.5 kPa) was obtained for

ESR-4, and the minimum (73.2 kPa) was obtained for ESR-1. During the undrained shear test, the pore water pressure was related to the tendency of contraction and expansion. The effective stress path of ESR is shown in Figure 3c. The slope of the ESR stress path to the left indicates that it generates a positive pore water pressure during shearing, which reflects the contraction behavior of the ESR (Figure 3c). The deviator stress-strain curve for the ESR in the 0-4% strain range was selected, as shown in Figure 3b. Figure 3b shows that under equal deviating stress, the strain of the rubber fiber-reinforced expansive soil was higher than that of the unreinforced expansive soil. This indicated that the rubber fibers reduced the initial stiffness of the expansive soil.

The undrained triaxial test results of the ESR at confining pressures of 100, 200, and 300 kPa are shown in Figures 4 and 5. The results reveal that the incorporation of rubber fibers increased ESR deviator stress, which was more significant at higher confining pressures (200 and 300 kPa). However, the effect on the deviator stress at 100 kPa was negligible. This is consistent with the results reported by Freilich et al. [30], who observed that the effectiveness of fibers for soil shear strength enhancement increases with an increase in confining pressure. Prior to shearing, the interaction between the soil particles and fibers improved owing to the increased confining pressure, and the fibers stretched gradually. As the presence of well-dispersed rounded particles destroys the sliding surface to increase the shear strength, the rougher surface of the rubber fibers makes the adhesion between the fibers and the soil play a significant role. The adhesion between the rubber fibers and the soil particles is greater than the friction force within the soil.

At the beginning of the shear phase, the fibers improved the distribution of the applied shear stress over a wider area. However, at lower confining pressures, the fibers may not interact effectively with the soil particles and may slip in shear. Thus, at relatively low confining pressures, fibers may not effectively confine soil particles or increase the strength of fiber-reinforced soil.

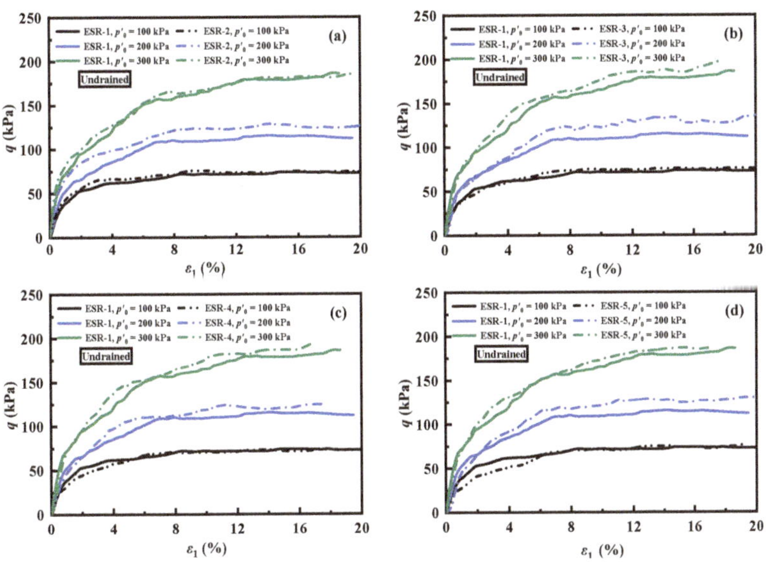

Figure 4. Deviator stress–strain of reinforced expansive soil and unreinforced soil under undrained conditions: (**a**) ESR-2, (**b**) ESR-3, (**c**) ESR-4, (**d**) ESR-5.

The pore water pressures of reinforced expansive soil and unreinforced soil under undrained conditions are shown in Figure 5. With the increase in rubber fiber content, the pore water pressure gradually increased and then decreased, and the maximum was obtained at ESR-4. The pore water pressures at 100, 200, and 300 kPa of ESR-1 were 40.6,

81.1, and 92.3 kPa, respectively, while the corresponding pore water pressures of ESR-4 were 44.3, 83.9, and 98.1 kPa, which were 9.1%, 3.5% and 6.3% higher, respectively. This indicated that the increase in pore water pressure is related to the distribution of fibers in the soil. It was noted that the effect of the confining pressure on the pore water pressure gradually increased as the confining pressure increased.

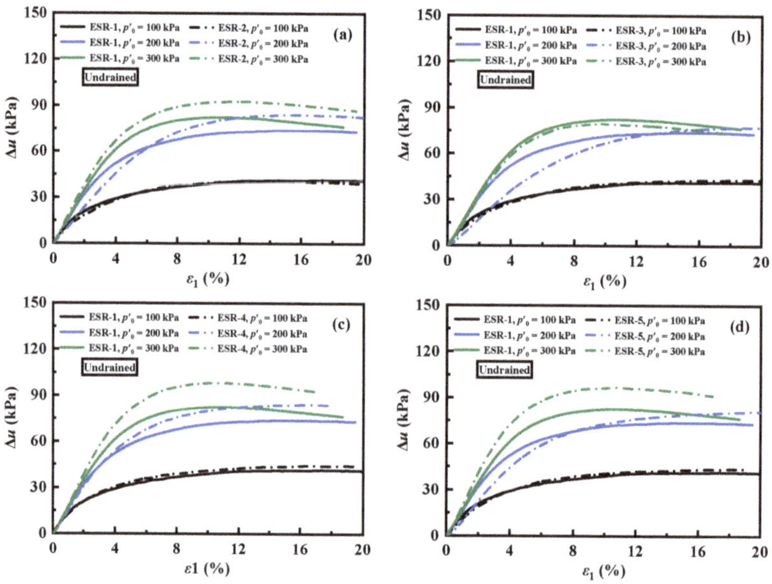

Figure 5. Pore water pressures of reinforced expansive soil and unreinforced soil under undrained conditions: (**a**) ESR-2, (**b**) ESR-3, (**c**) ESR-4, and (**d**) ESR-5.

3.2. Drained Behavior of ESR

The deviator stress–strain curves and volume strain–strain curves of the ESR with 0 and 10% rubber fiber content in the drained tests are shown in Figures 6 and 7, respectively. The tested ESR showed different damage patterns. The middle part of the unreinforced expansive soil (ESR-6) showed a significant uplift, whereas the uplift of the rubber fiber-reinforced expansive soil (ESR-7) was distributed uniformly along the height of the sample. This indicated that the rubber fibers played an important role in transferring the stress and strain inside the sample.

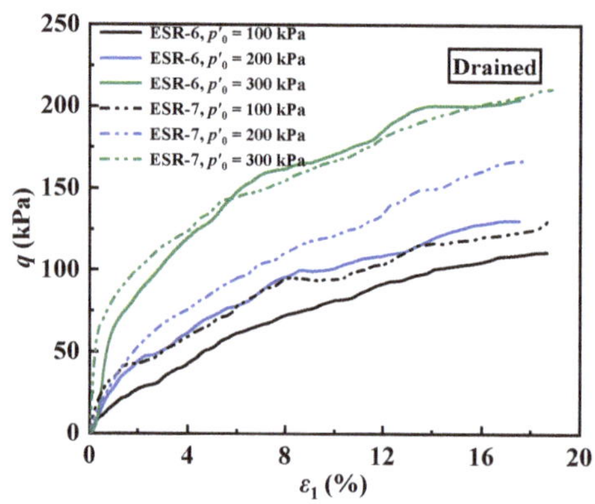

Figure 6. Deviator stress–strain curves of ESR-6 and ESR-7 in the drained tests.

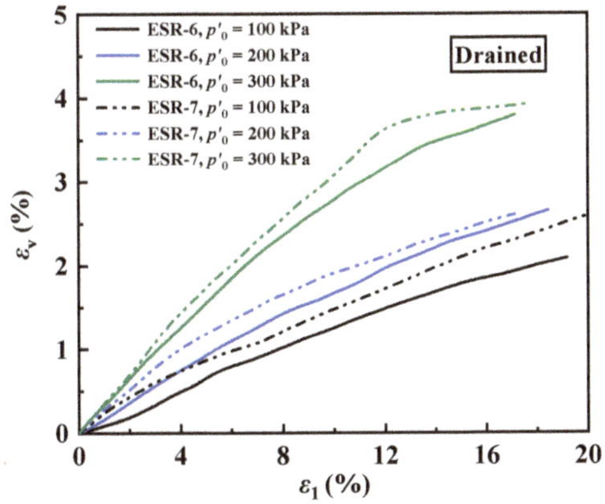

Figure 7. Volume strain of ESR-6 and ESR-7 in the drained tests.

As anticipated, the deviator stress and shrinkage increased as the rubber fiber content and confining pressure increased. Under equal confining pressure, the strain of the reinforced soil when attaining the maximum deviator stress was significantly higher than that of the unreinforced soil. This indicated that the rubber fiber significantly improved the ductility of the soil. This was in agreement with the study of the effect of plant roots on the mechanical properties of the soil by Xu et al. [31]. With an increase in strain, the deviator stress of ESR-6 attains a peak at 16% strain. After the plastic deformation of ESR-7, the resistance of the fibers was activated. There was no apparent peak deviator stress, which was manifested in the form of strain hardening. At confining pressures of 100, 200, and 300 kPa, the deviator stress of ESR-7 increased from 111.8, 130.6, and 204.3 kPa to 130.4, 177.8, and 210.8 kPa, respectively. This corresponded to increases of 16.6%, 36.1%, and 3.2%, respectively. Safdar et al. [32] indicated that this could be owing to frictional interactions between the particles and fibers, as well as an increase in tensile strength provided by the fiber modulus.

With regard to the volume strain–strain response, the volume strain–strain curves of ESR-6 and ESR-7 exhibit compressive behavior throughout the shear process until the end of the test, as shown in Figure 7. Under equal confining pressure, the volume variations of ESR-6 and ESR-7 were marginally different. ESR-7 exhibited a larger initial shrinkage than ESR-6 and was distributed over a larger strain range. This may have been owing to the compression of the rubber fiber. Correia et al. [33] also observed that fiber reinforced soil contracts more than unreinforced soil during shearing.

3.3. Strength Characteristics

The relationship between the shear stress and effective stress for the unreinforced expansive soil and 10% rubber fiber-reinforced expansive soil is shown in Figure 8. Here, the deviator stress of the undrained samples is lower than that of the drained samples. The difference decreases with an increase in the effective stress. Simultaneously, the effective stress of the undrained samples shifts to the left and the slope of the curve decreases owing to the pore water pressure.

For the drained and undrained conditions, the effective cohesion under the drained conditions was higher than that under the undrained conditions. The converse was true for the effective angle of internal friction. This is attributed to the effects of the pore water pressure and creep at the fiber–soil interface. Under drained conditions, owing to the lower strain rate, the test time increased significantly, and the influence of time on the interfacial strength (such as the time-varying slip or rearrangement of soil particles) was more evident. However, the pore water in the drained conditions was drained out of the soil, the stress was sustained by the soil skeleton and rubber fibers, and the occlusion capacity between the soil particles and fibers improved. Similarly, Freilich et al. [30] conducted consolidation undrained and consolidation drained triaxial tests on polypropylene fiber-reinforced soil. They observed that the effective shear strength parameters of the reinforced soil were higher under the undrained conditions. Özkul and Baykal [25] evaluated the shear strength of rubber fiber-reinforced soil under undrained and drained conditions. They observed that although the angle of internal friction was approximately equal for both conditions, the cohesion under the undrained conditions was higher than that under the under drained conditions.

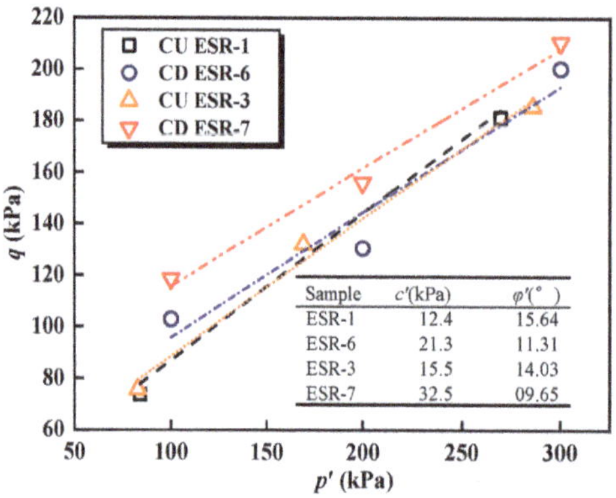

Figure 8. The relationship between the deviator stress and effective stress for the unreinforced expansive soil and 10% rubber fiber-reinforced expansive soil.

The addition of rubber fibers increased the effective cohesion of the samples from 12.4 kPa (ESR-1) to 15.5 kPa (ESR-3). This corresponded to an increase of 25%. From ESR-6 to ESR-7, the effective cohesion increased from 21.3 kPa to 32.5 kPa (an increase of 52.6%). Meanwhile, the effective internal friction angle varied marginally. This is similar to the results obtained by Valipour et al. [34]. This indicates that rubber fibers can effectively interlock soil particles, improve the interaction between the fiber surface and soil matrix, and increase the effective cohesion [35,36].

3.4. *Stiffness Characteristics*

3.4.1. Structural Characterization Based on Tangential Stiffness

Observing that ESR yields under shear, Malandraki and Toll [37] and Wang et al. [38] indicated that the yield point can be determined from the curve of the tangential stiffness $E_{\tan}$ and axial strain in double logarithmic coordinates. The formula for calculating $E_{\tan}$ refers to the generalized Hooke's law, as shown in Equation (2):

$$E_{\tan} = \left(\mathrm{d}\sigma_1{}' - 2\nu \mathrm{d}\sigma_3{}'\right)/\mathrm{d}\varepsilon_1 \quad (2)$$

where $\mathrm{d}\sigma_1{}'$ is the first effective principal stress increment, $\mathrm{d}\sigma_3{}'$ is the third effective principal stress increment, $\mathrm{d}\varepsilon_1$ is the axial strain increment, and ν is Poisson's ratio.

From the consolidated drained condition, $\sigma_3{}'$ is a constant, and $\mathrm{d}\sigma_3{}' = 0$. The shear tangential stiffness $E_{\tan}$, can be expressed as follows:

$$E_{\tan} = \frac{\mathrm{d}\sigma_1{}'}{\mathrm{d}\varepsilon_1} \quad (3)$$

To eliminate the influence of the average effective stress on the tangential stiffness during the shear process, $E_{\tan}$ was normalized. Figure 9 compares the $E_{\tan}/p'$–ε_1 relationship curves for different rubber fiber contents at a confining pressure of 100 kPa with that of rubber-reinforced soil determined by other scholars [36,39,40]. From the figure, when

$\varepsilon_1 < 1.2\%$, the ESR of E_{tan}/p' remained at approximately 20, although it reduced marginally. However, the magnitude of the reduction was not apparent. When $1.2\% < \varepsilon_1 < 15\%$, as the shear progressed, the samples began to yield structurally, and E_{tan}/p' decreased gradually with the increase in ε_1. When $15\% < \varepsilon_1$, the rate of variation in the normalized stiffness with the axial strain further increased, the curve was essentially close to being vertical, and the samples underwent overall failure. The relationship between E_{tan}/p' and ε_1 reveals an apparent yield point in the unreinforced soil. It corresponds to the destruction of the original arrangement and cementation of soil particles. After the addition of rubber, there was no apparent yield point in the reinforced soil. Yadav and Tiwari [38] indicated that the stretching of rubber would prevent the formation of face-to-face arrangement of soil particles.

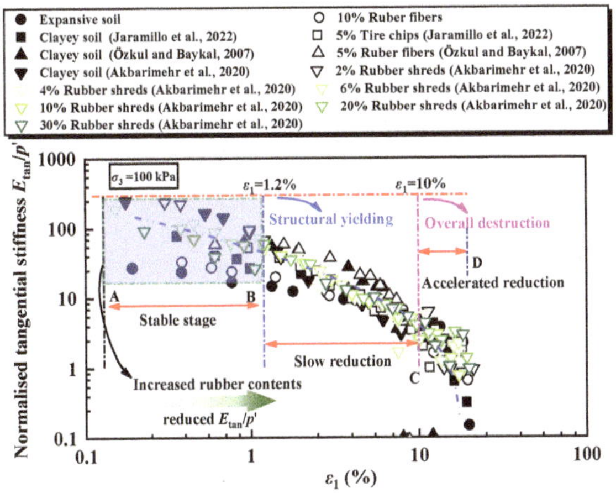

Figure 9. E_{tan}/p'–ε_1 relationship curves for different rubber contents [25,36,39].

Although the addition of rubber increased the shear strength of the mixture, E_{tan}/p' decreased gradually as the rubber content increased. This is attributed to the temporary compression of the rubber fibers when subjected to confining pressure or kinking. This is in agreement with the observations of Akbarimehr et al. [39] and Özkul and Baykal [25]. Similarly, Gray and Ohashi [41] determined that fibers with a higher modulus tended to increase the shear strength and failed to increase the stiffness of the mixture. Meanwhile, Michalowski and Čermák [42] demonstrated that fibers reduce the stiffness or modulus.

3.4.2. Maximum Dilatation Angle

In continuum mechanics, the most widely used parameter for quantifying expansion is the dilation angle ψ. It can be obtained from triaxial tests as the ratio of the volume strain increment to the axial strain increment. Zhao and Cai [43] defined the physical significance of ψ as frictional sliding along microcracks or particles.

The maximum dilatation angle under triaxial conditions was estimated according to the equation proposed by Vaid and Sasitharan [44]:

$$\psi_{max} = \sin^{-1}\left(\frac{2}{\frac{3}{|d\varepsilon_v/d\varepsilon_1|_{max}} - 1}\right) \tag{4}$$

where $|d\varepsilon_v/d\varepsilon_1|_{max}$ is the maximum rate of volume variation or expansion.

Figure 10 shows the variation in the maximum dilatation angle ψ_{max} with the rubber fiber content. It is positively correlated with the rubber fiber content for a given confining pressure [40,45]. It is demonstrated that the rubber affects the maximum dilatation angle of the ESR. This is because when the ESR sustains pressure, the rubber fiber sustains part of the stress transmission. Because of its deformable characteristics, the porosity ratio of the sample reduces. This considerably increases the compressibility of the sample. Perez et al. [46] determined that an increase in the size of the rubber also increases the maximum dilatation angle.

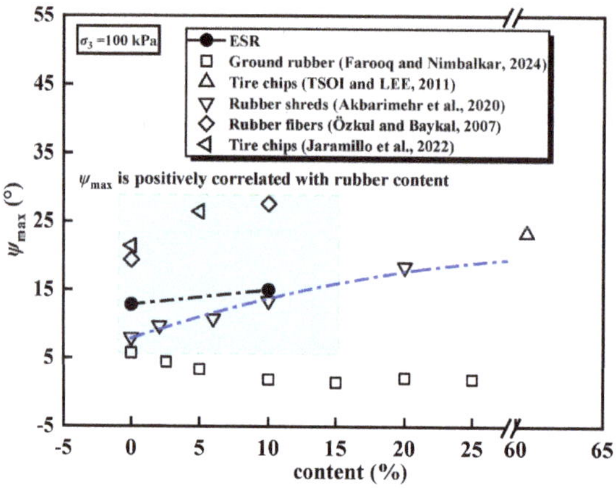

Figure 10. The variation in ψ_{max} with the rubber content [25,36,39,45,47].

3.5. Effect of Rubber Fibers on Pore Size Distribution

Relaxation refers to the process by which the spin of a proton is stabilized from an excited state to an equilibrium state. Transverse relaxation refers to the time required for the maximum excited state to approach zero. The transverse relaxation time is determined using the proton relaxation time T_{2B} and surface relaxation time T_{2S}.

$$\frac{1}{T_2} = \frac{1}{T_{2B}} + \frac{1}{T_{2S}} \approx \rho_2\left(\frac{S}{V}\right) \tag{5}$$

where ρ_2 is the surface relaxation rate, which is related to the physicochemical properties of the soil. S and V are the surface area and pore volume, respectively, of the porous medium.

The porosity and pore distribution characteristic parameters of the soil were calculated using the NMR transverse relaxation time T_2 spectrum. The relationship between the relaxation time and pore size is shown in Equation (6). From this equation, T_2 is linearly related to the pore size. Therefore, if the value of the surface relaxation rate ρ_2 is known, the pore-size distribution in the ESR can be obtained by measuring T_2. According to Lu [48], the ρ_2 of expansive soil is considered as 3 $\mu m\ s^{-1}$.

$$\frac{1}{T_2} \approx \rho_2\left(\frac{S}{V}\right) \approx \rho_2\left(\frac{2}{R}\right) \tag{6}$$

Figure 11 shows the relaxation curves of the ESR for different rubber fiber contents. The variation characteristics of the T_2 spectrum for different rubber fiber content were almost identical. This indicates that the distribution of internal pores was approximately identical. The T_2 spectra of the ESR with different fiber contents consist of two spectral peaks: a primary peak and secondary peak. The secondary peak is significantly smaller than the main peak. This indicates that the internal pore size had good continuity. Using ESR-8 as an example, the relaxation time corresponding to the vertex was 1.589 ms. Of this, the main peak accounted for 97.6%. To compare the data more visually, the spectral distributions of the other rubber fiber contents are presented in Table 4. As shown in Figure 11, the T_2 spectrum curve shifts to the left to a certain extent as the rubber fiber content increases. Comparing ESR-8 and ESR-11, the relaxation time corresponding to the main peak is equal (1.589 ms). However, the peak area of ESR-11 is larger than that of ESR-8.

This indicates that when the rubber fiber content attained 15%, the pores in the sample were destroyed because of the excessive rubber fiber content. The fibers were intertwined into a clump, and the porosity increased. This affected the mechanical properties of the expansive soil. When the rubber content is 5–10%, the relaxation time of the main peak of the spectrum is 1.383 ms. This is less than that of ESR-8 and ESR-11. This result reveals that when the rubber fiber content is in the range of 5–10%, the interface between the expansive soil particles and rubber fibers displays effective bonding. This can effectively inhibit the development of soil cracks and reduce the pore space around the cracks.

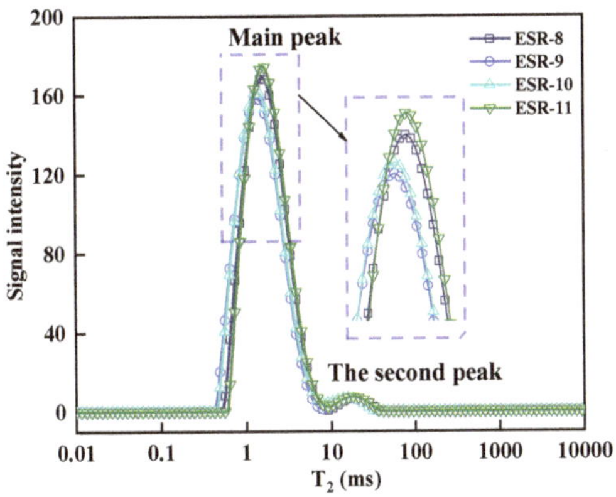

Figure 11. T_2 spectrum for different rubber fiber contents.

Table 4. T_2 spectrum distribution of the ESR for different rubber fiber content.

Sample	Beginning and Ending Relaxation Times of Spectrum Peaks (ms) (Peak Percentage)	
	Primary Peak	Secondary Peak
ESR-8	0.561–9.011 (97.6%)	9.659–36.123 (2.4%)
ESR-9	0.455–7.843 (97.6%)	8.407–33.701 (2.4%)
ESR-10	0.488–8.407 (97.4%)	9.011–31.440 (2.6%)
ESR-11	0.644–9.011 (97.3%)	9.659–36.123 (2.7%)

To quantitatively analyze the pore number distribution of ESR with different diameters, it is necessary to subdivide the pore size of the ESR. The pore diameter of the ESR at different rubber fiber contents can be obtained using Equation (6). According to Zhou et al. [49], the pore diameters of soil can be divided into five types: macropores (>3 μm), medium pores (1–3 μm), small pores (0.2–1 μm), micropores (0.01–0.2 μm), and ultra-micropores (<0.01 μm).

Figure 12a, b show the pore size distribution curves and pore content curves, respectively, of the ESR with different rubber fiber content. As shown in Figure 12a, the pore diameters of the ESR were less than 0.5 μm. This reflected the tight pore structure inside the sample. After the addition of an appropriate amount of rubber fiber (5–10%), Amplifying the ESR pore size distribution curve, it was found that with the increase of rubber content, the ESR pore size distribution curve shifted to the left, and the proportion of peak pore size distribution decreased first and then increased. This indicated that the porosity of expansive soil could be reduced by adding rubber fiber.

The pore content curves of the ESR with different rubber fiber contents are shown in Figure 12b. From Figure 12b, the ESR micropores accounted for the largest percentage of the number of pores. It was followed by the ultra-micropores, and the fewest number of small pores. The pore size inside the ESR was mainly distributed between 0.01 μm and 0.2 μm. The pore size of approximately 0.02 μm accounted for the largest proportion (approximately 4.2% of all the pores in the ESR). The addition of an appropriate amount of fiber (5–10%) decreased the proportion of small pores and micropores and increased the proportion of ultra-micropores. However, the addition of more rubber fibers (15%) increased the percentage of small pores and micropores. This weakened the internal structure of the sample.

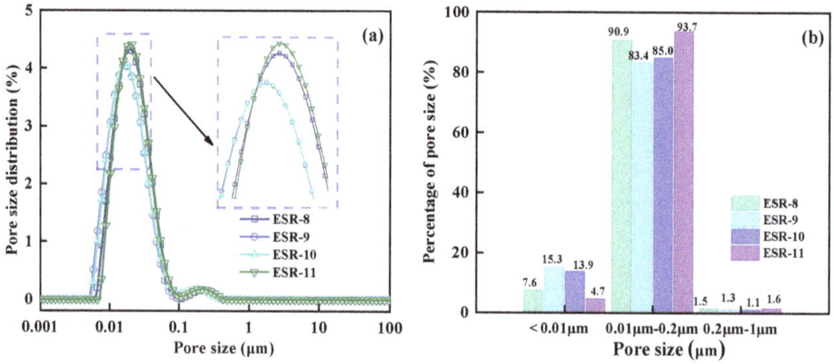

Figure 12. Pore size distribution and pore contents curves of the ESR with different rubber fiber content (**a**,**b**).

3.6. Shear Shrinkage Analysis Based on Classical Dilatation Equation

In recent years, scholars have proposed different forms of dilatation functions to describe the relationship between plastic volume strain and plastic strain. Table 5 summarizes the more commonly used stress–dilatancy relationship. Here, η is the stress ratio, and M is the critical stress ratio. The dilatancy rate and stress ratio of the triaxial rain test were orthogonalized according to the critical state parameters M (plotted in Figures 13 and 14) and compared with the dilatancy equation curves of the P.W. Rowe, Cambridge, and Modified Cambridge models.

Table 5. Several commonly used stress–dilatancy relationships.

Name of the Model	Formula
P.W. Rowe [50] model	$D = 9(M - \eta)/(9 + 3M - 2M\eta)$
Cambridge model	$D = M - \eta$
Modified Cambridge model	$D = (M^2 - \eta^2)/(2\eta)$

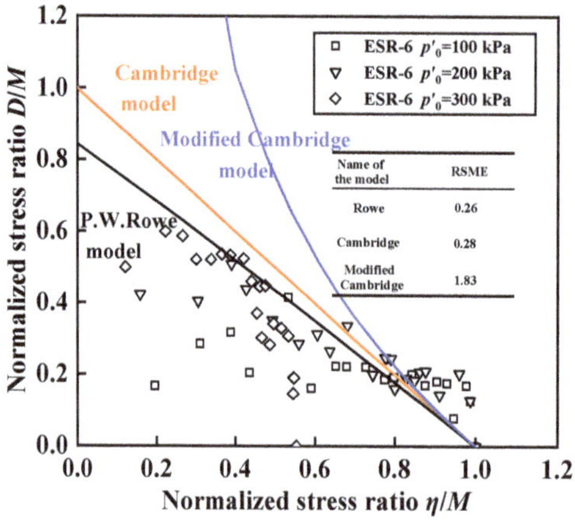

Figure 13. Comparison of dilatancy equations with test results (ESR-6).

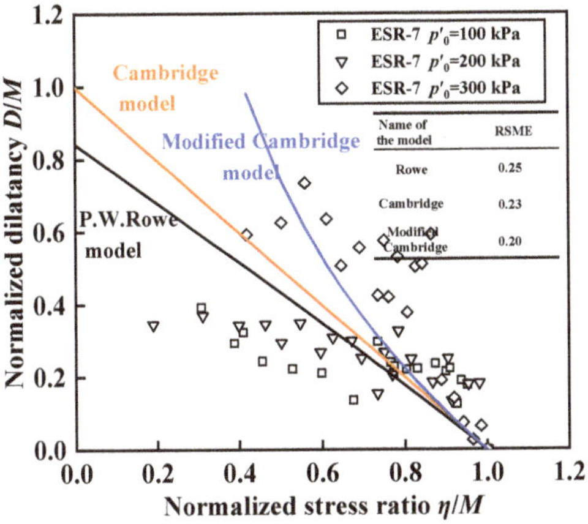

Figure 14. Comparison of dilatancy equations with test results (ESR-7).

The deviation between the model and the measured values is calculated using the root mean square error (RSME) method; the smaller the RSME, the more accurate the model is reflected. The relationship between the dilatancy rate and stress ratio for ESR-6 is shown in Figure 13. With an increase in the stress ratio, the dilatancy first increased to a peak and then decreased to zero. The Rowe model has a minimum RMSE of 0.26. The dilatancy equations of the Rowe model generally reflect the shear shrinkage pattern of ESR-6, whereas the Cambridge model and modified Cambridge model both overestimate its shear deformation. The relationship between the dilatancy rate and stress ratio of ESR-7 is shown in Figure 14. The figure shows that the confining pressure had a significant effect on the dilatancy rate of ESR-7. At low confining pressure (100 kPa and 200 kPa), it was found that the Rowe model could better describe the shear shrinkage of ESR-6 and ESR-7. Comparing the dilatancy rates of ESR-6 and ESR-7, the shear shrinkage deformation of ESR-7 was larger. This was more evident under a high confining pressure (300 kPa). At this time, the modified Cambridge model has the smallest RSME of 0.20; the dilatancy of ESR-7 was similar to that of the Modified Cambridge model. Similarly, To et al. [51] compared the dilatancy rate and dilatancy equation for fully weathered granite. They observed the Cambridge and Rowe models to be in better agreement with the test results. Meanwhile, the modified Cambridge model overestimated the dilatancy rate of granite.

4. Conclusions

To examine the effect of drainage conditions and rubber fibers on the mechanical properties of expansive soil, undrained triaxial, drained triaxial consolidation, and NMR ESR tests were performed in this study. The effects of rubber fibers on the ESR stress–strain relationship, tangential stiffness, maximum dilatation angle and pore size distribution were investigated. The main conclusions are as follows:

(1) With an increase in strain, the ESR deviator stress and pore water pressure for different rubber fiber contents increased nonlinearly. The ESR showed compressive behavior during the shearing process. The rubber fiber-reinforced expansive soil showed a higher initial contraction than the unreinforced expansive soil.

(2) The shear strength of ESR was affected significantly by the drainage conditions, which were influenced by the pore water pressure and creep of the fiber–soil interface. Under the consolidated drained conditions, the deviator stress and effective cohesion of the ESR were higher than those under the undrained conditions. The converse was true for the internal friction angle.

(3) After the addition of rubber fibers, the shear strength of the expansive soil improved. However, its stiffness reduced. After a significant addition of rubber fibers to the

unreinforced soil, the stretching effect of the rubber inhibited the formation of a face-to-face arrangement of soil particles. This alleviated the yield phenomenon of the ESR.

(4) The addition of an appropriate amount of rubber fiber (5–10%) reduced the proportion of small pores and micropores in the ESR, increased the proportion of ultra-micro pores, and inhibited the development of soil cracks. When the rubber content was excessively high (15%), the fibers interweaved into clusters. This resulted in an increased porosity and a weakened internal structure of the ESR.

This study provides a reference for the application of rubber fibers in the backfill, and the beneficial reuse of recycled tires provides a sound environmental alternative to the safe disposal problems associated with waste materials.

Author Contributions: Conceptualization, Z.Y.; data curation, writing—original draft, R.W.; methodology, W.S.; supervision, Z.S.; project administration, X.L. All authors have read and agreed to the published version of the manuscript.

Funding: This research was funded by the National Natural Science Foundation of China, grant number 42177153, and the National Major Scientific Research Instrument Development Project, grant number 41627801.

Institutional Review Board Statement: Not applicable.

Informed Consent Statement: Not applicable.

Data Availability Statement: The original contributions presented in the study are included in the article, further inquiries can be directed to the corresponding author/s.

Acknowledgments: The authors thank the School of Civil Engineering, Qingdao University of Technology for providing the NMR instrument.

Conflicts of Interest: The authors declare no conflicts of interest.

References

1. Ouyang, Q.; Xiao, H.; Li, Z.; Jiang, W.; Su, H.; Tian, X.; Ouyang, M. Experimental Study on the Influence of Microbial Content on Engineering Characteristics of Improved Expansive Soil. *Front. Earth Sci.* **2022**, *10*, 863357. [CrossRef]
2. Liang, C.; Yin, Z.Z.; Zhang, P. Relationship of resistivity with water content and fissures of unsaturated expansive soils. *J. China Univ. Min. Technol.* **2007**, *17*, 537–540. [CrossRef]
3. Fityus, S.; Smith, D.; Allman, M. Expansive soil test site near Newcastle. *J. Geotech. Geoenviron. Eng.* **2004**, *130*, 686–695. [CrossRef]
4. Rabab'ah, S.; Al Hattamleh, O.; Aldeeky, H.; Alfoul, B.A. Effect of glass fiber on the properties of expansive soil and its utilization as subgrade reinforcement in pavement applications. *Case Stud. Constr. Mater.* **2021**, *14*, e00485. [CrossRef]
5. Rezaei, M.; Ajalloeian, R.; Ghafoori, M. Geotechnical properties of problematic soils emphasis on collapsible cases. *Int. J. Geosci.* **2012**, *3*, 105–110. [CrossRef]
6. Dai, Z.; Chen, S.; Li, J. *The Failure Characteristics and Evolution Mechanism of the Expansive Soil Trench Slope*; PanAm Unsaturated Soils 2017: Dallas, TX, USA, 2017; pp. 196–205. [CrossRef]
7. Shi, B.; Jiang, H.; Liu, Z.; Fang, H. Engineering geological characteristics of expansive soils in China. *Eng. Geol.* **2002**, *67*, 63–71. [CrossRef]
8. Estabragh, A.; Rafatjo, H.; Javadi, A. Treatment of an expansive soil by mechanical and chemical techniques. *Geosynth. Int.* **2014**, *21*, 233–243. [CrossRef]
9. Mahedi, M.; Cetin, B.; White, D.J. Cement, lime, and fly ashes in stabilizing expansive soils: Performance evaluation and comparison. *J. Mater. Civ. Eng.* **2020**, *32*, 04020177. [CrossRef]
10. Indiramma, P.; Sudharani, C.; Needhidasan, S. Utilization of fly ash and lime to stabilize the expansive soil and to sustain pollution free environment–An experimental study. *Mater. Today Proc.* **2020**, *22*, 694–700. [CrossRef]
11. Lu, Y.; Liu, S.; Zhang, Y.; Li, Z.; Xu, L. Freeze-thaw performance of a cement-treated expansive soil. *Cold Reg. Sci. Technol.* **2020**, *170*, 102926. [CrossRef]
12. Mahedi, M.; Cetin, B.; White, D.J. Performance evaluation of cement and slag stabilized expansive soils. *Transp. Res. Rec.* **2018**, *2672*, 164–173. [CrossRef]
13. Tiwari, N.; Satyam, N. An experimental study on the behavior of lime and silica fume treated coir geotextile reinforced expansive soil subgrade. *Eng. Sci. Technol. Int. J.* **2020**, *23*, 1214–1222. [CrossRef]
14. Ikeagwuani, C.C.; Nwonu, D.C. Emerging trends in expansive soil stabilisation: A review. *J. Rock Mech. Geotech. Eng.* **2019**, *11*, 423–440. [CrossRef]

15. Al-Omari, R.R.; Fattah, M.Y.; Ali, H.A. Treatment of soil swelling using geogrid reinforced columns. *Ital. J. Geosci.* **2016**, *135*, 83–94. [CrossRef]
16. Dessouky, S.H.; Oh, J.; Ilias, M.; Lee, S.I.; Park, D. Investigation of various pavement repairs in low-volume roads over expansive soil. *J. Perform. Constr. Facil.* **2015**, *29*, 04014146. [CrossRef]
17. Malekzadeh, M.; Bilsel, H. Hydro-mechanical Behavior of Polypropylene Fiber Reinforced Expansive Soils. *KSCE J. Civ. Eng.* **2014**, *18*, 2028–2033. [CrossRef]
18. Chittoori, B.; Rahman, T.; Burbank, M. Microbial-facilitated calcium carbonate precipitation as a shallow stabilization alternative for expansive soil treatment. *Geotechnics* **2021**, *1*, 558–572. [CrossRef]
19. Wang, H.; Zhang, K.; Gan, L.; Liu, J.; Mei, G. Expansive soil-biochar-root-water-bacteria interaction: Investigation on crack development, water management and plant growth in green infrastructure. *Int. J. Damage Mech.* **2021**, *30*, 595–617. [CrossRef]
20. Hamza, M.; Nie, Z.; Aziz, M.; Ijaz, N.; Ijaz, Z.; Rehman, Z.U. Strengthening potential of xanthan gum biopolymer in stabilizing weak subgrade soil. *Clean Technol. Environ. Policy* **2022**, *24*, 2719–2738. [CrossRef]
21. Pooni, J.; Robert, D.; Gunasekara, C.; Giustozzi, F.; Setunge, S. Mechanism of enzyme stabilization for expansive soils using mechanical and microstructural investigation. *Int. J. Geomech.* **2021**, *21*, 04021191. [CrossRef]
22. Soon, N.W.; Lee, L.M.; Khun, T.C.; Ling, H.S. Improvements in engineering properties of soils through microbial-induced calcite precipitation. *KSCE J. Civ. Eng.* **2013**, *17*, 718–728. [CrossRef]
23. Thomas, B.S.; Gupta, R.C.; Panicker, V.J. Recycling of waste tire rubber as aggregate in concrete: Durability-related performance. *J. Clean. Prod.* **2016**, *112*, 504–513. [CrossRef]
24. Abbaspour, M.; Narani, S.; Aflaki, E.; Moghadas Nejad, F.; Mir Mohammad Hosseini, S. Strength and swelling properties of a waste tire textile fiber-reinforced expansive soil. *Geosynth. Int.* **2020**, *27*, 476–489. [CrossRef]
25. Özkul, Z.H.; Baykal, G. Shear behavior of compacted rubber fiber-clay composite in drained and undrained loading. *J. Geotech. Geoenviron. Eng.* **2007**, *133*, 767–781. [CrossRef]
26. Li, Z.; Tang, C.; Hu, R.; Zhou, Y. Research on model fitting and strength characteristics of critical state for expansive soil. *J. Civ. Eng. Manag.* **2013**, *19*, 9–15. [CrossRef]
27. Tajdini, M.; Nabizadeh, A.; Taherkhani, H.; Zartaj, H. Effect of added waste rubber on the properties and failure mode of kaolinite clay. *Int. J. Civ. Eng.* **2017**, *15*, 949–958. [CrossRef]
28. Soltani, A.; Deng, A.; Taheri, A.; Sridharan, A. Swell–Shrink–Consolidation Behavior of Rubber–Reinforced Expansive Soils. *Geotech. Test. J.* **2018**, *42*, 761–788. [CrossRef]
29. Alam, M.; Jiang, Y.J.; Umar, M.; Su, L.-J.; Rahman, M.; Ullah, F. Influence of drainage and root biomass on soil mechanical behavior in triaxial tests. *Acta Geotech.* **2022**, *17*, 2875–2893. [CrossRef]
30. Freilich, B.; Li, C.; Zornberg, J. Effective shear strength of fiber-reinforced clays. In Proceedings of the 9th International Conference on Geosynthetics, Guaruja, Brazil, 22–28 May 2010; pp. 1997–2000.
31. Xu, H.; Yuan, H.L.; Wang, X.Y.; Wang, D.; Chen, J.X.; Rong, C.Q. Study on the influence of root morphology and hierarchy on the mechanical properties of root-soil complexes. *J. Geotech. Eng.* **2022**, *44*, 926–935. (In Chinese)
32. Safdar, M.; Newson, T.; Shah, F. Consolidated drained (CID) behavior of fibre reinforced cemented Toyoura sand in triaxial loading conditions. *Int. J. Geo-Eng.* **2021**, *12*, 27. [CrossRef]
33. Correia, N.; Rocha, S.; Lodi, P.; McCartney, J. Shear strength behavior of clayey soil reinforced with polypropylene fibers under drained and undrained conditions. *Geotext. Geomembr.* **2021**, *49*, 1419–1426. [CrossRef]
34. Valipour, M.; Shourijeh, P.T.; Mohammadinia, A. Application of recycled tire polymer fibers and glass fibers for clay reinforcement. *Transp. Geotech.* **2021**, *27*, 100474. [CrossRef]
35. Tran, K.Q.; Satomi, T.; Takahashi, H. Improvement of mechanical behavior of cemented soil reinforced with waste cornsilk fibers. *Constr. Build. Mater.* **2018**, *178*, 204–210. [CrossRef]
36. Jaramillo, N.A.D.; dos Santos Ferreira, J.W.; Malko, J.A.C.; Casagrande, M.D.T. Mechanical behavior of clayey soil reinforced with recycled tire rubber using chips and fibers. *Geotech. Geol. Eng.* **2022**, *40*, 3365–3378. [CrossRef]
37. Malandraki, V.; Toll, D. The definition of yield for bonded materials. *Geotech. Geol. Eng.* **1996**, *14*, 67–82. [CrossRef]
38. Wang, H.B.; Zhou, Y.; Yu, G.; Zhou, B.; Zhang, A.J. Triaxial experimental study of structural granite residual soil. *Geotechnics* **2021**, *42*, 991–1002. (In Chinese)
39. Akbarimehr, D.; Eslami, A.; Aflaki, E. Geotechnical behaviour of clay soil mixed with rubber waste. *J. Clean. Prod.* **2020**, *271*, 122632. [CrossRef]
40. Yadav, J.; Tiwari, S. Effect of waste rubber fibres on the geotechnical properties of clay stabilized with cement. *Appl. Clay Sci.* **2017**, *149*, 97–110. [CrossRef]
41. Gray, D.H.; Ohashi, H. Mechanics of fiber reinforcement in sand. *J. Geotech. Eng.* **1983**, *109*, 335–353. [CrossRef]
42. Michalowski, R.L.; Čermák, J. Triaxial compression of sand reinforced with fibers. *J. Geotech. Geoenviron. Eng.* **2003**, *129*, 125–136. [CrossRef]
43. Zhao, X.; Cai, M. A mobilized dilation angle model for rocks. *Int. J. Rock Mech. Min. Sci.* **2010**, *47*, 368–384. [CrossRef]
44. Vaid, Y.; Sasitharan, S. The strength and dilatancy of sand. *Can. Geotech. J.* **1992**, *29*, 522–526. [CrossRef]
45. Tsoi, W.; Lee, K. Mechanical properties of cemented scrap rubber tyre chips. *Géotechnique.* **2011**, *61*, 133–141. [CrossRef]
46. Perez, J.L.; Kwok, C.; Senetakis, K. Micromechanical analyses of the effect of rubber size and content on sand-rubber mixtures at the critical state. *Geotext. Geomembr.* **2017**, *45*, 81–97. [CrossRef]

47. Farooq, M.A.; Nimbalkar, S. Static and cyclic performance of polyurethane foam adhesive bound soil-rubber mixtures under drained conditions. *Acta Geotech.* **2024**, *19*, 561–589. [CrossRef]
48. Lu, Z.C. Research on Dynamic Characteristics of Rubber Reinforced Expansive Soil under Freeze-Thaw Cycle. Master's Thesis, Qingdao University of Technology, Qingdao, China, 2022. (In Chinese)
49. Zhou, J.; Zhou, H.; Wang, C.; Guo, Z.; Pei, W. Analysis of moisture Migration, Temperature, and pore structure characteristics of the muddy clay column subject to artificial ground freezing Based on LF NMR. *Tunn. Undergr. Space Technol.* **2023**, *133*, 104948. [CrossRef]
50. Rowe, P.W. The stress-dilatancy relation for static equilibrium of an assembly of particles in contact. *Proc. R. Soc. Lond. Ser. A Math. Phys. Sci.* **1962**, *269*, 500–527. [CrossRef]
51. To, E.; Tham, L.; Zhou, Y. An elasto-plastic model for saturated loosely compacted completely decomposed granite. *Geomech. Geoengin. Int. J.* **2008**, *3*, 13–25. [CrossRef]

applied sciences

MDPI

Article

Mechanical Characteristics and Damage Constitutive Model of Fiber-Reinforced Cement-Stabilized Soft Clay

Tiecheng Yan [1,*], Xingyuan Zhang [1], Sutong Cai [2], Zefeng Zhou [3], Ran An [2] and Xianwei Zhang [2,*]

1 College of Civil Engineering, Longdong University, Qingyang 745000, China; 1018209084@tju.edu.cn
2 Wuhan Institute of Rock and Soil Mechanics, Chinese Academy of Sciences, Wuhan 430071, China; caisutong@outlook.com (S.C.); anran@wust.edu.cn (R.A.)
3 Advanced Modelling, Offshore Energy, Norwegian Geotechnical Institute (NGI), Sognsvn. 72, 0855 Oslo, Norway; zefeng.zhou@ngi.no
* Correspondence: 164902135@stu.cuz.com.cn (T.Y.); xwzhang@ucas.edu.cn (X.Z.); Tel.: +86-17874521240 (T.Y.); +86-13886084677 (X.Z.)

Abstract: Marine soft clays are prevalent in coastal regions of China, giving rise to engineering challenges such as salt swelling, corrosion, and load bearing in foundations with soft soil. This study is dedicated to enhancing the mechanical properties of fiber-reinforced cement-stabilized soft clay (FCSSC) and revealing its strengthening mechanism. Uniaxial compression tests are performed to explore the impact of fiber length, fiber amount, and curing ages on mechanical behavior. The stabilization mechanisms of cement and glass fibers are explored through damage analyses and microscopy. Based on the experimental results, a damage constitutive model is formulated for FCSSC, and its validity is established by comparing fitting curves with testing curves. The results demonstrate a significant improvement in the mechanical properties of the stabilized soil, attributed to the synergistic effects of the cement and glass fibers. The growth rate of the unconfined compressive strength decreased with increasing curing ages. Notably, the fiber length significantly impacted the strength index, with short-chopped fibers playing a crucial role in strength enhancement. The compressive strength exhibited an initial increase followed by a decrease with rising fiber content, reaching a maximum between 0.3% and 0.4%. The bridging effect of the glass fibers proved effective in inhibiting compression crack expansion and mitigating structural damage of the soil sample. However, excessive fiber content or length led to improved local porosity, resulting in the deterioration of strength and deformation properties. The stress–strain curves fitted using the proposed damage constitutive model accurately reflected the stress–strain relationship and deformation characteristics of the FCSSC.

Keywords: soft soil; glass fibers; compressive strength; microscopic mechanism; constitutive model

1. Introduction

Soft clay is extensively distributed in coastal bays, harbors, and inland salt lakes across China [1]. Sodium chloride (NaCl) in groundwater is acknowledged as a crucial factor contributing to the susceptibility of pavement, slopes, and foundations to failure in coastal regions [2]. Several types of soils either incorporate chloride ions in their chemical compositions or are influenced by NaCl due to their proximity to seawater sources. Within the expanse of the Port of Tianjin, the soft clay exhibits distinctive characteristics, including a high liquid limit, significant void ratio, elevated natural moisture content, low natural gravity, and diminished strength attributed to the hydrogeologic conditions [3–5]. These inherent properties render it unsuitable for direct utilization in engineering applications. The degradation of engineering properties leads to uneven deformation, cracking, and dissolution, necessitating the modification and treatment of soft clay for construction engineering [6].

The rapid development of infrastructure construction along the coastal areas of China has stimulated the development of major engineering projects, including high-speed railways, highways, airports, and nuclear power stations, all situated on soft clay terrain [7]. Consequently, the modification of soft clay has emerged as a pressing and crucial concern in geotechnical engineering. Presently, the widely adopted method for treating weak or problematic soils involves chemical curing, employing agents such as cement, lime, and other gelling agents. Utilizing cement as the curing agent, a profound mixture of soft clay and the curing agent can be achieved through techniques like grouting, anchoring, and mixed piling, among others [8]. In the context of pile foundations, this process transforms the soft clay into a robust pile characterized by integrity, water stability, and impressive bearing strength, facilitated by a series of physical and chemical reactions. Cemental soils significantly enhance the compressive strength of soft clay, mitigating the risk of excessive foundation settlement [9]. However, relying solely on traditional Portland cement as the curing agent to achieve a mixing pile that meets load bearing requirements presents practical challenges. Environmental factors and external loads can contribute to strength deterioration, cracking, and deformation in cement mixing piles, thereby posing risks to construction projects [10]. Therefore, there is an urgent imperative to develop high-performance stabilized soft clay capable of ensuring the stability of slopes and foundations [11,12].

The mechanical properties of soft clay wield significant influence on the load bearing capacity of cement mixing piles in geotechnical engineering. Critical elements such as the dosages of the gelling agent, material components, and admixtures all contribute substantially to its curing effectiveness [13–15]. Recent research endeavors have honed in on augmenting the strength of stabilized soil by incorporating various types and quantities of natural and artificial fibers [16,17]. With advancements in solid waste recycling, experiments have been undertaken to enhance the applicability of glass waste [18]. In practical applications, the incorporation of glass fibers in cement-based materials has exhibited promising outcomes. He et al. [19] delved into the influence of glass fibers on the mechanical behavior and microscopic morphology of filling slurry, highlighting the pronounced influence of fiber content on compressive and flexure strength. Via performing pullout tests and shear tests, Shi et al. [20] observed a significant improvement in the tensile strength of soft soil reinforced by glass fibers. Wang et al. [21] introduced high-titanium glass fibers into stabilized soft clay, revealing marked variations in fluidity, compressive strength, elastic modulus, and other physical parameters based on the fiber content. Schneider et al. [22] conducted mechanical experiments to scrutinize the reinforcement rules of glass fibers in cemental materials, delving into the reinforcement mechanism through microscopy. Collectively, soil reinforcement with glass fibers efficaciously elevates the bonding degree of stabilized soft clay, enhances mortar's interfacial adhesion, inhibits crack expansion, and maximizes the improvement in load bearing capacity [23]. However, the factors governing the strength of cemental soft clay with fiber reinforcement have not been comprehensively examined, underscoring the need to enrich the research in this regard.

In this study, uniaxial compression tests were conducted on fiber-reinforced cement-stabilized soft clay (FCSSC) to examine the changing patterns in its mechanical properties under different curing ages, lengths, and amounts of glass fibers. Microscopic image analysis was employed to unveil the glass fiber mechanism and identify the optimal content. Furthermore, we have established a damage constitutive model for specimens of FCSSC, verifying its accuracy and validity. The findings of this research offer theoretical and technical guidance for constructing infrastructure on grounds of marine soft clay.

2. Materials and Methods

Figure 1 presents an overview of the raw materials used and the testing procedure followed in this study. The materials employed for sample preparation, encompassing soft clay, cement, fly ash, and glass fibers, are presented in Figure 1a. The machine employed for mixing the materials is presented in Figure 1b, and a real photo of the cured soil sample is shown in Figure 1c.

Figure 1. Preparation of the specimen: (**a**) raw materials, (**b**) mixing machine, (**c**) specimen.

2.1. Testing Materials

2.1.1. Soft Clay

The experimental soft clay was sourced from a subgrade located in the coastal area of Zhuhai, Guangdong Province, China. The sampling site, situated in a coastal delta region, features a complex topography and abundant groundwater resources. Natural samples of soft clay were collected automatically using a medium-sized core drill. Based on basic ASTM standards, physical, strength, and mineralogical tests were conducted to characterize the physical properties of the material, the detailed results of which are provided in Table 1. The X-ray diffraction test, conducted following ASTM D4926-15 (2015) [24], indicated that the claystone consists of non-clay minerals including quartz (33.1%), feldspar (7.5%), chlorite (4.4%), and clay minerals including montmorillonite (12.9%) and kaolinite (42.1%). Following the ASTM standards (D4318-17e1, ASTM D6938-17ae1) [25,26], Table 1 details the physical properties of the soft clay, revealing characteristics such as its high natural water content, high compressibility, low load bearing capacity, and low permeability. The soft clay we collected falls within the category of a saturated cohesive soil based on its liquid limit value, transitioning between a soft plastic and fluid plastic state. Figure 2 depicts the particle size distribution of the soft clay, characterized by smaller particles, classifying it as a low-grade soil. The chemical composition results of the soft clay, detailed in Table 2, highlight that SiO_2 and Al_2O_3 are its dominant elements. The results of the mineral phase analysis indicate quartz to be the primary mineral phase in the soil, with sheet-like muscovite being the secondary mineral phase.

Table 1. Physical and chemical properties of the soft clay.

Properties	Moisture Content (%)	Liquid Limit (%)	Plastic Limit (%)	Plasticity Index	Relative Density	Clay Content (%)	Organic Matter Content (%)	pH Value
Value	29.40	41.54	21.01	20.53	2.71	91.10	3.71	7.91

Table 2. Chemical composition of the soft clay.

Ion Types	Cl^-	SO_4^{2-}	Al^{3+}	Mg^+	Na^+	Ca^+
Content (%)	9.4	2.3	5.0	2.4	15.1	1.6

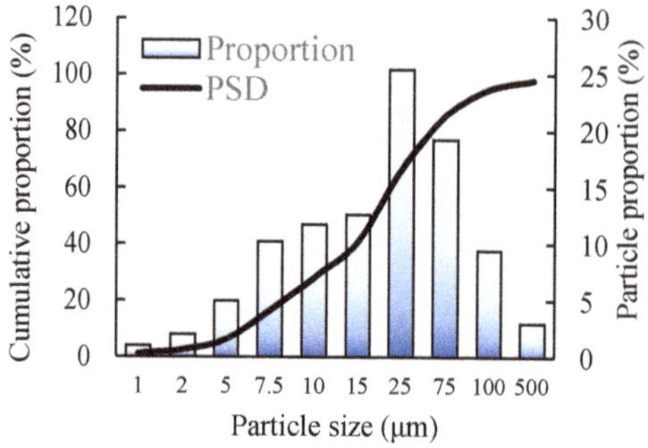

Figure 2. Particle size distribution of the soft clay.

2.1.2. Gelling Agent

The gelling agent employed for soil stabilization consists of cement and fly ash. The sulfoaluminate cement utilized here primarily comprises dicalcium silicate and anhydrous calcium sulfoaluminate as its main mineral components. The specific surface area of the cement particles is 349.5 m^2/kg, and it exhibits a final setting time of 90 min, which is considerably lower than conventional Portland cement, making it an exceptionally efficient gelling agent. The fly ash utilized herein was obtained from coal-fired power plants, with its particle specific surface area measuring 350 g/cmg. The dry density of the fly ash was recorded at 1.79 g/cm^3.

2.1.3. Glass Fibers

Technical indicators related to the glass fibers employed herein are listed in Table 3. The fibers exhibit high ductility, excellent mechanical properties, strong corrosion resistance, and non-toxicity, which is conducive to environmental protection.

Table 3. Major physical properties of the glass fibers.

Fiber Length (mm)	Single Filament Diameter (μm)	Density (g/cm^3)	Elongation (%)	Elastic Modulus (GPa)	Tensile Strength (MPa)	Corrosion Resistance	Dispersion
3–20	10	2.13	29	12.02	948	Strong	Good

2.2. *Experimental Method and Design Ratio*

Table 4 depicts the fundamental mix ratio of the stabilized soft clay. According to our preliminary experiments, the proportions of water, fly ash, cement, and clay in the research mixture were determined to be 22.5%, 3%, 12%, and 62.5%, respectively. The glass fibers were introduced through dry mixing. Initially, the weighed soft clay, fly ash, and cement were placed into the planetary mixer. Subsequently, glass fibers were added to the blender in batches, subjected to dry mixing for 120 s, followed by blending with water and stirring for 5 min. Subsequently, the mixture was transferred to the mold and compacted using the static compaction method, which has been widely employed for fine-grained soils. Finally, the mold was removed after the samples had reached a given curing time. Figure 1c illustrates the solidified samples of cement soil and fibers after pouring and curing, with the cylindrical samples having a diameter and length of 50 mm and 100 mm, respectively.

Table 4. The mix ratio of the stabilized soft clay.

Materials	Clay	Cement	Fly Ash	Water
Mass ratio (%)	62.5	12	3	22.5

To determine the optimal ratio for fiber-reinforced cemental soil, tests were conducted with three different fiber lengths and six different amounts of glass fibers. Mechanical tests were employed to ascertain the unconfined compressive strength (UCS) of the samples. Based on preliminary experiments, the proportions of water, fly ash, cement, and clay in the research mixture were determined to be 22.5%, 3%, 12%, and 62.5%, respectively. Furthermore, the fiber length was set at 5, 10, and 15 mm, denoted as short-cut, medium-cut, and long-cut fibers, respectively. We prepared stabilized soft clay samples with five different fiber contents, corresponding to dry masses of 0.1%, 0.2%, 0.3%, 0.4%, and 0.5% of the total sample mass, respectively. The level design of the 16 groups of factors under investigation is meticulously outlined in Table 5, with F0 serving as the reference group.

Table 5. Experimental factor design.

Test No.	Fiber Length (mm)	Fiber Content (%)
F0	-	0
F1, F2, F3, F4, F5	15	0.1, 0.2, 0.3, 0.4, 0.5
F6, F7, F8, F9, F10	10	0.1, 0.2, 0.3, 0.4, 0.5
F11, F12, F13, F14, F15	5	0.1, 0.2, 0.3, 0.4, 0.5

2.3. *Testing Method*

2.3.1. Unconfined Compressive Tests

The optimum water content was used for the test samples. Before performing a unconfined compressive test (UST), compaction tests were carried out on the cemented soft clays to obtain the maximum dry density and optimum water content. The compaction tests were conducted following the ASTM D698-12(2021) standard [27]. Subsamples with specified moisture contents were placed in three layers into a 101.6 mm diameter mold. Each layer was compacted with 25 blows from a 24.5 Newton ram dropped from a height of 305 mm. The compaction test was repeated five times to establish the compaction curve, determining the optimum moisture content and maximum dry density of the samples. Then, the tested samples were prepared under optimal water content conditions. UST experiments were executed using a uniaxial compression testing machine, with the loading rate set to 0.25 mm/min. Preloading was implemented before initiating the test to eliminate the gap between the bearing plate and the sample. The preloading force was set at 0.01 kN. Following sample failure, a high-definition camera was employed to capture images for analyzing the failure mode under compressive loading.

2.3.2. Scanning Electron Microscopy

The microstructure and morphology of the sample were examined using the results of scanning electron microscopy (SEM) tests. Prior to the experiment, a thin sheet sample measuring 5 mm × 5 mm × 2 mm was extracted from the center of the test block and immersed in absolute ethanol for 24 h. Subsequently, the sample underwent slicing and gold plating, and then morphologic observation under an SEM for measurement.

3. Results

3.1. *Stress–Strain Curves*

A uniaxial compression test was conducted on the fiber-reinforced cement-stabilized soft clay, with the sample dimensions measuring 50 mm in diameter and 100 mm in height, and with a compression rate of 0.25 mm/min. The stress–strain relationships and fracture morphology under different fiber content conditions were experimentally acquired. As depicted in Figure 3, the sample exhibited three stages of initial compaction with the increase in vertical strain, including elastic rise, post-peak stress drop, and residual distortion. Utilizing glass fibers resulted in an increase in the peak stress and a corresponding vertical strain increase. This suggests that fiber reinforcement enhances the plastic deformation of the sample compared to ordinary cemental soil and other solidified materials. In addition, introducing an appropriate amount fibers improves the strength and ductility of the materials.

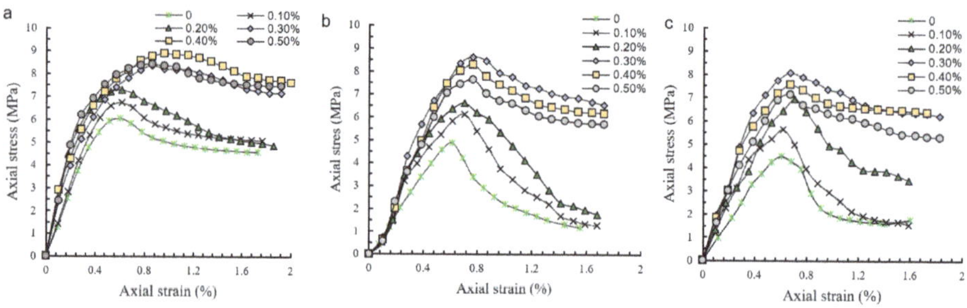

Figure 3. Stress–strain curves of the stabilized soft clay reinforced by fibers with different lengths: (**a**) $L = 5$ mm, (**b**) $L = 10$ mm, and (**c**) $L = 15$.

Compression failure realized mainly through three gradual stages, corresponding to the changes in the stress–strain curve. Figure 4 illustrates the morphology of the damaged sample observed through the uniaxial compression test. Figure 4a depicts the initial compression stage, where the native porosity within the sample is reduced. The vertical stress gradually rises with increasing strain, and small oblique cracks appear, accompanied by a few bumps on the surface. Figure 4b demonstrates the progression of damage as the sample undergoes plastic deformation, leading to a rapid increase in vertical stress with rising strain. Compression-induced cracks propagate towards the edge, resulting in oblique failure cracks, accompanied by a limited number of particles falling off the sample surface. In addition, Figure 4c displays the morphology of the sample in the fracturing stage. It is evident that as the vertical stress continues to increase, the compression cracks induced in the middle of the sample expand, and numerous articles progressively detach. In this stage, the vertical stress drops rapidly, leading to severe brittle failure of the FCSSC sample.

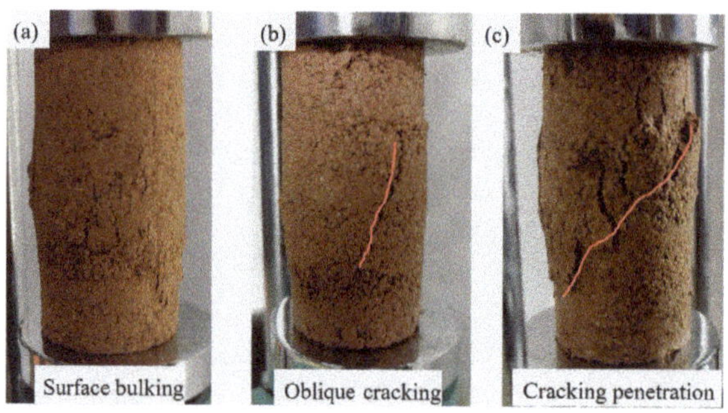

Figure 4. Fracture morphology of the soil specimens: (**a**) initial phase, (**b**) damage phase, and (**c**) destruction state.

3.2. *Effect of Curing Time, Fiber Length, and Fiber Amount on Strength*

The stress–strain curves for the specimens of FCSSC were obtained from the UCT results, encompassing conditions of varying fiber lengths and amounts. The pinnacle stress, indicative of the unconfined compressive strength, was extracted from these curves to quantify the static properties of the sample. These results are visually represented in Figure 5. Specifically, Figure 5a–c illustrate the significant impact of curing time, fiber content, and fiber length on the results of compressive strength testing. Table 6 presents a comprehensive overview of the uniaxial compression test results and the percentage growth in unconfined compressive strength. With the extension of curing time, there is a continuous increase in compressive strength observed across diverse groups of the FCSSC samples. The growth rate of the strength exhibits a higher value within the initial 1–7 days of curing, eventually stabilizing during the subsequent 7–14 days. In summary, the compressive strength of the cured samples demonstrates a higher magnitude in the early stages, reaching a peak at a curing age of 7 days. This experimental observation aligns with previous studies focusing on cement-based fiber-modified materials.

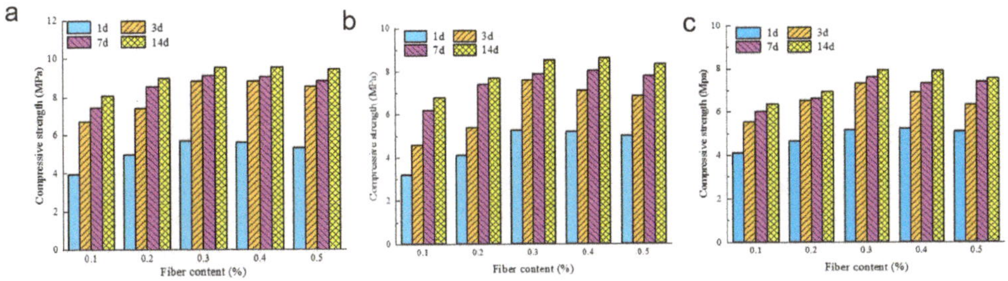

Figure 5. Compressive strength of the specimens modified by glass fibers: (**a**) short-cut fibers; (**b**) medium-cut fibers; (**c**) long-cut fibers.

Table 6. The values of fitting parameters *a* and *b*.

Fiber Content (%)	a	b
0	1.895	0.671
0.1	1.901	0.543
0.2	2.231	0.488
0.3	2.522	0.501
0.4	2.813	0.613
0.5	2.864	0.726

Under varying curing ages, the strength of FCSSC exhibits an initial increase followed by a decline with the rise in fiber content. The pivotal point for changes in the compressive strength of these specimens of FCSSC is found at amounts in the range of 0.3–0.4%. Notably, the length of the fibers plays a crucial role in influencing compressive strength, with short-cut fibers providing the most effective reinforcement. This effectiveness is attributed to the glass fiber monofilaments with a length of 5 mm (short-cut fibers) possessing the largest ratio of length to diameter. This characteristic ensures optimal contact with the hydration gel product and provides the ability to withstand greater compressive stress. The destruction of the sample is accompanied by the transfer and conversion of energy, and the emergence of failure cracks results from the accumulation and subsequent release of surface energy. The severity of fracturing damage correlates with the higher surface energy generated during compression failure. In samples without a fiber addition, cracks persistently expand under the load, concurrently enlarging the fracturing damage zone. As the fiber content increases, the intensity of compression damage to the FCSSC sample diminishes. However, excessive fiber length tends to counteract this, leading to poor deformation resistance in specific areas, consequently impacting the reinforcing effect.

The reinforcing effect of the fibers on the stabilized soft clay experiences an initial ascent followed by a decline with increases in fiber content, indicating that an excessive addition of fibers may yield diminishing returns. In comparison to the UCS values of the reference group, the maximum increase in compressive strength for the short-cut, medium-cut, and long-cut fiber-reinforced samples was achieved at fiber contents between 0.3% and 0.4%, showing values of 88.3%, 70.9%, and 77.1%, respectively. The addition of fibers promotes the formation of a three-dimensional network within the sample, amplifying adhesion and mechanical interlocking forces between the sand particles and hydration products, contributing to improving the load bearing capacity. Notably, at low fiber contents, the friction and cohesion between the fibers and clay grains are minimal, allowing for relatively easy sliding of the cemented particles. Furthermore, long fibers do not consistently arrest crack development, resulting in structured material damage. As the fiber content increases, the contact zones between the fibers and cemented particles expand, increasing the interfacial friction. Simultaneously, the fibers create intersections and form a connecting network, enhancing spatial confinement in the samples. Consequently, the unconfined compressive strength of fiber-reinforced cemental soil specimens gradually increases until it reaches a peak. However, excessive contents or lengths of glass fibers results in fiber interconnection, increasing the local porosity of the stabilized soft clay and reducing its ability to resist deformation under external loading.

3.3. Microscopic Characteristics of Fiber-Reinforced Cement-Stabilized Soft Clay

To elucidate the strengthening mechanism, SEM analyses were conducted on the samples of FCSSC with a curing age of 7 days, a fiber content of 0.3%, and a fiber length of 5 mm. Meanwhile, the morphology of both the intact and damaged samples after uniaxial compression was observed and compared using microscopic scanning. Previous studies have indicated that the active components in cement and fly ash, such as calcium silicate minerals, react with clay during the curing process, leading to the formation of cementitious products [1,7,8]. This reaction enhances the strength and stability of the soil. As depicted in Figure 6a, before failure, the stabilized soft clay contains C-S-H gel and $Ca(OH)_2$, with clay particles tightly wrapped, covered, and bonded by numerous hydrates, thereby enhancing the density of the stabilized soft clay. In Figure 6b, the hydration gel substance adheres to the fiber's surface, causing an overlap between the fibers and the hydration product, thereby improving the bonding force and mechanical interlocking force in the cement mortar. Simultaneously, the curing agent may reduce the water absorption of fibers and slow down the decay or degradation process of fibers, thereby enhancing the durability of fiber-rich soils. Figure 6c reveals the microscopic morphology of the fractured cement mortar sand. Notably, the fibers exhibited no breakage post sample failure but underwent pulling, suggesting fiber displacement or rotation. This signifies that the fibers possess high tensile strength and act as a bridge among the clay–cement particles inside the specimens of FCSSC. This inhibits the expansion of compression cracks, thus allowing cracks to expand on both sides of the fibers. Consequently, the overall severity of deformation and compression fracture is reduced in the sample. Regarding the mechanism of action, after reaching the peak compressive strength of the specimens of FCSSC, exceeding the optimal fiber content and length leads to overlapping and entanglement, causing non-uniform dispersion in the soil. This may result in weak interfaces, reducing the adhesive strength of the fibers and hydration products. Thus, glass fiber-reinforced cement-stabilized soft clay must contain an optimum combination of fiber content and length.

Figure 6. Microstructural morphology of cemental sand material at different states: (**a**) hydration products before destruction, (**b**) fibers in the sample before destruction, and (**c**) fiber pulling after failure.

4. Damage Constitutive Model

After decades of development, damage constitutive models for rock and soil under loading have been formulated by combining elastic and viscoplastic mesoscopic damage during deformation [28]. Particularly, the random distribution of micro-cracks in the specimen and the evolutionary damage in later stages are correlated with the initial damage [29]. Therefore, it is assumed herein that the axial strain of the specimen satisfies the following Equation (1) [30]:

$$P(\varepsilon) = \frac{m}{F}\left(\frac{\varepsilon}{F}\right)^{m-1} \exp\left[-\left(\frac{\varepsilon}{F}\right)^{m}\right] \tag{1}$$

where m and F are fitting constants, ε refers to the axial strain, and $P(\varepsilon)$ denotes the distribution function of probability.

As demonstrated in Figure 7, if N_D is designed to represent the number of microelements that have been destroyed inside a damaged specimen under a certain level of loading, the total number of microelements inside the specimen is signified by N_T, and the their ratio can be defined as the damage variable D, as follows [30]:

$$D = \frac{N_D}{N_T} \tag{2}$$

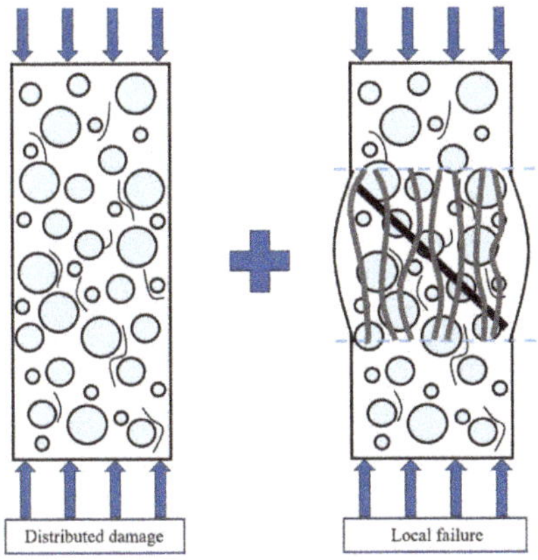

Figure 7. Schematic diagram of microelements under external loading.

The number of destroyed microelements N_t in the specimen is defined as follows:

$$N_t(\varepsilon) = \int_0^{\varepsilon} NP(x)dx = N\left\{1 - \exp\left[-\left(\frac{\varepsilon}{F}\right)\right]^{m}\right\} \tag{3}$$

The damage variable can be expressed by the following Equation (4).

$$D = 1 - \exp\left[-\left(\frac{\varepsilon}{F}\right)\right]^{m} \tag{4}$$

According to the principle of effective stress, the axial stress σ is defined as Equation (5):

$$\sigma_i = \frac{\sigma}{1 - D} = E\varepsilon(1 - D) \tag{5}$$

where σ_i is the effective stress, and E is elastic modulus.

The mechanical constitutive model plays a pivotal role in comprehending the mechanical behavior of reactive materials. Given that the stress–strain curves of stabilized soft clay exhibit linear rises and strain softening as their two most prominent features, a piecewise constitutive model based on mesoscopic damage mechanics is employed to accurately fit and address these curves. Drawing from previous research [31], the constitutive model equation is represented by Equation (6). The fitting parameters *a* and *b* of the constitutive model are determined by analyzing the specific UCS experimental data for five different fiber contents in the 5 mm short-fiber group using Equation (6), and these results are presented in Table 5. The correlation coefficient R^2 predicted by the constitutive model exceeds 0.95. These fitting outcomes affirm that the aforementioned constitutive equation adeptly captures the stress–strain relationship of stabilized soft clay samples.

$$\frac{\sigma}{s_u} = \begin{cases} a\frac{\varepsilon}{\varepsilon_u} + (3-2a)(\varepsilon/\varepsilon_u)^2 + (a-1.5)(\varepsilon/\varepsilon_u)^3 & \varepsilon \le \varepsilon_u \\ \frac{\varepsilon/\varepsilon_u}{b(\varepsilon/\varepsilon_u - 1)^2 + \varepsilon/\varepsilon_u} & \varepsilon > \varepsilon_u \end{cases} \quad (6)$$

where s_u is the peak stress, and ε_u represents the strain at the peak point.

The fitting parameters *a* and *b* in the proposed constitutive model carry specific physical significance [32]. Specifically, *a* in the rising section of the stress–strain curves reflects the ratio of the initial elastic modulus to the peak modulus of the materials, providing insights into the ductility characteristics of the materials. On the other hand, *b* in the descending section of the curves reveals the brittleness of the materials. As indicated in Table 6, with the increase in fiber content, *a* exhibits an upward trend, signifying an improved ductility of the fiber-reinforced cement-stabilized soil, which aligns with the test results. Simultaneously, *b* demonstrates a decrease followed by an increase, with the final value being lower than that of cement soil without fibers. This suggests that the brittleness damage degree of the materials becomes weaker after fiber modification [33,34]. Figure 8 compares the predicted results of the constitutive model and the measured results. It is evident that the stress of the samples follows a roughly complex linear elastic change law before reaching the peak value, while a structural fracture surface appears and the stress decreases after reaching the peak value [35,36]. In addition, Figure 8 illustrates highly similar shapes and geometric features between the measured curves and the predicted curves. Therefore, the constitutive model proposed in this research can reasonably and accurately reflect the mechanical behavior of specimens of FCSSC.

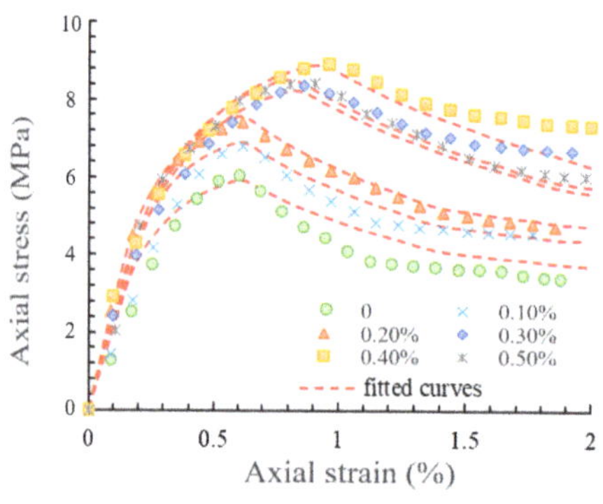

Figure 8. The measured and fitting results of the stress–strain curves.

5. Conclusions

In this investigation, glass fibers are utilized to reinforce and modify cement-stabilized soft clay. A comprehensive exploration of the mechanical strength is conducted through a uniaxial compression test, scrutinizing the influence of fiber length, fiber amount, and curing time on mechanical strength. Additionally, the mechanism of fiber reinforcement of FCSSC is analyzed based on microscopic images. The following conclusions have thus been drawn.

(1) The utilization of glass fibers enhances the performance of FCSSC in plastic deformation and compressive strength. At a curing age of 7 days, the stabilized soft clay

demonstrates heightened compressive strength, contributing to an improved early strength profile.

(2) The length of the glass fibers in FCSSC plays a pivotal role in determining the unconfined compressive strength of stabilized soft clay. The reinforcing effect is most pronounced for short-cut fibers, followed by medium-cut fibers and long-cut fibers. This is attributed to the intimate contact of shorter-chopped fibers with the hydration gel, enabling the sample to withstand greater compressive stress.

(3) The impact of increasing the fiber content on the strength index follows an initial increase and subsequently decreases. Considering both mechanical properties and engineering economy, the optimal amount is determined to be 0.3%. Excessive fiber content and excessively long fibers may result in cross-linking and entanglement, increasing porosity in the fiber aggregation area and diminishing the sample's ability to resist loading and deformation.

(4) Glass fibers function as a bridge in stabilized soft clay, effectively inhibiting the expansion of compression cracks and the formation of additional cracks. This action mitigates the overall severity of deformation and compression fractures in the sample. The synergistic interplay between glass fibers and hydration gel constitutes a significant factor involved in enhancing the mechanical properties of stabilized soft clay.

(5) The proposed constitutive model based on mesoscopic damage mechanics proves effective in describing the strain evolution of stabilized soft clay and accurately reflects the softening characteristics observed in the stress–strain curves under uniaxial compression.

Author Contributions: Conceptualization, T.Y.; methodology, T.Y. and S.C.; formal analysis, S.C.; validation, X.Z. (Xingyuan Zhang); investigation, X.Z. (Xingyuan Zhang); resources, X.Z. (Xianwei Zhang); writing—original draft preparation, T.Y.; writing—review and editing, T.Y. and R.A.; supervision, Z.Z. and X.Z. (Xianwei Zhang); funding acquisition, T.Y. All authors have read and agreed to the published version of the manuscript.

Funding: This research was funded by the Qingyang City Science and Technology Talent Special Plan Project (QY-STK-2022A-002); and the 2023 University Teacher Innovation Fund Project (2023A-145).

Institutional Review Board Statement: Not applicable.

Informed Consent Statement: Not applicable.

Data Availability Statement: The data presented in this study are available on request from the corresponding author. The data are not publicly available due to reasonable request.

Acknowledgments: This study was supported by the Qingyang City Science and Technology Talent Special Plan Project (QY-STK-2022A-002) and the 2023 University Teacher Innovation Fund Project (2023A-145). The authors also thank the anonymous reviewers for their comments on this manuscript.

Conflicts of Interest: The authors declare no conflicts of interest.

References

1. Rajasekaran, G. Sulphate attack and ettringite formation in the lime and cement stabilized marine clays. *Ocean Eng.* **2005**, *32*, 1133–1159. [CrossRef]
2. Xu, G.; Liu, J.; Liu, S.F.; Wang, Z.B.; Hu, G.; Kong, X.H. Modern muddy deposit along the Zhejiang coast in the East China Sea: Response to large-scale human projects. *Cont. Shelf Res.* **2016**, *130*, 68–78. [CrossRef]
3. Xiao, Z.; Yu, N.; An, J.; Zou, H.; Zhang, Y. Soil compressibility and resilience based on uniaxial compression loading test in response to soil water suction and soil organic matter content in northeast China. *Sustainability* **2022**, *14*, 2620. [CrossRef]
4. An, R.; Wang, Y.X.; Zhang, X.; Chen, C.; Liu, X.; Cai, S. Quantitative characterization of drying-induced cracks and permeabilityof granite residual soil using micron-sized x-ray computed tomography. *Sci. Total Environ.* **2023**, *876*, 163213. [CrossRef] [PubMed]
5. Zhou, J.; Tang, Y. Practical model of deformation prediction in soft clay after artificial ground freezing under subway low-level cyclic loading. *Tunn. Undergr. Space Technol.* **2018**, *76*, 30–42. [CrossRef]
6. Tang, Y.Q.; Zhou, J.; Zhang, M.D.; Liu, Y.D. Research on the thermal conductivity and moisture migration characteristics of Shanghai saline soil. I: Experimental modeling. *Bull. Eng. Geol. Environ.* **2015**, *74*, 577–593. [CrossRef]

7. Zhu, J.F.; Tao, Y.L.; Xu, R.Q.; Yang, H.; Pan, B.J. Investigation on the optimal formulation and mechanism of marine organic silt improved with magnesium-cement-based stabilizer. *Constr. Build. Mater.* **2022**, *341*, 127233. [CrossRef]
8. Yaghoubi, M.; Arulrajah, A.; Disfani, M.M.; Horpibulsuk, S.; Myint, W.B.; Stephen, D. Effects of industrial by-product based geopolymers on the strength development of a soft soil. *Soils Found.* **2018**, *58*, 716–728. [CrossRef]
9. Song, Y.; Cao, J.; Ding, W.; Song, Z.; Liu, H.; Huang, S.; Zhu, W. Influence of Peat Soil Environment on Mechanical Properties of Cement-Soil and Its Mechanism. *Sustainability* **2023**, *15*, 4580. [CrossRef]
10. Pheng, L.S.; Chuan, Q.T. Environmental factors and work performance of project managers in the construction industry. *Int. J. Proj. Manag.* **2006**, *24*, 24–37. [CrossRef]
11. Hu, J.F.; Chen, H.Q.; Zhang, Z.B. Mechanical properties of melamine formaldehyde microcapsules for self-healing materials. *Mater. Chem. Phys.* **2009**, *118*, 63–70. [CrossRef]
12. Savvilotidou, M.; Vassilopoulos, A.P.; Frigione, M.; Keller, T. Effects of aging in dry environment on physical and mechanical properties of a cold-curing structural epoxy adhesive for bridge construction. *Constr. Build. Mater.* **2017**, *140*, 552–561. [CrossRef]
13. Zhang, Y.; Luo, H.; Chen, P.; Liu, E.; Chen, Y. Mechanical Properties and Binary-Medium-Based Constitutive Model for Coral-Reef Limestone Samples Subjected to Uniaxial Loading. *Sustainability* **2022**, *14*, 12193. [CrossRef]
14. Somani, M.V.; Khandelwal, M.; Punia, V.; Sharma, V. The effect of incorporating various reinforcement materials on flexural strength and impact strength of polymethylmethacrylate: A meta-analysis. *J. Indian Prosthodont. Soc.* **2019**, *19*, 101. [CrossRef] [PubMed]
15. Tang, Y.Q.; Yan, J.J. Effect of freeze–thaw on hydraulic conductivity and microstructure of soft soil in Shanghai area. *Environ. Earth Sci.* **2015**, *73*, 7679–7690. [CrossRef]
16. Wille, K.; Naaman, A.E.; El-Tawil, S.; Gustavo, J. Ultra-high performance concrete and fiber reinforced concrete: Achieving strength and ductility without heat curing. *Mater. Struct.* **2012**, *45*, 309–324. [CrossRef]
17. Shi, X.Z.; Zhang, C.; Zhou, X.D. The statistical damage constitutive model of the mechanical properties of Alkali-resistant glass fiber reinforced concrete. *Symmetry* **2020**, *12*, 1139. [CrossRef]
18. Wang, D.H.; Ju, Y.Z.; Shen, H.; Xu, L.B. Mechanical properties of high performance concrete reinforced with basalt fiber and polypropylene fiber. *Constr. Build. Mater.* **2019**, *197*, 464–473. [CrossRef]
19. Schneider, K.; Lieboldt, M.; Liebscher, M.; Fröhlich, M.; Hempel, S.; Butler, M.; Schröfl, C.; Mechtcherine, V. Mineral-based coating of plasma-treated carbon fiber rovings for carbon concrete composites with enhanced mechanical performance. *Materials* **2017**, *10*, 360. [CrossRef] [PubMed]
20. He, G.S.; Li, J.; Zhang, F.S.; Wang, C.; Guo, S.Y. Effect of multistage tensile extrusion induced fiber orientation on fracture characteristics of high density polyethylene/short glass fiber composites. *Compos. Sci. Technol.* **2014**, *100*, 1–9. [CrossRef]
21. Liu, J.; Xie, X.; Li, L. Experimental study on mechanical properties and durability of grafted nano-SiO_2 modified rice straw fiber reinforced concrete. *Constr. Build. Mater.* **2022**, *347*, 128575. [CrossRef]
22. Wu, Z.M.; Khayat, K.H.; Shi, C.J. Effect of nano-SiO_2 particles and curing time on development of fiber-matrix bond properties and microstructure of ultra-high strength concrete. *Cem. Concr. Res.* **2017**, *95*, 247–256. [CrossRef]
23. Sun, J.F.; Zhao, F.; Yao, Y.; Jin, Z.; Liu, X.; Huang, Y.D. High efficient and continuous surface modification of carbon fibers with improved tensile strength and interfacial adhesion. *Appl. Surf. Sci.* **2017**, *412*, 424–435. [CrossRef]
24. *ASTM D4926-15*; Standard Test Method for Gamma Alumina Content in Catalysts and Catalyst Carriers Containing Silica and Alumina by X-Ray Powder Diffraction. ASTM: West Conshohocken, PA, USA, 2015.
25. *ASTM D4318-17e1*; Standard Test Methods for Liquid Limit, Plastic Limit, and Plasticity Index of Soils. ASTM: West Conshohocken, PA, USA, 2017.
26. *ASTM D6938-17ae1*; Standard Test Methods for In-Place Density and Water Content of Soil and Soil-Aggregate by Nuclear Methods (Shallow Depth). ASTM: West Conshohocken, PA, USA, 2017.
27. *ASTM D698-12*; Standard Test Methods For Laboratory Compaction Characteristics Of Soil Using Standard Effort (12 400 Ft-Lbf/Ft3 (600 KN-M/M^3)). ASTM: West Conshohocken, PA, USA, 2021.
28. Popovics, S. A numerical approach to the complete stress-strain curve of concrete. *Cem. Concr. Res.* **1973**, *3*, 583–599. [CrossRef]
29. Borodulina, S.; Kulachenko, A.; Nygårds, M.; Galland, S. Stress-strain curve of paper revisited. *Nord. Pulp Pap. Res. J.* **2012**, *27*, 318–328. [CrossRef]
30. Park, J.; Hyun, C.U.; Park, H.D. Changes in microstructure and physical properties of rocks caused by artificial freeze–thaw action. *Bull. Eng. Geol. Environ.* **2015**, *74*, 555–565. [CrossRef]
31. Zhang, D.W.; Cao, Z.G.; Fan, L.B.; Liu, S.Y.; Liu, W.Z. Evaluation of the influence of salt concentration on cement stabilized clay by electrical resistivity measurement method. *Eng. Geol.* **2014**, *170*, 80–88. [CrossRef]
32. Krajcinovic, D.; Basista, M.; Sumarac, D. Micromechanically Inspired Phenomenological Damage Model. *J. Appl. Mech.* **1991**, *58*, 305–310. [CrossRef]
33. Abbaspour, M.; Aflaki, E.; Nejad, F.M. Reuse of waste tire textile fibers as soil reinforcement. *J. Clean. Prod.* **2018**, *207*, 1059–1071. [CrossRef]
34. Owusu-Ansah, D.; Tinoco, J.; Correia, A.A.S.; Oliveira, P.J.V. Prediction of Elastic Modulus for Fibre-Reinforced Soil-Cement Mixtures: A Machine Learning Approach. *Appl. Sci.* **2022**, *12*, 8540. [CrossRef]

35. Zhao, Y.Y.; Ling, X.Z.; Gong, W.G.; Li, P.; Li, G.Y.; Wang, L.N. Mechanical Properties of Fiber-Reinforced Soil under Triaxial Compression and Parameter Determination Based on the Duncan-Chang Model. *Appl. Sci.* **2020**, *10*, 9043. [CrossRef]
36. Xie, S.; Lin, H.; Chen, Y.; Yong, R.; Xiong, W.; Du, S. A damage constitutive model for shear behavior of joints based on determination of the yield point. *Int. J. Rock Mech. Min. Sci.* **2020**, *128*, 104269. [CrossRef]

applied sciences

MDPI

Article

Predicting Swell in Clay-Sand Mixtures Used in Liners

Muawia Dafalla

Civil Engineering Department, King Saud University, Riyadh 11421, Saudi Arabia; mdafalla@ksu.edu.sa

Abstract: A method to predict the swellability of clay-sand mixtures is proposed. This model is a modified form of the Studds (1997) prediction model for clay-sand mixtures. The new proposed model uses the laboratory fall cone penetration technique to produce a characterization chart. This chart presents slope levels that can be used to obtain an equation for the final clay void ratio versus the vertical effective stress for clays. The porosity of the clay-sand mixtures was worked out based on a correction factor obtained from compression and porosity measurements in the laboratory. The porosity of the mixture was merged into the clay profile equation to compute the final clay void ratio at a specified stress level, which made it possible to predict the swelling behavior for different and variable stress levels. The swellability slope was obtained using fall cone tests conducted on the fine portion. Mixtures of kaolinite and bentonite were introduced to represent soils with different swell potentials. The fall cone measurements of a few points can be used to establish the swellability relationships for natural clay. Merging fall cone points with the swellability slope chart can define the profile of the vertical effective stress versus the clay void ratio.

Keywords: swelling; clay-sand mixtures; clay; effective stress; porosity

1. Introduction

Studies on the identification and quantification of the swell in expansive clays, which have been carried out by many researchers, are either based on direct measurements of the swell potential and the swelling pressure or estimated from index properties. Most of the formulae established, if not all, are considered of local validity. These variations are mainly due to the different clay origins and mineralogical composition [1–3].

Field measurements and tests for such soils are very expensive and require long-term monitoring, which can extend to years. There is a lack of a simple field or laboratory test that can give reliable and informative results within a reasonably short period.

Geotechnical engineers are concerned about the increase and decrease in the volume of clays following wetting and drying. Bentonite is the name for commercial products that are rich in montmorillonite. Sodium Wyoming bentonite has received the greatest attention due to its ability to absorb large quantities of water and swell more than other types.

Modeling studies on the mechanical behavior of swelling clays and/or bentonite have been introduced by many researchers [4–10]. Understanding the mechanisms of volume changes helps to properly quantify and predict swelling behavior. It is quite common in the literature to come across terms of interlayer swelling, interparticle swelling, and interaggregate swelling. The first refers to the condition in which water is introduced between two layers of montmorillonite, causing the basal spacing to increase. This increase is variable and depends on the state of moisture. For dry commercial bentonite, the basal spacing increase is estimated to be in the order of 12.5 to 20 Å [9,11]. The interlayer swellings can double the volume of dry clay [12]. Interparticle swelling or osmosis is computed based on the double-layer theory, which uses the charge and spaces between particles. Interaggregate swelling occurs when aggregations and clusters are formed within the clay fabric.

The model proposed by Studds [11] for the prediction of the swelling behavior of bentonite–sand mixtures is an experimental approach suggesting the swell curve to be presented in the form of the void ratio against the vertical effective stress. The model is generally based on three sets of data: the relationship between the clay void ratio (e_c) and the vertical effective stress (σ'_v) for bentonite with an appropriate permeant, the relationship between the sand porosity η_s and the vertical effective stress of the sand, and the sand porosity of the mixture in a dry state at an appropriate vertical effective stress. This approach is based on the concept of stress partitioning discussed by Mollins [9].

The clay void ratio concept used by Studds [11] and Mollins et al. [10] considers a two-phase sand–bentonite gel system (Figure 1). In this case, the clay is described in terms of the clay void ratio, defined as

$$e_c = \frac{Vw}{Vc} \tag{1}$$

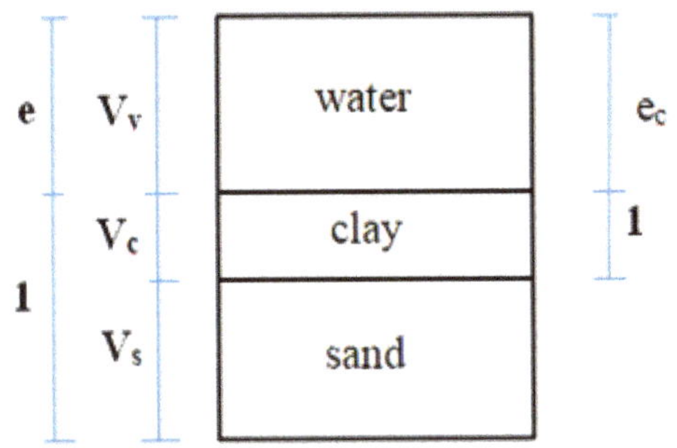

Figure 1. Phase diagram for saturated clay-sand mixture. *Vv* = volume of voids; *Vc* = volume of clay; *Vs* = volume of sand.

The conventional void ratio is shown on the left-hand side of the figure. The sand porosity is the total soil volume occupied by the paste and is given as

$$\eta s = \frac{(Vw + Vc)}{Vt} \tag{2}$$

where $Vt = Vw + Vc + Vs$.

Studds [11] assumed that the sand porosity of the dry mixture on a graph of η_s versus log σ'_v was parallel to the line of the pure sand but displaced from the compression line for the sand only. Studds proposed to obtain the sand porosity for a mixture at a given effective stress on the sand. This value was used to obtain the clay void ratio corresponding to the selected stress level via the following equation:

$$e_c = \frac{(\eta s(\text{mixture}) - x)}{x\,(1 - \eta s(\text{mixture})} \tag{3}$$

where x is the volumetric fraction of clay in the mixture. This equation was derived by combining the following two equations:

$$e = e_c \cdot x \tag{4}$$

and

$$\eta s(\text{mixture}) = \frac{(e + x)}{(1 + e)} \tag{5}$$

The effective vertical stress applied to the clay can be found when the clay void ratio related to the selected stress level is obtained. By adding up the vertical effective stress on the sand and the vertical effective stress of clay, we can obtain the vertical effective stress of the mixture.

This method enables the clay void ratio of the mixture versus the vertical effective stress to be plotted. The final clay void ratio can be used to compute the material void ratio using the relationship given in Equation (4). The percentage volume change or swellability at a given effective stress can be predicted accordingly. This model is designed to obtain quick assessment based on simple fall cone tests.

2. Materials and Methods

2.1. Materials

2.1.1. Natural Clay

The natural clay material used in this study was obtained from the Tabuk region, towards the northwest of Saudi Arabia. These soils belong to a semiarid region and are subjected to cycles of repeated wetting and drying. The problems related to swelling in this region are of great concern to residents and local government authorities. Although the Tabuk clay is categorized as low-expansion clay, it has severely harmed a number of engineering projects. Table 1 displays the soil's geotechnical parameters that were measured.

Table 1. The main physical properties of Tabuk clay.

Property	Average
Dry unit weight (kN/m^3)	20.4
Moisture content (%)	13.1
Liquid limit (LL) (%)	37
Plastic limit (PL) (%)	24
Plasticity index (PI)	13
Shrinkage limit (%)	21
Percent sand (%)	6.0
Passing 63-micrometer sieve (%)	81
Specific gravity	2.79
Swelling pressure (kN/m^2)	49
Compression index (Cc)	0.12
Angle of shearing resistance (φ)	22

2.1.2. Bentonite Material

Volclay Ltd. (london, UK) supplied the bentonite clay that was utilized in this study. It had a specific gravity of 2.76, a liquid limit of 405%, and a plastic limit of 48%. The laboratory index property tests were performed in accordance with BS1377 Part 2, 1990 [13]. The findings of the chemical examination of the bentonite material are shown in Table 2.

Table 2. Volclay bentonite chemical analysis.

Chemical Analysis	Percentage by Weight
SiO_2	63.02
Al_2O_3	21.08
Fe_2O_3	3.25
Na_2O	2.57
MgO	2.67
CaO	0.65
FeO	0.35
Trace elements	0.72
Loss of ignition	5.64
Total	99.95

2.1.3. Kaolinite Material

Processed kaolinite can have slightly variable chemical and physical properties depending on the nature of the processing applied to natural clay. Table 3 presents a typical chemical composition for a kaolinite sample.

Table 3. Typical chemical composition of kaolinite.

Chemical Analysis	Percentage by Weight
SiO_2	48 +/− 0.1
Al_2O_3	36 +/− 0.1
Fe_2O_3	<0.75
MgO	<0.3
Na_2O	<1.0
K_2O	1.5 +/− 0.5
TiO_2	<0.5
L.O.I.	**12.5 +/− 0.5**

Kaolinite can have a viscosity of more than 30 kg/m.s and a thixotropy of less than 1.5 kg/m.s. The kaolinite generally used in this research was a product of the ECC International (Europe) Company. The product was Spess-White China Clay. The size distribution of the Spess-White China Clay kaolin, as tested by the manufacturer, indicated 78% clay and 22% silt. The following tests were performed for the China Clay in the laboratory, as per ASTM and/or British standards. The liquid limit was reported as 50% and the plastic limit was measured at 27%. The specific gravity of the Spess-White kaolinite was 2.50 and the computed activity was reported as 0.28.

Kaolinite is known as a material that has negligible swelling. The index properties measured indicate a classification of low to medium swelling. This classification is not valid. This supports the claim of Chen that all swelling clays have index properties as per the established classification and that a material does not need to have a high liquid limit or plasticity index to be a swelling material [2].

2.1.4. Sand Material

Natural wadi sand material obtained from a valley in Riyadh was used to produce the test soils. The properties of this type of sand are as follows:

Fines: Less than 2%.
Particle shape: Rounded to subangular.
Maximum grain size: 0.6 mm.
Uniformity: Poorly graded.
Specific gravity: 2.65.

The particle size distribution curve is shown in Figure 2.

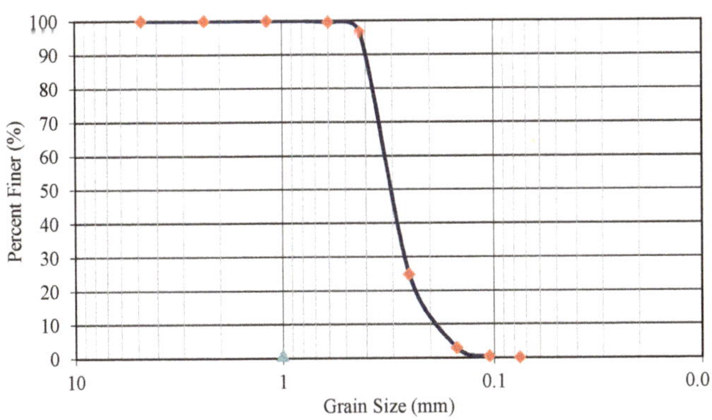

Figure 2. Grain size distribution for sand used in this study.

2.2. Testing Methods

2.2.1. The Fall Cone Test Method

This study's basic instrument was the penetrometer apparatus that is outlined in the British Standard (BS 1377: Part 2:1990:4.3) [13], which met the standards of BS 2000 Part 49.

The sample was cleansed of the particles that were retained on a 63 m test sieve. This was discovered to have an impact on the fall cone's penetration. For this test, the sample preparation technique was quite important. The fine-grained soil was vigorously combined using two palette knives for at least 10 min according to the British Standard to create a uniform paste. In order to avoid trapping air, a part of the mixed soil was forced into a normal cup or a ring with a 50 mm diameter and a 20 mm height. The cone was positioned such that its surface barely touched the sample, and it was then allowed to fall for five seconds. The original dial gauge readings before and after penetration were compared, and the difference was recorded to the nearest 0.1 mm.

The method used, as can be seen from the above description, is comparable to that for determining the liquid limit, with the exception of the preparation methods, which require that material with a particle size greater than 63 m be excluded. Additionally, when the penetration was less than 10 mm, the sample container was replaced with a 50 mm diameter by 20 mm high ring. Feng [14] discovered that a preparation method using specimen rings rather than cups was quicker and simpler. Additionally, he came to the conclusion that rings would lessen the possibility of the specimen accumulating air. The outcomes were found to be identical to those for the conventional cup.

For the fall cone method of determining the liquid limit, the test procedure was exactly as described above. A similar presentation of the data was used. A reduction in the impact of nonclay material on the penetration process can be achieved by rejecting material with a grain size greater than 63 m.

At various moisture contents, the test measured the penetration in mm. Semi-log charts can be used to display this relationship. Comparing the curve's trend with the curves of different materials is possible [15].

2.2.2. The Oedometer Testing Method

Using traditional oedometer techniques, tests for one-dimensional swelling were conducted. The sample was immersed in a container that enclosed a ring (the oedometer cell). Each sample had two porous disks at the bottom and top attached to Whatman filter sheets. This oedometer cell was housed inside a loading frame that transferred vertical loads via a uniquely crafted lever arm. Either a dial gauge with a precision of 0.01 mm or calibrated sensors that sense vertical movement were used to track changes in the sample height.

The utilized ring had a 50 mm diameter. A variable sample height was employed. The samples ranged in thickness from 3 to 8 mm. It was discovered that the initial swelling takes place more quickly in thin samples.

To estimate the swell parameters, one-dimensional swell testing was performed in accordance with ASTM D4546-96 method A [16]. The approach that deviates from this one uses a seating pressure of 7 kN/m^2. This technique involved submerging the sample and letting it swell vertically at the seating pressure for the duration of the principal swell. Similar procedures were used in the tests with various vertical effective stresses.

The ring's weight as well as the sample's weight were calculated. The sample's original height was noted. The initial moisture content and specific gravity were computed. These made it possible to calculate the height of the solids, Hs:

$$\mathrm{Hs} = \frac{\mathrm{Ws}}{\mathrm{A \cdot Gs \cdot \gamma_w}} \tag{6}$$

The initial void ratio e_o was computed using the following formula:

$$e_o = \frac{\mathrm{H} - \mathrm{Hs}}{\mathrm{Hs}} \quad (7)$$

The change in the void ratio can be measured at any stage of the test in a similar fashion. Thus, the final void ratios at the end of the test are given as

$$e_f = \frac{\mathrm{H} - \mathrm{Hs}}{\mathrm{Hs}} \quad (8)$$

where Gs is the specific gravity, W_s is the dry weight of the sample, and γ_w is the density of water. The degree of saturation is given as

$$\mathrm{Sr} \cdot e = \mathrm{wGs} \quad (9)$$

where w is the moisture content.

Similar procedures were used for the oedometer tests conducted on remolded samples, bentonite powder, and mixes.

The samples were given time to air dry. In order to create a paste with a known moisture level, distilled water was added. If the dry density is known, the amount of soil required to fill the consolidation ring can be calculated. The calculated amount was tamped into the ring in two or three lifts.

The final clay void ratio was obtained from the pretest conditions as above, e_1, and the post-test parameters, e_2, using the following equation:

$$1 + e_2 = \frac{\mathrm{Gs}\gamma_w}{\gamma_d} \quad (10)$$

where γ_d is the dry unit weight.

This confirmed the accuracy of the data observed and recorded.

The final void ratio was taken as the average of e_1 and e_2. The surcharge or vertical effective stresses used varied from 1.96 to 400 kPa. This range covered the needs of geotechnical practice.

2.2.3. Samples Selected for Clay-Sand Mixture Prediction

The clay paste, which included only 40% bentonite by weight, was tested for penetration for the characterization of clay using the fall cone method. The corresponding clay void ratio versus the vertical effective stress was obtained based on the slope of the fall cone test chart.

Artificial clay consisting of 40% bentonite and 60% kaolinite was added to natural sand to form three different clay-sand mixtures. The clay was taken as 20%, 40%, or 80% according to the weight of the sand.

2.3. *Modifications of the Studds Prediction Model*

The use of the fall cone to characterize swelling clays enabled the construction of a chart that defines different swelling zones. This chart introduces lines of similar slopes to the vertical effective stress versus the final clay void ratio. The slope of the swellability line can be found when the slope of the moisture versus penetration in the chart is known. This can replace the testing required by Studds to obtain the final clay void ratio versus the effective stress. Using direct measurements to obtain the final clay void ratio normally takes several days. The set of equations for the vertical effective stress versus the clay void ratio is related to the specific fall cone profile. These can be used and the intermediate points can be worked out by interpolation.

The second modification was made to adjust the porosity factor of the clay-sand mixture adopted by Studds [11], which is valid only for bentonite and cannot be used when kaolinite material is introduced. A new correction chart based on the inclusion of a non-swelling clay component was established. Figure 3 presents the new chart.

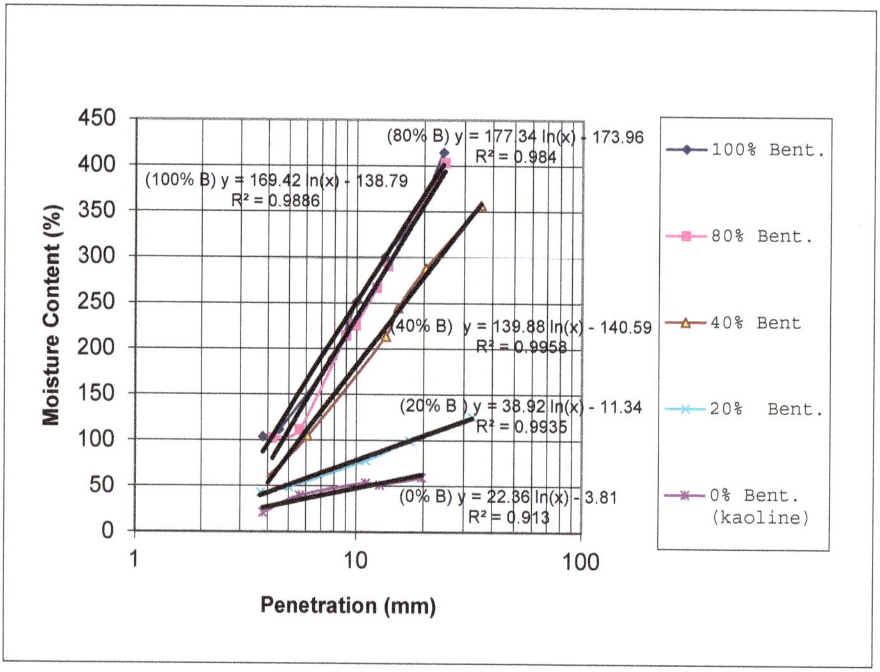

Figure 3. Idealized penetration chart for bentonite–kaolin mixtures.

The third modification of this system was to consider the silt material as part of the volumetric clay fraction (all material passing test sieve No. 200). This was implemented because, when measuring the shearing strength, the sample preparation technique for the fall cone includes this part. The attempt to include the silt within the clay for a specific purpose is not new. The authors of reference [17] considered all non-sand material as clay when investigating the liquid limit and the linear mixture law. The sand mixture porosity ($\eta_{s(mixture)}$) is given in Equation (5).

This value was calculated based on pretest dry conditions. When substituting for x, the silt material passing sieve No. 200 was included as a clay fraction.

Using the above modifications and assumptions, it was possible to predict the behavior of mixtures including sand, bentonite, and kaolin.

3. Results

Identification Tests

The test results obtained for the fall cone method were used to construct the chart presented in Figure 3. The clay void ratio versus the vertical effective stress for the selected bentonite–kaolin mixtures is presented in Figure 4. No prediction formula is given for the pure kaolinite in Figure 4. The laboratory porosity measurements carried out using the oedometer apparatus are presented in Figures 5 and 6. Trends in the artificial clay-sand mixtures are shown in Figure 7.

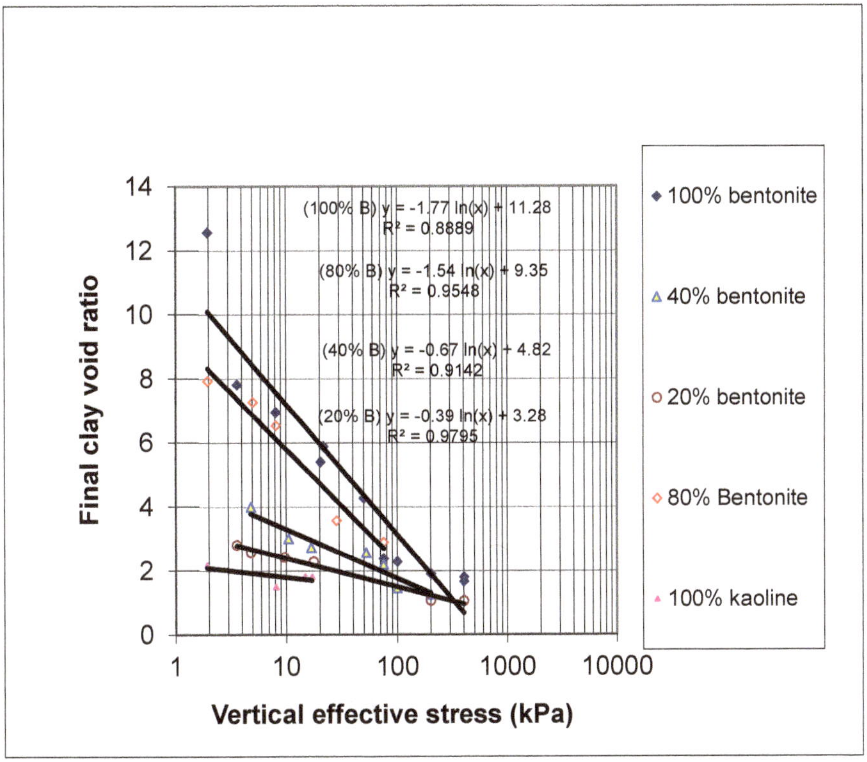

Figure 4. Final clay void ratio versus vertical effective stress for bentonite–kaolin mixtures.

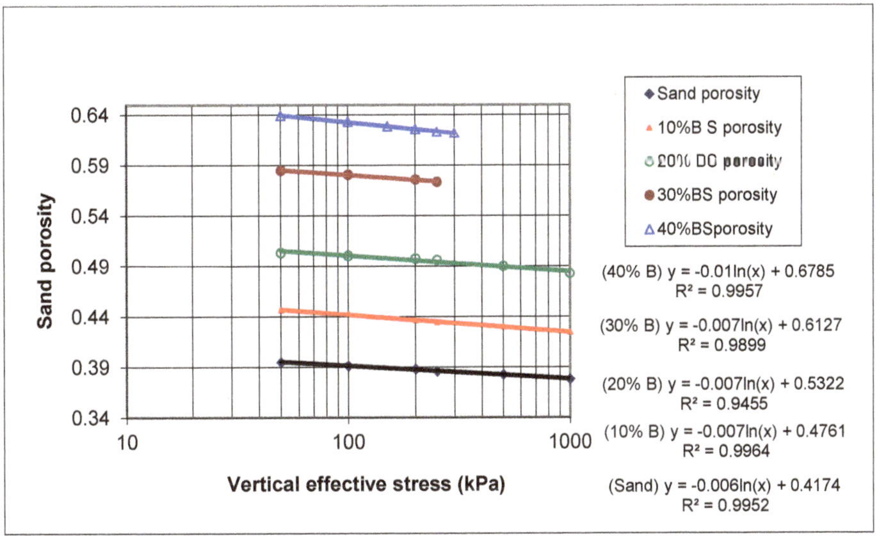

Figure 5. Compression and porosity of sand and sand with bentonite additives.

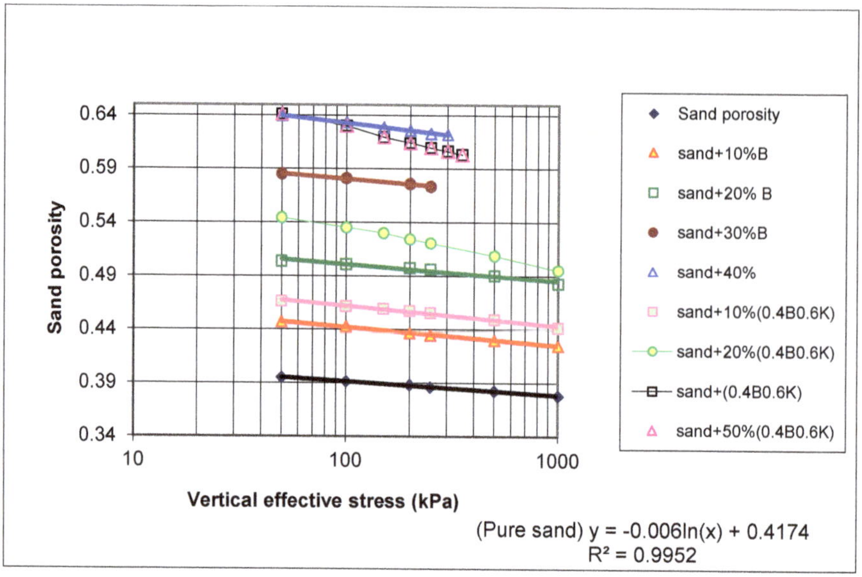

Figure 6. Compression and porosity of sand and sand with bentonite/kaolin additives.

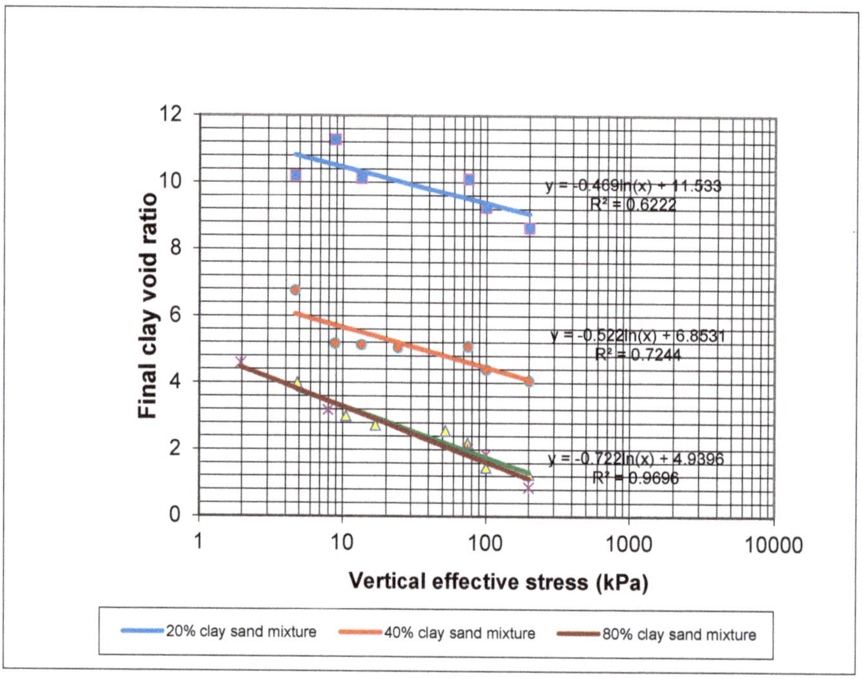

Figure 7. Vertical effective stress versus final clay void ratio for artificial clay-sand mixtures. (Green line is the reference bentonite material).

It is well known that index properties such as the liquid limit and plastic limit are used to roughly estimate the swelling of clay. These properties are strength-based parameters. The fall cone is a strength-measuring device ([18,19]). Therefore, it is expected to be very useful in serving the goals of this research. The authors of reference [20] stated that the act of penetration by the fall cone provides a strength measurement that can indicate the compressibility of the soil.

4. Discussion

4.1. Final Clay Void Ratio and Equilibrium

Equilibrium is the state after which no change in the void ratio occurs when swelling clays are in saturated or near-saturated conditions. Expansion problems associated with saturated soils or originally wet soils are minimal. Initially, dry or partially saturated swelling clays are likely to cause structural problems following a moisture increase. The prediction of the compressibility and expansion of partially saturated soils is a cumbersome and difficult analysis that requires knowledge of several parameters. Mollins [8] found that the compaction method does not affect the final clay void ratio of samples. This study presents the clay void ratio versus the vertical effective stress for selected artificial soils using bentonite and kaolin mixtures.

Fine-grained soil water systems are commonly defined by the liquid and plastic limits. These are strength-based moisture contents. The free swell limit is defined as the maximum water-holding capacity of clayey soils and corresponds to a stress-free reference state. In this research, the term final clay void ratio is frequently quoted. This is the state at which the clay reaches equilibrium in a state similar to a stress-free reference state.

4.2. Prediction of the Swelling Behavior of Clay

Many factors, including the mineralogical composition, fabric, stress history, placement conditions, and geologic mode of the formation of the clay, influence the swelling behavior of naturally occurring clays. This prediction method assumes that the active constituents of natural clay give similar results as for remolded clays. Material with a similar amount of active clay compacted to variable densities will swell to a maximum level and maintain the same slope of swellability. The actual swell profile of natural clay may indicate a shift or slight displacement. The slope of the clay void ratio versus the vertical effective stress for the natural clay is obtained as a function of the slope predicted for the fine clay constituents using the fall cone method. However, the study of artificial and remolded clay is necessary to understand how these minerals behave when subjected to a moisture increase, which forms a useful guide for predicting the swell behavior of mixtures. Artificial mixtures are more common in the waste containment and liner industries.

The proposed prediction is based on obtaining the clay void ratio versus the vertical effective stress from the slope of relevant artificial clay. To achieve this, it is necessary to run a single or a few fall cone tests to establish the slope of swellability.

The clay paste, which included only 40% bentonite by weight, was tested for penetration for the characterization of clay using the fall cone method. The profile of the clay void ratio versus the vertical effective stress was obtained for the same material. The predicted equation for this type of clay (shown in Figure 4) is given by the following:

$$e_c = -0.6656\, ln\, \sigma c' + 4.824 \qquad R^2 = 0.9142 \tag{11}$$

4.3. Prediction of the Swelling Behavior of Clay-Sand Mixtures

Clay can be found in nature in the form of pure clay or clay intermixed with sand, silt, or other nonclay material (usually discrete inclusions). A clay paste is formed when water in the mixture combines with the clay. The "clay paste" can be either too little in volume and not sufficient to fill up the pores between the coarse component or available in large quantities to fill all the available voids. Most of the stresses are taken by the sand or coarse part in the first condition, and are not likely to be of major concern. For the second condition, the stress is shared and the response of the mixture depends on many factors, such as the stress–strain properties of the assemblage of coarse particles, voids within the system, and the type of clay. It is well understood that adding sand to clay improves the shearing strength of the system. Additionally, it contributes indirectly by accelerating the consolidation of the slurry (Tan et al., 1994). It has also been found that the addition of sand to clay material affects the permeability of the mixture and helps with the quicker drainage of water.

Artificial clay consisting of 40% bentonite and 60% kaolinite was added to natural sand to form three different clay-sand mixtures. The clay was taken as 20%, 40%, or 80% according to the weight of sand.

The clay-sand mixture with 20% clay was tested for the effective stress and clay void ratio relations. The results measured in the laboratory are comparable to the data predicted by the Studds modified method. Table 4 presents how to calculate the final clay void ratio for different stress levels. The third column computes the porosity of the mixture at the given stress and as obtained from a direct test in the laboratory. The clay void ratio was computed and is presented in the fourth column. The stress of clay, which produces the same clay void ratio, is presented in a separate column. The stress exerted on the sand can be obtained by subtracting the stress on the clay from the total stress applied. The clay void ratio for the mixture was computed.

Table 4. Prediction table: clay-sand ratio of 0.2.

Stress Level (kPa)	Clay %	Porosity η	Void Ratio e_c	ln $\sigma'_{v\ clay}$	$\sigma'_{v\ clay}$ (kPa)	Stress on Sand $\sigma'_{v\ sand}$ (kPa)	e_c (mix)
2	0.2	0.597	4.931	−0.161186	0.851	1.1489	12.328
4	0.2	0.586	4.659	0.248376	1.281	2.7181	11.647
50	0.2	0.544	3.78	1.567824	4.796	45.204	9.451
100	0.2	0.533	3.567	1.888987	6.613	93.387	8.917
200	0.2	0.521	3.363	2.194884	8.979	191.021	8.408
300	0.2	0.515	3.248	2.367175	10.667	289.333	8.121
400	0.2	0.510	3.169	2.486579	12.020	387.980	7.922
500	0.2	0.507	3.108	2.577623	13.166	486.834	7.771
600	0.2	0.504	3.059	2.651015	14.168	585.832	7.649
700	0.2	0.501	3.019	2.71238	15.065	684.935	7.547
1000	0.2	0.495	2.926	2.852011	17.323	982.677	7.314

Table 4 indicates that sands contribute to taking some of the load from very low stresses. The portion taken by clay reduces upon increasing the applied load. This becomes insignificant for extremely high loads. The clay can fill up the voids of the sand and push the sand grains apart. This action causes the sand porosity to increase. When pressure is exerted on clay, it compresses to a smaller volume, making the sand less porous.

The clay-sand mixture with 40% clay by weight tested similarly, and yielded the results presented in Table 5.

Table 5. Prediction table: clay-sand ratio of 0.4.

Stress Level (kPa)	Clay %	Porosity η	Void Ratio e_c	ln $\sigma'_{v\ clay}$	$\sigma'_{v\ clay}$ (kPa)	Stress on Sand $\sigma'_{v\ sand}$ (kPa)	e_c (mix)
2	0.4	0.7062	2.606	3.332896	28.019	−26	6.514
4	0.4	0.6924	2.377	3.676886	39.523	−35.5	5.942
50	0.4	0.6422	1.692	4.705966	110.610	−60.6	4.229
100	0.4	0.6284	1.536	4.939705	139.730	−39.7	3.840
200	0.4	0.6146	1.392	5.156714	173.590	26.4	3.479
400	0.4	0.6008	1.257	5.358727	212.450	188	3.1430
500	0.4	0.5963	1.216	5.420823	226.070	274	3.040
600	0.4	0.5927	1.183	5.470554	237.59	362	2.957
800	0.4	0.587	1.132	5.547247	256.53	543	2.829
1000	0.4	0.5825	1.093	5.605286	271.86	728	2.733

It can be noted from Table 5 that the clay carried all the low stresses. The model shows that the stresses acting on the sand were negative. This results from the use of the clay profile equation over the full range. The points showing negative values reflect the theoretical figures so that all assumptions hold. This method can show when the sand starts to contribute to taking up stresses. This point is defined by Mollins as the threshold point. It can be graphically determined if the void ratio versus the effective stress profile of the clay-sand mixture intersects the profile of clay.

The clay-sand mixture with 80% clay was entirely governed by the behavior of the clay. The results of the measured clay void ratio were found to coincide with those of the clay alone (40% bentonite). For this case, the prediction model used the equation for the clay. This finding has been reported by many authors, including Mollins [8] and Studds [11].

Kumar and Wood [20] found that, for a clay content above 35%, the clay matrix alone controlled the mechanical behavior of the mixture. This value is for a clay–gravel mixture and for a particular clay material, which is kaolinite. However, when investigating the dry powder effect on the porosity of the coarse material, it can be observed that the shift in the porosity line is large compared with that in the original line of the coarse material only. Pure bentonite clay tested by Mollins [8] showed that a clay content as small as 10% could control the behavior of a mixture until a certain point of stress. This is the threshold point mentioned above. The clay content at which the entire mechanical behavior is controlled by the fine part depends on the type of clay, the porosity of the coarse material, and the stress level.

The authors of reference [21] regarded the sand as an inert filler and the clay phase as determining the swelling of clay-sand mixes. This is not necessarily true; the coarse material also has a role in the way clay can expand. The porosity and geometry of sand grains can influence the quantitative behavior of the clay paste.

Studies conducted to investigate the characteristics and shear strength behavior of clay-sand mixtures, in general, include the works in references [22–24]. In reference [25], it is stated that the clay content in clayey sand has a significant influence on the stress–strain profile of the soil mass. As the fine content increased, the dilatant behavior of the clayey sand decreased and the response gradually became similar to that of clay-like material at about 40% fine content. The authors of reference [26] experimentally studied the behavior of clay-sand mixtures under compression and extension triaxial loading paths. Reference [23] reports that the shear strength properties and the stress–strain characteristics of mixtures of sand–kaolin showed a significant change at a kaolin content of 20%. In a study in reference [27], the authors performed a series of unconsolidated, undrained triaxial tests to evaluate the effects of some factors on the compacted bentonite–sand mixtures. These included the moisture, the energy of compaction, and the confining pressure in addition to the clay content. Sridharan and Prakash [28] claim that the liquid limit and the plastic limits cannot represent the plasticity limits and they do not correspond to stress-free reference states. They suggested three characteristic water contents, namely the free swell limit, the settling limit, and the shrinkage limit.

5. Conclusions

The clay void ratio versus the vertical effective stress for a clay material can be obtained by comparing the fall cone penetration data of the clay with the slope obtained for artificial clay. The chart constructed for artificial clay using the fall cone can be used to characterize the swell potential of any given plastic to highly plastic clay.

The stress–strain behavior of the clay-sand mixture system is influenced by the behavior of each component. The independent responses of fine- and coarse-grained components can be useful for estimating stress portioning and predicting the overall behavior of the soil mass. The porosity compression relationship of the coarse part can be studied when the voids are partially filled up with fine material in a dry state. This was found to be comparable to the overall mixture behavior when wetted. The modified Studds model was suitable for predicting the clay void ratio versus the effective stress for 40% clay or less.

The clay-sand mixture with 80% clay was entirely governed by the behavior of the clay. The results of the measured clay void ratio were found to coincide with those of the clay alone.

Funding: This research was funded by the Deanship of Scientific Research of King Saud University. Researchers Supporting Project, number RSPD2023R1059.

Data Availability Statement: All data related to this manuscript are available upon request.

Acknowledgments: The author gratefully acknowledges the Researchers Supporting Project, number RSPD2023R1059, King Saud University, Riyadh, Saudi Arabia, for their financial support for the research work reported in this article.

Conflicts of Interest: The author declares no conflict of interest.

List of Symbols

Symbol	Description
e	Void ratio
e_c	Clay void ratio
e_f	Final void ratio
e_1	Pretest void ratio
e_2	Post-test void ratio
γ_d	Dry unit weight
σ'_v	Vertical effective stress
ηs	Sand porosity
$\eta s_{(mixture)}$	Sand porosity of mixture
x	Volumetric fraction of clay
Vw	Volume of water
Vc	Volume of clay
Vt	Total volume
H	Height of sample
Hs	Height of solids
Ws	Weight of solids
A	Area
Gs	Specific gravity
γ_w	Unit weight of water
Sr	Degree of saturation
w	Water content
LL	Liquid limit
PL	Plastic limit
PI	Plasticity index
Cc	Compression index
φ	Angle of shearing resistance
L.O.I.	Loss of ignition

References

1. Zein, A.K.M. Comparison of Measured and Predicted Swelling Behavior of a Compacted Black Cotton Soil. In Proceedings of the Sixth International Conference on Expansive Soils, New Delhi, India, 1–4 December 1987; pp. 121–126.
2. Chen, F.H. *Foundations on Expansive Soils*; Elsevier Scientific Publishing Company: Amsterdam, The Netherlands, 1988.
3. Dafalla, M.A.; Al-Sharmrani, M.A. Performance-Based Solutions For Foundations on Expansive Soils—Al Ghatt Region, Saudi Arabia. In Proceedings of the International Conference on Geotechnical Engineering, Chiangmai, Thailand, 10–12 December 2008.
4. Alonso, E.E.; Gens, A.; Josa, A. A constitutive model for partially saturated soils. *Géotechnique* **1990**, *40*, 405–430. [CrossRef]
5. Alonso, E.E.; Vaunat, J.; Gens, A. Modelling the mechanical behaviour of expansive clays. *Eng. Geol.* **1999**, *54*, 173–183. [CrossRef]
6. Gens, A.; Alonso, E.E. A framework for the behaviour of unsaturated expansive clays. *Can. Geotech. J.* **1992**, *29*, 1013–1032. [CrossRef]
7. Pusch, R.; Schomburg, J. Impact of microstructure on the hydraulic conductivity of undisturbed and artificially prepared smectitic clay. *Eng. Geol.* **1999**, *54*, 167–172. [CrossRef]

8. Mollins, L.H. The Design of Bentonite-Sand Mixtures. Ph.D. Thesis, University of Leeds, Leeds, UK, 1996.
9. Mollins, L.H.; Stewart, D.I.; Cousens, T.W. Predicting the properties of bentonite-sand mixtures. *Clay Miner.* **1996**, *31*, 243–252. [CrossRef]
10. Stewart, D.I.; Studds, P.G.; Cousens, T.W. The factors controlling the engineering properties of bentonite-enhanced sand. *Appl. Clay Sci.* **2003**, *23*, 97–110. [CrossRef]
11. Studds, P.G. The Effect of Ionic Solutions on the Properties of Soil. Ph.D. Thesis, University of Leeds, Leeds, UK, 1997.
12. Olphen, H.V. *An Introduction to Clay Colloid Chemistry, for Clay Technologists, Geologists, and Soil Scientists*; Interscience Publishers: Hoboken, NJ, USA, 1977.
13. *BS 1377-2:1990*; Methods of Test for Soils for Civil Engineering Purposes. Classification tests. British Standards Institution: London, UK.
14. Feng, T.W. Fall-cone penetration and water content relationship of clays. *Géotechnique* **2000**, *50*, 181–187. [CrossRef]
15. Dafalla, M.A. The Use of the Fall Cone Penetrometer in the Characterization and Assessment of Swelling Clays. Ph.D. Thesis, University of Leeds, Leeds, UK, 2004.
16. *ASTM D4546-96*; Standard Test Methods for One-Dimensional Swell or Settlement Potential of Cohesive Soils. ASTM International: West Conshohocken, PA, USA, 1996.
17. Tan, T.S.; Goh, T.-C.; Kuranaratne, G.P.; Lee, S.L. Shear strength of very soft clay-sand mixtures. *Geotech. Test. J.* **1994**, *17*, 27–34.
18. Wood, D.M. Some Fall Cone Tests. *Geotechnique* **1985**, *3*, 64–68. [CrossRef]
19. Wood, D.M.; Worth, C.P. The use of Cone Penetrometer to Determine the Plastic Limit of Soils. *J. Ground Eng.* **1978**, *11*, 37.
20. Kumar, G.V.; Muir Wood, D. Fall cone and compression tests on clay-gravel mixtures. *Geotechnique* **1999**, *49*, 727–739. [CrossRef]
21. Dixon, D.A.; Gray, M.N.; Thomas, A.W. A study of the compaction properties of potential clay-sand buffer mixtures for use in nuclear fuel waste disposal. *Eng. Geol.* **1985**, *21*, 247–255. [CrossRef]
22. Chalermyanont, T.; Arrykul, S. Compacted sand-bentonite mixtures for hydraulic containment liners. *Songklanakarin J. Sci. Technol.* **2005**, *27*, 313–323.
23. Olmez, M.S. Shear Strength Behavior of Sand-Clay Mixtures. Master's Thesis, Middle East Technical University, Ankara, Turkey, 2008.
24. Gueddouda, M.K.; Goual, I.; Lamara, M.; Aboubaker, N. Hydraulic Conductivity and Shear Strength of Compacted Dune Sand-Bentonite Mixtures. In Proceedings of the International Conference on Construction and Building Technology (ICCBT 2008), Kuala Lumpur, Malaysia, 16–20 June 2008; pp. 139–150.
25. Georgianou, V.; Burland, J.; Hight, D. The undrained behavior of clayed sands in triaxial compression and extension. *Geotechnique* **1990**, *41*, 383–393. [CrossRef]
26. Shafiee, A.; Tavakoli, H.R.; Jafari, M.K. Undrained behavior of compacted sand-clay mixtures under monotonic loading paths. *J. Appl. Sci.* **2008**, *8*, 3108–3118. [CrossRef]
27. Chen, Y.; Meehan, C.L. Undrained strength characteristics of compacted bentonite-sand mixtures. In *Geo-Frontiers 2011-ASCE, Geotechnical Special Publications (GSP) 211*; Han, A., Ed.; ASCE Library: Reston, VA, USA, 2011; pp. 2699–2708.
28. Sridharan, A.; Prakash, K. The characteristic water content of fine grained soil-water system. *Geotechnique* **1998**, *48*, 337–346. [CrossRef]

applied sciences

MDPI

Article

Determining the Macrostructural Stability of Compacted Wyoming Bentonites by a Disaggregation Method

José Manuel Moreno-Maroto [1,*], Óscar Merlo [2], Joel Torres-Serra [2], Jacinto Alonso-Azcárate [3], Mark Tyrer [4] and Vicente Navarro [2]

1 Department of Geology and Geochemistry, Faculty of Sciences, Autonomous University of Madrid, Cantoblanco, 28049 Madrid, Spain
2 Geoenvironmental Group, Universidad de Castilla-La Mancha, Avda. Camilo José Cela 2, 13071 Ciudad Real, Spain; oscar.merlo@uclm.es (Ó.M.); joel.torres@uclm.es (J.T.-S.); vicente.navarro@uclm.es (V.N.)
3 Department of Physical Chemistry, Faculty of Environmental Sciences and Biochemistry, Universidad de Castilla-La Mancha, Avenida Carlos III, s/n, 45071 Toledo, Spain; jacinto.alonso@uclm.es
4 Collegium Basilea (Institute of Advanced Study), Hochstrasse, 51, 4053 Basel, Switzerland; m.tyrer@mtyrer.net
* Correspondence: josemanuel.moreno@uam.es

Abstract: The use of compacted bentonites in radioactive waste repository barriers is a relevant application of geoenvironmental engineering. The on-site structural characteristics of the bentonites determine the performance and integrity of the barrier. The present work addresses the adaptation of the standardized sand equivalent shaking method for the controlled disaggregation of Wyoming bentonite specimens prepared at low, medium, and high compaction. The evolution of the macrostructural units' size distribution was determined by sieving at different shaking times. The stability of the compacted material increases with dry density. However, if enough energy is applied in the disaggregation process, the size distribution of the macrostructural units after disaggregation has the same characteristics as that of the uncompacted starting material, regardless of the applied degree of compaction. Since the disaggregation rate is a function of the aggregation level (compaction), it follows that compaction energy is not only spent on reducing porosity but also on generating more stable macrostructural units. These findings pave the way for future research with different materials and test conditions (compaction, moisture, etc.). In addition, the proposed shaking method is adaptable and could also be used in other sectors, such as agriculture, to determine the structural stability of natural soils.

Keywords: bentonite; disaggregation; macrostructural units; soil compaction; particle size distribution

1. Introduction

The word "bentonite" was first applied by Knight [1] to refer to Cretaceous clay deposits with highly plastic properties. The term was later redefined by Ross and Shannon [2] to designate those clays produced by in situ alteration of volcanic ashes, with smectite as the main component. Subsequently, Grim [3] defined bentonite as that clay with a significant smectite content (specifically montmorillonite), regardless of its origin; this is the most widely adopted definition, especially in the radioactive waste management community.

The problem of radioactive waste storage leads to the need to find safe containment systems that prevent the release and diffusion of radionuclides from inside the containers to the surrounding environment. In this sense, due to their physicochemical properties, such as low hydraulic conductivity, high specific surface, remarkable ion exchange capacity, swelling behavior, and high sorption capacity [4], the use of an Engineered Barrier System (EBS) integrating compacted bentonite has aroused a strong interest in the last decades as a sealing element [5]. Thus, the main function of the bentonite barrier is to protect and isolate the canister containing the residues from any type of geochemical or physical stress to maintain its integrity. In addition, the bentonite works as a hydraulic barrier that limits the mobilization of radioactive contaminants that could be released if the metallic container fails. These functions can be enhanced through blending with other materials, such as sand-bentonite mixtures [4].

Among the different types of bentonites, the application of MX-80 or Wyoming bentonites (in honor of their origin) has been the object of study in several countries in EBS for radioactive repositories [6]. Particularly well known are the studies being carried out to use this type of bentonite as a barrier in KBS-3-type repositories, a deposition concept considered in Sweden and Finland. The waste is sealed inside metal canisters, which are deposited in vertical holes constructed at great depth, surrounded by compacted MX-80 bentonite, which, either alone or mixed with other filler materials, is also used to seal the access tunnel [5].

Although the properties of EBS-bentonites have been extensively studied, no systematic procedure has been proposed to characterize the stability of their macrostructure. Stability is considered to critically affect the long-term integrity of the compacted bentonite, which has motivated its experimental characterization by in situ tests such as the BeLLT (Bentonite Large Scale Load Test) and the BLL (Bentonite Long-Term Load Test), both within the framework of the full-scale emplacement experiment [7,8]. These studies show a growing interest in defining a laboratory-scale protocol that ensures repeatability and contributes quantitative data toward the assessment of the effect of compaction on macrostructural stability.

Several methods are currently in use, the suitability of which depends on the processes to be studied, since each methodology can provide different information [9,10]. The most common methods are based on measuring the size distribution of the fragments or, more generally, macrostructural units (MU) generated after the application of a disaggregation effort. Such effort can be applied internally (e.g., by wetting) and also externally (mechanical agitation, tillage, etc.), resulting in the fracturing of the MU. Accordingly, the failure areas provide relevant information about the voids, pores, and solid phases of the soil and its MU [11].

Most of these fragmentation procedures are performed in two stages. First, the MU are generated, and subsequently, a sieving measurement is performed. Two different methods of sieving are considered: dry sieving (the process is carried out in the air) and wet sieving (the process is carried out in water or another liquid) [12,13]. In general, stability is determined by the relationship between the particle size distribution obtained between a short and long sieving time, so that the main source of disaggregation energy is that provided during the sieving process itself, particularly in dry methods [12]. This last procedure has a negative effect on the reproducibility and repeatability of the tests since, in most cases, it is not possible to know the disaggregation energy applied, to which should be added the fact that neither the equipment nor the protocol itself is subject to any standardization.

In the present study, a novel method is proposed using a standardized mechanical shaker, widely adopted in geotechnical laboratories, to perform a controlled disaggregation of the bentonite macrostructure. The MU size distribution (MUSD) generated for each shaking time is determined by dry sieving analysis, allowing the evolution of the disaggregation pattern of a soil to be studied as the disaggregation energy increases. Therefore, from a practical point of view, the objective is to develop a method and parameters that can be easily implemented to better understand the compaction and dis/aggregation processes in soils, which have a strong impact on their technological properties.

2. Materials and Methods

2.1. Basic Characterization of the Wyoming Bentonite

The same Wyoming bentonite used by Navarro et al. [14] was used in this work, which has similar properties to the material identified as Be-Wy-BT007-1-Sa-R by Kiviranta and

Kumpulainen [15]; see Table 1. The plastic limit (PL) was determined by the standard thread-rolling method, while the liquid limit (LL) was obtained by the Multipoint Method of the Casagrande cup percussion test [16]. The plasticity index (PI) and LL data were used to classify the material according to the USCS [17,18]. Moreover, the Moreno-Maroto and Alonso-Azcaráte [19,20] classification systems [21] were applied to estimate its maximum toughness, T_{max}.

Table 1. Properties of the tested bentonite [15].

Mineralogical Composition	wt.%
Smectite	87.6
Plagioclase	4.2
Quartz	4.1
K-feldspar	1.8
Rutile	0.9
Calcite	0.6
Pyrite	0.6
Illite	0.1
Other Properties	**eq/kg**
Cation Exchange Capacity	0.84
Na^+ (eq/kg)	0.58
Ca^{2+} (eq/kg)	0.25
Mg^{2+} (eq/kg)	0.08
K^+ (eq/kg)	0.02

2.2. New Method of Controlled Disaggregation

The objective of the present study is to compare how compaction affects the particle size distribution of the MU after shaking, in this case using compacted Wyoming bentonite specimens. However, the purpose is not only to develop a method applicable to this kind of clay but to any soil. For this reason, the protocol is based on basic equipment in geotechnical and materials laboratories, such as the mechanical shaker included in the sand equivalent test UNE-EN 933-8:2012+A1:2015/1M [22]. The procedure for the new disaggregation method is explained below.

Duplicate samples of height 17 mm and diameter 49.5 mm, with a water content of 10%, were prepared with target dry density values of 1.40 g/cm^3, 1.60 g/cm^3, and 1.80 g/cm^3. The compliance of the target density was checked by accurately determining the weight and dimensions of each specimen once prepared. Similarly, part of the excess material was used to check its moisture content by oven drying at 105 ± 5 °C.

Then, each cylindrical specimen (Figure 1a) was cut radially in successive subdivisions into 8 portions (Figure 1b). These portions were placed inside a plastic test tube presenting standardized dimensions of 430 mm in height and 32 mm in inner diameter [22] and closed with a rubber stopper. Next, the test tube containing the sample was coupled to the mechanical shaker (Figure 1c) and put into operation to start the agitation process. According to the sand equivalent test standard, the device used exerted a horizontal, periodic, and sinusoidal rectilinear motion of 200 ± 10 mm amplitude with a period of 1/3 s. To know the gradual evolution in the size of the MU, the process was interrupted after 5 s, 10 s, 20 s, 30 s, 1 min, 2 min, 5 min, and 10 min. For those specimens in which relatively large pieces were still observed, additional shaking times were applied, up to 50 min. At each shaking time, the material was extracted from the test tube and passed through sieves with apertures (in mm) of 10, 5, 2, 1.25, 0.63, 0.40, 0.16, and 0.080 (Figure 1d). Once the weights of each fraction were recorded, their particle size distribution was determined to analyze the change in the size of the MU as the shaking time increased and thus determine the disaggregation pattern.

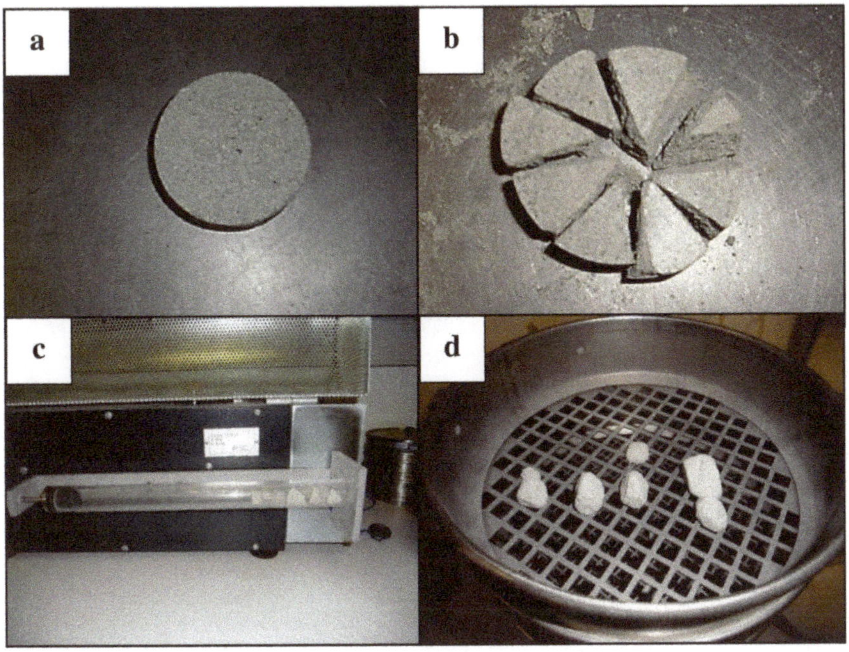

Figure 1. Method proposed in this study of controlled soil disaggregation from a compacted specimen: (**a**) initial compacted Wyoming bentonite specimen; (**b**) specimen cut into eight portions; (**c**) standardized shaking machine and test tube with the material inside; (**d**) passing through sieves after shaking.

3. Results and Discussion

3.1. Characterization of the Bentonite under Study

The particle size distribution of the total sample after disaggregation shows a trimodal differential curve (Figure 2), having a main peak with its maximum at approximately 5 µm, together with two secondary peaks with maxima at around 17 µm and 44 µm. The fineness of the bentonite studied is evident from the data in Table 2, with a d50 (the particle size below which 50% of the material is contained) of 5 µm and an average particle size of 13 µm.

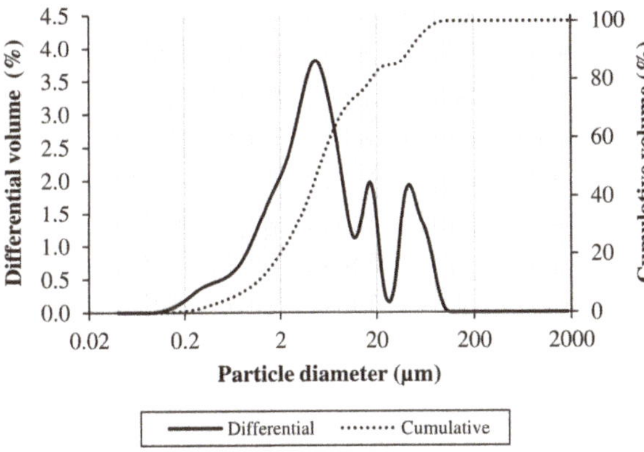

Figure 2. Particle size distribution of the Wyoming bentonite studied.

Table 2. Characterization results of the Wyoming bentonite under study.

Particle Size Distribution (μm)					Plasticity					
d10	d50	d90	Mean	>63 μm (%)	LL	PL	PI	PI/LL	T_{max} (kJ/m^3)	Class. Plasticity/Texture
1.1	5.1	44.9	13.1	3.8	459.7	43.4	416.3	0.91	73.5	CH/Clay

The water content of the Wyoming bentonite as received was w = 4.5%. The results of the Atterberg limits (Table 2) show the high plasticity of the material, which would be in accordance with its fineness (Figure 2 and Table 2), mineralogy, and exchange cations since the sample is predominantly sodic montmorillonite (see Table 1). The LL around 460 is consistent with the value of 400 usually found in the literature [23]. The PL value of 43.4 is also in the same order as those reported in the literature [24]. Although the PL has a relatively high value, the large difference with respect to the LL is evidenced by a very high PI value of 416.3. This causes a PI/LL ratio above 0.9, typical of materials with exceptional plastic and toughness properties, showing a T_{max} > 70 kJ/m^3 when values for ordinary clays are usually around 20–30 kJ/m^3 [20]. Therefore, as expected, the material studied is classified as CH according to both USCS [17,18] and Moreno-Maroto and Alonso-Azcárate [20] criteria, with a "clay" type texture.

The high toughness mentioned above is also demonstrated in the *flow index* of the LL test, the value of the slope obtained in the flow line after plotting the water content (w) versus the number of blows (N_B) required to close the groove. According to Casagrande [25], soils with high toughness (especially those rich in very active minerals, such as montmorillonite) have lower flow index values than most "normal" soils. In the case of the Wyoming bentonite studied, the flow line (Ln w = Ln 487.21 − 0.018·Ln N_B; coefficient of determination R^2 = 0.911) presented a flow index of −0.018, much lower than the average values reported in the literature based on the statistical study of a large number of soil samples. Thus, the ASTM D 4318-17e1 standard [16] considers that the most probable flow index is −0.121, while Spanish standard UNE 103103:1994 [26], based on the determinations of Eden [27], gave a value of −0.117. Therefore, the differences observed with respect to the Wyoming bentonite investigated here, together with the PI/LL ratio, demonstrate its high toughness.

3.2. Stability of the Macrostructure

The particle size distributions of the uncompacted bentonite, both as received (w = 4.5%) and with a water content of w = 10%, are presented in Table 3. The data are displayed in differential terms, whereby each numerical value represents the percentage by weight of MU with a size between the sieve aperture interval (class interval) indicated in each case. Nine class intervals and class marks are defined for the aggregate size ranges according to the sieve apertures where the different MU fractions are retained.

Table 3. Differential aggregate size distribution for the Wyoming bentonite studied both as received and at moisture content w = 10% prior to compaction.

Aggregate Size (mm)	Class Intervals	Class Marks (mm)	Bentonite (wt.%)	
			As Received	w = 10%
>10	1	15	0.0	0.0
5–10	2	7.5	0.0	0.0
2–5	3	3.5	0.5	0.5
1.25–2	4	1.63	2.3	2.6
0.63–1.25	5	0.94	5.4	11.2
0.40–0.63	6	0.52	26.7	24.3
0.16–0.40	7	0.28	39.5	37.4
0.08–0.16	8	0.12	13.8	13.1
<0.08	9	0.04	11.9	11.0

The evolution of MU size distribution (MUSD) as a function of shaking time and degree of compaction (dry density) is depicted in Figure 3 and detailed as Supplementary Material in Table S1.

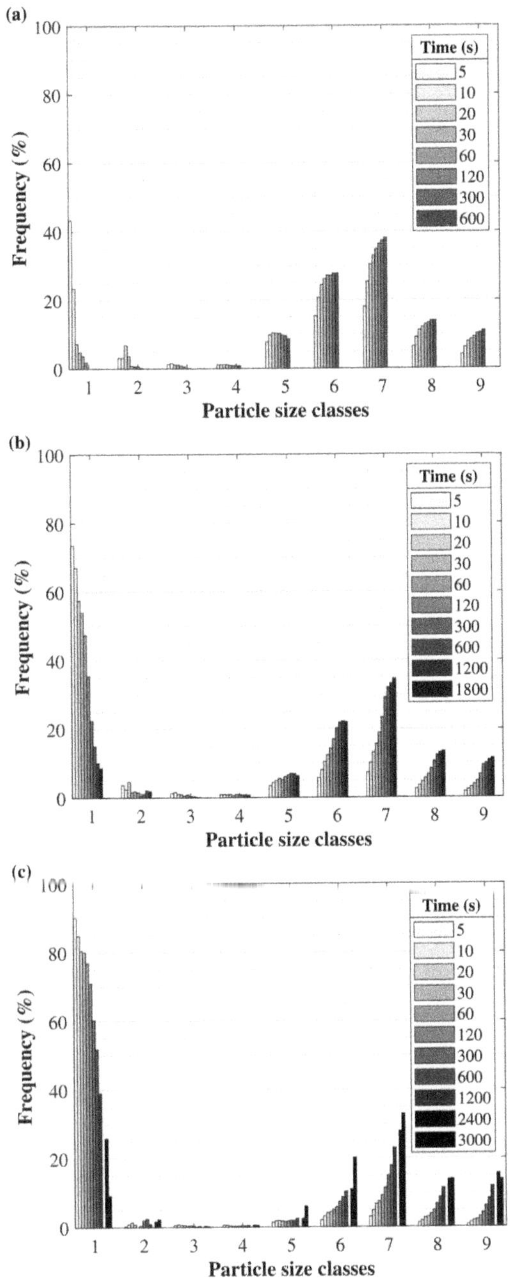

Figure 3. Evolution of the frequencies associated with the nine size intervals defined in Table 3 with respect to shaking time for average dry densities of: (**a**) 1.40 g/cm^3; (**b**) 1.57 g/cm^3; and (**c**) 1.74 g/cm^3.

Taking as reference the target dry densities for each degree of compaction, i.e., 1.40 g/cm^3 (low compaction), 1.60 g/cm^3 (medium compaction), and 1.80 g/cm^3 (high compaction), similar dry densities were obtained for the low and medium compaction energies for which the target density was reached (Table S1). In the case of the highest compaction, one specimen presented a dry density of 1.72 g/cm^3 and another of 1.75 g/cm^3, and it was not possible to obtain higher density values because the compressive capacity of the press was exceeded. However, both samples showed similar disaggregation behavior. Although more tests with different types of soils are advisable, the results in Table S1 point to the satisfactory repeatability of the shaking process. Taking this into account, to avoid redundancies in the discussion of the results, the average results for each pair of compaction conditions are analyzed, considering an average density of 1.74 g/cm^3 for the high compaction test.

The first aspect to be addressed is the analysis of how disaggregation of the samples evolves. The testing time is a first indicator of this parameter; therefore, in the less compacted specimens, it was 600 s (10 min), while in the medium and highly compacted samples, it was extended to 1800 s (30 min) and 3000 s (50 min), respectively. An appropriate and simple way to study the disaggregation evolution in quantitative terms is to focus on the fraction in class interval 1 (wt.% retained by the 10 mm sieve), which includes the original pieces prepared as shown in Figure 1d. Structural stability at low compaction (ρ_d = 1.40 g/cm^3) is relatively poor (fast disaggregation), since MU with a diameter equal to or greater than 10 mm crumbled completely before 5 min of shaking (Figure 3a). Conversely, when the compaction energy increases, the structural stability (slow disaggregation) increases as well. Specimens with a dry density of 1.57 g/cm^3 still had around 10 wt.% in class interval 1 after 30 min of shaking (Figure 3b). In the case of the dry density of 1.74 g/cm^3, after 40 min of shaking, there was still about 25 wt.% of MU with a size equal to or larger than 10 mm (Figure 3c), which demonstrates the outstanding mechanical stability of the highly compacted specimens. Shaking times over 3000 s (50 min) were required in this case for the disaggregation patterns to exhibit a stationary distribution of the MU sizes.

The differential MU size distribution curves (dMUSD) at the analyzed shaking times also explain the disaggregation evolution; see Figure 4, where frequencies are associated with the class marks defined in Table 3. With respect to dMUSD in uncompacted conditions, specimens with a higher degree of compaction showed slower time evolution of their dMUSD, as expected. Interestingly, as shown in Figure 4d, the dMUSD of all compacted specimens tended to conform to the uncompacted bentonite curves (both as received and with w = 10%) as shaking time increased.

Figure 5 at t = 5 s compares for the nine class intervals the associated frequencies to the dry uncompacted material and the three dry densities. MUSD evolved during the compaction process. Even though initially the main class interval was number 7 (0.40 mm–0.16 mm), its predominance decreased as the compaction loading (and dry density) increased. Class intervals 6 to 9 (between 0.63 mm and <0.08 mm) showed analogous behavior, whereas number 1 (>10 mm) became progressively dominant. For the bentonite considered, compaction developed for growing MU sizes larger than 0.40 mm, the maximum size of interval 7, which prevailed in uncompacted conditions.

Therefore, an aggregation index A is defined

$$A = \sum_{i=1}^{n} F_i \cdot fi \tag{1}$$

where n = 9 are the class intervals considered, F_i is the frequency associated with the i-th class, and f_i is a function defined by

$$f_i = \begin{cases} 1, \text{ if } d_{\min,i} \geq d_{REF} \\ 0, \text{ otherwise} \end{cases} \tag{2}$$

where $d_{min,i}$ is the minimum size of the i-th size interval (respectively equal to 10, 5, 2, 1.25, 0.63, and 0.40 mm) for class intervals 1 to 6, with classes with a minimum size larger or equal to d_{REF} = 0.40 mm taken as the reference size. The ability of index A to summarize the increase in aggregation with compaction is illustrated in Figure 6.

A is also a good estimator of the disaggregation process. This process, illustrated in Figure 3, is described in more detail in Figure 5, which shows for each dry density the evolution of the frequencies associated with the size intervals considered. The frequency of larger aggregates was reduced as shaking time advanced, eventually reaching a steady state. Remarkably, classes 2 to 5 (between 10 mm and 0.63 mm) showed low frequencies throughout the disaggregation process, which may be interpreted as metastable transition MU sizes.

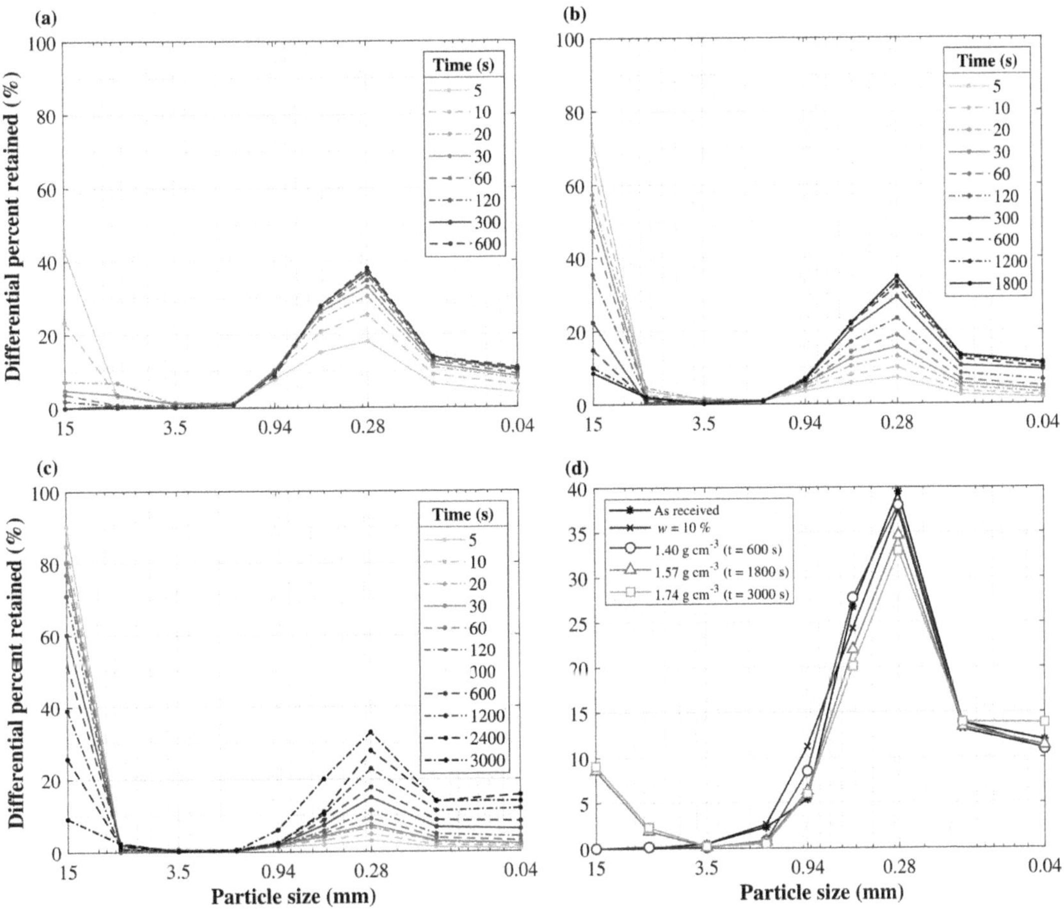

Figure 4. Differential aggregate size distribution, dMUSD (wt.% retained on sieve), for each shaking time and for each degree of compaction (average values) used in the bentonite studied: (**a**) 1.40 g/cm^3; (**b**) 1.57 g/cm^3; (**c**) 1.74 g/cm^3. Graph (**d**) compares the dMUSD for the maximum shaking time employed in each case with respect to the dMUSD of the samples before compaction, both as received and at moisture content w = 10 wt.%.

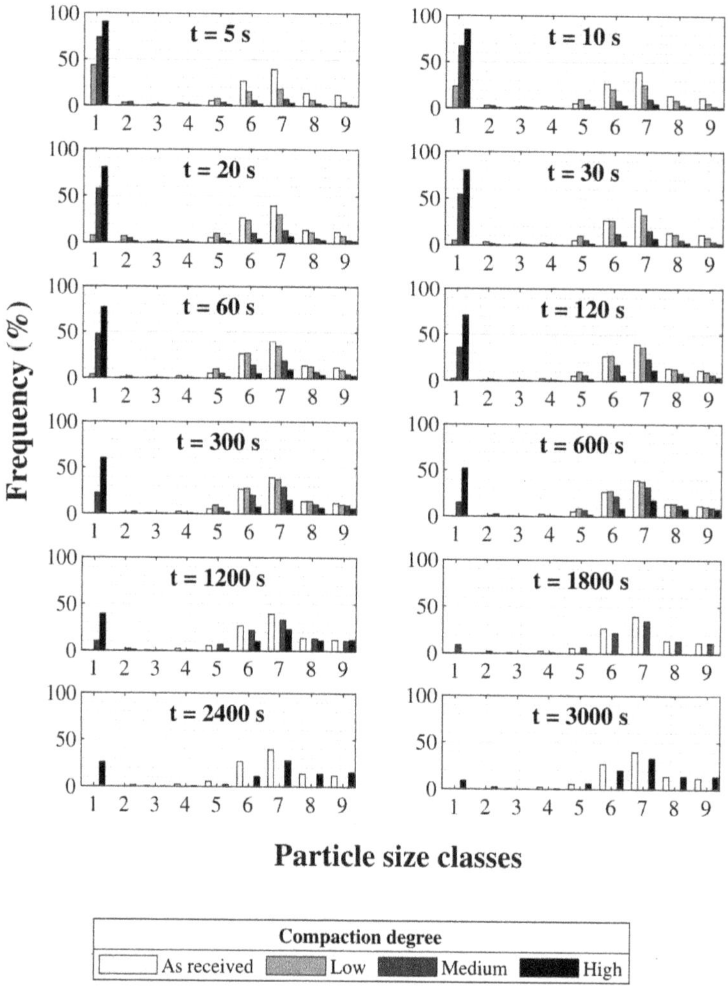

Figure 5. Associated frequencies to the dry uncompacted material and the three dry densities in the nine class intervals considered in this study as a function of shaking time, see Table 3.

The tendency of dMUSD to reach a stationary state around the initial size distribution in uncompacted conditions is depicted in Figure 7. The evolution of *A* during the shaking test is represented for each dry density, and the values of *A* for the uncompacted samples, dry and wet, are shown for reference. As the shaking progressed, *A* decreased.

For higher dry densities (when a higher amount of compaction energy is applied to the system), the aggregates formed seem to be more stable, resulting in a lower rate of disaggregation. The compaction process not only appears to involve the closure of the original interaggregate voids; however, the exerted energy is increasingly translated into the formation of structural elements with greater stability.

Based on all these results and the aggregation index, *A*, it is demonstrated that the proposed method is easily executable, standardizable, and useful to better understand the compaction and aggregation processes in soils, which ultimately generate a strong impact on their technological properties.

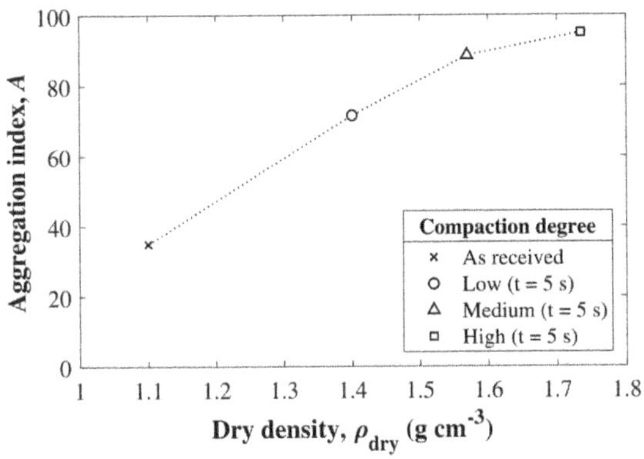

Figure 6. Evolution of the aggregation index, A, with the degree of compaction. The value of A at 1.10 g/cm^3 corresponds to the average dry density of the undisturbed sample.

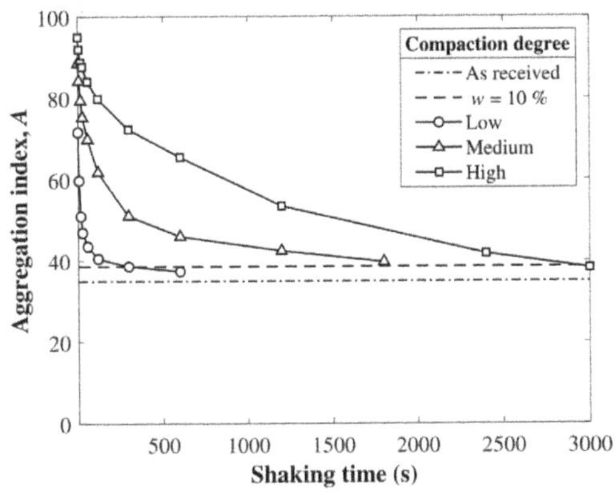

Figure 7. Evolution of the aggregation index, A, for each compaction degree during the shaking tests.

4. Conclusions

The present work has addressed the investigation of the macrostructural characteristics of compacted bentonites by means of a new method based on the adaptation of the equipment used in the standardized sand equivalent test. According to the results reflected in the previous sections, the following conclusions can be drawn:

- The characterization tests carried out not only confirmed the extraordinary plasticity but also demonstrated the toughness of the Wyoming bentonite.
- The use of the shaking method on specimens prepared at different degrees of compaction, ρ_d = 1.40 g/cm^3 (low compaction), ρ_d = 1.57 g/cm^3 (medium compaction), and ρ_d = 1.74 g/cm^3 (high compaction) has shown that there is a direct relationship between the dry density and the stability of the compacted bentonite, so that the higher the density, the longer the shaking time required to disaggregate the material.
- Although the required shaking times are different, regardless of the initial dry density, the final differential MU size distribution (dMUSD) obtained after disaggregation is equal to that of the initial uncompacted material.
- Although the same final dMUSDs are obtained, as the disaggregation rate is a function of the level of aggregation (compaction), it becomes clear that the compaction energy is not only spent on reducing porosity but also on generating more stable MU.

- An aggregation index, *A*, has been defined in Equations (1) and (2) to synthesize the aggregation information, favoring the comparison of states and the characterization of the soil disaggregation process.
- The satisfactory repeatability suggests that the new disaggregation method could potentially be standardized, especially considering that the equipment is very common in geotechnical laboratories.
- Although the disaggregation method developed in this work has been applied to compacted bentonite specimens, its use on other types of samples, such as natural or agricultural soils, could also be investigated in the future. Likewise, the method presents certain flexibility, being able to adapt the energy and/or the disintegration medium according to convenience.

Considering all the above points, this work paves the way for future research not only in the field of geotechnical engineering in bentonites and/or compacted soils but also in other fields where the proposed method may be beneficial.

Supplementary Materials: The following supporting information can be downloaded at: https://www.mdpi.com/article/10.3390/app132011159/s1. Table S1. Differential aggregate size distribution (wt.% retained on sieve) for each shaking time and for each degree of compaction used in the Wyoming bentonite studied.

Author Contributions: Conceptualization: J.M.M.-M. and V.N.; methodology: J.M.M.-M., Ó.M. and V.N.; validation, J.M.M.-M. and V.N.; formal analysis: J.M.M.-M. and V.N.; investigation: J.M.M.-M. and Ó.M.; resources: V.N.; writing—original draft preparation: J.M.M.-M. and V.N.; writing—review and editing: J.M.M.-M., J.T.-S., J.A.-A., M.T. and V.N.; visualization: J.M.M.-M., J.T.-S. and V.N.; supervision: J.M.M.-M. and V.N.; project administration: V.N.; funding acquisition: V.N. All authors have read and agreed to the published version of the manuscript.

Funding: This study was funded by the Junta de Comunidades de Castilla-La Mancha and the European Regional Development Fund (European Union) through project SBPLY/19/180501/000222.

Data Availability Statement: The data and material of this research will be made available to anyone who requests them by e-mail to the corresponding author.

Conflicts of Interest: The authors declare no conflict of interest.

References

1. Knight, W.C. Bentonite. *Eng. Min. J.* **1898**, *66*, 491.
2. Ross, C.S.; Shannon, E.V. The minerals of bentonite and related clays and their physical properties. *J. Am. Ceram. Soc.* **1926**, *9*, 77–96. [CrossRef]
3. Grim, R.E. *Clay Mineralogy*, 2nd ed.; International Series in the Earth and Planetary Sciences; McGraw-Hill Book Company: New York, NY, USA, 1968.
4. Wilson, J.; Savage, D.; Bond, A.; Watson, S.; Pusch, R.; Bennett, D. *Bentonite–A Review of Key Properties, Processes and Issues for Consideration in the UK Context*; Report QRS-1378ZG-1.1 for the Nuclear Decommissioning Authority; Quintessa Limited: Harwell, UK, 2011. Available online: https://rwm.nda.gov.uk/publication/bentonite-a-review-of-key-properties-processes-and-issues-for-consideration-in-the-uk-context/ (accessed on 20 October 2021).
5. Sellin, P.; Leupin, O.X. The use of clay as an engineered barrier in radioactive waste management—A review. *Clay Clay Miner.* **2013**, *61*, 477–498. [CrossRef]
6. Liu, L. Permeability and expansibility of sodium bentonite in dilute solutions. *Colloids Surf. A Physicochem. Eng. Asp.* **2010**, *358*, 68–78. [CrossRef]
7. Garitte, B.; Kober, F.; Müller, H.R.; Köhler, S.; Weber, H.; Blechschmidt, I. Stability of compacted bentonite blocks and block pedestals under changing climatic conditions in tunnels and long-term loads. In Proceedings of the 6th International Conference on Clays in Natural and Engineered Barriers for Radioactive Waste Confinement, Brussels, Belgium, 23–26 March 2015. [CrossRef]
8. Papafotiou, A.; Senger, R.; Li, C.; Singh, A.; Garitte, B.; Müller, H.; Marschall, P. A prediction-evaluation approach to the full-scale emplacement experiment (FE) in Mont Terri. In *Multiple Roles of Clays in Radioactive Waste Confinement*; Norris, S., Neeft, E.A.C., Van Geet, M., Eds.; Geological Society: London, UK, 2018; p. 482. [CrossRef]
9. Dexter, A.R. Physical properties of tilled soils. *Soil Till. Res.* **1997**, *43*, 41–63. [CrossRef]
10. Elliott, E.T.; Heil, J.W.; Kelly, E.F.; Monger, H.C. Soil structure and other physical properties. In *Standard Soil Methods for Long-term Ecological Research*; Robertson, G.P., Coleman, D.C., Bledsoe, C.S., Sollins, P., Eds.; Oxford University Press, Inc.: New York, NY, USA, 1999; pp. 74–88.

11. Kay, B.D.; Angers, D.A. Soil structure. In *Handbook of Soil Science*; Summer, M.E., Ed.; CRC Press: Boca Raton, FL, USA, 1999; pp. 229–276.
12. Díaz-Zorita, M.; Perfect, E.; Grove, J.H. Disruptive methods for assessing soil structure. *Soil Till. Res.* **2002**, *64*, 3–22. [CrossRef]
13. Moncada, M.P.; Gabriels, D.; Cornelis, W.; Lobo, D. Comparing aggregate stability tests for soil physical quality indicators. *Land Degrad. Develop.* **2013**, *26*, 843–852. [CrossRef]
14. Navarro, V.; Yustres, Á.; Asensio, L.; De la Morena, G.; González-Arteaga, J.; Laurila, T.; Pintado, X. Modelling of compacted bentonite swelling accounting for salinity effects. *Eng. Geol.* **2017**, *223*, 48–58. [CrossRef]
15. Kiviranta, L.; Kumpulainen, S. Quality Control and Characterization of Bentonite Materials; Posiva Oy, Working Report 2011-84. 2011. Available online: https://inis.iaea.org/collection/NCLCollectionStore/_Public/43/108/43108887.pdf (accessed on 20 October 2021).
16. *ASTM D 4318-17e1;* Standard Test Methods for Liquid Limit, Plastic Limit, and Plasticity Index of Soils. ASTM International: West Conshohocken, PA, USA, 2017. [CrossRef]
17. Casagrande, A. Classification and identification of soils. *ASCE Trans.* **1948**, *113*, 901–930. [CrossRef]
18. *ASTM D2487-17e1;* Standard Practice for Classification of Soils for Engineering Purposes (Unified Soil Classification System). ASTM International: West Conshohocken, PA, USA, 2017. [CrossRef]
19. Moreno-Maroto, J.M.; Alonso-Azcárate, J. Plastic Limit and Other Consistency Parameters by a Bending Method and Interpretation of Plasticity Classification in Soils. *Geotech. Test. J.* **2017**, *40*, 467–482. [CrossRef]
20. Moreno-Maroto, J.M.; Alonso-Azcárate, J. What is clay? A new definition of "clay" based on plasticity and its impact on the most widespread soil classification systems. *Appl. Clay Sci.* **2018**, *161*, 57–63. [CrossRef]
21. Moreno-Maroto, J.M.; Alonso-Azcárate, J.; O'Kelly, B.C. Review and critical examination of fine-grained soil classification systems based on plasticity. *Appl. Clay Sci.* **2021**, *200*, 105955. [CrossRef]
22. *UNE-EN 933-8:2012+A1:2015/1M;* Ensayos para Determinar las Propiedades Geométricas de los Áridos. Parte 8: Evaluación de los Finos. Ensayo del Equivalente de Arena. Asociación Española de Normalización y Certificación: Madrid, Spain, 2016.
23. Lajudie, A.; Raynal, J.; Petit, J.C.; Toulhoat, P. Clay-based materials for engineered barriers: A review. *MRS Online Proc. Libr.* **1996**, *353*, 221–230. [CrossRef]
24. Gómez-Espina, R.; Villar, M.V. Geochemical and mineralogical changes in compacted MX-80 bentonite submitted to heat and water gradients. *Appl. Clay Sci.* **2010**, *47*, 400–408. [CrossRef]
25. Casagrande, A. Research on the Atterberg limits of soils. *Public Roads* **1932**, *13*, 121–136.
26. *UNE103103:1994;* Determinación del Límite Líquido de un Suelo por el Método del Aparato de Casagrande. Asociación Española de Normalización y Certificación: Madrid, Spain, 1994.
27. Eden, W.J. Use of a One-Point Liquid Limit Procedure. In *Papers on Soils 1959 Meetings;* ASTM International: West Conshohocken, PA, USA, 1960. [CrossRef]

applied sciences

MDPI

Article

Research on the Accumulated Plastic Strain of Expansive Soil under Subway Loading

Lei Zhu [1], Ying Guo [2], Gan Cheng [1,*] and Xiangyang Liu [1]

[1] School of Urban Construction and Transportation, Hefei University, Hefei 230601, China; zhuleianhui@163.com (L.Z.); lxymm9632@163.com (X.L.)
[2] School of Civil Engineering, Anhui Jianzhu University, Heifei 230601, China; guoying626@std.ahjzu.edu.cn
* Correspondence: chenggan0713@163.com

Abstract: Expansive soil near Hefei Xinqiao International Airport was selected as the research focus. The effects of different loading modes on the dynamic strain of the saturated remodeled expansive soil was investigated by indoor dynamic triaxial tests and the effect of three factors, intermittent loading ratio, static deviatoric stress, and cyclic stress ratio, on the accumulated plastic strain of saturated remodeled expansive soil under intermittent subway cyclic loading was analyzed. Finally, the corresponding accumulated plastic strain calculation model was established based on the test results. The research results show that the loading modes do not affect the overall development mode of the strain and, under intermittent loading, the strain tends to be more stable. Under intermittent loading, when the number of cycles is the same, the larger the intermittent loading ratio, the smaller the accumulated plastic strain. The effect of static deviatoric stress and cyclic stress ratio on strain accumulation is significant. Based on the hyperbolic model, an accumulated plastic strain calculation model related to the number of cycles of remolded expansive soil under the load of the Hefei subway was proposed, which has a certain reference significance for the design, safety assessment, and safe operation of the Hefei subway.

Keywords: subway loads; expansive soil; dynamic strain; accumulated plastic strain model

1. Introduction

Expansive soil is widely distributed throughout the world and is known to exist in more than forty countries on six continents, of which China is one of the countries with the widest distribution and largest area of expansive soil [1,2]. Although these countries are geographically different and have different climatic environments, they all face the same problem of expansive soil hazards. Expansive soil poses formidable challenges in engineering construction and operational safety. Highways and railways constructed in expansive soil regions become "roadblocks" and "hard nuts" to crack. They epitomize the saying "Where there is a ditch, there is a slip; where there is a dike, there is a collapse". Similarly, working with river channels in these areas is akin to "Digging a foot during the day, only to find it stretches eight feet longer at night". Expansive soil not only causes great obstruction to above-ground projects such as buildings, highways, railroads, and canal slopes, but also poses a serious threat to underground engineering projects such as underground railways, pipelines, tunnels, and foundation structures [3–5]. The repetitive nature, complexity, and long-term potential hazards are hallmark features of expansive soil disasters, posing persistent challenges to the engineering community. The magnitude and widespread impact of the damages resulting from the unfavorable properties of expansive soil often necessitate significant human, material, and financial resources to address this unique geological hazard. According to statistics, the United States allocates billions of dollars each year towards addressing the issue of expansive soil hazards, surpassing the cumulative losses from disasters such as floods, hurricanes, earthquakes, and tornadoes.

Likewise, in China, the economic losses caused by expansive soil amount to over billions of dollars annually [6]. Hence, expansive soil has been recognized as a "cancer" that impedes engineering construction globally.

So far, researchers have made significant strides in studying the composition of expansive soil minerals, their macro–micro structure, swelling–shrinkage mechanisms, and physical–mechanical properties. However, there remains a relative scarcity of research concerning the mechanical properties of expansive soil under dynamic loading conditions.

Tang et al., investigated the attenuation patterns of a dynamic elastic modulus of saturated expansive soil under various influencing factors. Based on this, they established a model for the attenuation of a dynamic elastic modulus related to compaction degree, reinforcement layers, confining pressure, and frequency [7]. Yang et al., examined the dynamic shear modulus and damping ratio of expansive soil under freeze–thaw conditions. The results indicated that freeze–thaw cycles led to an increase in the dynamic elastic modulus of expansive soil, but a decrease in the damping ratio [8]. These researchers explored different experimental conditions to study dynamic modulus indices of expansive soil under dynamic loading. However, it is worth noting that under lower dynamic loads, the soil exhibits elastic deformation with closed hysteresis loops. In this scenario, dynamic shear modulus or dynamic elastic modulus can effectively describe the soil's ability to resist dynamic elastic deformation. Yet, under higher dynamic loads, the soil experiences irreversible strains, and hysteresis loops no longer close. Consequently, dynamic shear modulus or dynamic elastic modulus lose their significance.

Considering this, Chen and Zhang investigated the changing patterns of accumulated plastic strain under significant dynamic loading on expansive soil, proposing a corresponding strain model to predict strain growth during prolonged cyclic loading [9,10]. However, their simulation of dynamic loading in indoor experiments lacked meticulous detailing or a comprehensive rationale for employing sine wave simulation of dynamic loading.

In summary, the research on the strain development patterns of expansive soil under dynamic loading is still relatively limited, and a comprehensive and scientifically grounded theory has yet to be established. Furthermore, due to variations in soil properties across different regions, research outcomes also exhibit disparities. Therefore, this study employs indoor dynamic triaxial tests to investigate the impact of various factors on the strain behavior of expansive soil in the Hefei region under subway loading. The dynamic loading in the experiments is designed to closely emulate the authentic characteristics of subway loading to the greatest extent possible.

2. Dynamic Triaxial Test Plan

2.1. Test Equipment

The dynamic triaxial testing was performed utilizing the sophisticated electric motor-controlled dynamic triaxial testing system (DYNTTS) produced by GDS Company in the United Kingdom, as shown in Figure 1. The DYNTTS system finds extensive application in indoor experimental research on the dynamic properties of geotechnical materials. It not only facilitates dynamic testing of soils but also seamlessly integrates the entire suite of testing capabilities offered by static triaxial testing instruments. This system enables the execution of a range of experiments, such as soil sample consolidation, saturation, and dynamic loading tests. Moreover, during the testing process, it provides real-time dynamic visualization of experimental data.

2.2. Test Soil Samples and Remolded Soil Sample Preparation

The test soil was extracted from the expansive soil samples located near the subway station of Xinqiao International Airport in Hefei City. The diameter of the soil samples was 80 mm and the height was 200 mm, as shown in Figure 2.

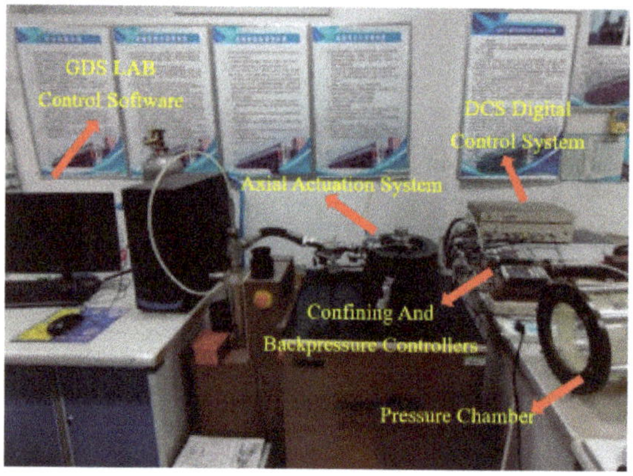

Figure 1. Dynamic triaxial testing system (DYNTTS).

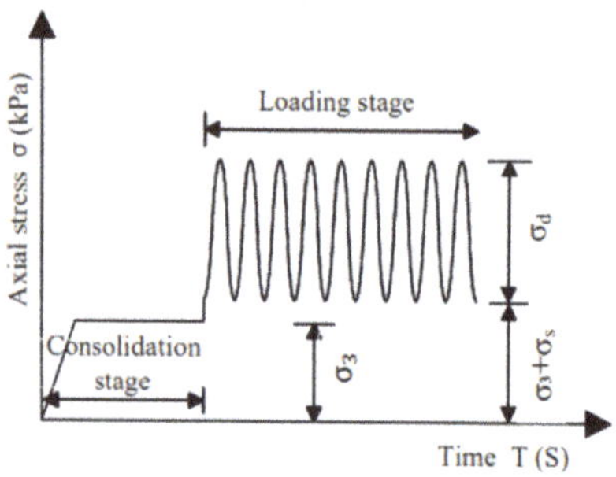

Figure 2. Schematic diagram of continuous cyclic loading waveform in the dynamic triaxial test.

The test soil was collected from the expansive soil at a subway station near Hefei Xinqiao International Airport in Hefei City. The soil samples were obtained with a diameter of 80 mm and a height of 200 mm. The retrieved soil samples of expansive soil were first subjected to a series of physical property index tests, with the experimental operations based on the "Geotechnical Testing Methods Standard" (GB/T50123-2019) [11]. The basic physical property indices of the undisturbed expansive soil samples are shown in Table 1, and according to the "Technical Code for Construction of Expansive Soil Areas" (GB 50112-2013) [12], the free swell ratio of the undisturbed soil samples falls within the range of 65% to 90%, therefore, the tested soil can be classified as moderately expansive.

Table 1. Basic physical property indices of soil samples.

ω (%)	ρ (g·cm^{-3})	W_L (%)	W_P (%)	I_p	δ_{ef} (%)
21.40	1.74	48.82	29.19	19.63	68

To ensure the uniformity of the soil texture and prevent the influence of physical parameter variations on the experiments, it was necessary to control other parameters and maintain consistency. Therefore, a series of studies was conducted using saturated remolded soil samples. The specific procedures for preparing the samples are as follows: (1) remove the soil sample from the sampling cylinder and cut it into slices, placing them on a tray. Then, put the tray into the drying oven and dry the soil at 110 °C for 24 h to ensure complete desiccation. (2) Crush and sieve the dried sample manually and mechanically, leaving soil particles with a diameter ranging from 1 mm to 2 mm. Then, thoroughly mix the sieved soil and measure the moisture content of the sample. (3) Utilize the wet method for sample preparation. Calculate the mass and moisture content of the dry soil required for the experiment based on the predetermined moisture content and maximum dry density. Add water to the dry soil container to reach the required moisture content. Additionally, cover the surface of the container with a damp cloth and allow it to rest undisturbed for 12 h to ensure uniform moisture distribution within the soil material. (4) Following the completion of wet soil sample preparation, employ a compaction method to compact the soil sample. Weigh the designated sample quantity, then pour it into the compaction mold in seven layers, ensuring an even distribution and leveling the soil surface. Throughout the testing process, it is essential to maintain consistent layer heights, and the layered surfaces should undergo trimming to remove excess soil. (5) Once the compaction is finished, use a cutting ring to extract a soil sample for weighing. Measure the moisture content of the soil samples cut from both ends of the ring, with an acceptable error range of ±1%. After completing the sample preparation, wrap it with cling film and allow it to rest undisturbed in preparation for the dynamic triaxial test.

2.3. Experimental Parameter Settings

The experimental conditions are crucial factors that impact the dynamic characteristics of the soil. To conduct this experiment reasonably, it is essential to make informed choices based on the research objectives and content, aiming to closely align with the actual field conditions and faithfully replicate the real-life scenario. The specific configuration of experimental parameters and conditions includes determining the influential factors of the experiment, establishing the parameters relevant to subway loading, and considering other control conditions for the experiment.

2.3.1. Determination of Experimental Influencing Factors

In general, the main influencing factors of the dynamic characteristics of expansive soil can be divided into two aspects: firstly, from the perspective of dynamic loads, factors such as the load waveform, stress amplitude, vibration frequency, and duration of action are crucial. Secondly, in relation to the experimental soil sample, factors mainly include the consolidation state of the soil (consolidation ratio, overconsolidation ratio) and the stress state of the soil (confining pressure, initial force conditions, etc.). Based on the research background and objectives of this paper, the experimental investigation focuses on three key factors as the influencing parameters: the intermittent loading ratio, the static deviatoric stress, and the cyclic stress ratio. The specific parameter settings are shown below.

(1) Intermittent Loading Ratio

Considering the intermittent nature of subway loading, the intermittent loading ratio is defined as the ratio of the intermittent loading duration ΔT and the functioning time of the subway loading T. The specific parameter settings for the intermittent loading ratio are determined as follows: since the typical subway train in Hefei City consists of six sections of Type B drum-shaped cars, with a length of approximately 120 m, and the train speed ranges from 30 to 80 km/h, the time taken for the subway to pass a specific point can be calculated as 5.4 s to 14.4 s. Taking 10 s as the time for the subway to pass a specific point, the intermittent duration is generally tens of times the running time [13]. The selected values for the intermittent loading ratio $\Delta T/T$ are 1, 5, 10, and 50, thus corresponding to intermittent durations of 10 s, 50 s, 100 s, and 500 s. A control experiment is also included with $\Delta T/T = 0$.

(2) Static Deviatoric Stress

The real loads imposed on the soil under train loading include a static load and cyclic dynamic loads, where the static load is caused by the instantaneous bias stress applied to the soil by the passing train. Researchers such as Wen et al. [14] have studied the effect of static deviatoric stress ratio on the deformation behavior of saturated clay. The loading procedure is as follows: a specific static deviator stress is first applied to a predetermined value, and then sinusoidal cyclic loads are superimposed on this static deviator stress. In the experiment, the static deviatoric stress ratios are set to 0, 0.1, and 0.2, with an effective consolidated confining pressure of 100 kPa. The results show that an increase in static deviator stress is detrimental to soil stability, and higher static deviator stress results in greater soil deformation. Similar experimental results were obtained by Chen et al. [15] and Wang et al. [16] regarding the influence of static deviator stress.

In addition, it has been found that the existence of static deviatoric stress may also cause preloading of the soil, which may increase the strength of the soil. Furthermore, the method of loading can impact the experimental outcomes, making it necessary to further study the influence of static deviator stress as a contributing factor. According to previous research results [17], the instantaneous stress produced by the tunnel surrounding soil under subway trains operation is 20~40 kPa. Chen et al. [18] considered the static load effects of the overlying structure (ballast, track, and rail sleeper) and used a static deviatoric stress of 15 kPa. In this paper, two different static deviatoric stresses of 15 kPa and 30 kPa are selected for comparison and analysis.

(3) Cyclic Stress Ratio

As one of the most important factors affecting the dynamic characteristics of soil, cyclic stress ratio can not be ignored in triaxial dynamic tests. The cyclic stress ratio η_d is defined as the ratio of the dynamic load amplitude (the difference between the peak and trough of the load) σ_d to the confining pressure σ_c. In the experiment, cyclic stress ratios η_d of 0.1, 0.2, and 0.4 were set.

2.3.2. Determination of Parameters Related to Subway Loads

(1) Load Waveform

Previous scholars have demonstrated that the subway running load has obvious characteristics of harmonic waves, and among various waveforms, the biased sinusoidal waveform closely approximates the actual subway train load. It can not only simulate the instantaneous bias stress generated by the train passing through the soil, but also meet the requirement that the soil experiences only compressive stress under subway loading conditions. Therefore, the biased sinusoidal wave is selected as the subway load waveform, as depicted in Figures 2 and 3.

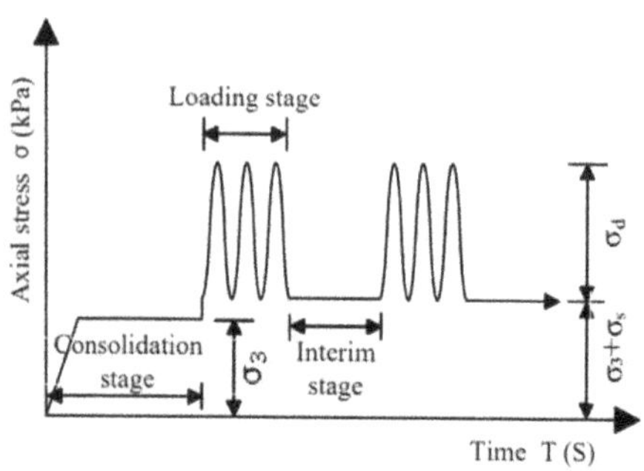

Figure 3. Schematic diagram of intermittent cyclic loading waveform in the dynamic triaxial test.

(2) Vibration Frequency

Zhang Yu'e [19] proposed that the main vibration caused by passing subway trains is of low frequency, and the high-frequency vibration has a minimal impact on structures. Based on research, Lei Shengyou et al. [20] found that the frequency of dynamic loads has little effect on the deformation of expansive soil. Zhou Xiaosheng [21] conducted experimental research on remolded expansive soil, and reached the conclusion that there is no significant regular change in the dynamic strain, dynamic elastic modulus, and damping ratio of the soil when the frequencies are 1 Hz and 2 Hz. Zhuang Jiahuang [22] determined the frequency of cyclic loading as 1 Hz based on measured data. Therefore, the influence of frequency on the dynamic loading tests for expansive soil is no longer considered, and a uniform value of 1 Hz is adopted.

2.3.3. Other Experimental Control Conditions

(1) Soil Consolidation Conditions

The actual consolidation state of the soil sample is not isotropic consolidation but is in a biased consolidation state. Therefore, the test adopts the biased consolidation condition of the soil, using the K_0 consolidation module provided by GDS, with an effective confining pressure of 120 kPa.

(2) Drainage Conditions and Test Termination Conditions

Both drained and undrained conditions are considered for the intermittent loading conditions of the subway, whereas only the undrained conditions are used for the continuous loading conditions of the subway.

The termination conditions for the continuous loading test are either when the axial strain reaches 10% or when the total number of cycles reaches 10,000, with the damage condition when the axial strain reaches 5%. For the intermittent loading test, the applied loads are relatively small, resulting in small strains ranging from 10^{-3} to 10^{-2}. The dynamic triaxial tests are performed with intermittent durations ranging from 10 s to 500 s. The instrument cannot automatically perform the cyclic loading and unloading operations, prolonged repeated loading and unloading following the termination condition for continuous loading can cause damage to the machine. Therefore, the termination condition for non-continuous loading tests are different from those for continuous loading tests. When the strain stabilizes within a certain range (order of magnitude 10^{-4}), the test is terminated, and the same number of cycles is controlled to avoid test errors caused by long-term manual operation.

In summary, the test scheme is shown in Table 2.

Table 2. Dynamic triaxial test plan.

Sample ID	Loading Method	Loading Time T (s)	Intermission Time ΔT (s)	Intermittent Loading Ratio $\Delta T/T$	Static Deviatoric Stress σ_s (kPa)	Cyclic Stress Ratio η_d
S1~S6	Continuous Loading	-	-	-		
S7~S12	Intermittent Loading	10	10	1	15	0.1
S13~S18			50	5	30	0.2
S19~S24			100	10		0.4
S25~S30			500	50		

3. Analysis of Test Results

3.1. Analysis of the Influence Law of Loading Methods on the Dynamic Strain of Expansive Soil

Figure 4a shows the dynamic strain–time curve of sample S3 under continuous loading conditions, while Figure 5a shows the dynamic strain–time curve of sample S21 under intermittent loading conditions (first 800 s). It can be observed from the figures that both loading modes exhibit dynamic strains comprising elastic strain ε_p and accumulated plastic strain ε_d. Unlike continuous loading, the dynamic strain–time curve under intermittent loading conditions develops in a "step-like" manner, including loading and intermittent phases. The strain value during the loading phase increases with time, while the strain value during the intermittent phase remains close to the strain value at the end of the loading phase, with a "strain platform" existing between the two phases. During each cycle, the strain experiences a period of growth followed by a period of nearly constant values. The reason for this is that under partially drained conditions, the axial strain is mainly composed of two parts: one is the volume change caused by the drainage of pore water, which leads to a decrease in the total volume of the soil sample and a corresponding decrease in the pore size between soil particles; the other is the deformation of the soil sample due to external loads. However, the deformation of the soil caused by the discharge of pore water accounts for the majority of the deformation. In addition, because the intermittent period stops loading, only a small amount of pore water can flow back into the soil sample. Therefore, the plastic strain during the quiescent period cannot be recovered and remains essentially unchanged.

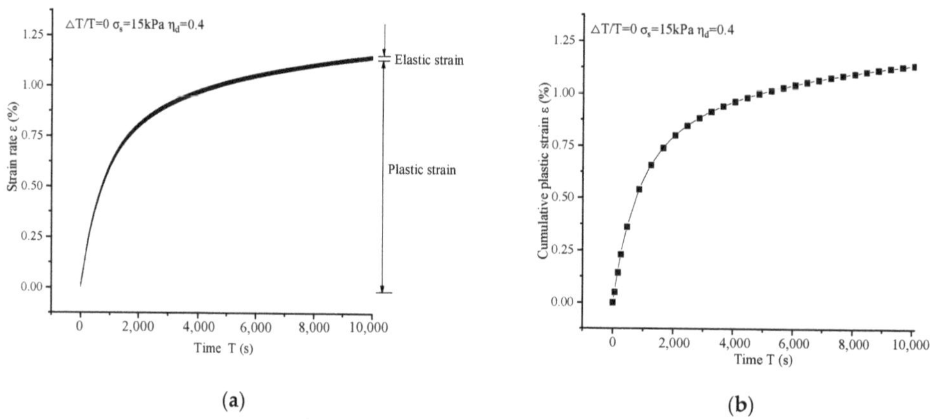

Figure 4. Strain–time curve of dynamic strain under continuous loading conditions. (**a**) Strain development curve; (**b**) cumulative strain development curve.

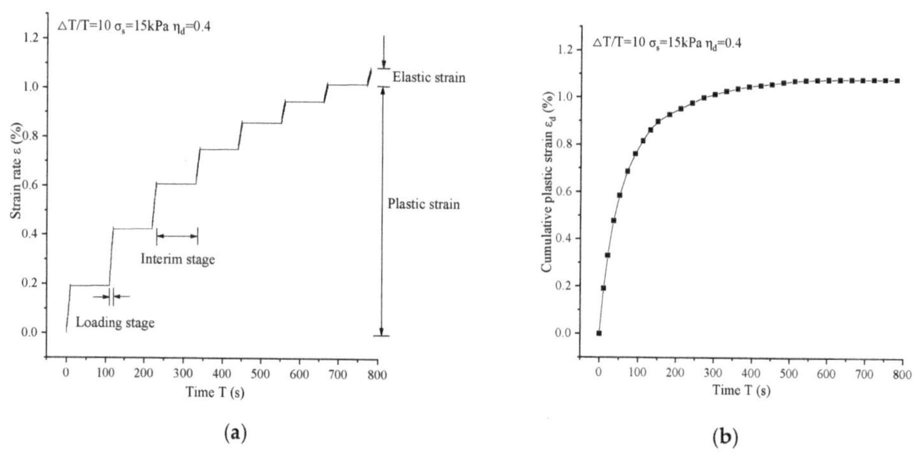

Figure 5. Strain–time curve of dynamic strain under intermittent loading conditions. (**a**) Strain development curve; (**b**) cumulative strain development curve.

The development curves of accumulated plastic strain under continuous and intermittent loading conditions are depicted in Figures 4b and 5b, respectively. With the increase in time, the development rate of the strain changes from fast to slow, and the dynamic strain still has an upward trend under the condition of continuous loading at the end, but the dynamic strain gradually approaches the stable value under the condition of intermittent loading. Moreover, it can be observed from the figures that the initial development rate of the plastic strain is significantly higher under intermittent loading conditions. The main reason for this phenomenon is that the accumulation of the strain under continuous loading conditions depends on the sustained action of external loads. As the number of loading cycles increases, the strain also increases accordingly. However, when the number of loading cycles reaches a certain threshold, the growth rate of the strain begins to decrease. Nevertheless, due to the sustained action of the subway load, the strain does not tend to stabilize until the test termination condition is reached, resulting in a continued upward trend. Under intermittent loading conditions, the dynamic strain is composed of two parts: the deformation induced by the external load during the loading stage and the deformation resulting from the volume reduction caused by pore water expulsion. The volume reduction caused by the release of pore water accounts for the majority of the effect. Thus, it can be inferred that the accumulation rate of plastic strain is initially lower under continuous loading conditions due to the undrained condition. As time progresses, the drainage volume gradually decreases, and the difference in strain between adjacent cycles also decreases. As a result, the development rate of accumulated plastic strain gradually decreases, and the rate of decrease in the development of accumulated plastic strain is faster under intermittent loading conditions compared to continuous loading conditions. Therefore, the accumulated plastic strain ultimately tends to stabilize under intermittent loading conditions.

3.2. Analysis of Accumulated Plastic Strain Behavior of Expansive Soil under Intermittent Loading

3.2.1. Analysis of the Influence of Intermittent Loading Ratio on the Accumulated Plastic Strain

Figure 6 shows the development curve of accumulated plastic strain with the number of cycles under different intermittent loading ratios. Under other conditions, including the same cyclic stress ratio and static deviatoric stress, the evolution of accumulated plastic strain under intermittent loading conditions follows a similar pattern. Initially, during the early stage of cyclic loading, the accumulated plastic strain increases rapidly with the number of vibration cycles. After a certain number of loading cycles, the rate of development curve of the accumulated plastic strain begins to slow down, and the slope of the curve decreases as the number of vibration cycles increases. Eventually, the cumulative plastic strain approaches a stable value.

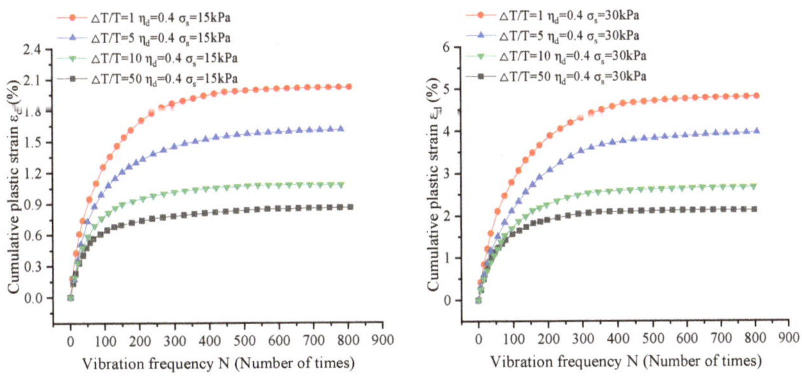

(a) (b)

Figure 6. Development curve of accumulated plastic strain with the number of cycles under different stress ratios of intermittent loading. (Sample numbers are S9, S12, S15, S18, S21, S24, S27, S30). (**a**) η_d = 0.4 σ_s =15 kPa; (**b**) η_d = 0.4 σ_s = 30 kPa.

However, it can be seen from Figure 6a,b that under the same number of cycles, the accumulated plastic strain decreases with the increase in the intermittent loading ratio. This is mainly because, for saturated soil samples, although the dissipation of pore pressure can promote strain growth, the amount of pore pressure dissipation does not continuously increase. Instead, it gradually diminishes over time. With the increase in the intermittent duration, the growth of accumulated plastic strain no longer depends on the strain growth transformed by the dissipation of pore pressure. At this time, because the recovery of strain when the loading is stopped accounts for the main part, the accumulated plastic strain will ultimately decrease with the increase in the intermittent loading ratio.

3.2.2. Analysis of the Influence Law of Static Deviatoric Stress on the Accumulated Plastic Strain

When the cyclic stress ratio η_d is 0.2, the development law of static deviatoric stress on accumulated plastic strain under intermittent cyclic loading in subway tunnels with σ_s = 15 kPa and σ_s = 30 kPa is analyzed as shown in Figure 7. The following conclusions can be drawn from the graph:

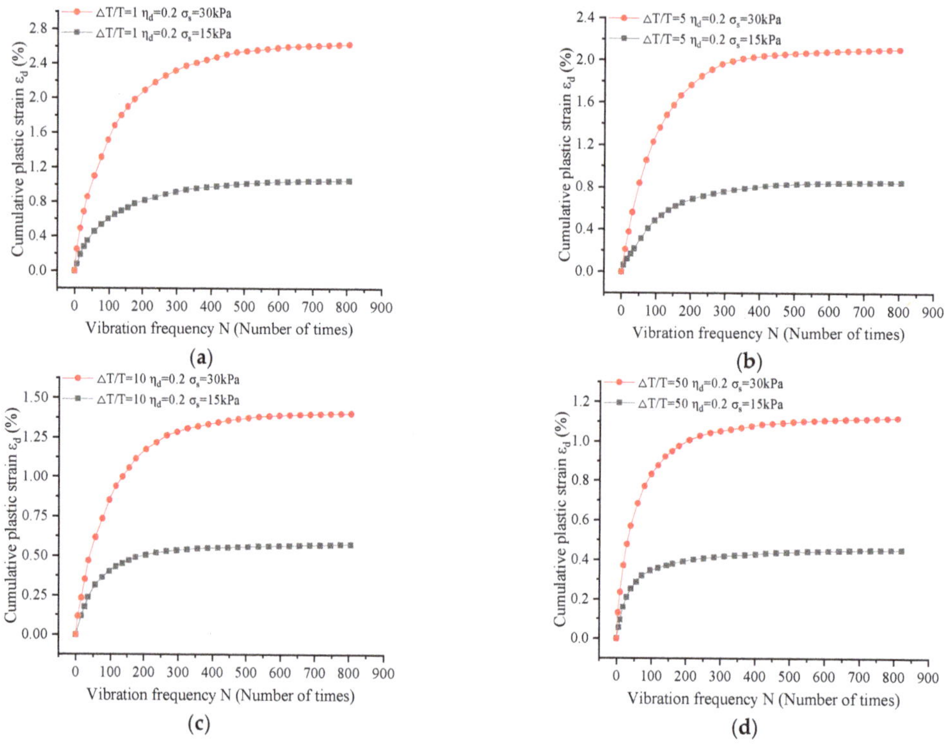

Figure 7. Translation: Development curve of accumulated plastic strain with the number of cycles under different levels of static deviatoric stress. (Sample numbers are S8, S11, S14, S17, S20, S23, S26, S29). (**a**) $\Delta T/T$ = 1; (**b**) $\Delta T/T$ = 5; (**c**) $\Delta T/T$ = 10; (**d**) $\Delta T/T$ = 50.

The accumulation of plastic strain development curves under two different static deviatoric stresses exhibited a similar pattern. Both displayed rapid initial growth followed by a deceleration in growth rate, ultimately reaching a stabilized state. Under the same cyclic loading ratio conditions, it was observed that static deviatoric stress notably influenced the accumulation of strain. The accumulated plastic strain increased with higher static deviatoric stress, leading to conclusions analogous to those from continuous loading scenarios. When the cyclic loading ratio $\Delta T/T$ was 1, the accumulated plastic strain at the point of stabilization under σ_s = 30 kPa conditions was approximately 150% higher compared to σ_s = 15 kPa conditions. For cyclic loading ratios of $\Delta T/T$ equal to 5, 10, and

50, the accumulated plastic strain increased by about 150%, 150%, and 149%, respectively. Consequently, doubling the static deviatoric stress resulted in an approximate 150% expansion in accumulated plastic strain under intermittent subway cyclic loading. Furthermore, the initial strain rate during cyclic loading was also influenced by the magnitude of static deviatoric stress. Larger static deviatoric stress led to higher initial strain rates.

This behavior primarily stems from the structural characteristics of the soil, characterized by weak inter-particle connections and significant instability. In dynamic triaxial testing, the soil experiences the combined effects of cyclic dynamic loading and static deviatoric stress. The substantial external loads lead to instantaneous rupture of the inter-particle connections, causing significant deformation in the initial stages. Consequently, the strain development rate is highest during this phase. When static deviatoric stress remains constant, the strain development rate diminishes, eventually reaching a stable state with minimal fluctuation within a small strain range as the cyclic cycles progress. With increasing static deviatoric stress, the instantaneous deformation of the soil also becomes more substantial, leading to a higher initial strain development rate. In summary, excessive static deviatoric stress significantly accelerates the sustained growth of soil strain.

3.2.3. Analysis of the Influence of Cyclic Stress Ratio on Accumulated Plastic Strain

Figure 8 shows the relationship curve between accumulated plastic strain and number of cycles under the influence of different cyclic stress ratios. As observed from the graph, under consistent conditions, the strain behavior for varying cyclic stress ratios follows a similar pattern. Specifically, in the early stages of cyclic loading, expansive soil experiences a rapid increase in strain. With an increase in the number of cycles, the strain gradually diminishes, eventually stabilizing its growth. As the cyclic stress ratio increases, the strain magnitude for the same number of cycles also increases. Towards the end of the loading process, when the strain development curve approaches stability, the accumulated plastic strain for a cyclic stress ratio of η_d = 0.2 is approximately 2.8 times that of a cyclic stress ratio of η_d = 0.1; similarly, for a cyclic stress ratio of η_d = 0.4, the accumulated plastic strain is roughly 5.4 times that of a cyclic stress ratio of η_d = 0.1. This demonstrates that an elevated cyclic stress ratio significantly promotes strain development.

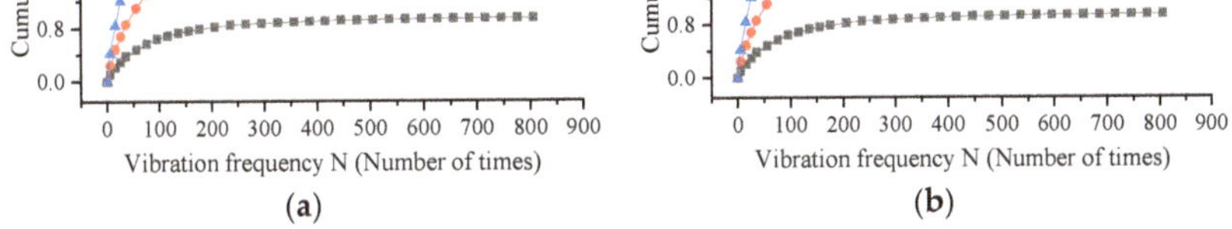

Figure 8. Development curves of accumulated plastic strain with number of cycles under different cyclic stress ratios. (The sample numbers are S10, S11, S12, S22, S23, S24). (**a**) $\Delta T/T = 1$ σ_s = 30 kPa; (**b**) $\Delta T/T = 10$ σ_s = 30 kPa.

This phenomenon is primarily due to the considerable influence of cyclic dynamic loading on soil deformation. Under smaller loads, the soil experiences elastic deformation without generating plastic strain. However, as external loads increase progressively, plastic strain accumulates within the soil. Consequently, an increase in the cyclic stress ratio leads to a noticeable rise in strain. In comparison to the effect of static bias stress, it is evident that a higher cyclic stress ratio has a slightly greater influence on promoting strain accumulation.

Additionally, it is evident from the graph that at lower cycles, the slope of strain at any given point increases with higher cyclic stress ratios. This can be attributed to the immediate surge in strain caused by initial instantaneous loading. Moreover, the cycles required for the accumulated plastic strain development curve to transition from rapid growth to gradual growth vary with different cyclic stress ratios. A larger cyclic stress ratio corresponds to a higher number of cycles needed for the strain development curve to enter the gradual growth phase. This trend arises from the fact that larger cyclic stresses make it more challenging for a strain to enter a stable growth phase, hence necessitating a higher number of cycles.

4. Classical Accumulated Plastic Strain Modeling Analysis

Currently, researchers from both domestic and international backgrounds have extensively investigated the accumulation of strain in soils under cyclic loading, and have proposed different strain models based on their research results, which can be summarized as explicit or implicit models related to accumulated plastic strain of soil samples and the number of cycles and experimental parameters. Classical accumulated plastic strain calculation models can roughly be divided into the following two types.

(1) Exponential Model represented by the Monismith Model

The exponential model proposed by Monismith et al. [23] is one of the widely used models, and subsequent scholars mostly used this model as a foundation and made refinements to develop improved models. The formula is expressed as follows:

$$\varepsilon_p = aN^b \tag{1}$$

In the equation: ε_p—accumulated plastic strain; a, b—related test parameters; N—number of cyclic vibrations.

Anand et al. [24] based their research on the Monismith exponential model, and considering factors such as confinement pressure and deviatoric stress, they proposed a modified model.

$$\varepsilon_p = a\left(\frac{\sigma_{oct}}{p_a}\right)^m N^b \tag{2}$$

In the equation: σ_{oct}—shear stress; p_a—atmospheric pressure, with a value of 101 kPa; a, b, m—test parameters; N—number of cyclic vibrations.

Chai et al. [25] developed a model that comprehensively considers the effects of initial static deviatoric stress, dynamic deviatoric stress, and static failure deviatoric stress. The formula is expressed as follows:

$$\varepsilon_p = a\left(\frac{\sigma_d}{\sigma_f}\right)^m \left(1+\frac{\sigma_{is}}{\sigma_f}\right)^n N^b \tag{3}$$

In the equation: ε_p—accumulated plastic strain; a, b, m, n—test parameters; σ_d, σ_f, σ_{is}—dynamic stress amplitude, static failure deviatoric stress, and initial static deviatoric stress, respectively; N—number of cyclic vibrations.

(2) Tang Yiqun Logit Model

Tang Yiqun [26] proposed using logarithmic relationship curves to reflect the strain accumulation characteristics of reinforced soils around tunnels under Shanghai subway loads. The model comprehensively considers the influences of dynamic stress amplitude, consolidation ratio, confining pressure, and soil properties. The specific calculation model is as follows:

$$\varepsilon_p = AlnN + B \tag{4}$$

In the equation: ε_p—accumulated plastic strain; A, B—parameters related to influencing factors; N—number of cyclic vibrations.

Fitting of the experimental results was carried out using both the Monismith index model and the Tang Yiqun model. However, due to the congruence of the conclusions drawn from both models, only specimens S8 and S29 were chosen for detailed illustration, with the omission of further elaboration on the remaining samples. This is demonstrated in Figures 9 and 10.

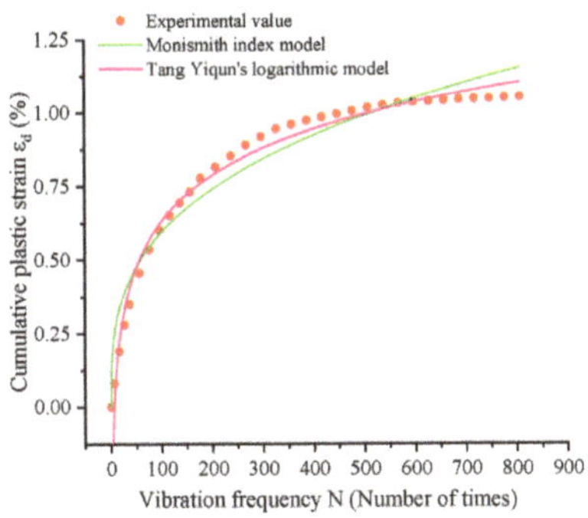

Figure 9. Fitting results of sample S8.

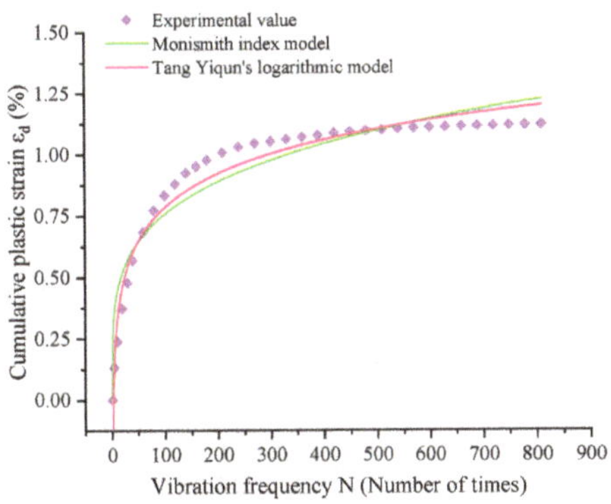

Figure 10. Fitting results of sample S29.

In comparison with the experimental data, prior to the inflection point of the fitting curves, the logarithmic model proposed by Tang Yiqun exhibits closer alignment with the experimental values. However, beyond this inflection point, both fitted curves deviate significantly from the experimental data. Consequently, for lower cycle numbers, Tang Yiqun's logarithmic model provides a better fit to the experimental data than the Monismith index model. Nevertheless, as the cycle numbers increase, both models continue to exhibit a rising trend, whereas the experimental values in this study tend to stabilize and show incremental growth. Ultimately, it can be determined that neither the Monismith index model nor the Tang Yiqun logarithmic model adequately captures the behavior observed in this study's experimental results.

5. Establishment of Accumulated Plastic Strain Model

Therefore, the development of accumulated plastic strain under intermittent subway loading is more suitable to be described by the hyperbolic model. The mathematical expression of the relationship between accumulated plastic strain and the number of cycles is shown in the following equation.

$$\varepsilon_d = \frac{N}{a + b \cdot N + c\sqrt{N}} \tag{5}$$

In the equation: ε_d—accumulated plastic strain; a, b, c—fitted parameters associated with different influencing factors; N—number of cycles.

The strain curve of the soil sample under intermittent loading conditions was fitted, and the fitting effect is shown in Figure 11. The related fitting parameters of the strain calculation model are shown in Table 3. From this analysis, it is evident that the coefficient of determination (R^2) for the specimens is consistently close to 1. A higher R^2 value indicates a stronger fit of the model to the data. As such, the experimental results can be effectively represented using the proposed cumulative plastic strain calculation model.

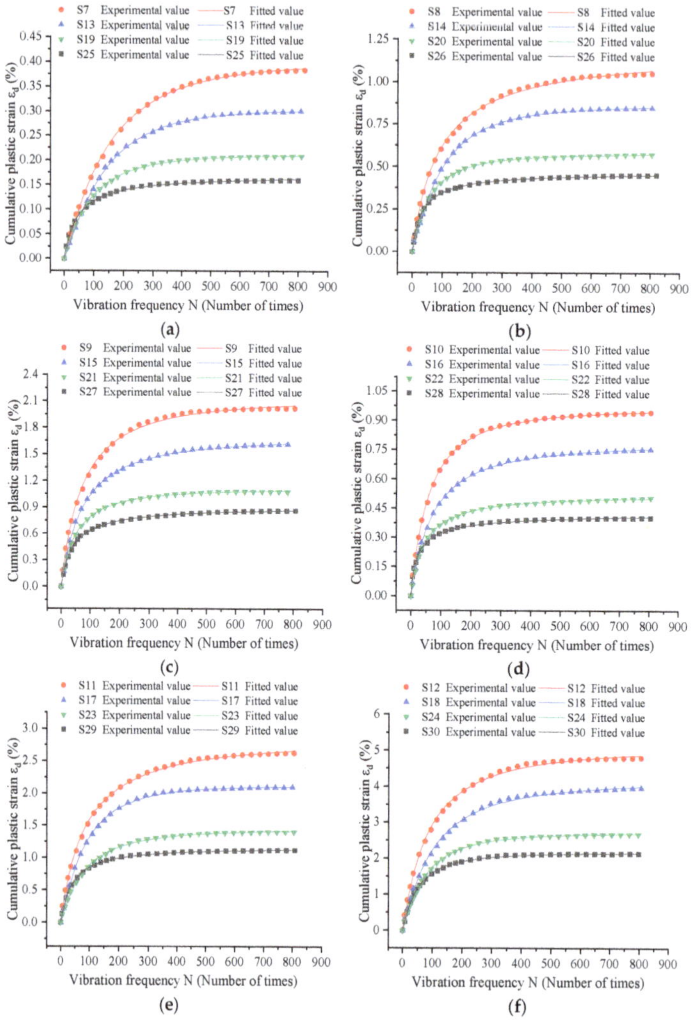

Figure 11. Comparison of test results and fitted values of soil samples under intermittent loading conditions. (**a**) Soil sample S7, S13, S19, S25; (**b**) soil sample S8, S14, S20, S26; (**c**) soil sample S9, S15, S21, S27; (**d**) soil sample S10, S16, S22, S28; (**e**) soil sample S11, S17, S23, S29; (**f**) soil sample S12, S18, S24, S30.

Table 3. Accumulated plastic strain model parameters under intermittent loading conditions.

Soil Sample Number	Fitting Parameter a	Fitting Parameter b	Fitting Parameter c	Correlation Coefficient R^2
S7	0.00182	0.00517	−0.04868	0.99925
S8	0.01350	0.01123	0.00458	0.99887
S9	0.02542	0.01213	−0.02809	0.99766
S10	0.01363	0.02225	−0.0058	0.99926
S11	0.03001	0.01060	−0.01619	0.99878
S12	0.05009	0.01008	−0.02974	0.99837
S13	0.00133	0.00534	−0.06115	0.99972
S14	0.00583	0.00764	−0.05782	0.9994
S15	0.01951	0.0118	−0.02916	0.99958
S16	0.01622	0.01264	−0.0248	0.99946
S17	0.01559	0.00849	−0.06604	0.99952
S18	0.03136	0.00798	−0.03901	0.99825
S19	0.00173	0.00897	−0.05603	0.9996
S20	0.00827	0.01494	−0.04905	0.99979
S21	0.01871	0.01713	−0.03461	0.99978
S22	0.01168	0.02225	−0.0058	0.99926
S23	0.01336	0.00987	−0.04763	0.99919
S24	0.02545	0.0103	−0.05811	0.99913
S25	0.00384	0.02234	0.00511	0.99869
S26	0.01486	0.03105	0.0196	0.99966
S27	0.02068	0.02263	0.00211	0.99929
S28	0.02723	0.06080	0.13971	0.99741
S29	0.02697	0.02311	−0.01206	0.9991
S30	0.03289	0.01593	−0.05258	0.99914

6. Conclusions

Taking the saturated remolded expansive soil in Hefei City as the research object, the dynamic characteristics of the soil under intermittent and continuous subway loading conditions were investigated through cyclic triaxial tests. The research aimed to explore the influence of intermittent and continuous loading modes on the dynamic behavior of saturated remolded expansive soil. The study also analyzed the effects of intermittent loading ratio, cyclic stress ratio, and static deviator stress on the variation of cumulative plastic strain with the number of cycles. Based on the test results, the following conclusions can be drawn:

(1) Under both intermittent and continuous subway loading, the accumulated plastic strain of saturated remolded expansive soil exhibited a similar development pattern of rapid growth–slow growth–eventual stabilization. It can be inferred that intermittent loading does not alter the overall trend of strain development but does show differences in the initial accumulation rate and final development of plastic strain. Comparative analysis revealed that the initial development rate of accumulated plastic strain was higher under intermittent loading conditions than under continuous loading conditions, and intermittent loading facilitated a quicker stabilization of cumulative plastic strain in the soil.

(2) Under intermittent subway loading with other test conditions being the same, a larger intermittent loading ratio resulted in smaller accumulated plastic strain at the same number of cycles, and the accumulated plastic strain value at stabilization was also smaller. Static deviatoric stress and cyclic stress ratio have a significant cumulative effect on plastic strain. When static deviatoric stress increases by one time, the accumulated plastic strain under intermittent subway cyclic loading increases by about 1.5 times. When the cyclic stress ratio increases by one time, the accumulated plastic strain increases by about 1~2 times.

(3) Based on the cumulative plastic strain development behavior under intermittent subway cyclic loading, an enhanced calculation model for cumulative plastic strain as a function of cycle count was proposed, building upon the Long–Yao hyperbolic

model. This model takes into account factors such as intermittent loading ratio, static bias stress, and cyclic stress ratio. The coefficient of determination (R^2) for this model is remarkably close to 1, affirming its high degree of fitness to the data. As a result, the proposed model demonstrates a notably favorable fitting performance.

Author Contributions: Writing—original draft, L.Z.; Writing—review & editing, Y.G., G.C. and X.L. All authors have read and agreed to the published version of the manuscript.

Funding: This research was funded by [2020 Natural Science Foundation of Anhui Province] grant number [2008085ME168] and by [2021 Project on Cultivation of Outstanding Top Talents in Universities].

Institutional Review Board Statement: Not applicable.

Informed Consent Statement: Not applicable.

Data Availability Statement: Not applicable.

Conflicts of Interest: The authors declare no conflict of interest.

References

1. Yang, H.; Qu, Y.; Zheng, J.; Liu, L. Preliminary Study on Engineering Geological Problems of Expansive Soil in Highway Construction in Western China. *J. Chang. Commun. Univ.* **2003**, *19*, 19–24.
2. Xie, Y.; Chen, Z.; Li, G. Study on law of three-dimensional expansive force of expansive soil in Nanyang. *J. Logist. Eng. Coll.* **2006**, 11–14+23.
3. Xu, Y.; Cheng, Y.; Xiao, J.; Lin, Y.; Qi, S. Research on Expansive Soil Landslide and New Prevention Technology for Engineering Slope. *J. Geotech. Eng.* **2022**, *44*, 1281–1294.
4. Liu, Q.; Xiang, W.; Wu, Y. *Study on Engineering Characteristics and Modification Theory of Expansive Soil*; China University of Geosciences Press: Wuhan, China, 2015; pp. 1–99.
5. Liao, S. *Expansive Soil and Railway Engineering*; China Railway Publishing House: Beijing, China, 1984; pp. 1–214.
6. Chen, S. *Research on Engineering Characteristics and Treatment Technology of Expansive Soil*; Huazhong University of Science and Technology: Wuhan, China, 2006.
7. Tang, X.; Wang, S.; Cheng, F. Experimental study on dynamic stress dynamic strain relationship and dynamic modulus of reinforced expansive soil. *IOP Conf. Ser. Earth Environ. Sci.* **2021**, *669*, 012006. [CrossRef]
8. Yang, Z.; Zhang, L.; Ling, X.; Li, G.; Tu, Z.; Shi, W. Experimental study on the dynamic behavior of expansive soil in slopes under freeze-thaw cycles. *Cold Reg. Sci. Technol.* **2019**, *163*, 27–33. [CrossRef]
9. Chen, H. *Experimental Study on Dynamic Characteristics of Expansive Soil under Cyclic Loading*; Hefei University of Technology: Hefei, China, 2021.
10. Zhuang, X. Experimental Study on Cumulative Deformation and Dynamic Strength Characteristics of Weak Expansive Soil under Cyclic Loading. *Geotech. Mech.* **2020**, *41*.
11. *GB/T50123-2019*; Standard Test Methods for Geotechnical Investigation. GB Standards: Taiwan, China, 2019.
12. *GB 50911-2013*; Technical Specification for Monitoring of Urban Rail Transit Project. GB Standards: Taiwan, China, 2013.
13. Zheng, Q.; Xia, T.; Zhang, M. Strain prediction model of intact silty clay under intermittent cyclic loading. *J. Zhejiang Univ.* **2020**, *54*, 889–898.
14. Wen, R.; Wang, C.; Chen, Y. Influence of static uniaxial stress caused by traffic load on deformation of saturated soft clay. *Rock Soil Mech.* **2009**, *30* (Suppl. S2), 119–122.
15. Chen, C.; Zhou, Z.; Zhang, X. Experimental study on dynamic cumulative characteristics of peat soil under cyclic loading. *Chin. J. Rock Mech. Eng.* **2017**, *36*, 1247–1255.
16. Wang, R.; Wang, L.; Hu, Z.; Zhao, Z.; Li, X.; Wang, Q. Influence of static uniaxial stress caused by traffic load on dynamic characteristics of compacted loess. *J. China Railw. Soc.* **2019**, *41*, 110–117.
17. Wang, Y. *Research on Dynamic Stability of Subway Tunnel-Soil under Cyclic Loading*; Tongji University: Shanghai, China, 2002.
18. Chen, X.; Nie, R.; Li, Y.; Guo, Y.; Dong, J. Resilient modulus of fine-grained subgrade soil considering load interval: An experimental study. *Soil Dyn. Earthq. Eng.* **2021**, *142*, 106558. [CrossRef]
19. Zhang, Y.; Bai, B. Simulation of excitation load of tunnel structure induced by subway train vibration. *J. Vib. Shock.* **2000**, *3*, 70–72+78+96.
20. Lei, S.; Hui, H. Comparative analysis of static and dynamic characteristics of expansive soil and its improved soil. *Chin. J. Rock Mech. Eng.* **2004**, *23*, 3003–3008.
21. Zhou, X. *Study on Dynamic Characteristics and Subgrade Response of Expansive Soil under Bi-Directional Cyclic Load*; Institute of Rock and Soil Mechanics, Graduate University of Chinese Academy of Sciences: Wuhan, China, 2010.
22. Zhuang, J. *Study on Dynamic Characteristics and Micro-Experimental of Soft Clay under Metro Circulatory Loads*; Zhejiang University: Hangzhou, China, 2019.

23. Monismith, C.L.; Ogawa, N.; Freeme, C.R. Permanent deformation characteristics of subgrade soils due to repeated loading. *Transp. Res. Rec.* **1975**, *537*, 1–17.
24. Anand, J.P.; Louay, N.M.; Aaron, A. Permanent deformation characterization of subgrade soils form RLT test. *J. Mater. Civ. Eng.* **1999**, *11*, 274–282.
25. Chai, J.; Miura, N. Traffic-Load-Induced Permanent Deformation of Road on Soft Subsoil. *J. Geotech. Geo-Environ. Eng.* **2002**, *128*, 907–916. [CrossRef]
26. Tang, Y.; Zhao, H.; Wang, Y.; Liu, R. Strain accumulation characteristics of surrounding soft clay reinforced under subway load. *J. Tongji Univ.* **2011**, *39*, 972–977.

applied sciences

MDPI

Article

The Effect of Different Fiber Reinforcement on Bearing Capacity under Strip Foundation on the Sand Soil: An Experimental Investigation

Bilge Aksu Alcan [1,*] and Semet Çelik [2]

[1] Department of Civil Engineering, Kafkas University, 36100 Kars, Turkey
[2] Department of Civil Engineering, Atatürk University, 25240 Erzurum, Turkey; scelik@atauni.edu.tr
* Correspondence: bilge.aksu@kafkas.edu.tr

Abstract: The aim of this study is to investigate the bearing capacity-settlement behavior of strip footing settling on sand soil randomly reinforced with glass fiber, basalt fiber, macromesh fiber, and four different hybrid fiber additives in which these fibers are used together. Model tests were carried out in the laboratory on the strip footing and placed on the unreinforced and reinforced sand with different fibers. In the study, model tests were carried out on seven types of randomly reinforced soils by using glass, basalt, macrame, and mixtures of these fibers as reinforcement. In the model tests, two different fiber contents, 1% and 2%, and two different fiber lengths, 24 mm and 48 mm, were used. Tests were carried out with Dr = 30% and 50% relative density, and reinforcement depths 1B, 2B, and 3B were selected. In addition, the photographs taken during the test were analyzed with the particle image velocimetry (PIV) method and the displacements on the soil were examined. As a result of the reinforced and unreinforced model tests, the highest ultimate bearing capacity was measured as 680 kPa from the tests with Dr = 50% relative density, 48 mm length, 2% contents, and 3B depth macromesh fiber reinforced. In hybrid fibers, the highest ultimate bearing capacity was measured as 495 kPa, with Dr = 50% relative density, 48 mm length, 2% contents, and 2D depth micromesh and basalt fiber-reinforced tests. In the reinforced tests, it was concluded that the most effective fiber on bearing capacity is macromesh fiber. It can be seen that in the PIV analysis, as the fiber additive increased, the settlements made by the foundation decreased under the same pressure. It has also been observed that adding reinforcement to the soil transfers the stresses occurring in the soil to a wider area.

Keywords: basalt fiber; glass fiber; macro-synthetic fiber; hybrid fiber; bearing capacity ratio; randomly reinforced soil; PIV

1. Introduction

The concept and principle of soil reinforcement were first developed by Vidal in 1969 when he found that adding reinforcements to a soil mass increases the shear strength of soil [1]. The reinforcing effect of fibers in soil reinforcement was first noticed by Waldron in 1977, when he observed that natural root fibers contribute to the improvement in soil strength due to their reinforcing effect [2]. In general, in fiber-reinforced soils, it has been observed that the added fiber can improve the mechanical properties of the soil due to the interface friction between the fiber and the soil. That is, the shear stresses in the soil actuate the tensile strength in the fibers and, as a result, give the soil more strength [3–5]. Randomly Distributed Fiber-reinforced Soil (RDFS) is a soil improvement technique in which fibers of desired properties and quantities are randomly ordered and compacted in place [6]. The RDFS method has become more popular in recent studies. Short staple fibers can be easily added and mixed into the soil such as cement, lime, or other additives. In this way, planes of weakness do not occur in the soil and an isotropic increase in the strength of the soil composite can be achieved [7,8].

Glass fiber improves the mechanical properties of the reinforced soil matrix such as flexural and tensile strength. They contribute to the flexibility of the soil matrix they are supplemented with and have a long period of use [9,10]. Glass fiber-reinforced sandy soil matrices have attracted great interest in many applications because of their relatively low weight, low cost, and good mechanical properties, which are desirable characteristics of natural fibers [11,12]. Philley et al. [13], in this study, first stabilized the soil with Class F fly ash and cement to improve the strength properties of the weak local soil and then reinforced it with glass fiber reinforcement. In the study, different mixtures were prepared with varying ratios such as fly ash (10, 30, 50), cement (0.5%, 1, 1.5, 2%), and glass fiber (0.25%, 0.5 and 1%). It was observed that there was a significant improvement in the pressure and shear strength parameters in the optimum additive percentage determined for all three materials. Benzian et al. [14] conducted direct shear box tests for their study to examine the effect of material properties and fiber content on the shear resistance of sand. In the study, glass fiber, polypropylene fiber, and sisal fiber were randomly reinforced into the soil mass at 50% relative density, 0.25%, 0.5, 0.75, and 1 ratio. It was observed that samples reinforced with sisal fibers had a more significant increase in strength parameters than samples reinforced with glass or polypropylene fibers. The reason one of the fiber types used in the experimental study was chosen as glass fiber is that it was thought to be suitable for comparing the results with basalt and macromesh fibers, as it is a familiar material that has been studied in the past.

Basalt fiber has high performance, natural compatibility, electrical insulation, corrosion, high temperature, and acid and alkali resistance [15,16]. When the studies in the literature are examined, it has been seen that it is widely used in thermoset resin and concrete as insulation, sound absorption, heat resistant materials, and reinforced materials [17,18]. In addition, the mechanical properties of basalt fiber such as tensile strength and modulus of elasticity, are much higher than polypropylene fiber [19,20]. Many studies have been carried out on concretes in which basalt fiber is used as an admixture. However, the use of basalt fiber additives in soil-related studies is relatively low [21]. Kale et al. [16] used three different lengths of fiber (12 mm, 18 mm, and 24 mm) to investigate the effect of basalt fiber additive on improving soil properties. They added 12 mm and 18 mm long fibers to the ground at a rate of 1.5%, 3%, 4.5%, 6%, and 24 mm long fibers at a rate of 1.5%, 2%, 2.5%, and 3% by weight. Xu et al. [22] reinforced loess soil with poor mechanical properties with basalt fiber additives in three different fiber lengths (6 mm, 12 mm, and 18 mm) and four different fiber contents (0.3%, 0.6%, 0.8%, and 1.0%). The maximum increase in strength was observed in the experimental group with a fiber content of 0.6% with a length of 12 mm, and it was determined that the shear strength of leucine of basalt fiber was improved. In their study, Mukhtar and Kumar [23] showed that the addition of basalt fiber to the local soil sample located in the Srinagar region increases the strength of the soil by decreasing the liquid limit and plastic limit, and improves the properties of the soil by decreasing the plasticity index by 62.64%. Chen et al. [24] concluded that the use of glass fiber and basalt fiber would be effective as reinforcement for higher bearing capacity and fewer cracks in granite residual soil. In their studies, fiber contents were chosen as 3%, 4%, and 5%, respectively, and fiber lengths were used as 6 mm, 9 mm, and 12 mm, respectively. While most of the studies on soil cover fine-grained soil groups, studies on cohesion-free sand soils are insufficient. Therefore, it is particularly important to study basalt fiber-reinforced cohesionless sand soil.

Macromesh fiber is a polypropylene (99%)-based macro synthetic fiber reinforcement in copolymer structure. It is variously named macro polypropylene fiber, spun polypropylene fiber, high-performance polypropylene fiber, macro synthetic fiber, and macro synthetic twisted bundle fiber [25]. They are environmentally friendly and new-generation reinforcements produced from high-performance polypropylene raw materials and have a longer life than traditional reinforcements [26,27]. In recent years, its effect on fiber-reinforced strengthening studies has begun to be investigated. Brasse et al. [28] strengthened the soil-cement composites made with cohesionless soil with the addition of single fibrillated-tapes polypropylene fibers (SFPF) and bundles of coiled fibrillated-tapes polypropylene fibers (BCFPF) and showed that the polypropylene fiber reinforcements used had a significant effect on the bending behavior. Daneshfar et al. [29], twisted, barchip, and fibrillated, three different macro polypropylene fibers were added to the concrete in three different volume ratios 0.2%, 0.4%, and 0.6%, and the reinforcement of macro PP fibers affected the ductility behavior and ductility of the concrete. investigated the effect on mechanical properties. According to the results, the twisted, and barchip fiber-reinforced concrete has higher compressive strength than the fibrillated fiber. It has been studied mostly on concrete and polymer composites but the studies on their effects on soil are few [30,31].

A single fiber type chosen as an additive in fiber additives can often significantly improve one or more of the material's properties. However, each type of fiber often has some weak points. For example, filament fiber can improve tensile strength but has the problem of poor dispersion. Therefore, recently researchers have mixed different fiber types to take full advantage of the reinforcing effect of each fiber especially in the field of concrete. Studies have been carried out in different combinations such as polymer-polymer hybrid fiber and polymer-natural hybrid fiber, steel-steel hybrid fiber, and steel-polymer hybrid fiber [32,33]. As a result of the studies, it has been seen that hybrid fibers can improve the toughness, ductility, energy dissipation capacity, and durability performance of concrete [30,34]. When the existing studies on improvement with fiber additives are evaluated, the effects on the soil are examined as a result of the addition of one or more fibers to the soil separately under various varying conditions [35–37]. However, studies on the effect of using two or more fibers together in a hybrid form on soil are very insufficient.

This study is aimed to contribute to the deficiencies in the literature mentioned above. In this context, an average of 300 laboratory model tests were carried out to investigate the effect of glass fiber, basalt fiber, macromesh fiber, and hybrid fiber-reinforced soils where these three fibers are used together in different combinations on the bearing capacity. In the experimental study, the scope of the study was expanded by investigating the effect of fiber length, fiber content, depth of reinforcement layer, and different relative density values, as well as fiber type. To investigate the effect of fiber length, two different fiber lengths of 24 mm and 48 mm were used. To see the effect of fiber content, fiber additives with two different contents of 1% and 2% were investigated. To see the effect of relative density, tests were carried out on soils with 30% and 50% relative density. In addition, soils were reinforced with fiber at depths of (B) 1, 2, and 3 times the foundation width, and the effect of three different reinforcement depths on soil stabilization was evaluated. Finally, to analyze the sand soil settlement behavior strengthened with different fiber types under varying environmental conditions within the scope of the study, the GeoPIV-RG program, an image analysis module running on MATLAB was used. For the image analysis of each test, a series of photographs taken at regular intervals during the test were used as data.

2. Materials and Methods

2.1. Test Materials

2.1.1. Soil Samples

Sieve analysis was performed in accordance with ASTM D2487 [38] standards to determine the grain diameter distribution of the soil used in experimental studies. The grain diameter distribution curve of the sand soil used in the experimental study is given in Figure 1. According to the Unified Soil Classification System (USCS), the soil class is determined as poorly graded (uniform) sand (SP). The granulometry parameters of the sand used in the tests are given in Table 1.

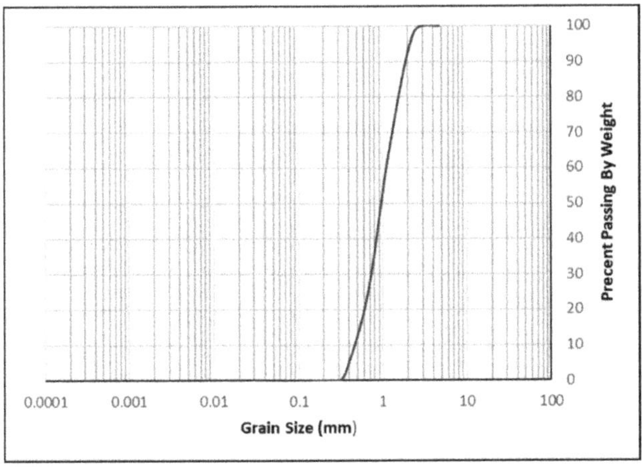

Figure 1. Grain size distribution of sand soil.

Table 1. Sieve analysis results of sand soil used in experimental study.

Granulometry Parameters	Value
D_{10}, Effective grain size (mm)	0.47
D_{30} (mm)	0.7
D_{60} (mm)	1.1
C_u (D_{60}/D_{10}), Coefficient of uniformity	2.34
C_c ($D_{30}^2/(D_{60} \cdot D_{10})$), Coefficient of curvature	0.95

In this study, model loading tests at approximately 30% and 50% relative density were performed. The index properties of sand for 30% and 50% relative densities are given in Table 2. Shear box (direct shear) tests were carried out in accordance with ASTM D 3080 [39] standard. The sand soil was placed in a Shear box cell with dimensions of 60 mm × 60 mm and a height of 38 mm at relative densities to be used in tests. Then, the soil sample was sheared horizontally by pushing on the bottom ring while holding the top ring under normal stresses of 30 kPa, 60 kPa, and 120 kPa, respectively. The sharing speed is set to 1 mm/min. According to the shear test results, the shear strength angle was obtained as $\phi = 32.1°$ for Dr = 30% relative density and as $\phi = 40.1°$ for Dr = 50% relative density.

Table 2. Index properties of poorly graded sand soil used in model tests.

Relative Density Parameters	Value
Specific gravity (G_s)	2.71
Maximum void ratio (e_{max})	0.84
Maximum dry unit weight (γ_{kmax}) (kN/m^3)	17.93
Minimum void ratio (e_{min})	0.47
Minimum dry unit weight (γ_{kmin}) (kN/m^3)	14.40
Void ratio (e) (Dr = 30%)	0.73
Void ratio (e) (Dr = 50%)	0.65
Dry unit weight (γ_k) (kN/m^3) (Dr = 30%)	15.30
Dry unit weight (γ_k) (kN/m^3) (Dr = 50%)	15.96

2.1.2. Fibers

Basalt fibers, glass fiber, and macromesh fibers used in the tests are shown in Figure 2. The physical and mechanical properties of the fibers used in the study are given in Table 3.

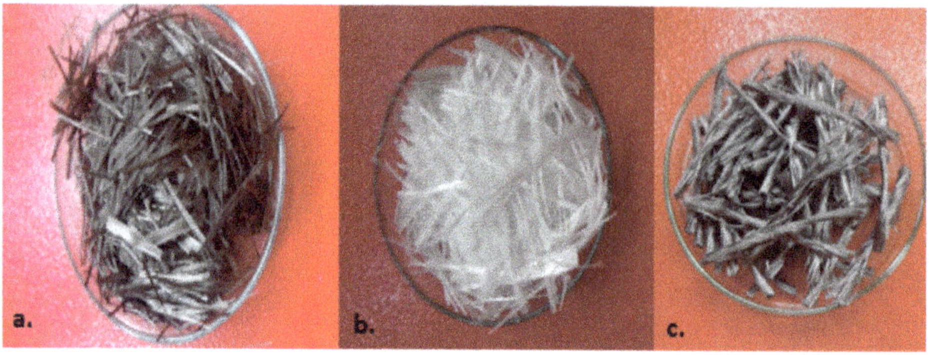

Figure 2. Material diagram; (**a**) Basalt fiber; (**b**) Glass fiber; (**c**) Macromesh fiber.

Table 3. Physical and mechanical properties of basalt fiber, glass fiber, and macromesh fiber.

Properties	Basalt Fiber	Glass Fiber	Macromesh Fiber
Fiber type	Micro	Micro	Macro
Length (mm)	24	24/48	24/48
Average diameter (μm)	16	14	850
Specific gravity (G)	2.65	2.60	0.91
Young's modulus (GPa)	89	77	5.75
Breaking tensile strength (MPa)	4840	3400	750

2.2. Experimental Setup

The model tank has a length of 1000 mm, a width of 300 mm, and a height of 1000 mm. The front is made of 20 mm thick glass fiber to display the displacement of the soil during loading, and the back and sides are made of sheet steel. Experimental studies were carried out under plane strain conditions using a strip foundation. For the model foundation, a rigid steel plate with a length of L = 300 mm, a height of H = 20 mm, and a width of B = 100 mm was used. The surfaces of the foundation in contact with the tank are covered with thick rubber material to reduce the friction coefficient. Based on the model, a hydraulic jack that can load up to 5000 kg according to the desired speed is used. To check the centric loading, two Linear Variable Differential Transformers (LVDT) were placed diagonally at the two corners of the footing plate and settlements were read. In the tests, an S-type load cell with a capacity of 5000 kg was used to determine the load values on the model base plate. During the tests, the settlements and stress values from the load cells and LVDTs were transferred to the data collection device and converted into numerical values. The schematic of the experimental setup is given in Figure 3.

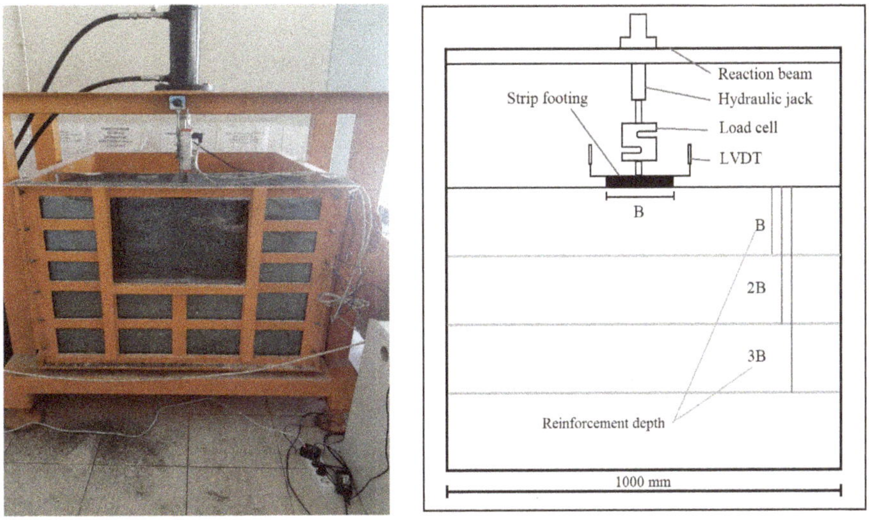

(a) (b)

Figure 3. (**a**) Photograph of the model tank, and (**b**) schematic shape of the model tank.

2.3. Test Program

Relative density has a great influence on the mechanical behavior of sands. Soils with low relative density are more likely to have settlement and bearing capacity problems. For this reason, loose (Dr = 30%) and medium-dense (Dr = 50%) soils were preferred to stabilize them with fiber additives. In the experimental study, firstly, loading tests were performed on model strip footing at 30% and 50% relative density in unreinforced soil, and ultimate bearing capacity was determined. Then, an experimental program was prepared for the reinforced soil tests. While determining the ultimate bearing capacity, there are options such as the 0.1B method (B is the width of model footing) [40], the tangential intercept method [41], the log-log method [42], and the hyperbolic method, apart from using load-deformation curves [43]. In the experimental studies, especially on reinforced soils, the 0.1B method was used. While determining the experimental study program, six parameters were taken into account. While determining the experimental study program, six parameters were taken into account. These parameters are two different relative densities (Dr = 30% and Dr = 50%), two different fiber contents (Fc = 1% and Fc = 2%), three different fiber additives (glass fiber, basalt fiber, and macromesh fiber), four different hybrid fiber additives (glass + basalt, basalt + macromesh, glass + macromesh, and glass + basalt + macromesh), two different fiber lengths (Fl = 24 mm and Fl = 48 mm), and three different reinforcement depths (B, 2B, and 3B). In the experimental study, 120 experimental groups were formed with the parameters given above. The representation of these groups is given in Table 4. When using hybrid fibers, fibers in each group were mixed in equal percentages. The randomly distributed fiber-reinforced sand soil with a relative density of 30% and 50% were placed in the tank in horizontal layers of 5 cm. To provide the desired relative density, the glass surface of the test tank was scaled from the outside, and the required sand and fiber weight for each layer was calculated in advance, mixed manually (by hand), and placed in the tank in a controlled manner.

Table 4. Test program.

		Fl = 24 mm		Fl = 48 mm		
Dr = 30%	Fc = 1%	7 group	B	3 group	B	(24 mm 7 group;
			2B		2B	1-glass
			3B		3B	2-macromesh
	Fc = 2%	7 group	B	3 group	B	3-basalt
			2B		2B	4-glass + macromesh
			3B		3B	5-glass + basalt
Dr = 50%	Fc = 1%	7 group	B	3 group	B	6-macromesh + basalt
			2B		2B	7-glass + macromesh + basalt
			3B		3B	48 mm 3 group;
	Fc = 2%	7 group	B	3 group	B	1-glass
			2B		2B	2-macromesh
			3B		3B	3-glass + macromesh)
Total		84 group		36 group		120 group

2.4. GeoPIV Analysis

GeoPIV-RG application was used for the analysis of the images obtained from the experimental studies. GeoPIV-RG is designed for geotechnical research applications and is used in the analysis of problems involving large displacements and deformations. It is capable of sub-pixel measurement resolution. At least a 10-fold improvement in measurement accuracy was achieved over the most widely used particle image velocimetry (PIV) approach for all deformation modes including rigid body displacements, rotations, and strains (compression and share) [44].

In our study, GeoPIV-RG software was used to analyze the settlement formed as a result of loading under model strip footing for each of the unreinforced and reinforced soil groups. GeoPIV-RG: It is an image analysis module designed for geotechnical research applications, running under the MATLAB R2019B program. To perform the analysis, a fixed camera was placed in front of the test tank and pictures were taken at regular intervals from the moment the soil was loaded. The basic principle of PIV is to look for changes in the analysis points of images taken at different times. The cross-correlation function can be calculated using the Fast Fourier Transform (FFT) to match the analysis points and obtain the motion states. The normalized cross-correlation function can be expressed as

$$R(m,n) = \frac{\sum_i \sum_j u(i,j) \cdot v(i+m,j+n)}{\sqrt{\sum_i \sum_j u^2(i,j) \sum_i \sum_j v^2(i+m,j+n)}} \quad (1)$$

Here, m, given in Equation 1 represents displacements in the horizontal direction, n represents the displacements in the vertical direction, $u(I,j)$ and $v(I,j)$ represent the gray distribution functions of the image before and after deformation [45]. The pictures taken during the tests were analyzed with the GeoPIV-RG software and the displacements in the soil were obtained. In the program, certain points are selected in the first picture. The program has divided the first image entered as data into very small areas. FFT was then used to compare each area with subsequent images in sequence. As a result of the subsequent analyses, the program obtained image processing results such as vector displacement, horizontal displacement, vertical displacement, general displacement, and stress distributions of the soil from the movement of soil particles (Figure 4).

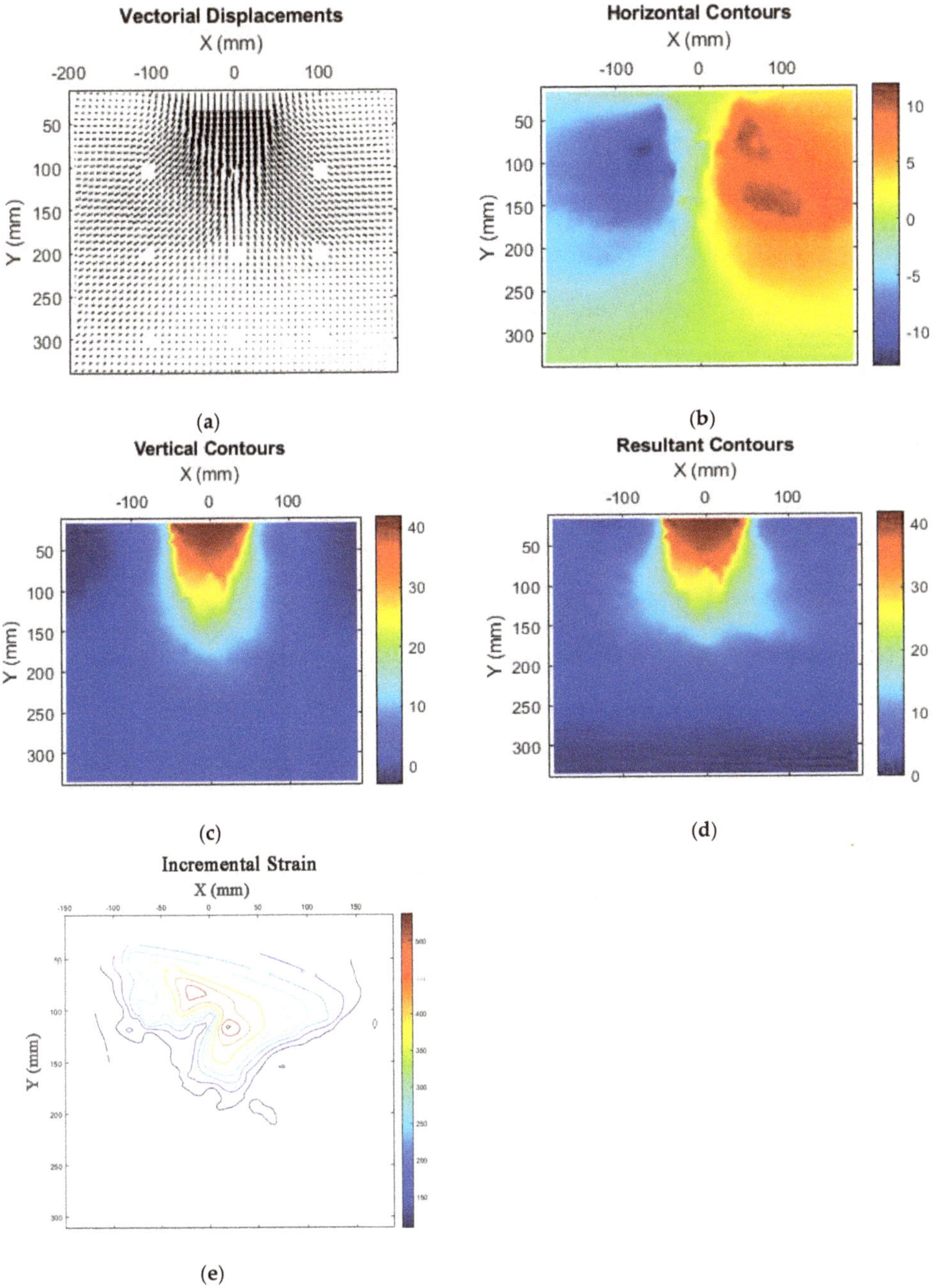

Figure 4. PIV results: (**a**) Vectorial displacements; (**b**) horizontal contours; (**c**) vertical contours; (**d**) resultant contours; (**e**) incremental strain.

3. Results Analysis and Discussion

The base pressure settlement curves obtained in this study as a result of the evaluation of other parameters according to fiber types and the corresponding bearing capacity ratio values were obtained as follows.

Glass Fiber;

The average base pressure-settlement curves obtained according to different fiber lengths, different fiber percentages, and different reinforcement layer depths as a result of model loading tests on sand soil randomly reinforced with glass fiber at Dr = 30% and Dr = 50% are shown in Figure 5, and the reinforcement depth-bearing capacity ratio relationship is given in Figure 6.

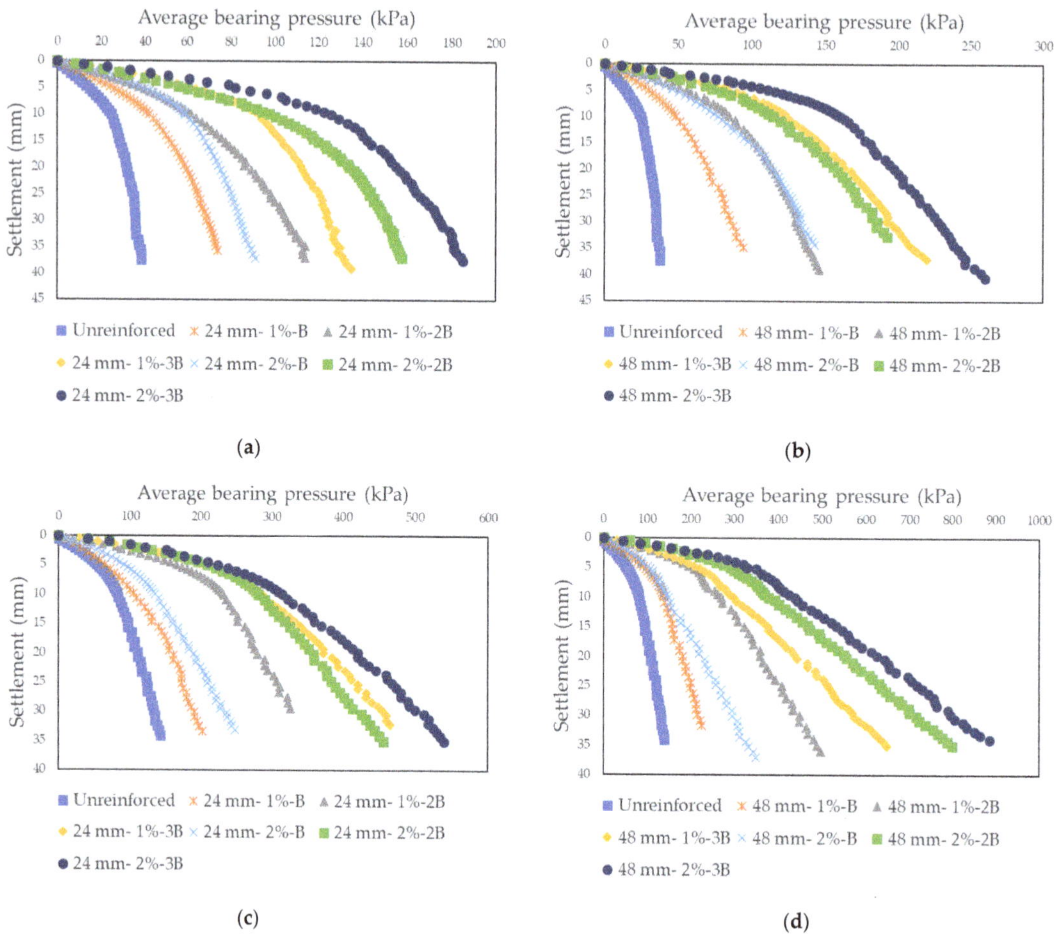

Figure 5. Average base pressure-settlement curves obtained as a result of model tests on sand soil randomly reinforced with glass fiber: (**a**) Dr = 30%, Fl = 24 mm; (**b**) Dr = 30%, Fl = 48 mm; (**c**) Dr = 50%, Fl = 24 mm; (**d**) Dr = 50%, Fl = 48 mm.

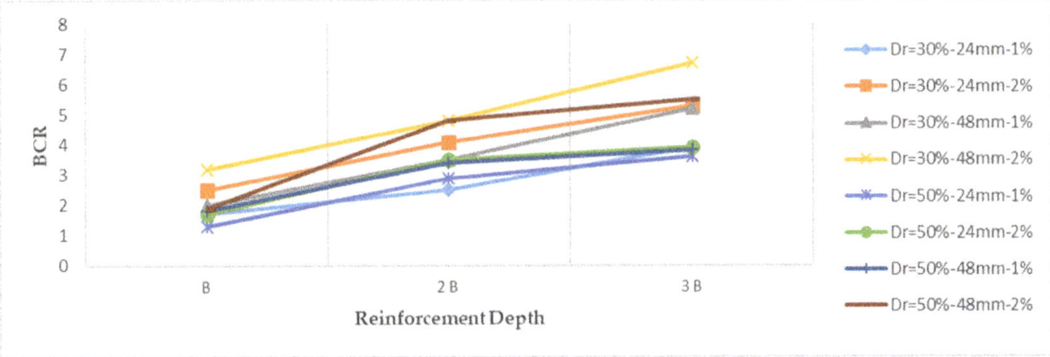

Figure 6. Variation of bearing capacity ratio (BCR) with reinforcement depth (Dr = 30–50%, Fl = 24–48 mm, Fc = 1–2%, and glass fiber).

- In glass fiber-reinforced soils, the highest ultimate bearing capacity (qu) was determined as 440 kPa in the case of adding glass fiber to the sand at Dr = 50%, Fc = 2%, Fl = 48 mm, and 3B reinforcement depth. The highest bearing capacity ratio (BCR = final bearing capacity of reinforced soil/final bearing capacity of unreinforced soil) was determined as 6.7 by adding glass fiber to sand at Dr = 30%, Fc = 2%, Fl = 48 mm, 3B reinforcement depth.
- For all fiber lengths and percentages at Dr = 30%, there was a 48–73% increase in BCR when reinforcement depth was changed from B to 2B, and there was a 30–52% increase in BCR when changed from 2B to 3B. At Dr = 50%, there was a 94–153% increase in BCR for all reinforcement lengths and percentages from depth B to 2B and an 11–23% increase in BCR from 2B to 3B. Changing the reinforcement depth from 2B to 3B at Dr = 50% had no effect on ultimate bearing capacity as much as the percentage increase seen at Dr = 30%.
- Changing the glass fiber length from 24 mm to 48 mm at all reinforcement depths and contents at Dr = 30% increased the BCR in the range of 16–39%. Changing the content of reinforcement from 1% to 2% at all reinforcement depths and lengths increased the BCR between 28% and 61%. Changing the glass fiber length from 24 mm to 48 mm at all reinforcement depths and percentages at Dr = 50% increased BCR in the range of 5–41%. Changing the percentage of reinforcement from 1% to 2% at all reinforcement depths and lengths increased the BCR between 6% and 45%. The increase in bearing capacity ratios obtained as a result of increasing the fiber length from 24 mm to 48 mm by increasing the fiber percentage from 1% to 2% in 50% relative density glass fiber-reinforced soils is similar. However, the increase in BCR obtained as a result of increasing fiber content in soils with 30% relative density glass fiber is higher than the increase in BCR obtained as a result of increasing fiber length.

Macromesh Fiber

The average base pressure-settlement curves obtained according to different fiber lengths, different fiber percentages, and different reinforcement layer depths as a result of model loading tests on sand soil randomly reinforced with macromesh fiber at Dr = 30% and Dr = 50% are shown in Figure 7, and the reinforcement depth-bearing capacity ratio relationship is given in Figure 8.

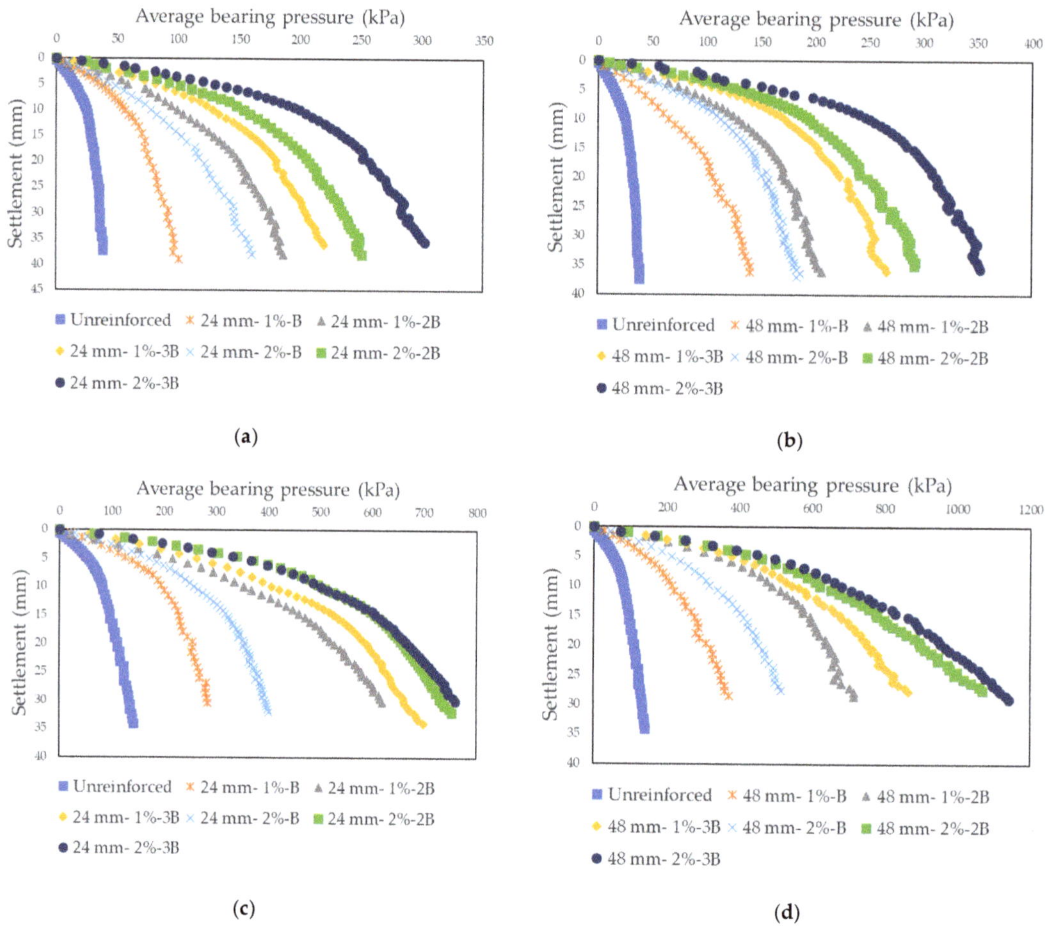

Figure 7. Average base pressure-settlement curves obtained as a result of model tests on sand soil randomly reinforced with macromesh fiber: (**a**) Dr = 30%, Fl = 24 mm; (**b**) Dr = 30%, Fl = 48 mm; (**c**) Dr = 50%, Fl = 24 mm; (**d**) Dr = 50%, Fl = 48 mm.

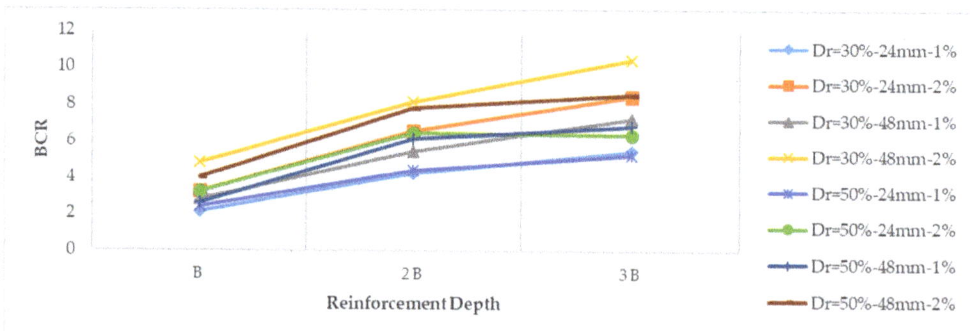

Figure 8. Variation of bearing capacity ratio (BCR) with reinforcement depth (Dr = 30–50%, Fl = 24–48 mm, Fc = 1–2%, and macromesh fiber).

- In macromesh fiber-reinforced soils, the highest ultimate bearing capacity (q_u) was determined as 680 kPa in the case of adding macromesh fiber to the sand at Dr = 50%, Fc = 2%, Fl = 48 mm, and 3B reinforcement depth. The highest BCR was determined as 10.4 by adding macromesh fiber to sand at Dr = 30%, Fc = 2%, Fl = 48 mm, 3B reinforcement depth.
- When the reinforcement depth was changed from B to 2B and from 2B to 3B for all fiber lengths and percentages at Dr = 30%, the BCR increased by 69–101% and 28–33%, respectively. For all reinforcement lengths and percentages, the BCR increased at Dr = 50% by 53–135% from depth B to 2B and by −1.5–18% from 2B to 3B. The ultimate bearing capacity did not rise by the same proportion when the reinforcement depth was increased from 2B to 3B at Dr = 50%. Especially in the experimental group with Dr = 50%, Fl = 48 mm, Fc = 2% macromesh fiber additives, when the depth was increased from 2B to 3B, there was no increase in the bearing capacity.

In both relative densities, an increase in bearing capacity was observed with increasing fiber length. The soil group, which is reinforced at depth B with Fc = 2% and examined at 30% relative density, has had the highest increase. In both relative densities, an increase in bearing capacity was observed with increasing fiber percentage.

Basalt Fiber

The average base pressure-settlement curves obtained according to different fiber percentages and different reinforcement layer depths as a result of model loading tests on sand soil randomly reinforced with basalt fiber at Dr = 30% and Dr = 50% are shown in Figure 9 and the reinforcement depth-bearing capacity ratio relationship is given in Figure 10.

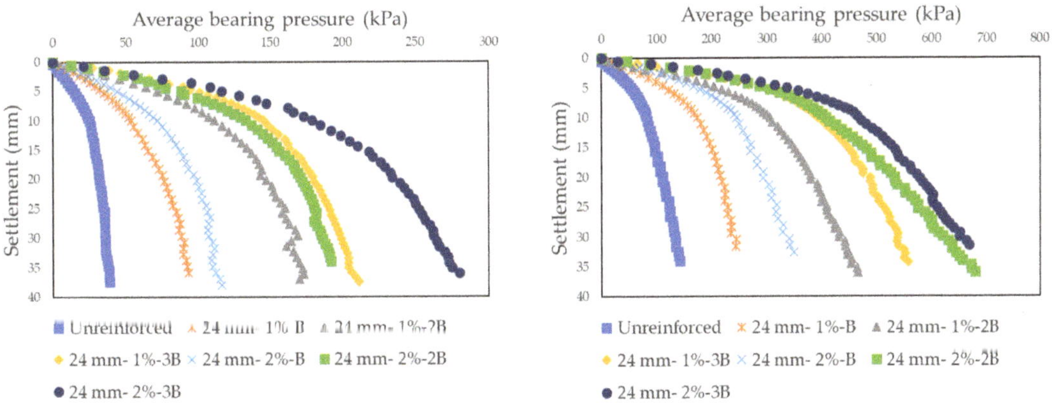

(**a**) (**b**)

Figure 9. Average base pressure-settlement curves obtained as a result of model tests on sand soil randomly reinforced with basalt fiber: (**a**) Dr = 30%, Fl = 24 mm; (**b**) Dr = 50%, Fl = 24 mm.

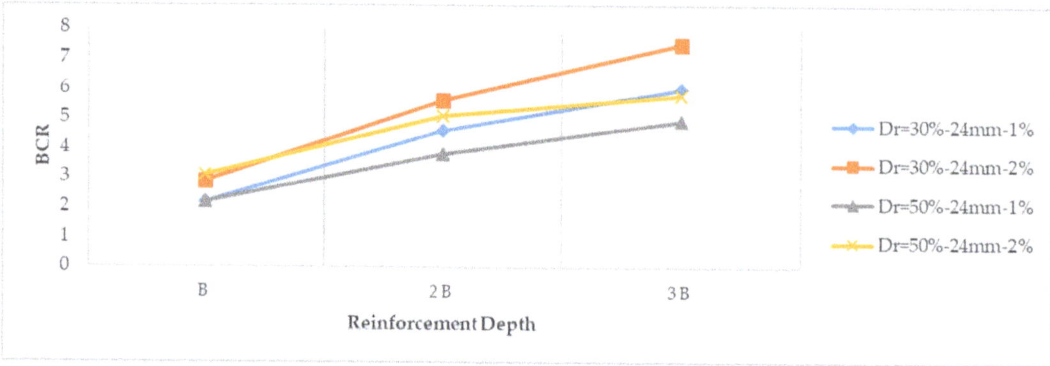

Figure 10. Variation of bearing capacity ratio (BCR) with reinforcement depth (Dr = 30–50%, Fl = 24 mm, Fc = 1–2%, and basalt fiber).

- In basalt fiber-reinforced soils, the highest ultimate bearing capacity (q_u) was determined as 467 kPa in the case of adding basalt fiber to the sand at Dr = 50%, Fc = 2%, Fl = 24 mm, and 3B reinforcement depth. The highest BCR was determined as 7.5 by adding basalt fiber to sand at Dr = 30%, Fc = 2%, Fl = 24 mm, 3B reinforcement depth.
- At both relative densities, the increase in bearing capacity is greater when reinforcement depth is increased from B to 2B. However, the impact on the ultimate bearing capacity of increasing the reinforcement depth from 2B to 3B at Dr = 50% is less significant than the percentage increase seen at Dr = 30%.

Glass + Macromesh Fiber

The average base pressure-settlement curves obtained according to different fiber lengths, different fiber contents, and different reinforcement layer depths as a result of model loading tests on sand soil randomly reinforced with glass + macromesh fiber at Dr = 30% and Dr = 50% are shown in Figure 11, and the reinforcement depth-bearing capacity ratio relationship is given in Figure 12.

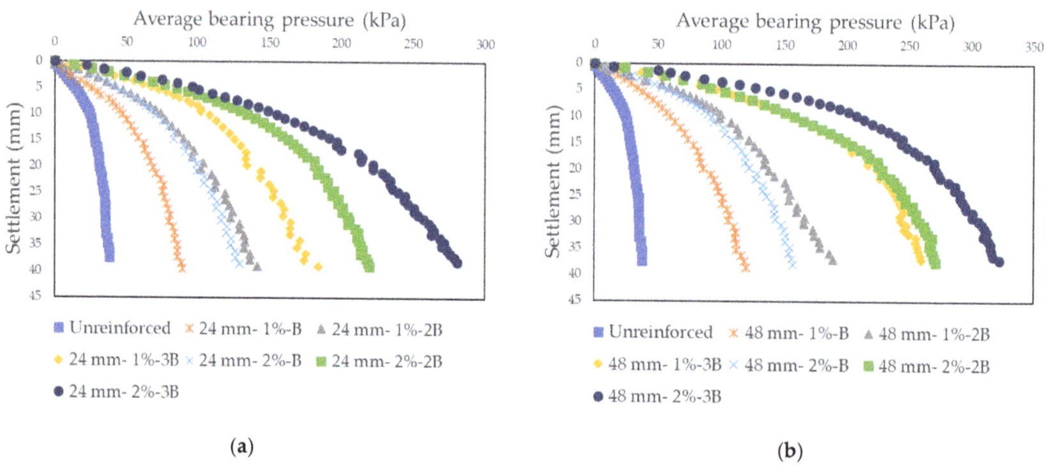

Figure 11. *Cont.*

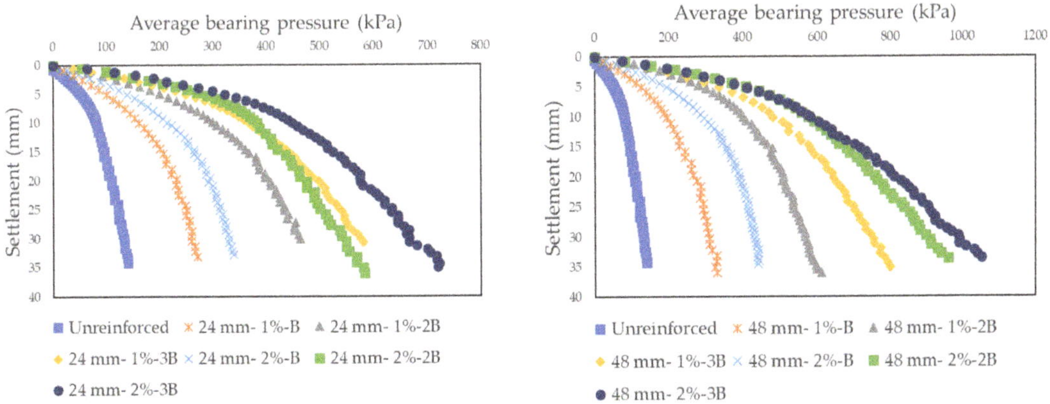

(c) (d)

Figure 11. Average base pressure-settlement curves obtained as a result of model tests on sand soil randomly reinforced with glass + macromesh fiber: (**a**) Dr = 30%, Fl = 24 mm; (**b**) Dr = 30%, Fl = 48 mm; (**c**) Dr = 50%, Fl = 24 mm; (**d**) Dr = 50%, Fl = 48 mm.

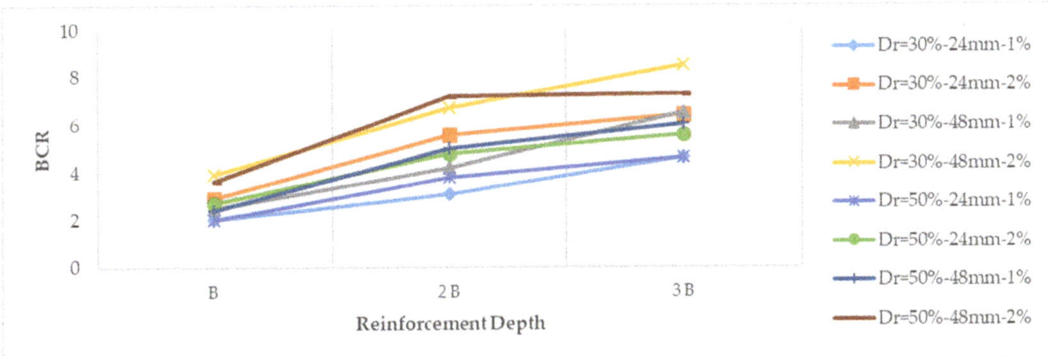

Figure 12. Variation of bearing capacity ratio (BCR) with reinforcement depth (Dr = 30–50%, Fl = 24–48 mm, Fc = 1–2%, glass and macromesh fiber).

- In glass and macromesh fiber reinforced soils, the highest ultimate bearing capacity (q_u) was determined as 580 kPa in the case of adding glass + macromesh fiber to the sand at Dr = 50%, Fc = 2%, Fl = 48 mm, and 3B reinforcement depth. The highest BCR was determined as 8.5 by adding glass + macromesh fiber to sand at Dr = 30%, Fc = 2%, Fl = 48 mm, 3B reinforcement depth.
- At 30% relative density, constant fiber content, and constant reinforcement depth, when the fiber length increases from 24 mm to 48 mm, the group with the highest increase in bearing capacity ratio is the group reinforced with glass + macromesh fiber in 3B reinforcement depth with 1% fiber content, and the rate of bearing capacity increased by 41%.
- When the soil layer depth reinforced at 30% relative density passes from 2B depth to 3B depth, the least increase in bearing capacity was observed in the 24 mm long 2% fiber-reinforced test. In both relative densities, the highest bearing strength was observed in the 48 mm long soil sample with 2% fiber reinforcement. There is almost no difference between the increase in bearing capacity as a result of improvement

in 2B depth and 3B depth in 2% fiber-reinforced soil with a length of 48 mm at 50% relative density.

Glass + Basalt Fiber

The average base pressure-settlement curves obtained according to different fiber percentages and different reinforcement layer depths as a result of model loading tests on sand soil randomly reinforced with glass + basalt fiber at Dr = 30% and Dr = 50% are shown in Figure 13 and the reinforcement depth-bearing capacity ratio relationship is given in Figure 14.

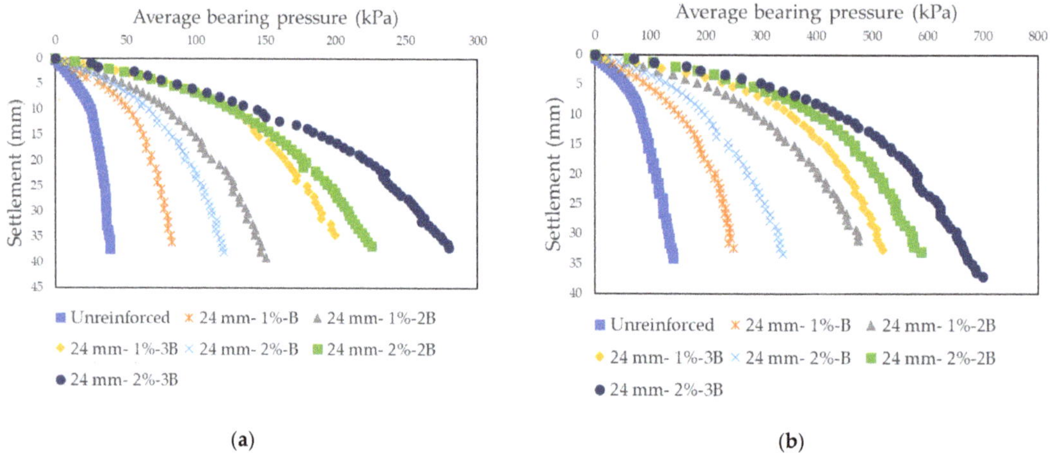

Figure 13. Average base pressure-settlement curves obtained as a result of model tests on sand soil randomly reinforced with glass + basalt fiber: (**a**) Dr = 30%, Fl = 24 mm; (**b**) Dr = 50%, Fl = 24 mm.

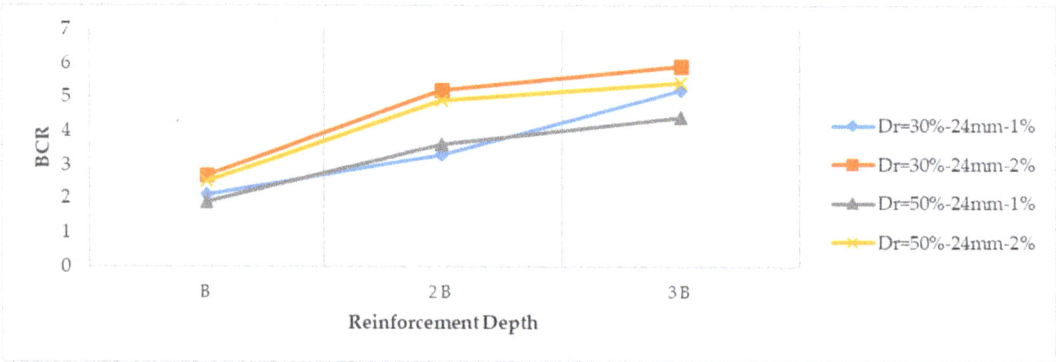

Figure 14. Variation of bearing capacity ratio (BCR) with reinforcement depth (Dr = 30–50%, Fl = 24 mm, Fc = 1–2%, glass and basalt fiber).

- In glass and basalt fiber-reinforced soils, the highest ultimate bearing capacity (q_u) was determined as 431 kPa in the case of adding glass + basalt fiber to the sand at Dr = 50%, Fc = 2%, Fl = 24 mm, and 3B reinforcement depth. The highest BCR was determined as 5.9 by adding glass + basalt fiber to sand at Dr = 30%, Fc = 2%, Fl = 24 mm, 3B reinforcement depth.
- The ultimate bearing capacity is significantly affected by increasing the reinforcement depth from B to 2B at Dr = 50%, However, the ultimate bearing capacity is little affected

by increasing the reinforcement depth from 2B to 3B. Reinforcement depth for soil reinforced with glass + basalt fiber at 50% relative density can be taken as 2B.

Macromesh + Basalt Fiber

The average base pressure-settlement curves obtained according to different fiber percentages and different reinforcement layer depths as a result of model loading tests on sand soil randomly reinforced with macromesh + basalt fiber at Dr = 30% and Dr = 50% are shown in Figure 15 and the reinforcement depth-bearing capacity ratio relationship is given in Figure 16.

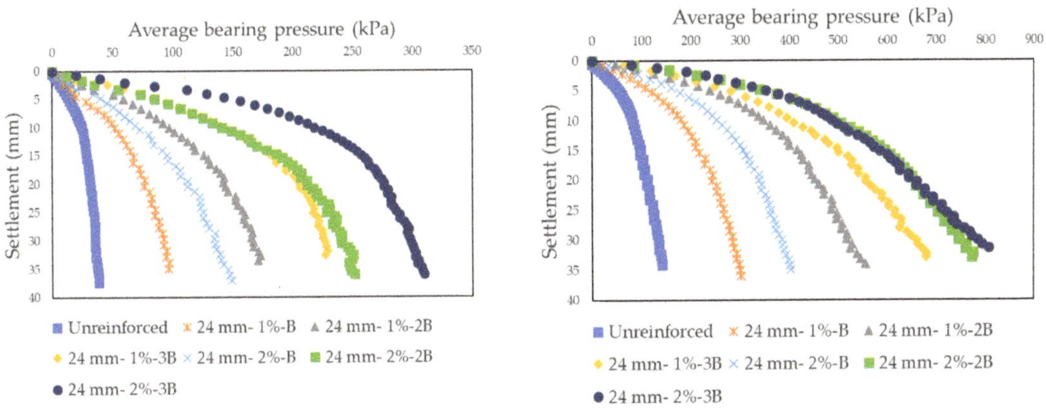

(a) (b)

Figure 15. Average base pressure-settlement curves obtained as a result of model tests on sand soil randomly reinforced with macromesh + basalt fiber: (**a**) Dr = 30%, Fl = 24 mm; (**b**) Dr = 50%, Fl = 24 mm.

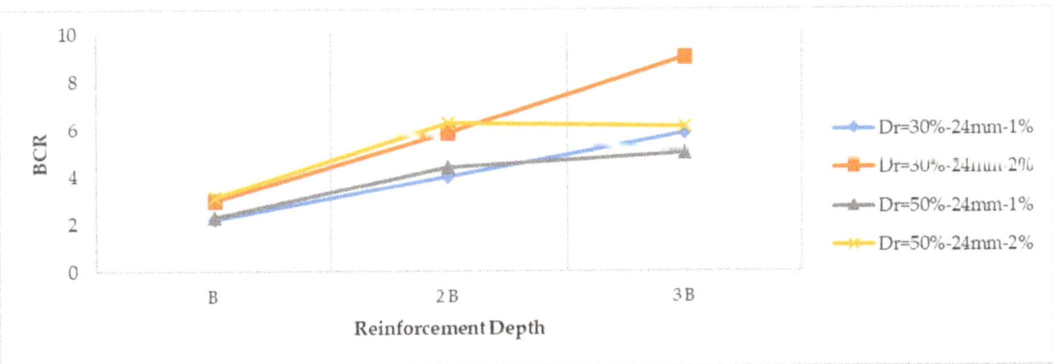

Figure 16. Variation of bearing capacity ratio (BCR) with reinforcement depth (Dr = 30–50%, Fl = 24 mm, Fc = 1–2%, macromesh and basalt fiber).

- In macromesh and basalt fiber-reinforced soils, the highest ultimate bearing capacity (q_u) was determined as 495 kPa in the case of adding macromesh + basalt fiber to the sand at Dr = 50%, Fc = 2%, Fl = 24 mm, and 3B reinforcement depth. The highest BCR was determined as 9 by adding macromesh + basalt fiber to sand at Dr = 30%, Fc = 2%, Fl = 24 mm, 3B reinforcement depth.
- For Dr = 50% Fc = 1% and 2%, basalt + macromesh fiber added experimental group showed an increase of 91% and 100% in BCR when the reinforcement depth was

increased from B to 2B, respectively, but when the reinforcement depth was increased from 2B to 3B, the increase in BCR was 14% and −1.6%, respectively. As can be seen here, when the reinforcement depth is increased from 2B to 3B in the 1% fiber added sample, the increase in the bearing capacity ratio was very low, while there was no increase in the bearing capacity ratio when the reinforcement depth was increased from 2B to 3B in the 2% fiber added sample. In fact, there was a very small decrease. Based on this, it was deemed more appropriate to choose 2B improving depth in experimental groups with 2% basalt + macromesh fiber at 50% relative density.

Glass + Macromesh + Basalt Fiber

The average base pressure-settlement curves obtained according to different fiber contents and different reinforcement layer depths as a result of model loading experiments on sand soil randomly reinforced with glass + macromesh + basalt fiber at Dr = 30% and Dr = 50% are shown in Figure 17, and the reinforcement depth-bearing capacity ratio relationship is given in Figure 18.

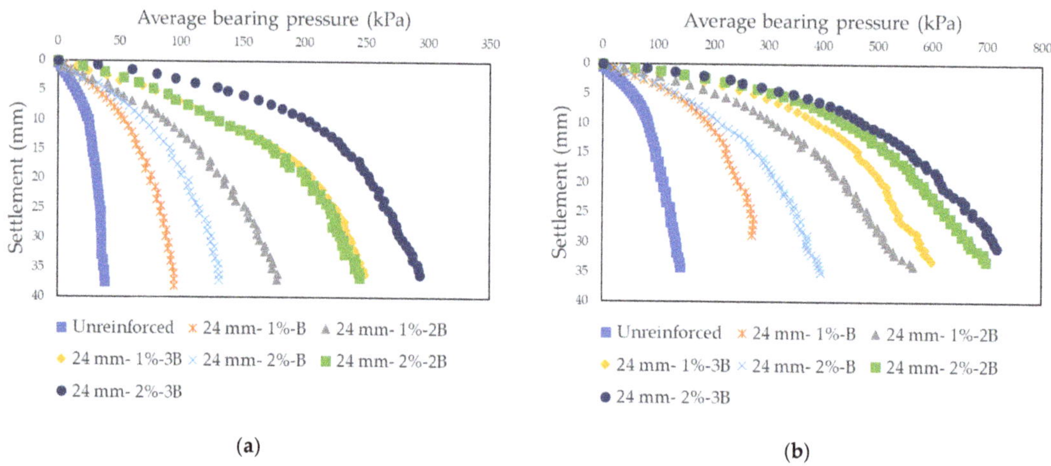

Figure 17. Average base pressure-settlement curves obtained as a result of model tests on sand soil randomly reinforced with glass + macromesh + basalt fiber: (**a**) Dr = 30%, Fl = 24 mm; (**b**) Dr = 50%, Fl = 24 mm.

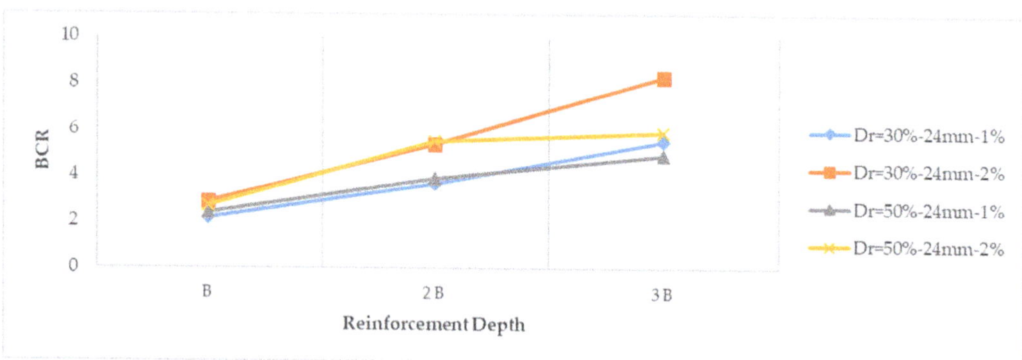

Figure 18. Variation of bearing capacity ratio (BCR) with reinforcement depth (Dr = 30–50%, Fl = 24 mm, Fc = 1–2%, glass + micromesh + basalt fiber).

- In glass + macromesh + basalt fiber-reinforced soils, the highest ultimate bearing capacity (q_u) was determined as 475 kPa in the case of adding glass + macromesh + basalt fiber to the sand at Dr = 50%, Fc = 2%, Fl = 24 mm, and 3B reinforcement depth. The highest BCR was determined as 8.3 by adding glass + macromesh + basalt fiber to sand at Dr = 30%, Fc = 2%, Fl = 24 mm, 3B reinforcement depth.
- Depending on the fiber additive ratio, a decrease in the amount of increase in the bearing capacity ratio was seen as the depth of layers grew. In tests with glass + basalt + macromesh fiber added with Dr = 50% relative density, 2B reinforcement depth can be preferred.

3.1. The Effect of Different Fiber Additives on Soil Behavior

The graphs obtained within the scope of the study as a result of evaluating together the different fiber types (glass, basalt, and macromesh fiber), hybrid fiber (glass + basalt, glass + macromesh, basalt + macromesh, and glass + basalt + macromesh fiber), fiber length (Fl = 24 mm and Fl = 48 mm), fiber content (Fc = 1% and Fc = 2%), relative density (Dr = 30% and Dr = 50%) and reinforcement depth (B, 2B and 3B) is given in Figures 19 and 20. The BCR values obtained as a result of the tests are also given in Table 5.

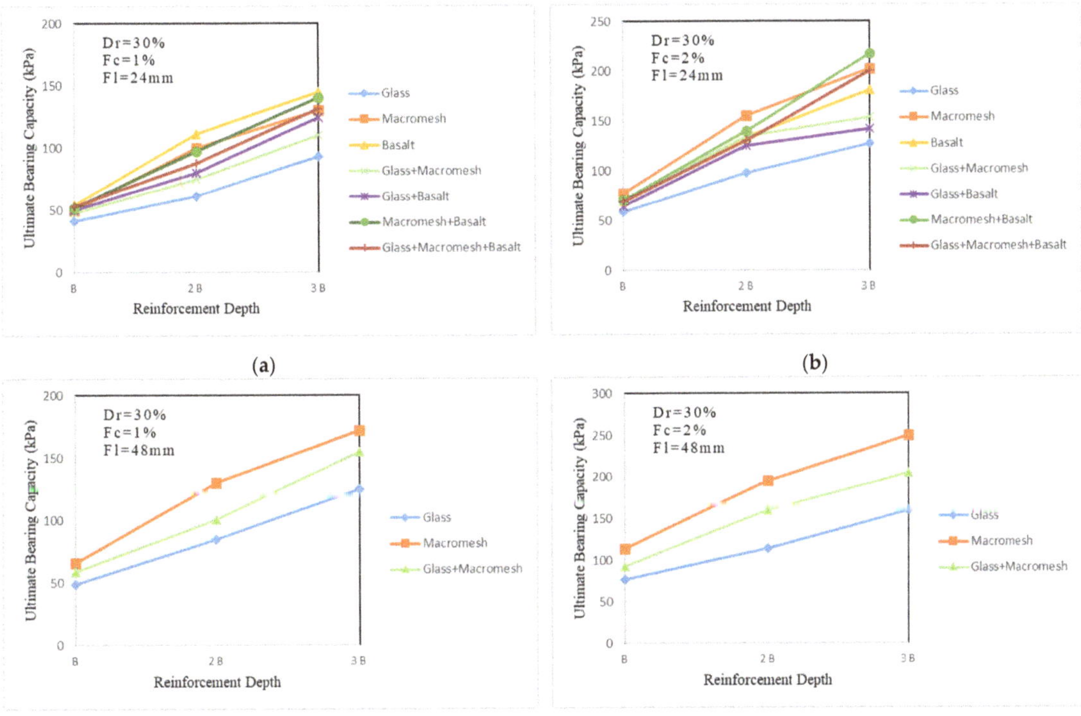

Figure 19. Reinforcement depth-ultimate bearing capacity relationship corresponding to fiber type in tests with Dr = 30% relative density: (**a**) fiber content 1%, fiber length 24 mm; (**b**) fiber content 2%, fiber length 24 mm; (**c**) fiber content 1%, fiber length 48 mm; (**d**) fiber content 2%, fiber length 48 mm.

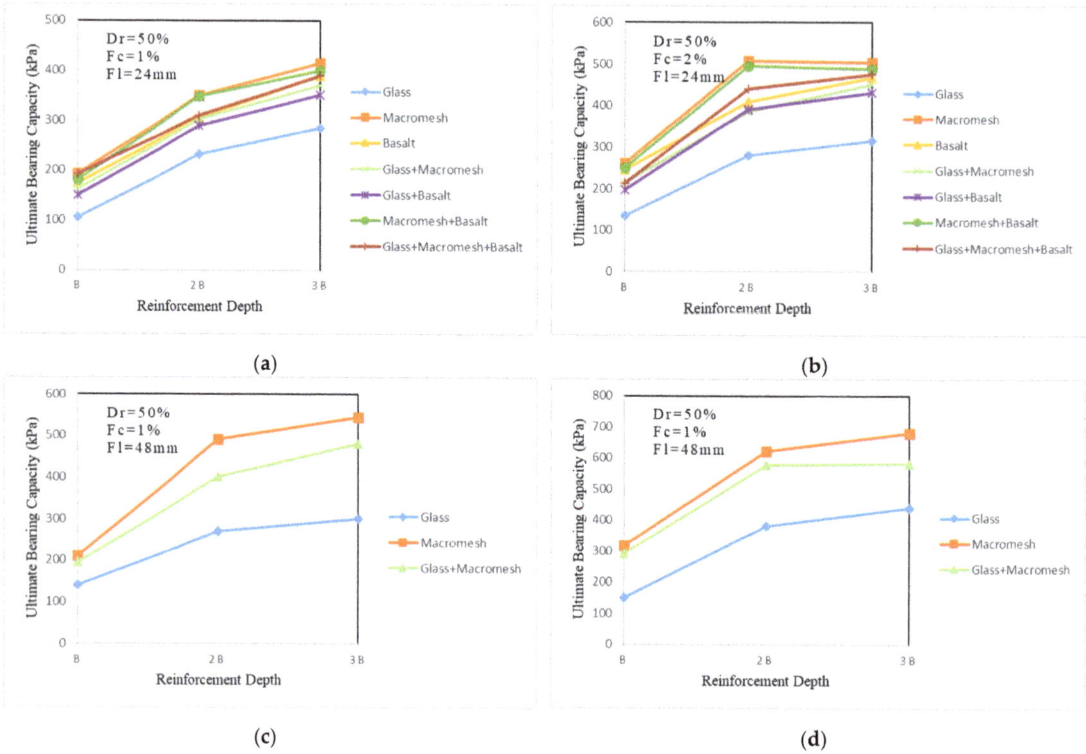

Figure 20. Reinforcement depth-ultimate bearing capacity relationship corresponding to fiber type in tests with Dr = 50% relative density: (**a**) fiber content 1%, fiber length 24 mm; (**b**) fiber content 2%, fiber length 24 mm; (**c**) fiber content 1%, fiber length 48 mm; (**d**) fiber content 2%, fiber length 48 mm.

Table 5. BCR values obtained as a result of tests.

			Glass		Macromesh		Basalt		Glass + Macromesh		Glass + Basalt		Macromesh+ Basalt		Glass + Basalt + Macromesh	
			30%	50%	30%	50%	30%	50%	30%	50%	30%	50%	30%	50%	30%	50%
24 mm	1%	B	1.7	1.3	2.1	2.4	2.2	2.2	2.0	2.0	2.1	1.9	2.2	2.3	2.2	2.4
		2B	2.5	2.9	4.2	4.4	4.6	3.8	3.1	3.8	3.3	3.6	4.0	4.4	3.7	3.9
		3B	3.9	3.6	5.4	5.2	6.0	4.9	4.6	4.6	5.2	4.4	5.8	5.0	5.5	4.9
	2%	B	2.5	1.7	3.2	3.2	2.9	3.1	2.9	2.7	2.7	2.5	3.0	3.1	2.9	2.7
		2B	4.1	3.5	6.5	6.4	5.6	5.1	5.6	4.8	5.2	4.9	5.8	6.2	5.4	5.5
		3B	5.3	3.9	8.4	6.3	7.5	5.8	6.4	5.6	5.9	5.4	9.0	6.1	8.3	5.9
48 mm	1%	B	2.0	1.8	2.8	2.6			2.5	2.4						
		2B	3.5	3.4	5.4	6.1			4.2	5.0						
		3B	5.2	3.8	7.2	6.8			6.5	6.0						
	2%	B	3.2	1.9	4.8	4.0			3.9	3.6						
		2B	4.8	4.8	8.1	7.8			6.7	7.2						
		3B	6.7	5.5	10.4	8.5			8.5	7.3						

A thorough review of the literature revealed that the use of fibers in geotechnical engineering is applicable in six fields, including pavement layers (road construction), retaining walls, railway embankments, protection of slopes, earthquake, and soil-foundation engineering. According to an existing study, the Bradshaw Field Training Area in Australia's Northern Territory stabilized aprons, taxiways, and a helipad using Portland cement and PP fibers. According to some researchers, the use of fibers in conjunction with geogrids can make it easier and more affordable to build tall vertical walls for railway embankments in low-lying residential areas. Additionally, because they can accept the irregular shape of failed slopes, soil mixtures containing randomly distributed fibers can be employed as patches in the localized repair of slope failures. The fiber reinforcement can eliminate the necessity for an anchorage that is present with planar reinforcement and also lessens gully erosion in the reinforcement of soil veneer, such as landfill covers [46].

In the case of using random fiber reinforcement to improve the soil properties, the differences in the results obtained according to the type of fiber selected as reinforcement are significant [47]. Three different fiber types were used in experimental studies. These fibers are glass fiber, basalt fiber, and macromesh fiber, respectively. Glass fiber and macromesh fiber were used in two different lengths of 24 mm and 48 mm, while basalt fiber was used only in 24 mm length. Therefore, in the comparison of these fibers, experimental groups made with a length of 24 mm were evaluated. As can be seen in Figure 19a, using basalt fibers as reinforcement with Dr = 30%, Fl = 24 mm, and Fc = 1% yielded higher ultimate bearing capacity values compared to the use of glass fiber and macromesh fiber at all reinforcement depths. In other groups, the ultimate bearing capacity values of macromesh fiber-added soils are higher than basalt and glass fiber-added soils. Strength increases are related to the shear stresses that occur in the soil as a result of the incorporation of fibers into the soil, revealing the tensile strength of the fibers. Liu et al. [48] investigated the effect of fiber additives of different types (polypropylene, basalt, and glass) on the mechanical performance of sand, similar to our study. When the results were examined, polypropylene fiber with the same content gave more strength to the treated floor due to its good flexible structure and strength properties compared to other fibers. Malidarreh et al. [49] reported that the better interconnection properties of polypropylene fiber also contribute to the ductility gain. In addition, the superiority of macromesh fiber-reinforced samples compared to glass and basalt fiber-reinforced samples is due to the higher surface roughness. Tagnit-Hamou et al. [50] and Tang et al. [51] studies, they reported that the fiber reinforcement mechanism is based on friction and the roughness of the surface of the fibers strongly affects the sand fiber interaction. A fiber's original surface roughness has an impact on how well its strength is mobilized by nearby soil particles. Hard sand particles that touch the fiber surface during the cutting process cut abrade it, causing plastic deformation and even the loss of a portion of the surface layer. The development of holes and grooves on the surface of the fiber increases the interactions between that surface and the sand matrix by forming an interlock. Compared to basalt fiber and glass fiber macromesh fiber, their positive effects on strength increase were more limited as they have a smoother surface.

When the experimental results with hybrid fibers are examined, as observed in the separate use of the fibers, the increase in the relative density, the increase in the depth of reinforcement, the increase in the fiber additive ratio from 1% to 2%, and the increase in the fiber length have positively affected the increase in the ultimate bearing capacity value. In Zhang et al. [36], Ghadakpour et al. [52] and Soltani et al. [53]'s studies, it has been stated that the simultaneous use of fibers with different lengths and aspect ratios (hybrid) improves soil properties more effectively than a single fiber. However, in the experiments, the ultimate bearing capacity values obtained from the soils with macromesh fiber additives are mostly higher than the ultimate bearing capacity values obtained from the soils with hybrid fiber additives. Among the hybrid experimental groups, the highest ultimate bearing capacity was obtained from the group reinforced at 3B depth, Dr = 30%, Fc = 2%, and in which basalt and macromesh fiber were used together. In addition, the ultimate bearing capacities obtained from the glass fiber-reinforced tests were lower than all the other fibers used. However, much higher qu values were obtained when basalt and macromesh were mixed with fiber. When hybrid fiber-reinforced soils are evaluated, it has been observed that when glass fiber is used together with basalt or macromesh fiber, dispersibility is better than when only glass fiber is added to the soil. With sufficient content and length, two or more fibers form a good spatial network structure in the soil [30,34]. Due to the differences in fiber diameters, large-diameter fiber forms a large-scale spatial skeleton structure, while small-diameter fiber created a smaller scale spatial structure, and the combination of the two gave better results in soil reinforcement [37,54].

3.2. The Effect of Fiber Length on Soil Behavior

In this study, glass fiber and macromesh fiber were used in two different lengths of 24 mm and 48 mm, and the effect of the length of the fibers on the bearing capacity was investigated. In order to examine the effect of fiber length on ultimate bearing capacity, 36 experimental groups were evaluated. Looking at the results, it was observed that the increase in fiber length increased the ultimate bearing capacity in both glass fiber and macromesh fiber added tests and glass + macromesh fiber added tests. Chen et al. [24] and Tran et al. [55], in their previous studies, stated that the increase in fiber length positively affects the carrying capacity. Fiber length affects the reinforcing effect by affecting the contact area, but this effect is relatively small compared to the fiber content. In addition, Moghal et al. [56] reported that the increase in carrying capacity with the increase in fiber length was associated with the surface morphology of the fiber. When glass fiber and macromesh fiber are compared among themselves, it was observed that the macromesh fiber-reinforced soil bearing capacity increased more as a result of the increase in fiber length. As seen in the experimental study, since the outer surface of the macromesh fiber is rougher than the glass fiber, the resistance created on the surface of the fiber increased as the fiber length increased. As the fiber length increases, the bonding and friction force between the fiber and soil grains increases, and the bearing capacity increases. The shear strength of the soil increases due to the tensile force formed as a result of the friction force between the fiber and soil grains. Tabakouei et al. [57] said that the bridging effect of the fibers increases as the fiber length increases. According to Liang et al. [58], their study reported that when exposed to an external force, the effect between fibers of different lengths and the soil takes different forms of failure. When the fiber length is short, the failure patterns are characterized by shearing or slipping of the fiber. As the fiber length increases, the failure patterns are characterized by fiber breakage. However, excessively long fiber makes uniform fiber distribution in soil difficult. Because, in long fibers soil, this can cause agglomeration and uneven distribution. As a result, when the length of the fibers used to strengthen the soil is extended, more energy must be spent for a certain deformation to occur in the soil. When the experimental groups were examined among themselves, the effect of fiber length was seen mostly in the tests (49%) reinforced with Dr = 50%, Fc = 2%, glass + macromesh fiber at 2B depth, and it was seen in the tests (48%) reinforced with Dr = 30%, Fc = 2%, macromesh fiber at B depth, and it was seen in the tests (40%) rein-forced with Dr = 50%, Fc = 2%, glass fiber at 3B depth.

3.3. The Effect of Fiber Content on Soil Behavior

Within the scope of the study, fiber reinforcement was added to the soil at two different additive rates 1% and 2% by weight, and the effect of the fiber additive ratio on the bearing capacity was investigated. The fiber contents determined in the experimental study were selected as a result of the studies in the literature. When the fiber content is increased from 1% to 2%, it has been observed that the ultimate bearing capacity increases. When the fiber content does not exceed a certain amount, the fiber is effectively dispersed. As the fiber content increases, it affects the contact area between the fiber and the soil, affecting the cohesion and friction between the fiber and the soil. Similar observations were made by Zafar et al. [59] on natural and synthetic fiber-reinforced soil. Therefore, when the fiber content was increased from 1% to 2%, a larger contact area occurred in the soil and improved mechanical properties. When the tests were examined among themselves, the effect of fiber content was mostly obtained from tests with glass + macromesh fiber reinforcement (80%) with Dr = 30% Fl = 24 mm, reinforcement depth 2B, and it was obtained from tests (73%) with macromesh fiber reinforcement with Dr = 30% Fl = 48 mm, reinforcement depth B. However, as the fiber ratio increased, difficulties began to be experienced in placement. In addition, as the fiber content increased, it was observed that the amount of increase in the bearing capacity ratio decreased, especially in model tests performed at Dr = 50% relative density. If the fiber content is increased too much, the fibers may begin to show a tendency to agglomerate. As a result, the fibers do not effectively increase the contact area. They can reduce the effectiveness of the fiber by creating a weak surface in the soil. These results are consistent with those found by Tabakouei et al. [57]. According to Prabakar and Sridhar [60], increasing the fiber content above a certain point lessens the interaction between soil particles and, as a result, results in less cohesion being mobilized, which lowers the shear strength of the fiber soil composite. Therefore, the contribution rate was not increased above 2% in the tests. In addition, when the fibers were evaluated among themselves, the increase in the bearing capacity with the increase in fiber content was observed in the macromesh fibers the most. This is mostly related to the dispersibility of the fiber [61]. Even after the glass fibers are added to the soil and mixed well, they are mostly dispersed in the soil in bundles. This reduces the benefit to be gained from increased fiber content. Although basalt fibers showed better dispersibility than glass fiber, they were not as effective as macromesh fiber.

3.4. The Effect of Reinforcement Depth on Soil Behavior

Within the scope of the experimental study, the soils were equipped with 1, 2, and 3 times the width of the foundation under the foundation plate. To examine the effect of the reinforced soil layer depth on the bearing capacity, the determined experimental groups were evaluated. Determining the optimum reinforcement depth in soils is an important parameter to design and build a more economical structure. As the depth of the reinforced soil from the foundation base increased, the ultimate bearing capacity of the soil increased has also been reported before by other researchers such as Li et al. [62], Doğan Küçükçongar et al. [63]. In the model loading tests performed at Dr = 30%, the ultimate bearing capacity increased proportionally as the reinforcement depth increased. At Dr = 50%, changing the depth of reinforcement from B to 2B resulted in significant increases in bearing capacity, but when changed from 2B to 3B, the rate of increase in bearing capacity decreased. When compared according to fiber types, as the reinforcement depth increases, the increase in ultimate bearing capacity was highest in glass fiber reinforced tests and least in macromesh-reinforced tests. In model tests with Dr = 50% Fl = 24 mm Fc = 2% macromesh fiber reinforced and Dr = 50% Fl = 24 mm Fc = 2% basalt + macromesh fiber-reinforced tests, there was no increase in ultimate bearing capacity when the reinforcement depth was increased from 2B to 3B. Therefore, at Dr = 50%, it is considered sufficient to reinforce the 2B depth from the foundation base. It has been observed that the soil under strip footing and the ideal reinforcement depth are between 2B and 3B according to the varying relative density values [63,64].

3.5. The Effect of Relative Density on Soil Behavior

In the study, unreinforced and reinforced tests were carried out in loose and medium-dense sand with Dr = 30% and Dr = 50%. Tests to examine the effect of relative density on bearing capacity were evaluated. In all tests, increases in ultimate bearing capacity occurred when Dr = 30% increased to 50%. Different relative densities mean different dry densities. In previous studies on fiber-reinforced sand with different dry densities, it was stated that the stiffness and bearing capacity of the fiber-reinforced soil with higher dry density were higher [65]. In addition, the interfacial movement between the fiber and the soil increases significantly with increasing relative stiffness [54]. A larger relative density means smaller porosity and smaller porosity means increased cohesion and friction as a result of increased contact area between fiber and soil particles. Similar observations were made by Belkhatir et al. [66] on fiber-reinforced soil. The authors reported that a high relative density causes a further increase in the effective contact area between soil particles and fiber. Yixian et al. [67] and Li-Hua et al. [68] indicated that increasing the relative density and increased fiber deformation increased the mechanical interaction between fiber and soil. Thus, it can be noted that the relative density positively affects the soil's mechanical behavior. In the reinforced tests, the bearing capacity ratios obtained in the tests performed at Dr = 30% relative density at all fiber types, fiber length, fiber percentage, and reinforcement depths were higher than the bearing capacity ratios obtained from the tests performed at Dr = 50% relative density. Maximum bearing capacity ratios according to relative density were: BCR = 6.7 at Dr = 30%, BCR = 5.5 at Dr = 50% for glass fiber; BCR = 7.5 at Dr = 30%, BCR = 5.8 at Dr = 50% for basalt fiber; BCR = 10.4 at Dr = 30%, and BCR = 8.5 at Dr = 50% for macromesh fiber. It is seen that the effect of reinforcement is more effective in loose soils [69]. The highest increase in ultimate bearing capacity was determined as 280% in glass fiber-reinforced tests with Fl = 24 mm Fc = 1% reinforcement depth 2B; as 277% in the tests with macromesh fiber reinforcement with Fl = 48 mm Fc = 1% reinforcement depth 2B, and as 286% from the tests with macromesh fiber reinforcement with Fl = 24 mm Fc = 1% reinforcement depth B. Especially in soils reinforced with 2B and 3B depths, as the relative density increased, the ultimate bearing capacities increased significantly.

3.6. GeoPIV Analysis Results

The analysis of pictures taken while the tests were carried out to monitor the deformation at the time of collapse and the formation of the slip band in the unreinforced and reinforced model tests in the soil under different conditions were analyzed using the PIV method. To use this method, GeoPIV-RG, an image analysis module designed for geotechnical research applications running under the MATLAB R2019B program was used [70]. First, the image analysis of the tests carried out on unreinforced soil was made. Afterward, PIV analyses of the experimental groups with 1% and 2% fiber content at 30% relative density were performed, and the PIV analyses of the experimental groups with 1% and 2% fiber content at 50% relative density were performed, respectively. Although the images obtained according to the fiber type are similar, the analysis results obtained from the macromesh fiber-added soils, which have the greatest effect on the bearing capacity in general, are given. The shear strain images resulting from the PIV analyses are shown in Figures 21 and 22.

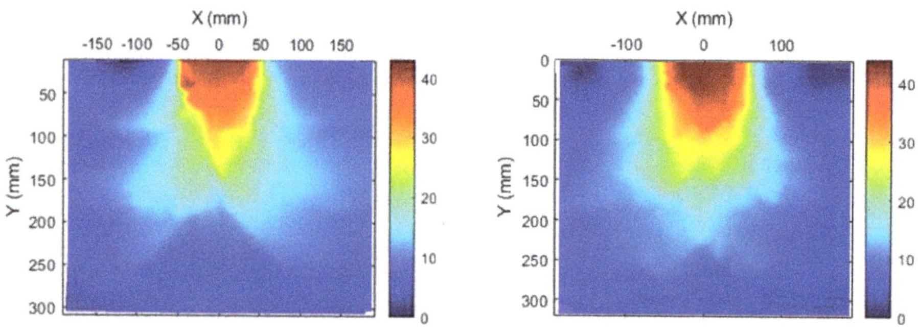

(a) (b)

Figure 21. *Cont.*

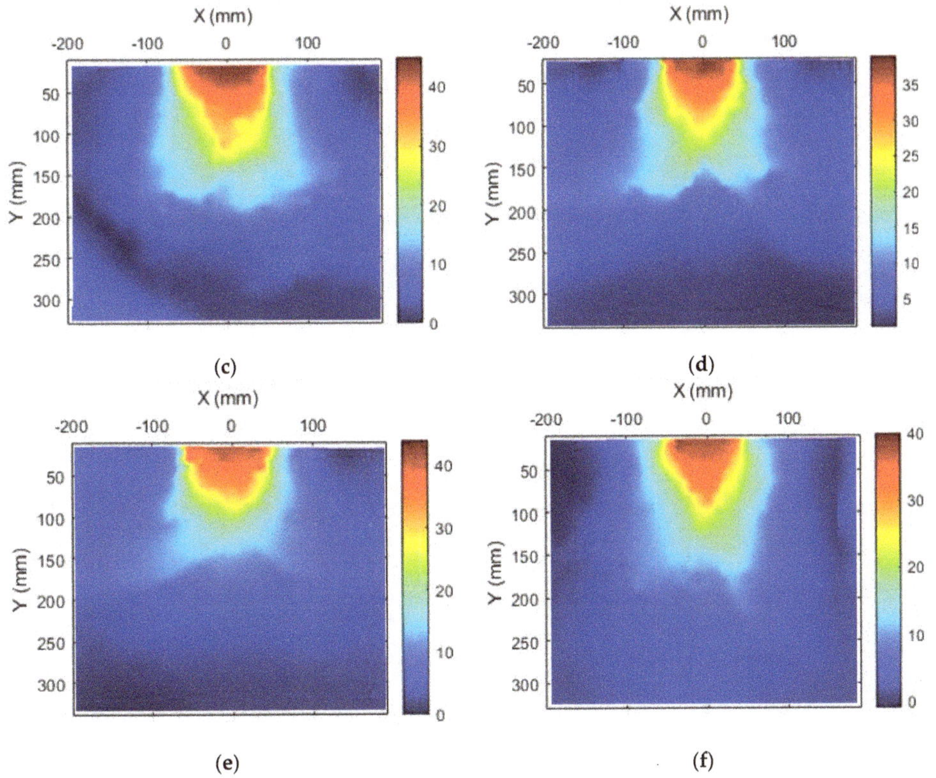

Figure 21. PIV results with Dr = 30%: (**a**,**b**) Unreinforced; (**c**,**d**); Fiber content 1%; (**e**,**f**) Fiber content 2%.

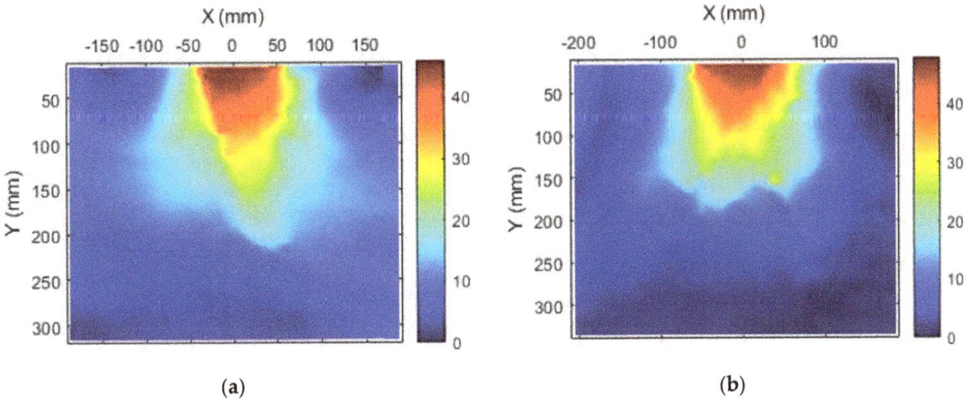

Figure 22. *Cont.*

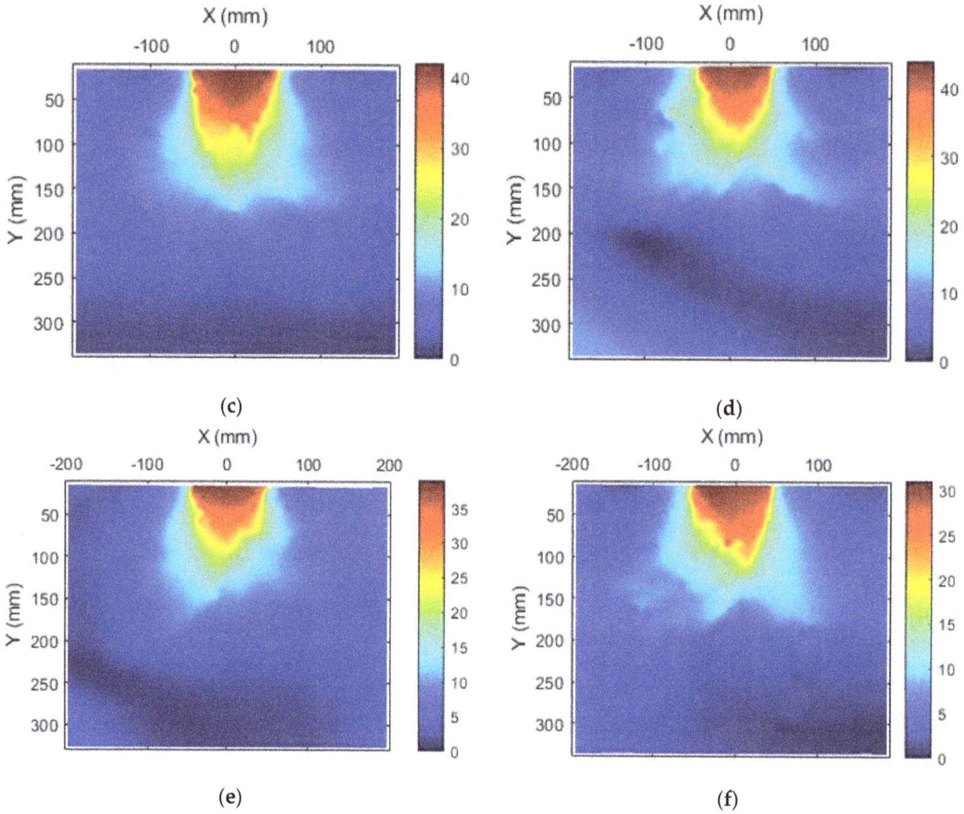

Figure 22. PIV results with Dr = 50%: (**a**,**b**) Unreinforced; (**c**,**d**); fiber content 1%; (**e**,**f**) fiber content 2%.

When the analyses were examined, it was seen that the grain movement in the unreinforced soil was deeper than in the reinforced soil. When the data on reinforced soils were evaluated, it was observed that the displacement decreased as the relative density increased. In general, it can be said that with the increase in the relative density and increasing the fiber reinforcement ratio from 1% to 2%, the displacement decreases and the effect of the loads from the foundation is not transmitted to deeper [45]. In unreinforced model tests, shear surfaces at Dr = 30% relative density were observed to form as punching shear failure, and slip surfaces at Dr = 50% relative density were observed to form as a local shear failure. When reinforcement is added to the unreinforced soil, the slip surfaces of the soil go from a punching shear failure to a local shear failure at Dr = 30% soil. When the percentage of reinforcement was increased from 1% to 2%, the stresses were distributed over a larger area, and a larger area was affected on the right and left sides of the foundation. In general, the addition of reinforcement to the soil transferred the stresses in the soil to a larger area that was observed in the current paper in the case of fiber-reinforced specimens has also been reported before by other researchers such as Çelik et al. [71], Yazdandoust et al. [72]. In this way, the area over which the stress is distributed has increased and it has been observed that the ultimate bearing capacity is higher in reinforced soils than in unreinforced soils.

Limitations and Future Work

Since basalt fiber with a length of 48 mm could not be supplied in the study, the effect of fiber length could not be investigated in the basalt fiber-added experimental groups. In hybrid fiber groups, the fibers were limited by mixing them at the same length and in equal proportions. In the future, the fibers used in the studies to be carried out with random hybrid fibers can be added to the soil at different percentages, and model tests can be made. The effects of hybrid fibers, which will be prepared by using fibers of different lengths, on soil behavior can be investigated by laboratory and field tests. In addition to the fibers used in the study, the use of different fibers as hybrid reinforcement in the ground can be investigated.

4. Conclusions

1. It has been observed that the effect of macromesh fiber on improving the strength properties of the soil is higher than basalt and glass fiber.
2. It was observed that the increase in fiber length and fiber percentage increased the ultimate bearing capacity.
3. When the situation where glass fiber and basalt fiber are added separately to the soil and the use of macromesh fiber in the hybrid form are compared, it is seen that the hybrid fiber has positive effects on soil bearing capacity.
4. BCR values obtained as a result of reinforcement with fiber are higher in soils with low relative density. It is seen that the positive effect of reinforcement with fiber reinforcement is greater in loose soils.
5. It has been observed that the soil under strip footing and the ideal reinforcement depth are between 2B and 3B according to the varying relative density values.
6. As a result of the PIV analyses, when the fiber is added to the unreinforced soil, it was seen that the settlement in the soil decreases and the effect of the loads coming from the foundation is not transmitted to deeper.

Author Contributions: Conceptualization, B.A.A. and S.Ç.; methodology, B.A.A. and S.Ç.; software, B.A.A.; validation, B.A.A. and S.Ç.; formal analysis, B.A.A. and S.Ç.; investigation, B.A.A. and S.Ç.; resources, B.A.A.; data curation, B.A.A.; writing—original draft preparation, B.A.A.; writing—review and editing, B.A.A.; visualization, B.A.A.; supervision, S.Ç.; project administration, B.A.A. and S.Ç.; funding acquisition, B.A.A. All authors have read and agreed to the published version of the manuscript.

Funding: This research received no external funding.

Institutional Review Board Statement: Not applicable.

Informed Consent Statement: Not applicable.

Data Availability Statement: The data presented in this study are available upon request from the corresponding author.

Conflicts of Interest: The authors declare no conflict of interest.

References

1. Vidal, H. The principle of reinforced earth. *Highw. Res. Rec.* **1966**, *282*, 1–16.
2. Waldron, L. The shear resistance of root-permeated homogeneous and stratified soil. *Soil Sci. Soc. Am. J.* **1977**, *41*, 843–849. [CrossRef]
3. Jamshidi, R.; Towhata, I.; Ghiassian, H.; Tabarsa, A. Experimental evaluation of dynamic deformation characteristics of sheet pile retaining walls with fiber reinforced backfill. *Soil Dyn. Earthq. Eng.* **2010**, *30*, 438–446. [CrossRef]
4. Ghiassian, H.; Jamshidi, R.; Tabarsa, A. Dynamic performance of Toyoura sand reinforced with randomly distributed carpet waste strips. In *Geotechnical Earthquake Engineering and Soil Dynamics IV*; ASCE: Reston, WV, USA, 2008; pp. 1–13.
5. Gray, D.H.; Ohashi, H. Mechanics of fiber reinforcement in sand. *J. Geotech. Eng.* **1983**, *109*, 335–353. [CrossRef]
6. Yetimoglu, T.; Salbas, O. A study on shear strength of sands reinforced with randomly distributed discrete fibers. *Geotext. Geomembr.* **2003**, *21*, 103–110. [CrossRef]
7. Hataf, N.; Rahimi, M. Experimental investigation of bearing capacity of sand reinforced with randomly distributed tire shreds. *Constr. Build. Mater.* **2006**, *20*, 910–916. [CrossRef]
8. Yetimoglu, T.; Inanir, M.; Inanir, O.E. A study on bearing capacity of randomly distributed fiber-reinforced sand fills overlying soft clay. *Geotext. Geomembr.* **2005**, *23*, 174–183. [CrossRef]
9. Mukherjee, K.; Mishra, A.K. Impact of glass fibre on hydromechanical behaviour of compacted sand–bentonite mixture for landfill application. *Eur. J. Environ. Civ. Eng.* **2021**, *25*, 1179–1200. [CrossRef]

10. Patel, S.K.; Singh, B. Shear strength response of glass fibre-reinforced sand with varying compacted relative density. *Int. J. Geotech. Eng.* **2019**, *13*, 339–351. [CrossRef]
11. Patel, S.K.; Singh, B. A comparative study on shear strength and deformation behaviour of clayey and sandy soils reinforced with glass fibre. *Geotech. Geol. Eng.* **2020**, *38*, 4831–4845. [CrossRef]
12. Sujatha, E.; Atchaya, P.; Darshan, S.; Subhashini, S. Mechanical properties of glass fibre reinforced soil and its application as subgrade reinforcement. *Road Mater. Pavement Des.* **2021**, *22*, 2384–2395. [CrossRef]
13. Philley, P.D.; Kalita, A.; Omo, Y. Improvement of Strength Behaviour of Fly Ash and Cement Stabilized Soil with Glass Fibre Reinforcement. In *Advances in Geo-Science and Geo-Structures*; Springer: Singapore, 2022; pp. 133–142.
14. Benziane, M.M.; Della, N.; Sert, S.; Denine, S.; Nouri, S.; Bol, E.; Elroul, A.B. Shear behaviour of sandy soil from Chlef river reinforced with different types of fibres. *Mar. Georesources Geotechnol.* **2022**, *40*, 1232–1241. [CrossRef]
15. Lv, X.; Zhou, H.; Liu, X.; Song, Y. Experimental study on the effect of basalt fiber on the shear behavior of cemented sand. *Environ. Earth Sci.* **2019**, *78*, 688. [CrossRef]
16. Kale, R.; Neha, D.; Sakshi, M.; Gaurav, A.; Pranit, J.; Vaishnavi, R.; Chetan, R. Soil stabilization by using basalt fibers. *Int. J. Sci. Res. Eng. Trends* **2020**, *6*, 1931–1935.
17. Gobinath, R.; Mahesh, V.; Shyamala, G.; Rajesh, A. Strength and Settlement studies on basalt fiber reinforced marginal soil. In *IOP Conference Series, Proceedings of the International Conference on Recent Advancements in Engineering and Management (ICRAEM-2020), Warangal, India, 9–10 October 2020*; IOP Publishing: Philadelphia, PA, USA, 2020; p. 032083.
18. Gao, C.; Du, G.; Guo, Q.; Zhuang, Z. Static and dynamic behaviors of basalt fiber reinforced cement-soil after freeze-thaw cycle. *KSCE J. Civ. Eng.* **2020**, *24*, 3573–3583. [CrossRef]
19. Wang, S.; Chen, F.; Xue, Q.; Zhang, P. Splitting tensile strength of cement soil reinforced with basalt fibers. *Materials* **2020**, *13*, 3110. [CrossRef]
20. Boz, A.; Sezer, A.; Özdemir, T.; Hızal, G.E.; Azdeniz Dolmacı, Ö. Mechanical properties of lime-treated clay reinforced with different types of randomly distributed fibers. *Arab. J. Geosci.* **2018**, *11*, 122. [CrossRef]
21. Ndepete, C.P.; Sert, S.; Beycioğlu, A.; Katanalp, B.Y.; Eren, E.; Bağrıaçık, B.; Topolinski, S. Exploring the effect of basalt fibers on maximum deviator stress and failure deformation of silty soils using ANN, SVM and FL supported by experimental data. *Adv. Eng. Softw.* **2022**, *172*, 103211. [CrossRef]
22. Xu, J.; Wu, Z.; Chen, H.; Shao, L.; Zhou, X.; Wang, S. Triaxial shear behavior of basalt fiber-reinforced loess based on digital image technology. *KSCE J. Civ. Eng.* **2021**, *25*, 3714–3726. [CrossRef]
23. Mukhtar, M.; Kumar, E.A. To Study the Behavior of Compacted Soil Using Basalt Fibers. *Int. J. Innov. Res. Eng. Manag.* **2022**, *9*, 360–366.
24. Chen, W.; Zhao, J.; Fan, L.; Li, J.; Yuan, B.; Li, H.; Jiang, G.; Li, H.; Chen, T. The effect of length and content of fiber on glass fiber and basalt fiber-reinforced granite residual soil. *Adv. Civ. Eng.* **2022**, *2022*, 7803002. [CrossRef]
25. Latifi, M.R.; Biricik, Ö.; Mardani Aghabaglou, A. Effect of the addition of polypropylene fiber on concrete properties. *J. Adhes. Sci. Technol.* **2022**, *36*, 345–369. [CrossRef]
26. Wang, Q.; Ding, Y.; Zhang, Y.; Castro, C. Effect of macro polypropylene fiber and basalt fiber on impact resistance of basalt fiber-reinforced polymer-reinforced concrete. *Struct. Concr.* **2021**, *22*, 503–515. [CrossRef]
27. Rostami, R.; Zarrebini, M.; Abdellahi, S.B.; Mostofinejad, D.; Abtahi, S.M. Investigation of flexural performance of concrete reinforced with indented and fibrillated macro polypropylene fibers based on numerical and experimental comparison. *Struct. Concr.* **2021**, *22*, 250–263. [CrossRef]
28. Brasse, K.; Tracz, T.; Zdeb, T. The Effect of the Type and Amount of Synthetic Fibers on the Effectiveness of Dispersed Reinforcement in Soil-Cements. *Materials* **2020**, *13*, 3917. [CrossRef] [PubMed]
29. Daneshfar, M.; Hassani, A.; Aliha, M.; Berto, F. Investigating Flexural Performance of Fiber-Reinforced Concrete with Different Contents and Types of Macrosynthetic Fiber. *Strength Mater.* **2022**, *54*, 650–661. [CrossRef]
30. Badri, M.Z.A.; Khalid, F.S. Performance of concrete containing combination of macro mesh PP and nylon fibers. *Recent Trends Civ. Eng. Built Environ.* **2021**, *2*, 475–481.
31. Akın, S.K.; Kartal, S.; Müsevitoğlu, A.; Sancıoğlu, S.; Zia, A.J.; Ilgün, A. Macro and micro polypropylene fiber effect on reinforced concrete beams with insufficient lap splice length. *Case Stud. Constr. Mater.* **2022**, *16*, e01005. [CrossRef]
32. Khan, M.; Cao, M.; Xie, C.; Ali, M. Hybrid fiber concrete with different basalt fiber length and content. *Struct. Concr.* **2022**, *23*, 346–364. [CrossRef]
33. Khan, M.; Cao, M.; Xie, C.; Ali, M. Efficiency of basalt fiber length and content on mechanical and microstructural properties of hybrid fiber concrete. *Fatigue Fract. Eng. Mater. Struct.* **2021**, *44*, 2135–2152. [CrossRef]
34. Ding, Y.; Wang, Q.; Pacheco-Torgal, F.; Zhang, Y. Hybrid effect of basalt fiber textile and macro polypropylene fiber on flexural load-bearing capacity and toughness of two-way concrete slabs. *Constr. Build. Mater.* **2020**, *261*, 119881. [CrossRef]
35. Zhang, J.; Xu, W.; Gao, P.; Su, L.; Kun, B.; Yueyuan, L.; Bohan, Y. Integrity and crack resistance of hybrid polypropylene fiber reinforced cemented soil. *J. Eng. Fibers Fabr.* **2022**, *17*, 15589250211068428. [CrossRef]
36. Zhang, J.; Xu, W.; Gao, P.; Yao, Z.; Su, L.; Qiu, N.; Huang, W. Compressive strength characteristics of hybrid fiber-reinforced cemented soil. *Int. J. Pavement Eng.* **2022**, 1–27. [CrossRef]
37. Shu, Y.; Zhang, J. Effect of Basalt Fiber Content and Length on the Strength and Crack Development of Polyvinyl Alcohol/Basalt Hybrid Fiber-Reinforced Cement Soil. *Polymers* **2023**, *15*, 2146. [CrossRef]

38. *ASTM D2487-17*; Standard Practice for Classification of Soils for Engineering Purposes (Unified Soil Classification System). ASTM International: West Conshohocken, PA, USA, 2020.
39. *ASTM D3080-04*; Standard Test Method for Direct Shear Test of Soils Under Consolidated Drained Conditions. ASTM International: West Conshohocken, PA, USA, 2012.
40. Briaud, J.-L.; Jeanjean, P. Load settlement curve method for spread footings of sand. In *Vertical and Horizontal Deformations of Foundations and Embankments*; ASCE: Reston, WV, USA, 1994; pp. 1774–1804.
41. Trautmann, C.H.; Kulhawy, F.H. Uplift load-displacement behavior of spread foundations. *J. Geotech. Eng.* **1988**, *114*, 168–184. [CrossRef]
42. De Beer, E. Experimental determination of the shape factors and the bearing capacity factors of sand. *Geotechnique* **1970**, *20*, 387–411. [CrossRef]
43. Ateş, B.; Şadoğlu, E. Kum zeminlerdeki kazıklı radye temellerin optimum kazık aralığının deneysel olarak incelenmesi. *Tek. Dergi* **2021**, *32*, 10477–10494. [CrossRef]
44. Stanier, S.A.; Blaber, J.; Take, W.A.; White, D. Improved image-based deformation measurement for geotechnical applications. *Can. Geotech. J.* **2016**, *53*, 727–739. [CrossRef]
45. Xu, L.; Chen, H.-b.; Chen, F.-q.; Lin, Y.-j.; Lin, C. An experimental study of the active failure mechanism of narrow backfills installed behind rigid retaining walls conducted using Geo-PIV. *Acta Geotech.* **2022**, *17*, 4051–4068. [CrossRef]
46. Hejazi, S.M.; Sheikhzadeh, M.; Abtahi, S.M.; Zadhoush, A. A simple review of soil reinforcement by using natural and synthetic fibers. *Constr. Build. Mater.* **2012**, *30*, 100–116. [CrossRef]
47. Kravchenko, E.; Liu, J.; Niu, W.; Zhang, S. Performance of clay soil reinforced with fibers subjected to freeze-thaw cycles. *Cold Reg. Sci. Technol.* **2018**, *153*, 18–24. [CrossRef]
48. Liu, J.; Bai, Y.; Song, Z.; Kanungo, D.P.; Wang, Y.; Bu, F.; Chen, Z.; Shi, X. Stabilization of sand using different types of short fibers and organic polymer. *Constr. Build. Mater.* **2020**, *253*, 119164. [CrossRef]
49. Malidarreh, N.; Shooshpasha, I.; Mirhosseini, S.; Dehestani, M. Effects of reinforcement on mechanical behaviour of cement treated sand using direct shear and triaxial tests. *Int. J. Geotech. Eng.* **2018**, *12*, 491–499. [CrossRef]
50. Tagnit-Hamou, A.; Vanhove, Y.; Petrov, N. Microstructural analysis of the bond mechanism between polyolefin fibers and cement pastes. *Cem. Concr. Res.* **2005**, *35*, 364–370. [CrossRef]
51. Tang, C.; Shi, B.; Gao, W.; Chen, F.; Cai, Y. Strength and mechanical behavior of short polypropylene fiber reinforced and cement stabilized clayey soil. *Geotext. Geomembr.* **2007**, *25*, 194–202. [CrossRef]
52. Ghadakpour, M.; Choobbasti, A.J.; Kutanaei, S.S. Investigation of the Kenaf fiber hybrid length on the properties of the cement-treated sandy soil. *Transp. Geotech.* **2020**, *22*, 100301. [CrossRef]
53. Soltani, A.; Deng, A.; Taheri, A. Swell–compression characteristics of a fiber–reinforced expansive soil. *Geotext. Geomembr.* **2018**, *46*, 183–189. [CrossRef]
54. Li, J.; Tang, C.; Wang, D.; Shi, B.; Pei, X. Single fiber pullout tests on interfacial shear strength of wave-shape fiber-reinforced soils. *Chin. J. Geotech. Eng.* **2014**, *36*, 1696–1704.
55. Tran, K.Q.; Satomi, T.; Takahashi, H. Effect of waste cornsilk fiber reinforcement on mechanical properties of soft soils. *Transp. Geotech.* **2018**, *16*, 76–84. [CrossRef]
56. Moghal, A.A.B.; Chittoori, B.C.; Basha, B.M.; Al-Shamrani, M.A. Target reliability approach to study the effect of fiber reinforcement on UCS behavior of lime treated semiarid soil. *J. Mater. Civ. Eng.* **2017**, *29*, 04017014. [CrossRef]
57. Tabakouei, A.R.; Narani, S.; Abbaspour, M.; Aflaki, E.; Siddiqua, S. Coupled specimen and fiber dimensions influence measurement on the properties of fiber-reinforced soil. *Measurement* **2022**, *188*, 110556. [CrossRef]
58. Liang, S.; Xiao, X.; Wang, J.; Wang, Y.; Feng, D.; Zhu, C. Influence of fiber type and length on mechanical properties of MICP-treated sand. *Materials* **2022**, *15*, 4017. [CrossRef] [PubMed]
59. Zafar, T.; Ansari, M.A.; Husain, A. Soil stabilization by reinforcing natural and synthetic fibers–A state of the art review. *Mater. Today Proc.* **2023**, *in press*. [CrossRef]
60. Prabakar, J.; Sridhar, R. Effect of random inclusion of sisal fibre on strength behaviour of soil. *Constr. Build. Mater.* **2002**, *16*, 123–131. [CrossRef]
61. Duong, N.T.; Tran, K.Q. Estimation of seepage velocity and piping resistance of fiber-reinforced soil by using artificial neural network-based approach. *Neural Comput. Appl.* **2023**, *35*, 2443–2455. [CrossRef]
62. Li, J.; Zhou, C.; Xin, G.; Long, G.; Zhang, W.; Li, C.; Zhuang, P.; Yao, Z. Study on reinforcement mechanism and reinforcement effect of saturated soil with a weak layer by DC. *Appl. Sci.* **2022**, *12*, 9770. [CrossRef]
63. Doğan Küçükçongar, Ş.; Çelik, S.; Karimi Ghalehjough, B. Experimental and Numerical Investigation for Estimating Optimal Depth-Bearing Capacity of Randomly Fiber-Reinforced Sandy Soils. *J. Sci. Ind. Res.* **2022**, *81*, 1037–1049.
64. Li, L.; Xiao, H.; Gao, M.; Liu, Y.; Hu, Z.; Yao, K. Experimental investigations on effect of geocell, waste tire chips, and geocell–tire chips on foundation reinforcement. *J. Perform. Constr. Facil.* **2019**, *33*, 04019074. [CrossRef]
65. Brahimi, A.; Bouri, D.E.; Boutaraa, Z.; Nougar, B.; Krim, A.; Chemmam, M.; Arab, A. Numerical and experimental study on the effect of fiber reinforcement on the shear strength and hydraulic conductivity of Chlef soil. *Innov. Infrastruct. Solut.* **2023**, *8*, 40. [CrossRef]
66. Belkhatir, M.; Schanz, T.; Arab, A.; Della, N.; Kadri, A. *Insight into the Effects of Gradation on the Pore Pressure Generation of Sand-Silt Mixtures*; ASTM International: Oxford, UK, 2014.

67. Yixian, W.; Panpan, G.; Shengbiao, S.; Haiping, Y.; Binxiang, Y. Study on strength influence mechanism of fiber-reinforced expansive soil using jute. *Geotech. Geol. Eng.* **2016**, *34*, 1079–1088. [CrossRef]
68. Li, L.; Liu, S.; Xiao, H.; Feng, W.; Li, J.; Wu, P. Experimental investigation on reinforcement effect of sustainable materials for different subgrades. *J. Clean. Prod.* **2022**, *343*, 130944. [CrossRef]
69. Fazeli Dehkordi, P.; Ghazavi, M.; Karim, U.F. Bearing capacity-relative density behavior of circular footings resting on geocell-reinforced sand. *Eur. J. Environ. Civ. Eng.* **2022**, *26*, 5088–5112. [CrossRef]
70. Stanier, S.; Dijkstra, J.; Leśniewska, D.; Hambleton, J.; White, D.; Wood, D.M. Vermiculate artefacts in image analysis of granular materials. *Comput. Geotech.* **2016**, *72*, 100–113. [CrossRef]
71. Çelik, S.; Ghalehjough, B.K.; Majedi, P.; Akbulut, S. Effect of randomly fiber reinforcement on shear failure surface of soil, behind flexible retaining walls at different conditions. *Indian J. Geo-Mar. Sci.* **2017**, *46*, 2097–2104.
72. Yazdandoust, M.; Taimouri, A.B.B. Lateral Pressure Distribution and Shear Band Development in Two-Tiered MSE Walls under Footing Loading. *Int. J. Geomech.* **2023**, *23*, 04023038. [CrossRef]

applied sciences

MDPI

Article

Mechanical Behaviors and Numerical Simulation Analysis of a New Isolation System

Bo Liu [1,2], Danguang Pan [1,*], Wanying Zhang [2], Yankai Lu [2] and Guowei Ni [2]

[1] School of Civil and Resource Engineering, University of Science and Technology Beijing, Beijing 100083, China; liubo19840311@ncst.edu.cn

[2] School of Architectural Engineering, North China University of Science and Technology, Tangshan 063210, China; zwy_sep03@163.com (W.Z.); 17331755699@163.com (Y.L.); tingw@126.com (G.N.)

* Correspondence: pdg@ustb.edu.cn; Tel.: +86-13519157048

Abstract: Finite element analysis is one of the key steps in evaluating the effectiveness of an isolation system. The aim of this paper is to research the mechanical properties of sliding and rolling friction systems and their application for isolating frame structures. A hybrid seismic isolation system is proposed, which aims to protect the superstructure by reducing dynamic response. The effect of different parameters on the friction coefficient of the isolation bearing was tested using a compression-shear testing machine, and the results showed that various parameters significantly affected the friction coefficient. In addition, a numerical analysis finite element model was established using SAP2000 software to compare the seismic isolation performance of the hybrid isolation system with that of traditional seismic-resistant structures. The results showed that the composite isolation system not only effectively reduced inter-story shear and acceleration, but also slightly enhanced the seismic isolation effect as the friction coefficient decreased.

Keywords: friction coefficient; mechanical properties; isolation performance; response analysis

1. Introduction

Earthquakes can trigger high-intensity vibrations in a short time, producing tremendous amounts of energy that may result in structural damage or even destruction. To reduce the degree of damage to superstructures under seismic excitation, researchers have developed various types of isolation devices, including rubber, friction, and composite isolation bearings.

The most typical passive isolation device is the rubber isolation system (RIS). The RIS is composed of rubber and steel plates laminated together, which is widely used in practical projects because of the elasticity of rubber and the bearing capacity of steel plates with excellent isolation performance. Over the past two decades, researchers have conducted experiments and finite element studies on the stiffness, shear deformation, temperature, and hysteresis curve properties of RIS from multiple perspectives, e.g., [1–5]. Taking a steel frame structure with FRP rubber bearings as the research object, WU [6] obtained the ground vibration response of the base-isolated structure under different vertical loads, different structural height and width ratios, and different site factors by shaking table testing. Meanwhile, they established a finite element model of the seismic isolated structure with OpenSees to compare the ground vibration response analysis, ensuring the correctness of the results, which showed that FRP rubber isolation bearings can effectively reduce the influence of earthquakes on the superstructure. SUN [7] verified the influence of temperature on the seismic isolation effect by analyzing the correlation between temperature and the bearing performance and stiffness of the rubber bearing, providing a theoretical reference for the engineering application and development of rubber seismic isolation bearings.

Friction isolation bearings take friction as the damping force of the isolation layer and have good isolation performance, among which the friction pendulum has a bright development prospect because of its automatic reset capability. On this basis, researchers have developed various forms of friction pendulum bearings and conducted mechanical model derivations and experimental testing to promote the development of friction isolation bearings [8–10]. For example, Harvey [11] proposed a coupled two-way dynamic nonlinear model to numerically analyze the dynamic performance of horizontal two-way systems based on considering the rotation degree of freedom factor. Zhang [12] studied the effects of two types of vibration isolation bearings, ball and roller, on the rolling friction coefficient and analyzed the effects of the friction coefficient on the vibration isolation effect by modeling with SAP2000 software. The results showed that the seismic isolation performance is most effective when the rolling friction coefficient is within a certain range.

In recent years, composite isolation systems have become a research focus in the field of engineering earthquake resistance due to their advantages of gathering various types of seismic isolation devices and innovative combinations to improve seismic isolation performance to the maximum extent [13–19]. In terms of innovative combinations, Gao [20] studied the mechanical properties and seismic isolation performance of a disc spring-high damping rubber three-dimensional composite seismic isolation bearing. Numerical simulations were conducted to analyze the mechanical properties of the disc spring using ABAQUS finite element software, and test data were used to analyze the mechanical properties of the high-damping rubber seismic isolation bearing. Li [21] conducted horizontal and vertical seismic isolation performance tests on proposed disc-spring-stacked rubber three-dimensional composite seismic isolation bearings under different working conditions. In the horizontal direction, the hysteresis performance of the bearing under simple harmonic excitation was tested, and the effects of shear strain, vertical pressure, and loading frequency on its horizontal mechanical properties were investigated. In the vertical direction, the equivalent stiffness and equivalent damping ratio of the bearing were investigated, and the effects of loading amplitude, pressure, and loading frequency on its vertical hysteresis performance were investigated, as well as how changes in vertical stiffness affect the horizontal mechanical properties of the laminated rubber bearing. Hui developed a new type of variable-frequency rolling pendulum bearing consisting of a rolling bearing and an additional viscous damper to protect acceleration-sensitive equipment. The authors conducted comparative analyses using computer software and vibration table tests on different types of ground motion, including short-period (high-frequency), medium-frequency, and long-period (low-frequency) ground motion. They tested the effects of different parameters on device acceleration and relative displacement responses and the accuracy of the model.

With the advancement of science, technology, and engineering design, composite isolation bearings are constantly innovating and developing in the field of seismic resistance. In this study, a new type of composite seismic isolation system is proposed that is based on the flexibility of ball bearings and solves the limitations of traditional sliding isolation bearings. In order to analyze the dynamic friction performance of sliding and rolling, compression-shear tests were conducted. By controlling variables such as vertical pressure, horizontal velocity, number of pads, and number of rolling balls, the influence of these factors on the dynamic friction coefficient of the composite isolation bearing was studied. The obtained data can be used to analyze the structural response under earthquake excitation.

2. Design of the Isolation System

2.1. Description of the Designed Isolation System

The hybrid isolation system designed to isolate vibration comprises rolling friction and sliding friction. The physical diagram of the isolation device is shown in Figure 1, while its individual components are displayed in Figure 2. The bearing capacity is distributed by compressing the spring.

Figure 1. Physical diagram of the seismic isolation device.

Figure 2. Individual components.

When the lower surface of the upper plate of this composite vibration isolation bearing contacts the friction surface of the upper surface of the lower plate, the spring in the upper part of the ball roller group receives extrusion, as per Hooke's law:

$$f = -kx$$

where x is the compression of the spring, mm; k is the coefficient of elasticity of the spring, kN/mm; and f is the pressure of a single spring, KN.

Then, the upper load borne by the rolling ball group is

$$F = nf$$

where n is the number of springs in the upper part of the ball group.

The friction sliding surface of the upper load is

$$N = M - F$$

where M is the total upper structure load.

The distribution of the load-bearing capacity is controlled by the bearing by changing the number of gimbals and shims. Without balls, the system relies solely on sliding friction, which is equivalent to a planar sliding bearing. With the addition of balls, rolling friction and sliding friction work together to reduce structural damage by decreasing the upward transmission path of earthquakes.

2.2. Design of the Test Bearing

This experiment uses a micro-controlled hydroelectric servo compression shear tester and a combination of two isolation systems and an intermediate steel plate. The arrangement is a double shear combination, with the isolation system located in the middle, as shown in Figure 3.

Figure 3. Bearing design of the test.

This experiment studied the variation law of the friction coefficient of the composite seismic isolation bearing under different variables. The calculation of the friction coefficient was carried out using the following formula:

$$F = \mu N$$

To facilitate the determination of the tested friction coefficient, a pair of composite isolation bearings with two friction surfaces was used. The assembly of the isolation bearing is shown in Figure 4. Therefore, the friction coefficient of the composite isolation bearing is given by the test results obtained from this setup.

$$\mu = F_1/2N_1 \quad (1)$$

where μ is the friction coefficient of the composite seismic isolation bearing.

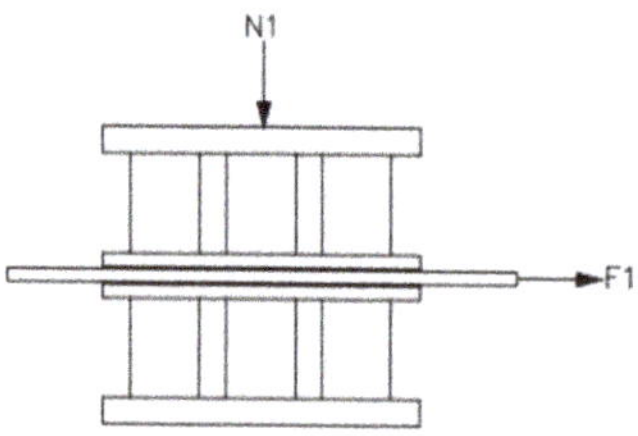

Figure 4. Sketch of the test device.

The upper connecting plate has a diameter of 250 mm and a thickness of 20 mm. Its lower surface is reserved by five bolt holes of diameter 20 mm and depth 7.5 mm. The middle column has a diameter of 54 mm and a total height of 92 mm, with a bolt column on the upper surface of 20 mm diameter and 7 mm depth, and two cylindrical grooves of diameter 45 mm and depth 32 mm, and diameter 32 mm and depth 43 mm on the lower surface. The lower connecting plate has a diameter of 250 mm and a thickness of 10 mm, with five cylindrical grooves of diameter 54 mm and depth 5 mm on its upper surface, and five cylindrical grooves of diameter 45 mm and depth 5 mm on its lower surface. The two cylindrical axes of the upper and lower grooves are aligned. Bearings will be equipped with different numbers of balls and shims according to the designed working conditions. To ensure the stability of the bearing, the number of balls is set to 1, 4, and 5. The materials of the parts of the bearings are different; the upper and lower connecting plates and intermediate columns are made of 45# steel; the shims, springs, and friction plates are made of 65 mn spring steel; the universal ball is made of turned heavy duty universal ball SP-25; the internal ball material is bearing steel; and the friction material is selected as PTFE.

2.3. Test Loading Procedure

A total of 75 operating conditions were established in this project, mainly focused on four parameters: vertical pressure, horizontal velocity, number of shims, and number of rolling balls. In the loading process, the vertical pressure is applied slowly at a loading rate of 0.04 MPa/s, and the application is stopped when the predetermined loading condition is reached. Then, the horizontal servo actuator drives the middle steel plate to provide horizontal force to the support friction plate with an action speed of 2 mm/s, 5 mm/s, and 10 mm/s, respectively, and ensures the horizontal uniform motion. The effect of different factors on the friction coefficient is calculated by the friction formula.

2.4. Test Results

2.4.1. Friction Performance Test under the Action of No Ball

When no ball is placed, the friction characteristics of the bearing under different pressures and speeds are investigated by a standard compression shear test, which is conducted under the action of vertical pressures of 39, 59, 102, and 204 KN and speeds of 2 mm/s, 5 mm/s, and 10 mm/s, respectively.

Figure 5 shows the relationship between the sliding speed and the sliding friction coefficient under different pressures. At a certain vertical pressure, the friction coefficient tends to decrease and then increase as the speed increases. It can be seen that the minimum sliding friction range is 0.04548–0.0516.

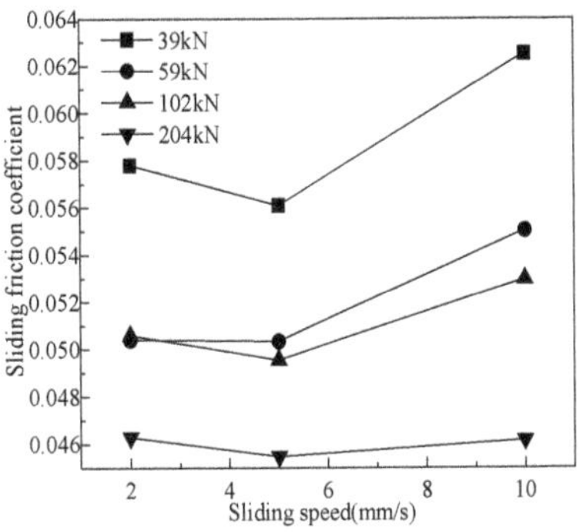

Figure 5. Relationship between sliding speed and pure sliding friction coefficient under different vertical pressures.

Figure 6 shows the relationship between the vertical pressure and the sliding friction coefficient at different velocities. The results show that at a certain speed, the friction coefficient shows a decreasing trend with an increase in vertical pressure. Obviously, the optimal friction parameters are 0.04548–0.04629, and the minimum friction coefficient is 0.04548.

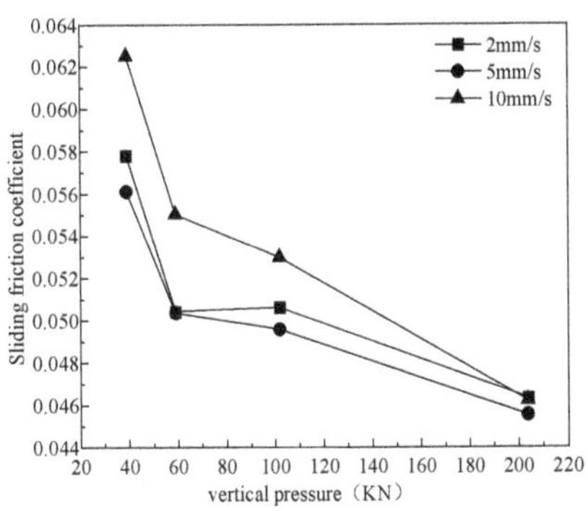

Figure 6. Relationship between vertical pressure and pure sliding friction coefficient under different sliding speeds.

2.4.2. Friction Performance Test under the Action of Balls

The ball in the system, combined with the spring and shim, plays a role. The main role played by the spring and shim is to adjust the force of the ball so that the total upper load is proportionally distributed to the ball group and PTFE.

This part of the study focuses on the effect of the bearing ratio on the dynamic friction coefficient. The effect of the number of shims and balls on the dynamic friction coefficient is shown in Figures 7 and 8. The relationship between the number of shims and friction coefficient for four-roller ball and five-roller ball isolation bearings with horizontal velocities of 2 mm/s, 4 mm/s, and 10 mm/s, respectively, under different vertical pressures.

In Figure 7a, when the horizontal speed is 2 mm/s, the friction coefficients under the action of three bearing ratios show different changes with an increase in the number of shims. When the ball group occupies 25%, the friction coefficient increases first and then decreases, and the minimum is 0.5516; when the ball group occupies 50%, the friction coefficient increases continuously, and the minimum is 0.05805; when the ball group occupies 100%, the friction coefficient decreases gradually, and the minimum is 0.06757.

In Figure 7b,c, when the horizontal velocity is 5 mm/s and 10 mm/s, the friction coefficient has a tendency to increase and then decrease with the increase in the number of shims, and the minimum friction range occurs when the number of shims is 2 and the pressure-bearing ratio is between 25% and 50%, which are 0.05514–0.05614 and 0.05759–0.05804, respectively.

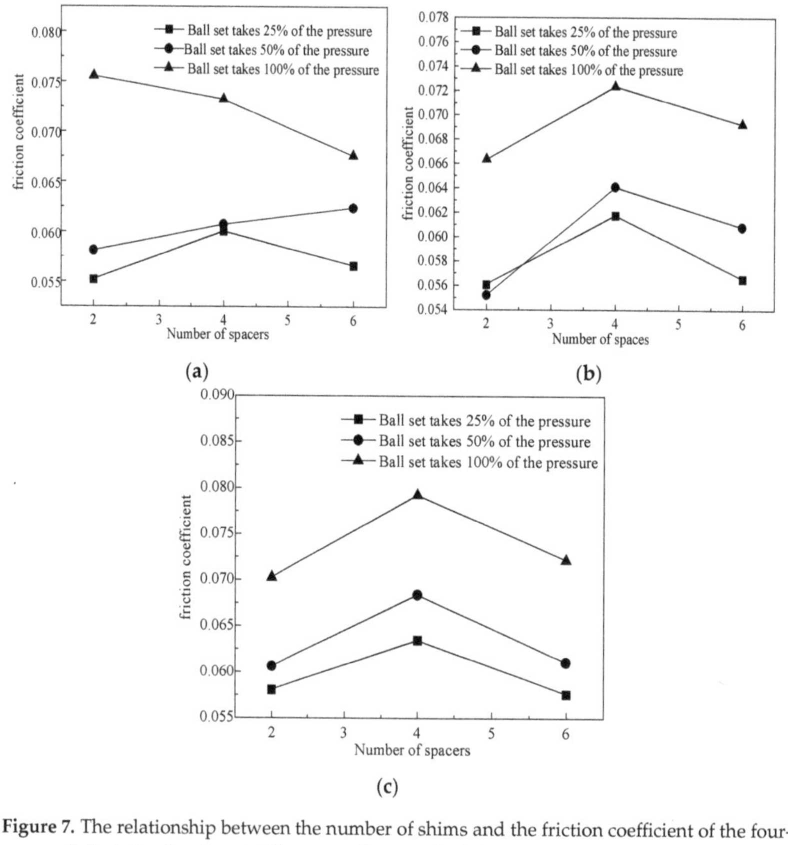

Figure 7. The relationship between the number of shims and the friction coefficient of the four-ball composite isolation bearing at different motion speeds: (**a**) 2 mm/s; (**b**) 5 mm/s; (**c**) 10 mm/s.

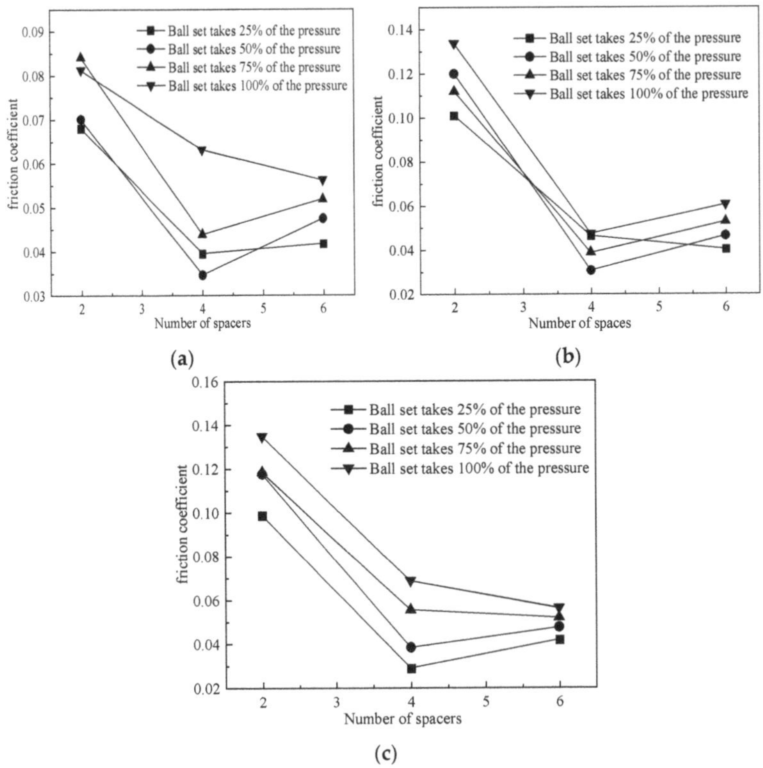

Figure 8. The relationship between the number of shims and the friction coefficient of the five-ball composite isolation bearing at different motion speeds: (**a**) 2 mm/s; (**b**) 5 mm/s; (**c**) 10 mm/s.

In Figure 8, at a certain horizontal velocity, the friction coefficients under the action of different bearing ratios showed a trend of decreasing and then increasing, and the minimum ranges, respectively, were 0.03458–0.03947, 0.03047–0.03869, and 0.02883–0.04169. Most of them were found when the number of shims was 4 and the pressure-bearing ratio was between 25 and 50%.

In Figures 7 and 8, comparing the data of a four-roller ball and a five-roller ball, it can be found that the coefficient of friction decreases with the increase in the number of balls, and there is a difference in the influence of the parameters of the minimum coefficient of friction; the four-roller ball mostly appears when the number of shims is 2, the five-roller ball mostly appears when the number of shims is 4, and the pressure-bearing ratio is 25–50%.

3. Numerical Analysis of a Based Isolated System and Conventional Structure

To evaluate the effectiveness of the isolation system, we used the finite element software SAP2000 to perform numerical simulations on a four-story reinforced concrete frame structure. Through finite element analysis, we compared the performance of the isolation system with isolated supports at the bottom and traditional seismic-resistant structures in terms of base shear, inter-story displacement, and floor acceleration, and evaluated the isolation effect based on these comparisons.

3.1. Structure Model

As shown in Figure 9, the superstructure of the isolation bearing is a reinforced concrete structure. According to the engineering design requirements, the constant load of the structure, which includes the self-weight of the structure and the uniform load of the floor slab, is 3 kN/m^2, the live load is 2 kN/m^2, the total height of the building is 14.4 m, the floor height is 3.6 m, and the floor size is 12 m $\times$ 30 m. The length and width of the column section are 600 mm, the beam section length and width are 600 mm and 300 mm, respectively, and the thickness of the slab is 120 mm. The structural plan is shown in Figure 9, with longitudinal reinforcement of HRB335 and hoop reinforcement of HPB300 to meet the minimum reinforcement rate of the structure. The seismic isolation bearings are placed on the lower surface of each column and connected to the foundation. In SAP2000, beams and columns are modeled using frame elements. Non-bottom nodes are constrained using diaphragms, while the bottom nodes of non-seismic structures are all fixed support nodes. The bottom nodes of composite isolation systems use the Friction Isolator element in the Connection/Support Property with an infinite radius setting. Figure 10 shows a three-dimensional simulation model for a non-seismic structure.

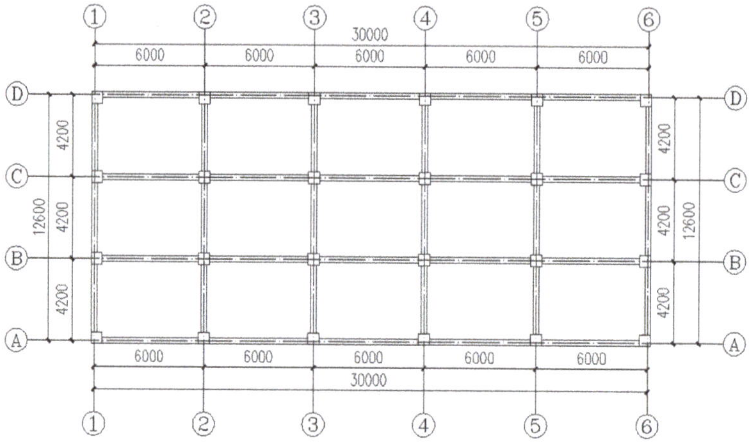

Figure 9. Structure plan.

Figure 10. Seismic isolation finite element model diagram.

3.2. Test Setup

Three seismic records were selected from the Pacific seismic station network: El Centro, Kobe, with two strong seismic records, and Shanghai artificial waves, which were used as the input excitation signals. The peak ground accelerations of 200 gal and 400 gal were selected to correspond to the seismic intensities specified in the China Code at the 8-degree preparedness level (the seismic intensity level considered in the design process). These acceleration response spectra are shown in Figure 11.

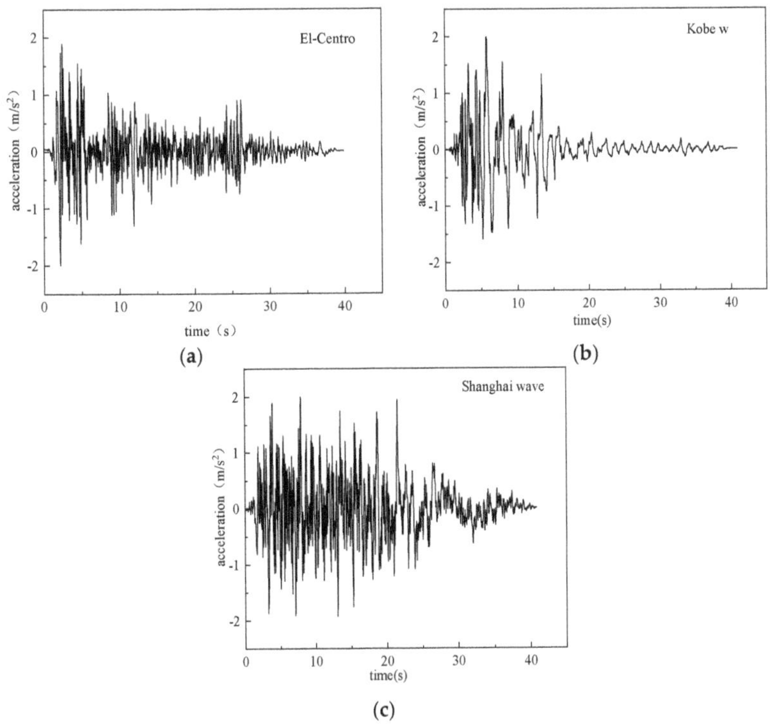

Figure 11. Seismic waveform diagram. (**a**) EI-Centro wave; (**b**) Kobe wave; (**c**) Shanghai wave.

3.3. *Model Results Analysis*

3.3.1. Basement Shear

Based on the friction coefficient of 0.05 obtained from previous tests, Figure 12 shows the base shear curves of the composite isolation structure and the seismic-resistant structure under the same working conditions. It can be seen from the figure that the base shear of the structure varies under different earthquake waves and increases with an increase in seismic intensity.

Compared with traditional seismic-resistant structures, the composite isolation structure exhibits good isolation effects. Regardless of the input ground motion, the base isolation model can significantly reduce the structure's shear force.

3.3.2. Floor Acceleration

Through sliding friction tests without rolling balls as well as compound friction tests with four and five rolling balls, friction coefficients of 0.04 and 0.05 were obtained. Based on Table 1 and Figure 13, acceleration response spectra for the isolation structure and seismic-resistant structure with friction coefficients of 0.04 and 0.05 under three types of earthquake waves in the X direction were obtained. The friction coefficient of the isolation layer under different earthquake waves and intensities can affect the acceleration response of each floor. However, compared to the seismic-resistant structure, the isolation structure can achieve good isolation effects under various working conditions. Overall, the Shanghai artificial wave has the greatest impact on the acceleration response of the structure, followed by the Kobe wave and the El Centro wave.

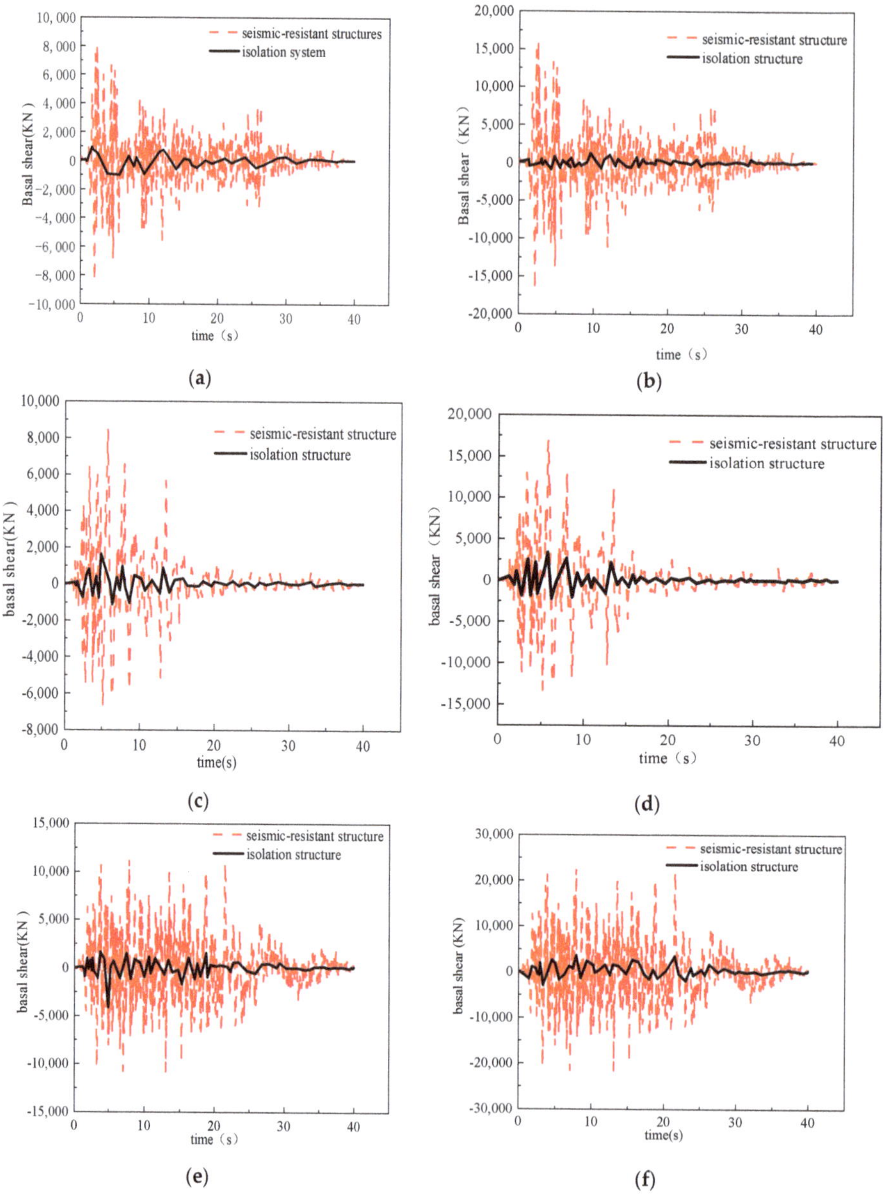

Figure 12. Time curve of basal shear force of seismically isolated structures and seismically resistant structures. (**a**) El-Centro wave 200 gal; (**b**) El-Centro wave 400 gal; (**c**) Kobe wave 200 gal; (**d**) Kobe wave 400 gal; (**e**) Shanghai wave 200 gal; (**f**) Shanghai wave 400 gal.

Table 1. Inter-story displacement of based isolated structures and conventional structures.

Friction Coefficient	X Acceleration/gal	Floor	El Centro Wave Convention m/s^2	El Centro Wave Based Isolated m/s^2	Kobe Wave Convention m/s^2	Kobe Wave Based Isolated m/s^2	Shanghai Wave Convention m/s^2	Shanghai Wave Based Isolated m/s^2
0.04	200	1	2.12	0.99	2.81	1.02	3.41	1.79
		2	2.62	0.82	2.89	0.93	3.82	1.51
		3	3.25	0.76	3.51	0.98	4.23	1.28
		4	3.48	0.84	4.22	0.84	4.96	1.37
	400	1	7.63	1.81	6.39	1.92	7.03	2.68
		2	8.41	1.42	7.84	1.86	8.42	2.16
		3	9.29	1.21	9.01	1.58	9.11	2.08
		4	10.54	1.34	11.46	1.49	10.96	2.43
0.05	200	1	2.43	1.12	3.11	1.37	3.88	1.97
		2	3.06	1.04	3.95	1.22	4.21	1.69
		3	3.94	0.86	4.26	1.02	4.86	1.55
		4	4.71	0.95	5.02	1.21	5.32	1.72
	400	1	9.11	2.83	9.82	3.84	7.23	3.25
		2	10.93	2.42	9.46	3.19	9.46	3.12
		3	11.16	2.15	11.59	2.93	12.77	2.97
		4	14.83	2.36	15.23	3.54	16.84	2.51

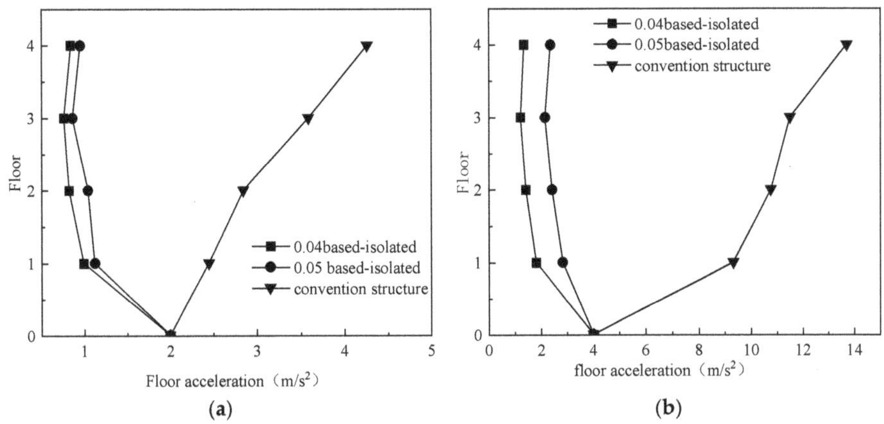

Figure 13. Comparison Diagram of Floor Acceleration under El Centro wave. (**a**) El Centro wave 200 gal; (**b**) El Centro wave 400 gal.

In addition, under the same working conditions, there is generally a positive correlation between the friction coefficient and the acceleration response of the structure. As the friction coefficient increases, the acceleration of the structure also increases correspondingly, but the increase in amplitude is not very significant. This indicates that increasing the friction coefficient of the isolation layer will have a certain impact on the seismic response of the overall structure, but the degree of influence is relatively small.

3.3.3. Inter-Story Displacement

Through sliding friction tests without rolling balls as well as compound friction tests with four and five rolling balls, friction coefficients of 0.04 and 0.05 were obtained. Using three types of earthquake waves to analyze the interlayer shear force in the X direction, interlayer displacement data for the isolation structure were obtained, as shown in Table 2. It can be seen that the El Centro wave, Kobe wave, and Shanghai artificial wave have different degrees of influence on the interlayer displacement of the structure. Among them, the Shanghai artificial wave has the greatest influence on the interlayer displacement of the structure, followed by the Kobe wave, while the El Centro wave has a relatively small influence. In addition, at an earthquake intensity of 200 gal, as the friction coefficient increases, the interlayer displacement of the first, second, third, and fourth layers of the structure will increase, but the interlayer displacement of the isolation layer will decrease. When the earthquake intensity is 400 gal, the interlayer displacement of the first, second, and third layers of the structure will also increase with the increase in the friction coefficient, but the interlayer displacement of the isolation layer will decrease with the increase in the friction coefficient. Generally speaking, the greater the earthquake intensity, the larger the corresponding structure's interlayer displacement. In addition, as the number of floors increases, the interlayer displacement gradually decreases.

Table 2. Comparison of displacement of seismically isolated structures under different seismic waves.

Friction Coefficient	X Acceleration/gal	Floor	Inter-Story/mm		
			El Centro	Kobe	Shanghai Wave
0.04	200	isolated	27.4	34.1	56.4
		1	15.1	18.4	33.9
		2	12.4	15.3	26.9
		3	10.4	13.2	21.5
		4	9.1	11	16.8
	400	isolated	98.1	118.1	153.7
		1	50.2	55.6	63.8
		2	32.8	42	49.6
		3	25.4	31.8	39.2
		4	16.2	18.3	21.8
0.05	200	isolated	25.3	31.4	57.2
		1	16.1	20.2	35.4
		2	11.8	15.9	28.6
		3	10.4	14.1	23.7
		4	9.3	12.8	17.2
	400	isolated	90.2	109.4	149.5
		1	45.6	57.9	68.6
		2	36.2	42.3	48.4
		3	27.5	34.5	36.2
		4	18.4	20.6	21.8

4. Conclusions

This article introduces a new type of isolation system that takes into account the potential impact of parameters, such as the number of spheres, on the friction coefficient. Based on this, numerical simulations were conducted for both the base isolation structure and traditional seismic-resistant structures under earthquake excitation to determine their dynamic characteristics.

Firstly, in the experimental test data, we found that the number of spheres and pads has a certain influence on the friction coefficient. Without spheres, the minimum friction coefficient occurs when the horizontal velocity is 5 mm/s and the vertical pressure is 204 KN. With the involvement of spheres, there is little difference in the friction coefficient range between five and four spheres. In addition, the minimum friction coefficient appears in the bearing ratio of 25–50%, indicating that the number of spheres and pads can reduce the friction coefficient within a certain range. These results indicate that the number of spheres and pads is an important factor affecting the friction coefficient of the isolation system.

Secondly, after conducting numerical simulations for the base isolation structure and traditional seismic-resistant structures under earthquake excitation, we found that the new composite isolation system can effectively reduce the dynamic response of the model compared to traditional seismic-resistant structures. This indicates that the new isolation system has good seismic performance.

Finally, it should be pointed out that this experiment mainly studied the influence of parameters such as the number of balls on the friction coefficient of the isolation support. However, in order to better apply the new isolation support in practical engineering, we still need to further study the shear performance of the isolation support using ABAQUS finite element software and explore its applicability under different parameter conditions. These research results will help provide a theoretical basis for the practical application of the new isolation support.

Author Contributions: Conceptualization, B.L. and D.P.; validation, B.L.; formal analysis, B.L.; investigation, B.L.; resources, B.L.; data curation, W.Z.; writing—original draft preparation and writing, W.Z.; review and editing, D.P.; visualization, Y.L.; supervision, G.N.; project administration, G.N.; funding acquisition, B.L. All authors have read and agreed to the published version of the manuscript.

Funding: This research received no external funding.

Data Availability Statement: The data can be obtained from the author by contacting them at liubo19840311@ncst.edu.cn.

Acknowledgments: We extend our appreciation to all individuals who participated in this study and provided us with their time and cooperation.

Conflicts of Interest: The authors declare no conflict of interest.

References

1. CNIPA. A New type of Friction damping isolated Rubber Bearing. *Rubber Sci. Technol.* **2022**, *20*, 515. (In Chinese)
2. Zhang, Q.; Sang, G.R.; Li, S.S. Analysis on Damping effect of laminated rubber cushion foundation isolation Structure. *Jiangxi Build. Mater.* **2022**, 290–291. (In Chinese)
3. Huang, Z.S. *Study on the Stability of Three Dimensional Vibration Isolation Rubber Bearing*; Guangzhou University: Guangzhou, China, 2022. (In Chinese)
4. Fiore, A.; Marano, G.C.; Natale, M.G. Theoretical prediction of the dynamic behavior of rolling-ball rubber-layer isolation systems. *Struct. Control Health Monit.* **2016**, *23*, 1150–1167. [CrossRef]
5. Cao, S.; Ozbulut, O.E.; Wu, S. Multi-level SMA/lead rubber bearing isolation system for seismic protection of bridges. *Smart Mater. Struct.* **2020**, *29*, 055045.
6. Sun, M.M.; Li, Z.W.; Wang, C.S. Review on the influence of temperature on rubber isolation bearing. *Build. Saf.* **2020**, *35*, 75–78. (In Chinese)
7. Ma, Z.X.; Wen, W.L.; Guan, Q.S.; Yu, W.Z.; Pan, W. Research on equivalent Linearization parameters and isolation design method of Friction pendulum Bearing. *J. Disaster Prev. Mitig. Eng.* **2022**, *42*, 751–760. (In Chinese)
8. Calhoun, S.J.; Tehrani, M.H.; Harvey, P.S. On the performance of double rolling isolation systems. *J. Sound Vib.* **2019**, *449*, 330–348. [CrossRef]
9. Jampole, E.; Eeri, S.; Deierlein, G. Full scale dynamic testing of a sliding seismically isolated unibody house. *Earthq. Spectra* **2016**, 3, 16. [CrossRef]
10. Harvey, P.S. Vertical Accelerations in Rolling Isolation Systems: Experiments and Simulations. *Eng. Struct.* **2016**, *142*, 04015091. [CrossRef]
11. Zhang, H. *Experimental Study on Friction Performance of Rolling Isolation Bearing*; North China University of Science and Technology: Qinhuangdao, China, 2018.
12. Zhang, H.; Liang, X.; Gao, Z.; Zhu, X. Seismic performance analysis of a large-scale single-layer lattice dome with a hybrid three-directional seismic isolation system. *Eng. Struct.* **2020**, *214*, 110627. [CrossRef]
13. Pourmasoud, M.M.; Lim, J.B.P.; Hajirasouliha, I. Multi-Directional Base Isolation System for Coupled Horizontal and Vertical Seismic Excitations. *J. Earthq. Eng.* **2020**, *26*, 1145–1170. [CrossRef]
14. Jing, X.G. *Nonlinear Mechanical Properties Analysis and Experimental Study of Series Isolation Components*; Hebei University of Engineering: Handan, China, 2021. (In Chinese)
15. Gao, J.; Xue, Y.T.; Xiao, C.Z.; Zhou, X.Y.; Han, X. Experimental study on Series laminated Rubber bearing with variable Stiffness. *Build. Struct.* **2020**, *50*, 109–113. (In Chinese)
16. Wu, Q.; Yan, H.; Zhu, H.; Bai, X. Shaking table test study on the seismic isolation effect of a hybrid passive control System. *Measurement* **2020**, *164*, 108125. [CrossRef]
17. Yuan, Y.; Zhou, A.H.; Yang, S.B.; He, G.F. Effect of Friction Pressure Ratio on Isolation Performance of Parallel Composite Isolation System. *Ind. Archit.* **2014**, *44*, 24–28+153. (In Chinese)
18. Wu, D.; Li, J.J.; Tan, P.; Xiong, Y. Seismic vulnerability analysis of series isolation structural system. *Eng. Mech.* **2017**, *34*, 227–232. (In Chinese)

19. Wei, B.; Wang, P.; Liu, W.; Yang, M. The impact of the concave distribution of rolling friction coefficient on the seismic isolation performance of a spring-rolling system. *Int. J. Non-Linear Mech.* **2016**, *83*, 65–77. [CrossRef]
20. Gao, J.Y. *Research on Disc Spring-High Damping Rubber Isolation Bearing*; Beijing University of Technology: Beijing, China, 2018. (In Chinese)
21. Li, X.Y.; Liang, S.Z.; Xue, S.D.; Gao, J.Y. Experimental study on mechanical properties of disc-spring-stacked rubber three-dimensional composite vibration isolation bearings. *Build. Struct.* **2021**, *51*, 14–20+104. (In Chinese)

applied sciences

MDPI

Review

Research Progress on the Influence of Thermo-Chemical Effects on the Swelling Pressure of Bentonite

Jinjin Liu, Chuanqin Yao *, Wenbo Su and Yizhe Zhao

School of Civil Engineering, Shanghai Normal University, Shanghai 201418, China; yy21_anhuibbu@163.com (J.L.)
* Correspondence: cqyao@shnu.edu.cn

Abstract: The swelling pressure of bentonite changes dramatically due to diffused nuclear radiation heat and underground osmosis, causing the failure of the buffer isolation layer in deep geological repositories for the disposal of high-level radioactive waste. A detailed overview of the relevant research results on the swelling pressure variation of bentonite under thermo-chemical effects is presented in this paper. The results showed that the values of the swelling pressure obtained by different test methods are dissimilar. The swelling pressure of bentonite decreased with the increasing pore solution concentration; nevertheless, the effect of temperature on the swelling pressure is still controversial. At the micro-level, crystal layer swelling and double- layer swelling are generally considered to be the main factors affecting the swelling pressure; the pore structure and water distribution of bentonite will change owing to thermo-chemical effects. At the macro-level, involving intergranular stress, a mechanical parameter was proposed to explain the mechanism of the changes in the swelling pressure of bentonite. Finally, future research directions for the study of the evolution of bentonite swelling properties under thermo-chemical effects are proposed, based on the current research results.

Keywords: thermo-chemical; swelling pressure; test methods; microscopic properties; intergranular stress

1. Introduction

With the rapid evolution of nuclear energy, a large amount of high-level radioactive waste (HLW) has been produced [1]. Disposing of this high-level radioactive waste safely, in an environmentally friendly way, and effectively for a long time has become a major problem that requires urgent solutions worldwide. At present, the most effective and feasible method of disposal of HLW is deep geological disposal, which requires choosing a suitable site (crust with poor water content, good stability, far from human activity zones, etc.) to set up an engineering barrier, buffering and sealing the tanks containing HLW in a deep geological repository at a depth of 500–1000 m from the surface to prevent the waste from being transported to the biosphere in case of a spill [2]. Bentonite, which contains a large amount of montmorillonite, is internationally recognized as a buffer/backfill material for the deep geological disposal of HLW due to its characteristics of low permeability, large swelling, strong adsorption, and good thermal conductivity [3,4]. Furthermore, bentonite not only has the important characteristics of buffering and backfilling, but also has the advantages of low cost, wide distribution, easy mining, and easy processing; so, it has become the preferred buffering backfilling material around the world. In China, Gaomiaozi bentonite produced in Xinghe County, Inner Mongolia, was selected preliminarily as the base material for buffering/backfill material [5,6].

Bentonite is in an unsaturated state after the construction of deep geological repositories. As the groundwater level rises, the compacted bentonite absorbs water (solution) undergoing an expansive deformation that can tightly enclose the waste tank and fill the gap between the buffer material and the surrounding rock mass, fully blocking the surrounding rock and the waste tank. In the meantime, a sufficiently large swelling pressure under almost constant volume conditions can improve the sealing efficiency of bentonite [7–9], which is the main governing factor for the stability of subsurface structures. However, the near-field chemical conditions are complex during the operation of an HLW repository [10], as bentonite draws water from the surrounding rock and becomes gradually saturated due to subsurface infiltration. This will result in changes in the chemical concentration of bentonite pore water. Moreover, the temperature in the repository will be changed on account of the diffuse nuclear radiation heat. In the long term, problems such as the loss of bentonite's swelling properties may arise. Therefore, the study of bentonite's swelling performance under coupled chemical and temperature gradients field conditions is a crucial issue for the safe operation of nuclear waste repositories.

The swelling characteristics of compacted bentonite have been extensively investigated by scholars in the past decades. Focusing on this theme, this paper systematically describes the progress and limitations of the studies on bentonite swelling pressure under the effect of thermochemistry, with regard to the experimental test methods, the law and mechanism of swelling pressure variations under the influence of pore solution concentration and temperature, etc. On this basis, an outlook of future research developments to understand the influence of thermo-chemical effects on bentonite swelling pressure is provided.

2. Test Methods for Swelling Pressure

The swelling pressure can be measured by different methods, but different test methods produce different swelling pressure values [9,11,12]. It should be emphasized that it is ideal to think of the swelling pressure as the pressure measured when bentonite is wetted under the condition of limited volume [13]. The constant volume method, constant load method, and paired swell test (PS-test) were proposed to measure the swelling pressure [14].

The constant volume method for measuring the swelling pressure is based on strain control techniques [15,16] including the zero-swell test and the direct method. The main difference between the two approaches is that the latter measures the swelling pressure directly using a pressure transducer, such as the S-type load cell shown in Figure 1. The zero-swell method prevents the specimen from expanding by applying a vertical stress, which is generally around the year 2000 [11,17]. It is challenging to guarantee optimal swelling readings and constant volume conditions during the test, and the soil pressure is very sensitive to load increments during the test [9,18], especially in the zero-swell test [14]. Furthermore, the swelling pressure will be overestimated due to the friction between the specimen and the consolidator ring during the recovery of the specimen to the initial volume [9]. In spite of this, the value measured with the zero-swell method is only one-third of the value measured with the swelling reloading test [9,17]. The reason is that the soil particles in the swelling reloading test are sufficiently hydrated, especially under low load conditions, whereas soil particle swelling is restricted and hydration is not sufficient in the constant volume method. Large errors with the direct method are due to the device's characteristics, including the stiffness of the sensor [19,20], and the corresponding swelling pressure error can reach a maximum of 1–2 MPa [9].

The constant load test methods include the swell-consolidation test and the double consolidation test. The swell-consolidation method involves compressing the specimen after free swelling for 24 h until it returns to its initial state of porosity or height, which is typical of the constant load method [14]. The swelling pressure is the pressure necessary for the sample to return to its initial height [11], similarly to the swelling pressure of the free-swelling method. The difference between the swell-consolidation method and the swell-reload method is that the latter applies compression to the specimen in steps, with an interval of 24 h between each step [9], which ensures the full swelling of the specimen and also increases related errors such as friction; the swelling pressure is the vertical stress required to restore the specimen to its original state. The "extra" macroscopic void fraction generated during the initial swelling at low pressure results in higher swelling pressure measured by the swell-consolidation method compared to the constant volume method [21]. The double consolidation test was performed on two samples, one of which was wetted, swollen and completely submerged, and its maximum expansion was recorded after 24 h of expansion, while the other was compressed at natural moisture content, and the amount of compression was equal to the swelling of the wetted sample; the vertical stress was defined as the swelling pressure [14]. Apparently, the constant load method is constrained by the free-swelling method [14]. The values measured by the double consolidation test are significantly higher than the values obtained by the free swelling test [22].

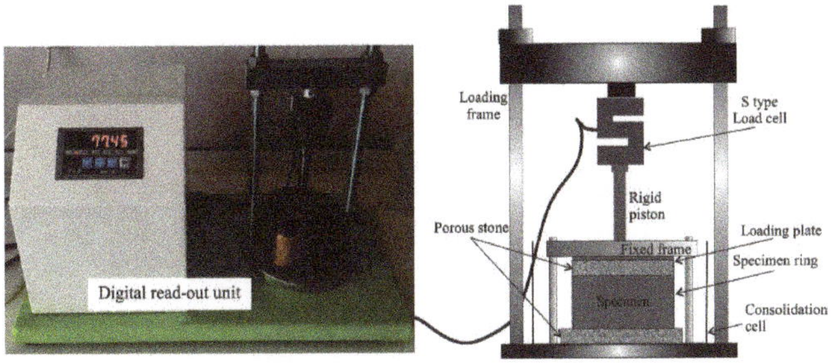

Figure 1. Schematic view of the experimental setup using a constant-volume cell, according to [14].

The PS-test is a simple, reasonable, and reliable swelling test based on the improvement of the loading of the tested specimen; its validity has also been confirmed by other numerous tests. Its most important advantages are that it does not require a complicated setup or design, uses conventional consolidation equipment, and does not need the provision of constant volume conditions. On the other hand, the disadvantage is that two samples are needed at the same time. The principle is that the response of one specimen is used to refine the load step of the second specimen and that a small incremental load-controlled swelling does not cause excessive water absorption in the soil specimens [14].

Many optimization test schemes were proposed according to the different properties of the above-mentioned test methods. For example, as early as in 1975, Bishop et al. [23] proposed a triaxial instrument that could directly measure the vertical swelling of a sample in a hydraulic triaxial stress path cell; it could predict the swelling potential of the soil under test better than the consolidation test method [11]. A new constant-volume sensor, developed by Tang et al. [9], could measure the swelling pressure without both any strain adjustment and the influence of test equipment stiffness, thereby reducing the measurement errors (Figure 2). To reduce diaphragm deformation under external pressure, a mercury-filled pore gap is introduced below the diaphragm in the BER-A-58S pressure sensor utilized in the upper portion. The upper-pressure transducer is in contact with the upper surface of the sample for the direct measurement of the swelling pressure. Figure 3 shows the diagram equipped with this pressure transducer used with the constant volume method without manual load adjustment, which makes it very convenient for monitoring the long-term changes in swelling pressure.

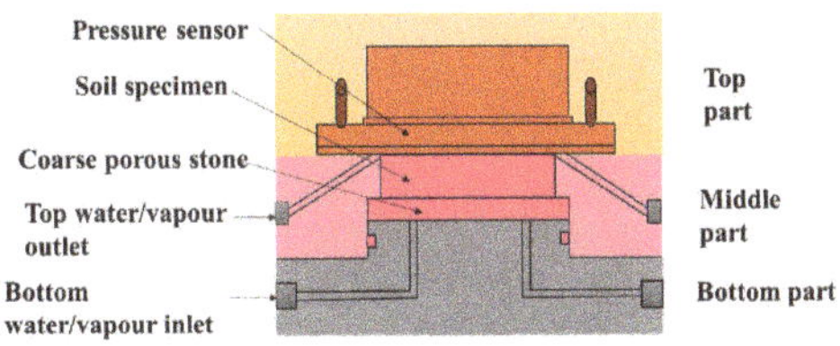

Figure 2. Schematic view of the developed constant-volume cell [9].

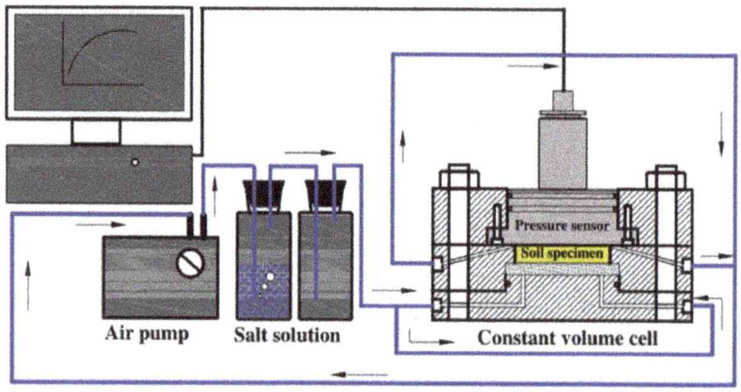

Figure 3. Schematic view of the experimental setup using the constant-volume cell [9].

Accuracy, directness, simplicity, and long-term monitoring have become important criteria for selecting a test method. In addition, the different loading and wetting conditions used by a method will also cause the achieved swelling pressure to behave differently [11]. For instance, at constant volume, the swelling pressure is surprisingly reduced after wetting in soil subjected to a high initial vertical stress [9].

In recent years, microstructural investigations of the bentonite swelling pressure have clarified the causes of swelling pressure variation in relation to the test method. It was shown that similar swelling pressures obtained by different tests depended on their similar micro- or macro-porosity. The swelling pressure will show a good consistency, independent of the test method, when the samples are wetted under a sufficiently high vertical stress [21]. In general, the constant volume test is still the most commonly used method in current research [24–26].

3. Influence of Thermo-Chemical Effect on the Swelling Pressure

This review mainly discusses the changes of the swelling pressure from the perspective of crystal layer swelling and double electric layer swelling in the presence of salt solutions and in different temperature conditions, almost ignoring the influence of mineral alteration during this process.

Generally, the crystalline-layer swelling mechanism involving the internal minerals and the diffusive double-layer swelling mechanism involving the mineral particles are considered to be the main swelling mechanisms of bentonite [27] and are commonly considered to analyze the changes in swelling pressure in complex environments [28]. As shown in Figure 4, the wedging of water molecules into the crystalline layers leads to the crystalline layer swelling. The swelling changes from crystalline-layer swelling to double-electric-layer swelling when the number of water layers in the crystalline layers reaches a certain value [27,29,30]. Both swelling mechanisms are referred to as interlayer swelling. Interlayer swelling above 22 Å is associated with double-layer swelling, and interlayer swelling between 10 and 22 Å is associated with crystalline layer swelling caused by the adsorption of water molecules and is determined by the layer charge of clay minerals, exchange cations, and the attraction between water molecules and polar surface groups [31–34]. The crystalline layer interval and the thickness of the diffusion double layer are essential factors in the study of the magnitude of the swelling pressure, and changes in temperature and pore solution can lead to changes in both, thus affecting the magnitude of the swelling pressure.

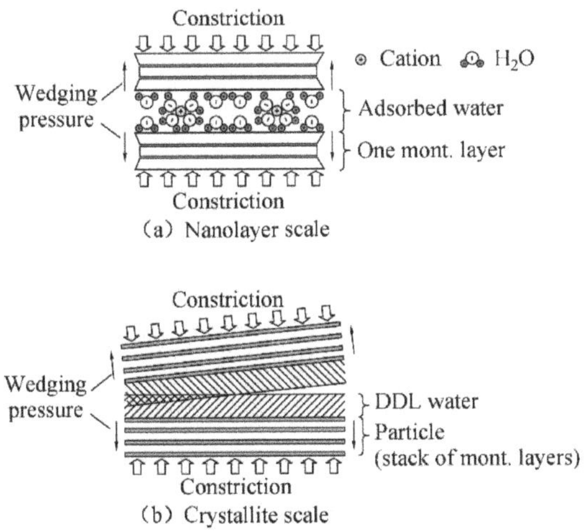

Figure 4. Formation of the wedging pressure at (**a**) a nanolayer scale and (**b**) a crystallite scale [35].

3.1. Effect of the Pore Salt Solution Concentration on the Swelling Pressure

Studies have shown that the increase of the pore solution concentration leads to a decrease in bentonite expansibility [3,36,37]. Different types of solutions have different effects on the swelling pressure of bentonite.

The concrete material in an HLW repository is gradually degraded during the interaction with pore water producing an alkaline solution, while montmorillonite, the main mineral component of bentonite, is dissolved gradually by the alkaline solution [38,39].

As shown in Figure 5, GMZ01 bentonite with a montmorillonite content up to 75.4% was tested for its constant-volume swelling pressure in the presence of different saline solution concentrations and types of infiltration, while maintaining the temperature at 20 ± 1 °C [26]. According to the classical double- layer theory, increasing the solution concentration will inhibit the swelling of the double layer [40], and the thickness of the diffusion double layer is inversely proportional to the square root of Na^+ concentration [16]. Therefore, high-concentration saline solutions will reduce the repulsive force between the particles, leading to a decrease in the swelling pressure. It is obvious that the ion types, Ca^{2+} and Na^+, had similar effects on the swelling pressure at low concentrations, but when the concentrations were greater than 0.5 M, the inhibition of the swelling pressure of bentonite by Ca^{2+} was not so obvious compared to that induced by Na^+. A greater base spacing of Ca-based montmorillonite under constant volume conditions was reported [41], and Chen et al. [42] suggested that Ca^{2+} replacement changed the pore fabric characteristics, leading to a greater crystalline-layer swelling deformation. However, with regard to the diffusive double-layer swelling, the Gouy-Chapman model [43,44] suggested that the higher the electrolyte valence type and the electrolyte concentration, the smaller the double-layer thickness.

Several researchers attribute the greater swelling pressure in the presence of Ca^{2+} to the fact that its action produces larger and thicker particles [45], and the bilayer effect is not so pronounced; so, bentonite has greater mechanical strength and greater swelling pressure.

As shown in Figure 6, where the effect of temperature was ignored, the constant-volume swelling pressure test was performed on GMZ Na–bentonite with 75.4% bentonite, initial water content of 11.92%, dry density of 1.7 Mg/m^3; the swelling pressure curve showed three phases [24]. Analogously, when the dry density was high, 1.7 Mg/m^3, and the initial water content was relatively low, i.e., 10.76%, the swelling process under the action of the saline solution comprised three stages: firstly, crystalline swelling, followed by the breakup of quasicrystals, and then double-layer swelling [26]. The first stage was characterized by a rapid increase in the swelling pressure, marked by increased interlayer cation hydration, and as the cation concentration continued to increase, water molecules within the crystals left the crystal layer, which gradually inhibited the swelling of the crystal layer [46]. The second stage was the transition from crystalline to permeable swelling, with a small decrease in the swelling pressure due to the swelling of soil particle agglomerates and the collapse of the soil skeleton, which was manifested by the splitting of thick quasicrystals into fine quasicrystals and the filling of large pores (inter-agglomerate pores). The third stage was the double-layer swelling stage, where the swelling pressure increased again, and the crystalline swelling became less pronounced; for specimens infiltrated with the salt solution or distilled water, the final swelling pressure in this stage was maintained in equilibrium [47].

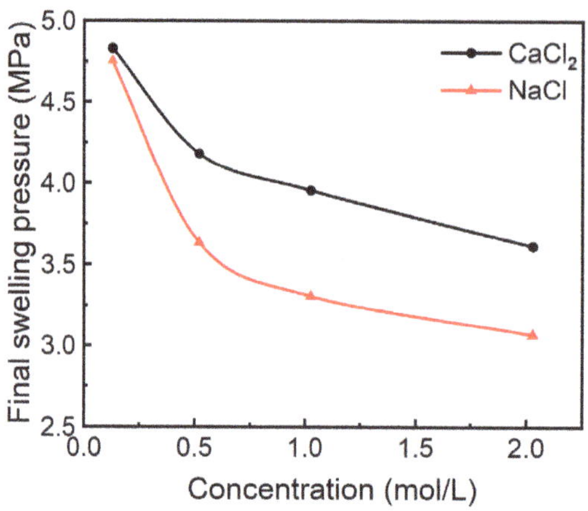

Figure 5. Comparison of the influence of the cation type on the swelling pressure of GMZ01 bentonite [26].

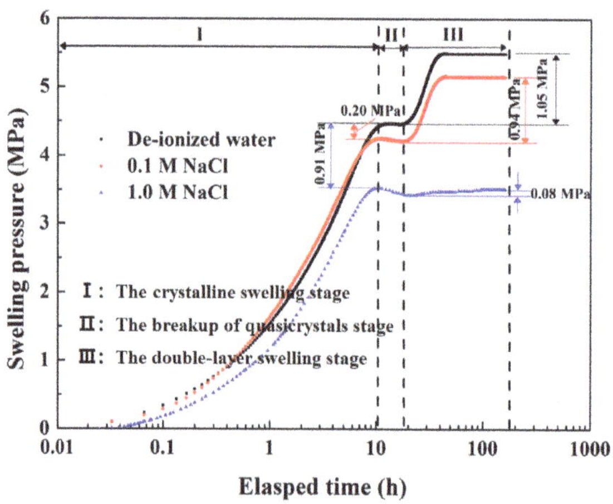

Figure 6. Effect of a salt solution at different concentration on the swelling pressure [24].

Zhang et al. [48] proposed to distinguish the two swelling mechanisms by the pore ratio. Similarly, ignoring the temperature change, the swelling pressure of GMZ01 bentonite with 75% montmorillonite content was investigated, and the combined effect of double-layer swelling and crystal-layer swelling was distinguished from the swelling caused by the crystal-layer swelling effect with a critical pore ratio of 0.41. In the case of low porosity (less than 0.41), a diffusion double layer could not form [49], and the swelling pressure increased with the increasing concentration but was not particularly pronounced. Thus, the swelling pressure could be explained by the theory of crystalline-layer swelling with increasing cation valence, on the basis of which, Ca^{2+} will lead to a greater crystalline-layer swelling compared to Na^{+} [42]. The swelling pressure increased with the increasing pore solution concentration for pore ratios less than 0.5 [50]. Combined with the above analysis, this stage was considered to be the crystal-layer swelling control stage, and the swelling pressure was almost generated by the swelling between the crystal layers.

The above study investigated the variation of the swelling pressure from the perspective of the pore ratio under the action of two swelling mechanisms, while some studies further analyzed the reasons for the variation of the swelling pressure from the perspective of the variation of soil–water interfacial forces. For illustration purposes, the concepts of accumulated wedging pressure (AWP) and dissipated wedging pressure (DWP) were used to determine the development of the swelling pressure in terms of the competition between the two pressures [35].

The wedging pressure results from the wedging of water molecules into the interlayer of montmorillonite, interparticle swelling, and the competition between the dissociation of the initially accumulated montmorillonite layer and aggregated particles, due to the continuous entry of water molecules and the deformation, damage, and dislocation of the bentonite block, leading to pressure dissipation. It manifests as small pore formation and large pore collapse when the latter process predominate [51]. The concept also links the flatness of the pore structure to the ultimate stabilization of the swelling pressure, the essence of which remains inseparable from the swelling mechanism of the mineral crystal layer and the interparticle double layer.

3.2. *Effect of Temperature on the Swelling Pressure*

Elevated temperatures can have an impact on bentonite and its engineering properties. For example, simulations of high-discharge waste disposal in China have shown that thermal swelling and saturation processes may affect a tank stability for long periods [52], as higher temperatures can provoke montmorillonite dehydration [53]. The effect of temperature on the swelling pressure is highly controversial. Liu et al. [25] used distilled water and performed constant-volume swelling pressure tests on GMZ bentonite particle mixtures at different temperatures. It was shown that the swelling pressure changed faster and gradually increased with higher temperatures (Figure 7). A similar phenomenon was also indicated by some related studies [5,54]. Nevertheless, other tests also showed that the swelling pressure decreased with the increasing temperature [55–58].

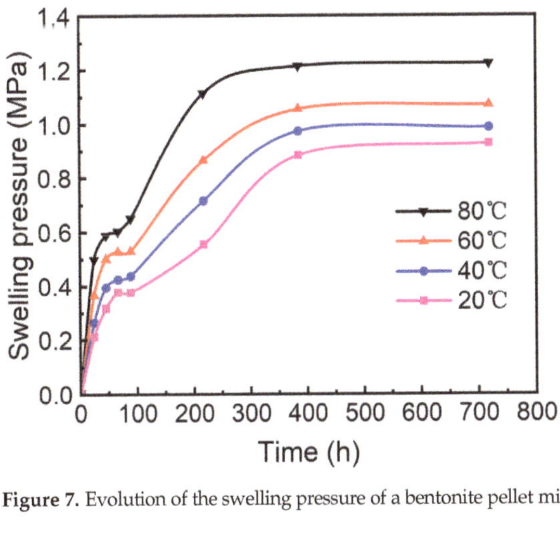

Figure 7. Evolution of the swelling pressure of a bentonite pellet mixture at different temperatures [25].

In contrast, some studies have shown opposite effects of temperature on different bentonite types, such as heating leading to an increase in the swelling pressure of Na-based bentonite and a decrease in the swelling pressure of Ca-based bentonite [59]. This can be attributed to two main mechanisms: osmotic pressure and the net effect of crystalline lattice shrinkage. Lattice shrinkage caused by the dehydration of the interlayer space and the increase in osmotic pressure at the interlayer contact is different for the two clays, with lattice shrinkage dominating in Ca-based bentonite, and the increase of osmotic pressure dominating in Na-based bentonite. Different bentonite types produce different responses to temperature variation; therefore, combined with the swelling mechanism, it can be concluded that the effect of temperature on hydration greatly influences the variation of the final swelling pressure of bentonite.

Hydration causes an increase in the interlayer pore volume of clay minerals, leading to the swelling of clay particles and, thus, to the formation of clay aggregates [60]. The bentonite swelling potential is influenced by the water content due to the coupling of micro- and macro-structures caused by hydration [61]. As shown in Figure 7, a test was conducted by injecting distilled water at the bottom of the specimen, and the results were analyzed. Firstly, the water entering from the bottom caused the soil particles to swell and densify, and thus the swelling pressure increased rapidly; however, the swollen particles prevented further water penetration, which slowed down the increase in the swelling pressure. Finally, the reduced strength of the wetted particles was counteracted by the rearrangement of the upper particles, and thus a plateau in the development of swelling pressure occurred [25]. The dual-structure model may explain the reduction of the particle strength in the wetted state [62,63]: on the one hand, wetting increased the interlayer distance between the clay flakes leading to the swelling of the clay aggregates, and on the other hand, wetting weakened the macro-structural resistance.

Liu et al. [25] provided a comprehensive explanation of the relationship between swelling pressure and temperature, indicating that the variation of the swelling pressure with temperature depends on the competition between hydration pressure, hydraulic pressure, and osmotic pressure thermo-tropic response. The Na-based GMZ bentonite with a dry density of 1.45 mg/m^3 in the study exhibited an enhanced swelling pressure as it possesses a large number of pores between the particles, and the temperature-dependent increase of its swelling pressure was affected by the enhanced water pressure and osmotic pressure masking the limited decrease in hydration force. Therefore, the effect of temperature on the swelling pressure depends not only on hydration but also on the pore water pressure and electrolyte properties.

3.3. Coupled Thermo-Chemical Effect on the Swelling Pressure

Chen et al. [28] conducted alternating wetting tests using distilled water and NaCl solutions of different concentrations on Na–GMZ bentonite with an initial dry density of 1.7 g/cm^3 at 20 °C and 60 °C, as shown in Figure 8. The results showed that when the thermo-chemical effect was coupled, the increase of temperature and electrolyte concentration was not conducive to the development of swelling pressure, and in the multi-step salination–desalination process, the initial swelling pressure of the sample could not be fully restored after saline circulation. The possible causes were the irreversible interpore collapse and the further homogenization of the clay structure at the macrostructural level.

Similar results were obtained with the swelling pressure test of monovalent bentonite with the same dry density, in India [58]. In addition, that study showed that a high temperature can promote the expansion of bivalent bentonite. In general, the swelling pressure of the divalent bentonite increased monotonically and then stabilized, while the collapse occurred in the middle of the monovalent bentonite and increased again after equilibrium. When the thermo-chemical effect was coupled, the swelling pressure of monovalent bentonite was greatly reduced, but that of divalent bentonite was very small. Compared with the equilibrium swelling pressure when using distilled water wetting at 25 °C, the inhibition effect of a 0.1 M NaCl solution was less than the promotion effect of

95 °C temperature on the swelling pressure. Therefore, under the condition of 95 °C and 0.1 M NaCl solution infiltration, the equilibrium swelling pressure of bivalent bentonite still increased. However, when the sample was soaked in a $CaCl_2$ solution with the same concentration and at the same temperature, the equilibrium swelling pressure of bivalent bentonite was reduced. The authors attributed this to the combined effect of high temperature, leading to the expansion of the soil particles, and the increase of microscopic pore water pressure. At the same time, when low-concentration Na^+ was exchanged with Ca^{2+}, the exchanged Na^+ was hydrated until the exchange process was saturated, which promoted permeability and expansion [46]. However, this study did not take into account the phenomenon of mineral alteration occurring in bentonite. The microscopic mechanism of bentonite swelling pressure variation under coupled thermo-chemical effects needs to be further refined and studied.

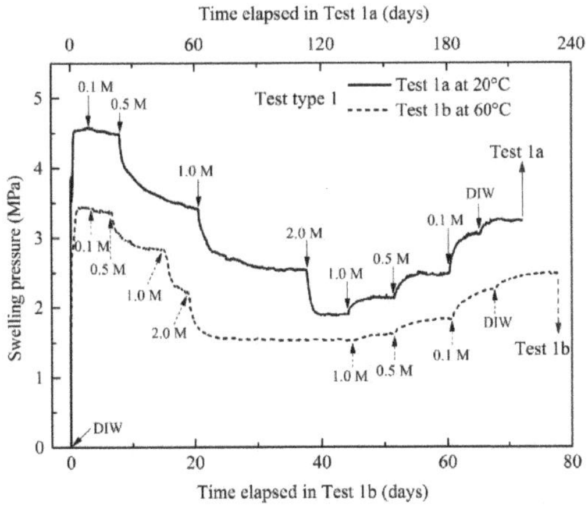

Figure 8. Evolution of the swelling pressure under thermo-chemical effects [28].

4. Effect Mechanism of Swelling Properties

Generally, the mechanistic explanation of macroscopic phenomena ultimately requires microscopic-scale evaluations. Thermo-chemical effects will have a great impact on the microstructure of the soil, and as a consequence, it is essential to consider the variations occurring in the microstructure of bentonite with the hydration process when describing the expansion behavior of bentonite. Aside from pore distribution, another important factor affecting the microstructure of soil is water distribution in the pores, which has a strong control effect on the mechanical and seepage characteristics of unsaturated soil.

4.1. Pore Distribution

Clay represents a complex porous structural body, with clay layers and interlayer pores composing a basic unit of clay particles, such as montmorillonite lamellae and interlayer pores, as shown in Figure 9 [60]. Water in clay soils includes microstructural water and macro-structural water [64]. As shown in Figure 9, water in interparticle and interlayer pores is microstructural water, i.e., adsorbed water, and water in interaggregate pores is macrostructural water. The surface of the clay particle unit carries a negative charge, and a diffuse double layer is formed between the clay particles, such as the diffused double-electric-layer water layer at the crystallite scale, shown in Figure 4b. The clay particles and the inter-clay pores form the basic clay aggregate, then the aggregates and the inter-aggregate pores develop a representative basic clay unit. Consequently, the clay block can be roughly divided into interlayer pores, intergranular pores, and inter-aggregate pores at three levels of pore space, and fractures visible to the naked eye develop from the inter-aggregate pores.

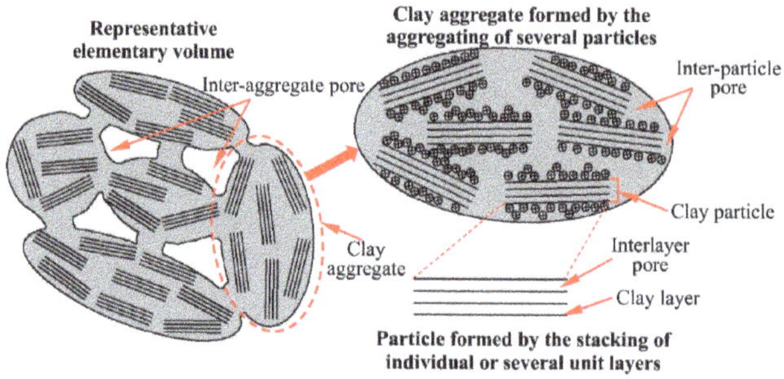

Figure 9. Fabric units and pore spaces of compacted clay (modified after [2,65]).

Investigating bentonite microstructure entails the study of the pore structure of bentonite as well as its mineral composition. Freeze–drying techniques [64,66] and mercury intrusion pore size measurement (MIP technique) [67,68] were used to obtain PSD distribution curves and thus distinguish between macro- and micropores. Nevertheless, the limitations of the MIP technique itself [69–71], combined with the differences in clay types, led to different pore size measures. For example, when the specimen was a mixture of bentonite and Callovo–Oxfordian (COx) claystone with low dry density, the average pore sizes of its three main pore groups were 10–22 μm for macro-pores, 0.17–0.52 μm for mesopores, and about 25 nm for small pores [60].

Ma et al. [72] investigated the microstructure of saturated bentonite samples with a moisture content of 0.46 in the wet–dry cycle and found that micro-pores and macro-pores corresponded to 12 nm and 6 μm; they considered the size of 500 nm as a threshold to distinguish between intra-aggregate and inter-aggregate pores. Zhang et al. [21] performed MIP analysis of compacted MX80 Wyoming bentonite samples with 80–92% montmorillonite content and identified small and large pore groups with major pore sizes of 17 nm and 21 μm, respectively; pores with a diameter of less than 100 nm were considered micro-pores [35]. In addition, scanning electron microscopy (SEM) tests can provide high-precision clay images to study the structural changes of clays more visually, and X-ray diffraction (XRD) can evaluate the changes in the composition of clay minerals to determine the causes of changes in swelling pressure [24,73], as well as measure variations in mineral layer spacing as a support condition for changes in the swelling pressure [74]. The structure determines the properties, and the complex structure of clays dictates that variations in swelling properties are significantly influenced by changes in pore distribution and pore water status.

4.2. Pore Water Status

Pore water refers to water that is confined to soil pores or clay minerals, and the properties of pore water are changed by the interaction between soil particles and pore water. The pore water status of soils also affects their mechanical properties to some extent. Studies on soil moisture characteristics suggested that soil moisture is controlled by two mechanisms: capillary and adsorption [75,76]. Adsorption predominates under high suction, and capillary dominates under low suction [77,78]. Ma et al. [72] tested the wet–dry cycle with different suction values in powder and compacted Ningming expansive soils. When the suction reached 24 MPa, the SWCCs curves of the two kinds of soil samples coincided. Therefore, the suction pressure of 24 MPa was defined as the threshold for capillary and adsorption of the soil samples, distinguishing capillary and adsorption effects in expansive soil with a quantitative suction magnitude. Meanwhile, some constant-volume experiments provided similar results [79]. The variation of absorbed water content in the soil with temperature is controlled by two mechanisms: pore water potential and soil structure. The pore water potential of adsorbed water is smaller than that of capillary water and decreases with the decrease in temperature [80]. According to Le Chatelier's principle [81], an increase in temperature facilitates the desorption process and inhibits the exothermic process. According to the thermodynamic theory Equation (1), both cation and surface hydration processes are exothermic because the adsorbed water is adsorbed on the crystal surface, the entropy of the system decreases, $\Delta S_a < 0$, and then the enthalpy of adsorption becomes $\Delta H_a < 0$; therefore, adsorption is exothermic [82]. The decrease in temperature enhances hydration, influencing the entropy of the hydration reaction [83].

$$\Delta G_a = \Delta H_a - T\Delta S_a = 0 \tag{1}$$

The absorbed water content increases with the decrease in temperature, which can be attributed to the decrease in pore water potential and the increase in micro-pore volume, as shown in Figure 10. Depending on the Yang–Laplace equation, capillary suction can be described by Equation (2) [82]. As the temperature increases, under a given suction Pc, the water-saturated pore diameter d decreases, with the decrease of surface tension (σ) and contact angle (θ) the faster the speed decreases. This results in a reduction in the amount of capillary water [84,85]. Thus, the conversion of adsorbed water to capillary water with inreased temperature and the further loss of capillary water will be detrimental to the retention of water in clays. Liu et al. [82] demonstrated that the water retention capacity of three bentonite samples decreased with increasing temperature for a given suction condition. In both restricted- and free-swelling conditions, bentonite water retention capacity will decrease with the increasing temperature (20–80 °C), and this is especially evident when the suction is low [57].

$$P_c = \frac{4\sigma \cos\theta}{d} \tag{2}$$

The adsorbed water content is simultaneously controlled by the pore solution concentration, which increases to the detriment of the adsorbed water [80]. It is clear that an increase in the pore solution concentration will inhibit the development of the bilayer thickness, according to Equation (3). However, the role of salt concentration on crystal layer swelling needs to be further investigated. As regards the changes in pore structure, the adsorbed water content is affected by the salt concentration in the same way as the micro-pore volume. Norrish et al. [86] concluded that, although salt concentrations less than 1 M have little effect on the swelling of crystalline layers, the increased electrolyte concentration leads to the dehydration of montmorillonite by osmosis, which is detrimental to the occurrence of inter-montmorillonite hydration. Thus, high salt concentrations have an inhibitory effect on hydration.

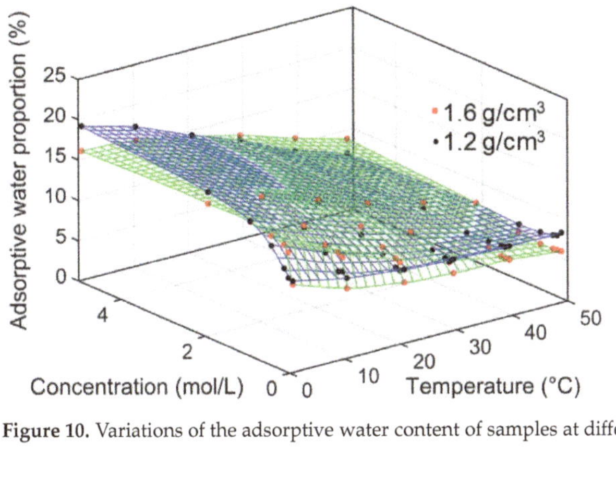

Figure 10. Variations of the adsorptive water content of samples at different dry densities [80].

The thickness of the double electric layer is calculated by [40]:

$$\frac{1}{K} = \left(\frac{\varepsilon^0 DkT}{2n^0 e^2 v^2}\right)^{\frac{1}{2}} \quad (3)$$

$\frac{1}{K}$ is the thickness of the double layer, ε^0 is the vacuum coefficient, D is the dielectric constant, k is the Boltzmann constant, T is the temperature, n^0 is the pore solution concentration, e is the unit charge, and v is the ionic valence.

Significantly, because of the thermo-chemical effects on pore water, as shown in Figure 10 [80], the volume of micro-pores decreased, and the volume of macro-pores increased with increasing salt concentrations, and micro-pores were more frequent at low temperatures. In addition, it was observed that more water was adsorbed at low temperatures and in high-salt-concentration environments, which seems to be contrary to the theory. By investigating the effective activation energy [87], which is known as the energy barrier for water molecule conversion, Tian et al. [87] found that elevated salt concentrations accelerated the water molecule relaxation and produced a higher effective activation energy compared to samples saturated with distilled water, enhancing water molecule adsorption to some extent. In reality, such water with a short relaxation time is somewhat different from adsorbed water or exists in the form of hydrated ions surrounded by water molecules, an effect that is more pronounced in the presence of saline solutions, masking the inhibition of the double electric layer and the reduction of the micro-pore volume by the increase of the pore solution concentration; hence, the higher measured adsorbed water content [80]. Bentonite exposed to a concrete-leaching groundwater solution introduced at elevated temperatures showed a decrease in macro-pore volume and an increase in micro-pores, with the average pore size decreasing from 6 nm to 4 nm [88]. This phenomenon suggests that a high-temperature situation may also lead to an increase in micro-pores; so, the study of adsorbed water content should take into account the pore structure in the case of coupled thermo-chemical effects.

In conclusion, dry and wet cycles destroy the soil pore structure and weaken the water retention capacity of bentonite; elevated temperatures reduce the water retention performance of bentonite; increased salt concentrations are detrimental to the development of micro-pores and the adsorption of adsorbed water, thereby reducing the swelling capacity of bentonite. When the thermo-chemical effects act simultaneously, combining with changes in soil structure, thermodynamic properties, and ion hydration effects, their influence on the pore water status becomes more complex. As a consequence, more in-depth studies are essential.

4.3. Intergranular Stress

At present, studies on the swelling mechanism of bentonite generally remain at the microscopic level. A few scholars have proposed a new formula to determine the effective stress in unsaturated soil from a macroscopic perspective, which can describe the stress state in unsaturated soil well [89,90]. In 2014, Wei [91] proposed a theory of intergranular stress acting between soil particles, which comprehensively characterized the physicochemical interactions between soil particles from a macroscopic perspective.

Wei [91] introduced the concept of equilibrium solution and pointed out that the measured pore pressure is not the same as the true pore pressure within the soil. For bentonite, when the soil is saturated with an aqueous solution, the magnitude of the water pressure within the pore space is not uniform due to the physicochemical interaction between pore solution and soil water. The equilibrium solution is defined as the water in a measurement vessel that has reached equilibrium with a pore solution without surface action of the soil particles. The soil is in equilibrium with the sensor, when measuring the pore pressure. In other words, we determine the pore pressure by measuring the pressure of the equilibrium solution, whereas the instrument measures only the potential energy of the water and not the mechanical pressure.

Wei [91] deduced the expression of true pore pressure considering that the water potential energy in the pore is the same as that in the container at an equilibrium state:

$$p^l = p^l{}_w + \Pi \quad (4)$$

where $p^l{}_w$ is the actual pore water pressure in the soil, and there is a difference Π from the measured pore pressure p^l; Π is called generalized osmotic pressure. Intergranular stress is defined as:

$$\sigma'' = \sigma - (p^l{}_w + \Pi) \quad (5)$$

where σ'' is the intergranular stress; σ is the total stress.

This generalized osmotic pressure reflects the interaction between soil particles and pore water and is divided into two components: the Donnan osmotic pressure and the pressure derived from surface forces. The Donnan osmotic pressure relates to the concentration difference between the pore solution and the equilibrium solution. The second component of generalized osmotic pressure is the pressure caused by surface forces, including the surface tension of the interface, the electrostatic force, Van der Waals gravity, electric double-layer repulsion, etc. The equation is:

$$\Pi = \Pi_D - \rho_{\oplus}^{l_{H_2O}} \Omega^l \quad (6)$$

In the equation, $\rho_{\oplus}^{l_{H_2O}}$ is the mass density of pure water, equal to 1.0 g/cm^3, Π_D is the Donnan osmotic pressure, and Ω^l is the surface potential energy caused by microscopic surface forces. Donnan osmotic pressure is the pressure caused by the pressure difference betwe en the internal and the external concentration of the soil related to the porosity and solution concentration in the soil. It is expressed by:

$$\Pi_D = \frac{RT\rho_{\oplus}^{l_{H_2O}}}{M_{H_2O}} \ln\left(\frac{a_A^{l_{H_2O}}}{a^{l_{H_2O}}}\right) \quad (7)$$

where M_{H_2O} is the molar mass of water, equal to 18 g/mol, $a_A^{l_{H_2O}}$ is the activity of water in the equilibrium solution, which is equal to the molar fraction of water in the equilibrium solution in the ideal solution, $a^{l_{H_2O}}$ is the activity of water in the soil pore solution, which is equal to the molar fraction of water in the soil pore solution in the ideal solution. Ω^l is expressed as:

$$n^l \rho^l \Omega^l(T, n^l) = n_0^l \rho^l \Omega^l(T, n_0^l) + \int_{n^l}^{n_0^l} \left[s_M(T, n^l) - \Pi_D(T, n^l)\right] dn^l \quad (8)$$

where n^l is the volume fraction of water, defined as the volume of water in the pore as a proportion of the total volume, p^l is the density of water, for dilute solutions, $\rho^l \approx \rho_{\oplus}^{l_{H_2O}}$, s_M is the measured (or controlled) matrix suction, p_W^l is the measured pore pressure, n_0^l is the volume fraction of pore water at saturation, equal to the porosity n.

The intergranular stress expression can consider the physicochemical effects between particles and unitize the infiltration, capillary, and adsorption effects caused by physicochemical interactions, which can better describe the behavior of bentonites in terms of strength and deformation. Moreover, the capillary and physicochemical pressure at the microscopic scale can be better applied to the intergranular stress expression at the macroscopic scale, in unsaturated soils.

Ma et al. [92] experimentally researched the effect of pore solution on the swelling pressure of expansive soils, pointed out that the variation of the swelling pressure is determined by the intergranular stress, described the swelling pressure variation from the perspective of macro-mechanics, and proposed a model of swelling force prediction. The superiority of this model for swelling pressure prediction was confirmed by validating it on several bentonites with different properties (Figure 11). Overall, studies of the effect of intergranular stress on the swelling pressure of bentonite under thermo-chemical effects are still relatively rare and lack extensive experimental validation; therefore, further investigation is necessary.

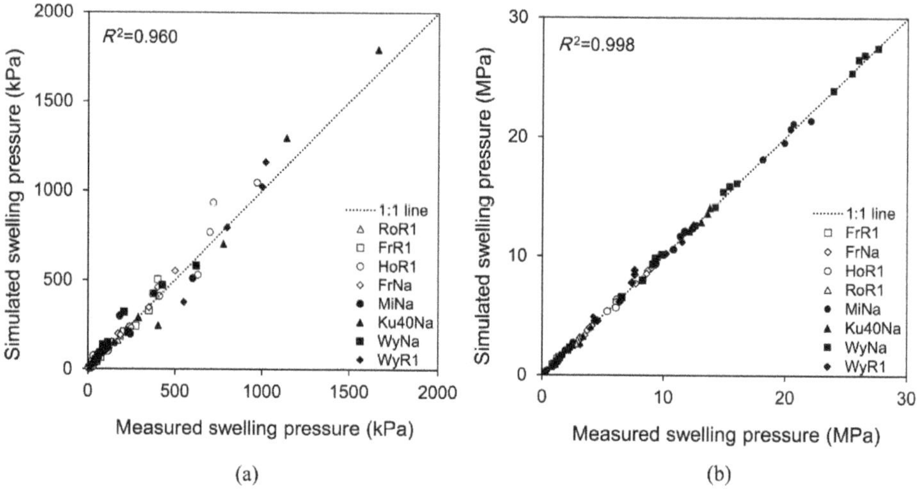

Figure 11. Comparison of measured and predicted swelling pressures for eight soils [93]: (**a**) low swelling pressure; (**b**) high swelling pressure [92].

5. Conclusions and Prospects

In recent years, scholars have conducted elaborate studies on the swelling pressure of bentonite in the deep geological repositories of HLW, focusing on experimental methods, influencing factors, change patterns, and intrinsic influencing mechanisms. However, considering the coupled thermochemistry and hydrodynamics phenomena involved in the actual nuclear waste containment process, the existing research findings still have certain limitations. Additional research is desired in the future in the following areas:

(1) Various test methods will result in different values of measured swelling pressure. A comparative analysis of different test methods should be conducted, specifically considering the coupled environmental effects in the nuclear waste containment process to identify the optimal test method.

(2) Diffuse nuclear radiation heat is a major influencing factor in nuclear waste containment. Considering the effect of temperature, studies on the hydrodynamic properties of bentonite remain to be conducted.

(3) Analysis of the mechanism of swelling pressure changes: from the microscopic perspective, the process and mechanism of swelling pressure changes still lack profound recognition. In addition, from the macro-mechanics perspective, there are only sporadic studies which generally lack experimental verifications. In the future, the analysis of the swelling pressure variation law and the internal mechanism of swelling pressure evolution needs to be further refined.

Author Contributions: Conceptualization, J.L. and C.Y.; validation, J.L., W.S. and Y.Z.; investigation, J.L., W.S. and Y.Z.; resources, J.L., W.S. and Y.Z.; writing—original draft preparation, J.L.; writing—review and editing, J.L. and C.Y.; supervision, C.Y.; funding acquisition, C.Y. All authors have read and agreed to the published version of the manuscript.

Funding: This research was funded by National Natural Science Foundation of China (Grant No. 52109133), Shanghai Sailing Program (Grant No. 21YF1432700).

Institutional Review Board Statement: The study did not involve humans or animals.

Informed Consent Statement: The study did not involve humans or animals.

Data Availability Statement: All data, models, or codes that support the findings of this study are available from the corresponding author upon reasonable request.

Acknowledgments: This research was supported by the National Natural Science Foundation of China (Grant No. 52109133), Shanghai Sailing Program (Grant No. 21YF1432700).

Conflicts of Interest: The authors declare no conflict of interest.

References

1. Prongmanee, N.; Chai, J.-C.; Shen, S. Hydraulic Properties of Polymerized Bentonites. *J. Mater. Civ. Eng.* **2018**, *30*, 04018247. [CrossRef]
2. Liu, L. Prediction of swelling pressures of different types of bentonite in dilute solutions. *Colloids Surf. A Physicochem. Eng. Asp.* **2013**, *434*, 303–318. [CrossRef]
3. Castellanos, E.; Villar, M.; Romero, E.; Lloret, A.; Gens, A. Chemical impact on the hydro-mechanical behaviour of high-density FEBEX bentonite. *Clays Nat. Eng. Barriers Radioact. Waste Confin.* **2008**, *33*, S516–S526. [CrossRef]
4. De la Morena, G.; Asensio, L.; Navarro, V. Modelling the hydro-mechanical behaviour of GMZ bentonite. *Eng. Geol.* **2018**, *239*, 195–205. [CrossRef]
5. Ye, W.M.; Wan, M.; Chen, B.; Chen, Y.G.; Cui, Y.J.; Wang, J. Temperature effects on the swelling pressure and saturated hydraulic conductivity of the compacted GMZ01 bentonite. *Environ. Earth Sci.* **2013**, *68*, 281–288. [CrossRef]
6. Schanz, T.; Tripathy, S. Swelling pressure of a divalent-rich bentonite: Diffuse double-layer theory revisited. *Water Resour. Res.* **2009**, *45*. [CrossRef]
7. Komine, H.; Ogata, N. Prediction for swelling characteristics of compacted bentonite. *Can. Geotech. J.* **1996**, *33*, 11–22. [CrossRef]
8. Lloret, A.; Villar, M.V.; Sánchez, M.; Gens, A.; Pintado, X.; Alonso, E.E. Mechanical behaviour of heavily compacted bentonite under high suction changes. *Géotechnique* **2003**, *53*, 27–40. [CrossRef]
9. Tang, C.; Tang, A.; Cui, Y.; Delage, P.; Schroeder, C.; De Laure, E. Investigating the swelling pressure of compacted crushed-Callovo-Oxfordian claystone. *Clays Nat. Eng. Barriers Radioact. Waste Confin.* **2011**, *36*, 1857–1866. [CrossRef]
10. Gawin, D.; Pesavento, F.; Schrefler, B.A. Hygro-thermo-chemo-mechanical modelling of concrete at early ages and beyond. Part I: Hydration and hygro-thermal phenomena. *Int. J. Numer. Methods Eng.* **2006**, *67*, 299–331. [CrossRef]
11. Al-Mhaidib, A.I. Swelling Behaviour of Expansive Shales from the Middle Region of Saudi Arabia. *Geotech. Geol. Eng.* **1998**, *16*, 291–307. [CrossRef]
12. Zhang, F.; Cui, Y.-J. Microstructure-based insight into different swelling pressure determination methods. *Eng. Geol.* **2022**, *307*, 106777. [CrossRef]
13. Wang, H.; Ito, D.; Shirakawabe, T.; Ruan, K.; Komine, H. On swelling behaviours of a bentonite under different water contents. *Geotechnique* **2022**, *in press*. [CrossRef]
14. Selçuk, L. A rational method for determining the swelling pressure of expansive soils: Paired swell test (PS-test). *Acta Geotech.* **2022**, *17*, 3161–3180. [CrossRef]
15. Hoffmann, C.; Alonso, E.; Romero, E. Hydro-mechanical behaviour of bentonite pellet mixtures. *Clays Nat. Eng. Barriers Radioact. Waste Confin.-Part 2* **2006**, *32*, 832–849. [CrossRef]
16. Tripathy, S.; Sridharan, A.; Schanz, T. Swelling pressures of compacted bentonites from diffuse double layer theory. *Can. Geotech. J.* **2004**, *41*, 437–450. [CrossRef]
17. Al-Shamrani, M.A.; Dhowian, A.W. Experimental study of lateral restraint effects on the potential heave of expansive soils. *Eng. Geol.* **2003**, *69*, 63–81. [CrossRef]
18. Pejon, O.; Zuquette, L. Effects of strain on the swelling pressure of mudrocks. *Int. J. Rock Mech. Min. Sci.* **2006**, *43*, 817–825. [CrossRef]
19. Tang, C.-S.; Tang, A.M.; Cui, Y.; Delage, P.; Schroeder, C.; Shi, B. A study of the hydro-mechanical behaviour of compacted crushed argillite. *Eng. Geol.* **2011**, *118*, 93–103. [CrossRef]
20. Kayabali, K.; Demir, S. Measurement of swelling pressure: Direct method versus indirect methods. *Can. Geotech. J.* **2011**, *48*, 354–364. [CrossRef]
21. Zhang, F.; Cui, Y.-J.; Chen, B. Investigation of Suction Effects Due to Stress Release with Compacted MX80 Bentonite. *J. Geotech. Geoenviron. Eng.* **2022**, *148*, 04022070. [CrossRef]
22. Weston, D.J. Expansive Roadbed Treatment for Southern Africa. In Proceedings of the Conference on Expansive Soils ASCE, Denver, CO, USA, 16–18 June 1980. Report Number RR 288.
23. Bishop, A.W.; Wesley, L.D. A hydraulic triaxial apparatus for controlled stress path testing. *Géotechnique* **1975**, *25*, 657–670. [CrossRef]
24. Liu, L.-N.; Chen, Y.-G.; Ye, W.-M.; Cui, Y.-J.; Wu, D.-B. Effects of hyperalkaline solutions on the swelling pressure of compacted Gaomiaozi (GMZ) bentonite from the viewpoint of Na^+ cations and OH− anions. *Appl. Clay Sci.* **2018**, *161*, 334–342. [CrossRef]

25. Liu, Z.-R.; Ye, W.-M.; Cui, Y.-J.; Zhu, H.-H.; Wang, Q. Water infiltration and swelling pressure development in GMZ bentonite pellet mixtures with consideration of temperature effects. *Eng. Geol.* **2022**, *305*, 106718. [CrossRef]
26. Zhu, C.-M.; Ye, W.-M.; Chen, Y.-G.; Chen, B.; Cui, Y.-J. Influence of salt solutions on the swelling pressure and hydraulic conductivity of compacted GMZ01 bentonite. *Eng. Geol.* **2013**, *166*, 74–80. [CrossRef]
27. Saiyouri, N.; Hicher, P.Y.; Tessier, D. Microstructural approach and transfer water modelling in highly compacted unsaturated swelling clays. *Mech. Cohesive-Frict. Mater.* **2000**, *5*, 41–60. [CrossRef]
28. Chen, Y.-G.; Dong, X.-X.; Zhang, X.-D.; Ye, W.-M.; Cui, Y.-J. Cyclic thermal and saline effects on the swelling pressure of densely compacted Gaomiaozi bentonite. *Eng. Geol.* **2019**, *255*, 37–47. [CrossRef]
29. Villar, M.; Gómez-Espina, R.; Gutiérrez-Nebot, L. Basal spacings of smectite in compacted bentonite. *Appl. Clay Sci.* **2012**, *65–66*, 95–105. [CrossRef]
30. Liu, Z.-R.; Ye, W.-M.; Zhang, Z.; Wang, Q.; Chen, Y.-G.; Cui, Y.-J. Particle size ratio and distribution effects on packing behaviour of crushed GMZ bentonite pellets. *Powder Technol.* **2019**, *351*, 92–101. [CrossRef]
31. Olphen, H.V. *An Introduction to Clay Colloid Chemistry*; Interscience Publishers: New York, NY, USA, 1964.
32. Yong, R. Soil suction and soil-water potentials in swelling clays in engineered clay barriers. *Eng. Geol.* **1999**, *54*, 3–13. [CrossRef]
33. Saiyouri, N.; Tessier, D.; Hicher, P.Y. Experimental study of swelling in unsaturated compacted clays. *Clay Miner.* **2004**, *39*, 469–479. [CrossRef]
34. Villar, M.V. Water retention of two natural compacted bentonites. *Clays Clay Miner.* **2007**, *55*, 311–322. [CrossRef]
35. Liu, Z.-R.; Ye, W.-M.; Cui, Y.-J.; Zhu, H.-H.; Wang, Q.; Chen, Y.-G. Development of swelling pressure for pellet mixture and compacted block of GMZ bentonite. *Constr. Build. Mater.* **2021**, *301*, 124080. [CrossRef]
36. Alawaji, H.A. Swell and compressibility characteristics of sand–bentonite mixtures inundated with liquids. *Appl. Clay Sci.* **1999**, *15*, 411–430. [CrossRef]
37. Sun, D.; Zhang, L.; Li, J.; Zhang, B. Evaluation and prediction of the swelling pressures of GMZ bentonites saturated with saline solution. *Appl. Clay Sci.* **2015**, *105–106*, 207–216. [CrossRef]
38. Rao, S.M.; Thyagaraj, T.; Rao, P.R. Crystalline and Osmotic Swelling of an Expansive Clay Inundated with Sodium Chloride Solutions. *Geotech. Geol. Eng.* **2013**, *31*, 1399–1404. [CrossRef]
39. Bauer, A.; Velde, B. Smectite transformation in high molar KOH solutions. *Clay Miner.* **1999**, *34*, 259–273. [CrossRef]
40. Mitchell, J.K.; Soga, K. *Fundamentals of Soil Behavior*; John Wiley & Sons: New York, NY, USA, 2005; Volume 3.
41. Pusch, R.; Yong, R.N. *Microstructure of Smectite Clays and Engineering Performance*; CRC Press: London, UK, 2006; ISBN 0-429-18022-5.
42. Chen, Y.-G.; Jia, L.-Y.; Li, Q.; Ye, W.-M.; Cui, Y.-J.; Chen, B. Swelling deformation of compacted GMZ bentonite experiencing chemical cycles of sodium-calcium exchange and salinization-desalinization effect. *Appl. Clay Sci.* **2017**, *141*, 55–63. [CrossRef]
43. Gouy, M. Sur la constitution de la charge électrique à la surface d'un électrolyte. *J. Phys. Thèor. Appl.* **1910**, *9*, 457–468. [CrossRef]
44. Chapman, D.L. LI. A contribution to the theory of electrocapillarity. *Lond. Edinb. Dublin Philos. Mag. J. Sci.* **1913**, *25*, 475–481. [CrossRef]
45. Mesri, G.; Olson, R.E. Shear Strength of Montmorillonite. *Géotechnique* **1970**, *20*, 261–270. [CrossRef]
46. Lee, J.O.; Lim, J.G.; Kang, I.M.; Kwon, S. Swelling pressures of compacted Ca-bentonite. *Eng. Geol.* **2012**, *129–130*, 20–26. [CrossRef]
47. Herbert, H.-J.; Kasbohm, J.; Sprenger, H.; Fernández, A.M.; Reichelt, C. Swelling pressures of MX-80 bentonite in solutions of different ionic strength. *Clays Nat. Eng. Barriers Radioact. Waste Confin.* **2008**, *33*, S327–S342. [CrossRef]
48. Zhang, F.; Ye, W.-M.; Wang, Q.; Chen, Y.-G.; Chen, B. An insight into the swelling pressure of GMZ01 bentonite with consideration of salt solution effects. *Eng. Geol.* **2019**, *251*, 190–196. [CrossRef]
49. Karnland, O.; Muurinen, A.; Karlsson, F. Bentonite swelling pressure in NaCl solutions—Experimentally determined data and model calculations. In Proceedings of the Clays in Natural and Engineered Barriers for Radioactive Waste Confinement, Reims, France, 9–12 December 2005; pp. 241–256.
50. Zhang, F.; Ye, W.-M.; Chen, Y.-G.; Chen, B.; Cui, Y.-J. Influences of salt solution concentration and vertical stress during saturation on the volume change behavior of compacted GMZ01 bentonite. *Eng. Geol.* **2016**, *207*, 48–55. [CrossRef]
51. Liu, Z.-R.; Cui, Y.-J.; Ye, W.-M.; Chen, B.; Wang, Q.; Chen, Y.-G. Investigation of the hydro-mechanical behaviour of GMZ bentonite pellet mixtures. *Acta Geotech.* **2020**, *15*, 2865–2875. [CrossRef]
52. Chen, L.; Liu, Y.; Wang, J.; Cao, S.; Xie, J.; Ma, L.; Zhao, X.; Li, Y.; Liu, J. Investigation of the thermal-hydro-mechanical (THM) behavior of GMZ bentonite in the China-Mock-up test. *Eng. Geol.* **2014**, *172*, 57–68. [CrossRef]
53. Liang, S.-Y.; Lin, W.-S.; Lin, G.-F.; Liu, C.-W.; Fan, C. The Effect of Porosity Change in Bentonite Caused by Decay Heat on Radionuclide Transport through Buffer Material. *Appl. Sci.* **2021**, *11*, 7933. [CrossRef]
54. Jadda, K.; Bag, R. Thermochemical Effect on Swelling Pressure and Microstructure of Monovalent and Divalent Indian Bentonites. *J. Hazard. Toxic Radioact. Waste* **2022**, *26*, 04021039. [CrossRef]
55. Kim, M.; Lee, S.; Cheon, E.; Kim, M.; Yoon, S. Thermochemical changes on swelling pressure of compacted bentonite. *Ann. Nucl. Energy* **2021**, *151*, 107882. [CrossRef]
56. Kanazawa, S.; Kobayashi, S.; Ichikawa, N.; Yanai, M.; Ishiyama, K. The swelling behaviour of bentonite buffer material considering the effect of temperature. *Int. J. GEOMATE* **2020**, *19*, 216–221. [CrossRef]
57. Villar, M.; Lloret, A. Influence of temperature on the hydro-mechanical behaviour of a compacted bentonite. *Clays Nat. Eng. Barriers Radioact. Waste Confin.* **2004**, *26*, 337–350. [CrossRef]

58. Daniels, K.; Harrington, J.; Sellin, P.; Norris, S. Closing repository void spaces using bentonite: Does heat make a difference? *Appl. Clay Sci.* **2021**, *210*, 106124. [CrossRef]
59. Pusch, R.; Karnland, O.; Hökmark, H. *GMM—A General Microstructural Model for Qualitative and Quantitative Studies of Smectite Clays*; No. SKB-TR–90-43; Swedish Nuclear Fuel and Waste Management Co.: Solna, Sweden, 1990.
60. Zeng, Z.; Cui, Y.-J.; Talandier, J. Evaluation of swelling pressure of bentonite/claystone mixtures from pore size distribution. *Acta Geotech.* **2022**, *18*, 1671–1679. [CrossRef]
61. Chen, B.; Peng, F.; Zhang, L.; Sun, D. Investigation on swelling characteristics of GMZ bentonite with different initial water contents. *Ann. Nucl. Energy* **2023**, *181*, 109565. [CrossRef]
62. Sánchez, M.; Gens, A.; Guimarães, L.D.N.; Olivella, S. A double structure generalized plasticity model for expansive materials. *Int. J. Numer. Anal. Methods Géoméch.* **2005**, *29*, 751–787. [CrossRef]
63. Gens, A.; Alonso, E.E. A framework for the behaviour of unsaturated expansive clays. *Can. Geotech. J.* **1992**, *29*, 1013–1032. [CrossRef]
64. Delage, P.; Marcial, D.; Cui, Y.J.; Ruiz, X. Ageing effects in a compacted bentonite: A microstructure approach. *Géotechnique* **2006**, *56*, 291–304. [CrossRef]
65. Mašín, D.; Khalili, N. Swelling phenomena and effective stress in compacted expansive clays. *Can. Geotech. J.* **2016**, *53*, 134–147. [CrossRef]
66. Diamond, S. Pore Size Distributions in Clays. *Clays Clay Miner.* **1970**, *18*, 7–23. [CrossRef]
67. Romero, E. A microstructural insight into compacted clayey soils and their hydraulic properties. *Spec. Issue Res. Appl. Unsaturated Soil Mech.* **2013**, *165*, 3–19. [CrossRef]
68. Delage, P.; Audiguier, M.; Cui, Y.-J.; Howat, M.D. Microstructure of a compacted silt. *Can. Geotech. J.* **1996**, *33*, 150–158. [CrossRef]
69. Romero, E.; Simms, P.H. Microstructure Investigation in Unsaturated Soils: A Review with Special Attention to Contribution of Mercury Intrusion Porosimetry and Environmental Scanning Electron Microscopy. *Geotech. Geol. Eng.* **2008**, *26*, 705–727. [CrossRef]
70. Bian, X.; Cui, Y.-J.; Li, X.-Z. Voids effect on the swelling behaviour of compacted bentonite. *Géotechnique* **2019**, *69*, 593–605. [CrossRef]
71. Wei, X.; Hattab, M.; Fleureau, J.-M.; Hu, R. Micro–macro-experimental study of two clayey materials on drying paths. *Bull. Eng. Geol. Environ.* **2013**, *72*, 495–508. [CrossRef]
72. Ma, T.; Wei, C.; Yao, C.; Yi, P. Microstructural evolution of expansive clay during drying–wetting cycle. *Acta Geotech.* **2020**, *15*, 2355–2366. [CrossRef]
73. Lee, J.O.; Kang, I.M.; Cho, W.J. Smectite alteration and its influence on the barrier properties of smectite clay for a repository. *Adv. Smectitic Clay Res.* **2010**, *47*, 99–104. [CrossRef]
74. Jadda, K.; Bag, R. Effect of initial compaction pressure and elevated temperature on swelling pressure of two Indian bentonites. *Environ. Earth Sci.* **2020**, *79*, 197. [CrossRef]
75. Lebeau, M.; Konrad, J.-M. A new capillary and thin film flow model for predicting the hydraulic conductivity of unsaturated porous media. *Water Resour. Res.* **2010**, *46*. [CrossRef]
76. Lu, N.; Dong, Y. Correlation between Soil-Shrinkage Curve and Water-Retention Characteristics. *J. Geotech. Geoenviron. Eng.* **2017**, *143*, 04017054. [CrossRef]
77. Tuller, M.; Or, D. Water films and scaling of soil characteristic curves at low water contents. *Water Resour. Res.* **2005**, *41*. [CrossRef]
78. Lu, N.; Khorshidi, M. Mechanisms for Soil-Water Retention and Hysteresis at High Suction Range. *J. Geotech. Geoenviron. Eng.* **2015**, *141*, 04015032. [CrossRef]
79. Romero, E.; Della Vecchia, G.; Jommi, C. An insight into the water retention properties of compacted clayey soils. *Géotechnique* **2011**, *61*, 313–328. [CrossRef]
80. Yao, C.; Wei, C.; Ma, T.; Chen, P.; Tian, H. Experimental Investigation on the Influence of Thermochemical Effect on the Pore–Water Status in Expansive Soil. *Int. J. Géoméch.* **2021**, *21*, 04021080. [CrossRef]
81. Campbell, J.A. Le Châtelier's principle, temperature effects, and entropy. *J. Chem. Educ.* **1985**, *62*, 231. [CrossRef]
82. Liu, Z.-R.; Ye, W.-M.; Cui, Y.-J.; Zhu, H.-H.; Wang, Q. Temperature effects on water retention behaviour and structural evolution of GMZ bentonite pellet mixtures. *Appl. Clay Sci.* **2022**, *222*, 106492. [CrossRef]
83. Svensson, P.D.; Hansen, S. Combined Salt and Temperature Impact on Montmorillonite Hydration. *Clays Clay Miner.* **2013**, *61*, 328–341. [CrossRef]
84. Romero, E.; Gens, A.; Lloret, A. Temperature effects on the hydraulic behaviour of an unsaturated clay. *Geotech. Geol. Eng.* **2001**, *19*, 311–332. [CrossRef]
85. Grant, S.A.; Salehzadeh, A. Calculation of Temperature Effects on Wetting Coefficients of Porous Solids and Their Capillary Pressure Functions. *Water Resour. Res.* **1996**, *32*, 261–270. [CrossRef]
86. Norrish, K. The swelling of montmorillonite. *Discuss. Faraday Soc.* **1954**, *18*, 120–134. [CrossRef]
87. Tian, H.; Wei, C.; Yan, R. Thermal and saline effect on mineral-water interactions in compacted clays: A NMR-based study. *Appl. Clay Sci.* **2019**, *170*, 106–113. [CrossRef]
88. Morozov, I.; Zakusin, S.; Kozlov, P.; Zakusina, O.; Roshchin, M.; Chernov, M.; Boldyrev, K.; Zaitseva, T.; Tyupina, E.; Krupskaya, V. Bentonite–Concrete Interactions in Engineered Barrier Systems during the Isolation of Radioactive Waste Based on the Results of Short-Term Laboratory Experiments. *Appl. Sci.* **2022**, *12*, 3074. [CrossRef]

89. Lu, N.; Godt, J.W.; Wu, D.T. A closed-form equation for effective stress in unsaturated soil. *Water Resour. Res.* **2010**, *46*. [CrossRef]
90. Lu, N.; Likos, W.J. Suction Stress Characteristic Curve for Unsaturated Soil. *J. Geotech. Geoenviron. Eng.* **2006**, *132*, 131–142. [CrossRef]
91. Wei, C. A Theoretical Framework for Modeling the Chemomechanical Behavior of Unsaturated Soils. *Vadose Zone J.* **2014**, *13*, 1–21. [CrossRef]
92. Ma, T.; Yao, C.; Dong, Y.; Yi, P.; Wei, C. Physicochemical approach to evaluating the swelling pressure of expansive soils. *Appl. Clay Sci.* **2019**, *172*, 85–95. [CrossRef]
93. Karnland, O.; Olsson, S.; Nilsson, U. *Mineralogy and Sealing Properties of Various Bentonites and Smectite-Rich Clay Materials*; Technical Report SKB-TR-06-30; Swedish Nuclear Fuel and Waste Management Co.: Stockholm, Sweden, 2006.

applied sciences

MDPI

Article

Experimental Study on Maximum Dynamic Shear Modulus of Yangtze River Overconsolidated Floodplain Soft Soils

Yifeng Zhou, Xing Xiao, Zhenglong Zhou and Qi Wu *

Institute of Geotechnical Engineering, Nanjing Tech University, Nanjing 211899, China
* Correspondence: qw09061801@163.com

Abstract: This study conducted experimental tests on the undisturbed Nanjing Yangtze River floodplain soft soil using the bender element instrument to determine the maximum dynamic shear modulus of the Yangtze River floodplain overconsolidated soft soil. The G_{max} of floodplain soft soil with different overconsolidated ratio OCR, initial effective confining pressure σ_{3c}', and void ratio e are discussed. The results indicated that G_{max} reduced as e rose for given σ_{3c}' and OCR. In addition, an increase in OCR contributed to a gradual decrease in the decay rate of G_{max}, while the G_{max} decay rate is insensitive to the change of σ_{3c}'. The void ratio-normalized maximum shear modulus $G_{max}/F(e)$ improved with the increase in the stress-normalized initial effective confining pressure σ_{3c}'/P_a, whereas the growth rate gradually drops, and a power relationship is then obtained between $G_{max}/F(e)$ and σ_{3c}'/P_a. Based on the regression analysis, a G_{max} prediction method is established for reasonably characterizing Yangtze River floodplain soft soils with various over-consolidation states, initial stress conditions, and compactness levels, with a prediction error of less than 10%.

Keywords: floodplain soft soil; overconsolidated ratio; Initial effective confining pressure; void ratio; bender element testing

1. Introduction

Floodplain soils formed by unstable sedimentary environments are widespread in rivers, lakes, coasts, and other landforms. Foundations of bridges, offshore projects, and sub-sea tunnels will inevitably pass through this type of soil. However, under the reciprocal influence of the water flow or wave, the floodplain soils exhibit distinctive horizontal stratification and depositional rhythm. They are characterized by a high water content, large porosity ratio, high compressibility, and an evident over-consolidation state. The floodplain soils are susceptible to residual deformation or strength damage caused by waves and earthquakes, essential to soil subsidence and building instability. G_{max} (defined as the dynamic shear modulus G when the shear strain level is less than 10^{-5}, with the soil deformation belonging to the elastic range) is a fundamental parameter for describing the dynamic properties of soils, and it plays a vital role in geotechnical problems such as soil deformation prediction, potential liquefaction evaluation, seismic site response analysis, and dynamic foundation design parameters [1–4]. Therefore, conducting a systematic study on the G_{max} of overconsolidated floodplain soft soils is necessary.

Extensive studies have been conducted to investigate the G_{max} of soils using resonant column (RC) and bender element (BE) tests [5–7]. The level of densification (void ratio e or relative density D_r) and the initial effective confining pressure σ_{3c}' is surely the most fundamental effects on G_{max} [8–10]. However, the effect of frequency f on G_{max} is controversial. Kim and Stokoe [11] proposed that an increase in f will lead to an increase in G of clayey soils in the range 0.001 Hz < f < 200 Hz, while Irfan [12] insisted that the f range of 10~100 Hz did not have a significant effect on G. In this paper, the effect of e and σ_{3c}' was considered instead of f. For sandy soils, Hardin and Black [13] indicated, based on a large number of RC tests, that G_{max} rises as σ_{3c}' increases and decreases as e improves.

Kim and Novak [14] confirmed the effect of σ_{3c}', e, and other factors on G_{max}. Payan and Chenari [15] and Liu et al. [16] demonstrated that stress anisotropy and the particle shape of soils can still affect G_{max}.

Gu et al. [17] conducted a series of combined cyclic triaxial and bender element tests regarding clayey soils. They revealed that cyclic stress and strain history lead to a decrease in G_{max}, while the reduction in G_{max} in the strain-controlled tests is less than that in the stress-controlled tests. Kokusho and Yoshida [18] reported that factors such as stress history (OCR) and plasticity index (I_p) also have some influence on the G_{max}. Vucetic and Dobry [19] exhibited that G_{max} does not change with I_p for normally consolidated clays (OCR = 1), and for overconsolidated clays (OCR > 1), G_{max} increases with I_p. Marika et al. [20] also explored the effect of the initial consolidation path on the G_{max} of Boston blue clay. Laureano et al. [21] examined the impact of moisture content (w) and consolidation time (T) on the G_{max} of Texas-expanded clay and established an empirical equation considering these two factors. Lin et al. [22] investigated the influence of montmorillonite to kaolinite (M-K) ratio and T on the G_{max} of clay soils. Sadeghzadegan et al. [23] discovered that the G_{max} initially decreases as clay content increases from 0 to 20% and then rises slightly as clay content ranges from 20 to 30%. In addition, under the same clay content, the G_{max} enhances when the degree of saturation drops from 100 to 95%. Kantesaria and Sachan [24] indicated that the G_{max} of compacted high-plasticity clay is related to the mean effective stress and slightly depends on deviatoric stress. Simultaneously, G_{max} decreases as the strain level increases during shearing. The above studies provide valuable research values and scientific insights for investigating soft soil's dynamic properties.

As regional soft soils, research on the dynamic properties of the Yangtze River floodplain soft soil is still limited, particularly because the knowledge of shear modulus remains inadequate. G_{max} is the most fundamental parameter for describing the dynamic soil properties, essential for analyzing the seismic response and soil-structure dynamic interaction. This study conducts a series of BE tests to investigate the G_{max} of overconsolidated soft soils in the Yangtze River floodplain. A G_{max} prediction method is developed for reasonably characterizing Yangtze River floodplain soft soils with various over-consolidation states, initial stress conditions, and compactness levels. This study can provide primary data for the engineering and construction requirements of the Yangtze River floodplain.

2. Maximum Dynamic Shear Modulus Tests

2.1. Tested Materials

Floodplain soft soil is extensively dispersed clay formed in the late Quaternary period that is commonly characterized by high moisture content, high void ratio, and low permeability due to many factors, including environment, stress history, and structural properties and can be quaking when the vibration level is high. As depicted in Figure 1, the samples tested in this study belong to typical Yangtze River floodplain soft soils, which are gray-brown and have an apparent horizontal layer texture and sand trap structure. The original soil is air-dried, milled, and sieved, resulting in the formation of loose particles. The soft soils in the diffuse phase are composed entirely of clay with grain sizes less than 0.075 mm, except for the sand layer, which is fine sand.

Figure 1. Undisturbed Yangtze River floodplain facies soft soil.

Figure 2 illustrates the SEM image of the soft soil in the Yangtze River floodplain. It indicates that the original soft soil is mostly agglomerate or flaky aggregates microscopically. Its surface is frequently covered with a loose arrangement of clay particles in the form of stacked flakes. In addition to agglomerates and clumps of mucilage, several highly angular and irregular powder particles are visible. In the relationship between the structure and these agglomerates or granular aggregates, direct contact linkage, bonding material linkage, edge-surface, and point-surface contact play the most critical roles.

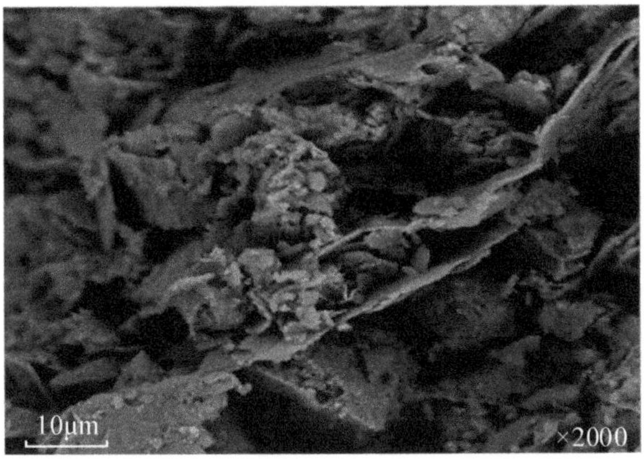

Figure 2. Microstructure of undisturbed Yangtze River floodplain facies soft soil.

Long-term hydraulic transport dominates the sedimentation of Yangtze River floodplain soft soil, which has high dispersibility, hydrophilicity, moisture content, void ratio, compressibility, sensitivity, and generally creeping and thixotropic properties. The physical analysis also demonstrated that the clay minerals species are single, mainly with illite (70%) and chlorite (21.8%), and the clastic minerals are mainly quartz (5.5%) and feldspar (2.7%), and the material composition is unified.

2.2. Specimen Preparation, Saturation, and Consolidation

The Yangtze River floodplain's soft soil is mainly distributed between 5 and 25 m below the surface. The typical original floodplain soft soils, 51 in total, were extracted using an open-end thin-wall sampler. Because the underground water level is about 1–2 m, all the undisturbed samples are saturated soils. Table 1 lists the fundamental physical and mechanical properties of the soft soils in each drill hole. The density ρ, moisture content w, void ratio e, and plasticity index I_p are measured using ASTM D4318 [25]. It indicates that the w values of the floodplain soft soils are between 37 and 42%, and I_p is distributed between 14.8 and 17.8. The samples are classified into four groups based on the depth H of the samples: Group A: H = 5–10 m; Group B: H = 10–15 m; Group C: H = 15–20 m; Group D: H = 20–25 m. σ_{3c}' was set to 50, 85, 120, and 150 kPa for samples in groups A, B, C, and D, respectively. The consolidation of each group was performed with OCR = 1, 2, and 3 in combination with the e distribution in each group. Table 1 details the test conditions.

Table 1. Basic physical properties of test soil samples.

No.	H	ρ	w	e	LL	PL	I_p	Grain size			$\sigma_{3c}{}'$	OCR
	(m bgs)	(g·cm^{-3})	(%)	(-)	(%)	(%)	(%)	Sand [a] (%)	Silt [b] (%)	Clay [c] (%)	(kPa)	(-)
A1	8.7–8.9	1.39	39.32	0.95	38.1	21.8	16.3	25.8	59.8	14.4	50	1
A2	6.8–7.0	1.27	40.11	1.10	49.7	32.5	17.2	16.6	64	19.4	50	1
A3	6.3–6.5	1.31	41.08	1.04	37.8	20.4	17.4	12.6	63.7	23.7	50	1
A4	8.3–8.5	1.27	40.14	1.13	35.4	18.5	16.9	12.5	70.3	17.2	50	1
A5	8.2–8.4	1.33	38.81	1.03	34.4	17.6	16.8	12.5	66.9	20.6	50	2
A6	7.7–7.9	1.29	42.61	1.08	36.9	20.1	16.8	19.9	59	21.1	50	2
A7	9.2–9.4	1.26	41.43	1.12	39.4	22.0	17.4	46.8	39.3	13.9	50	2
A8	7.4–7.7	1.26	40.43	1.15	37.1	20.1	17.0	19.9	59	21.1	50	2
A9	7.3–7.5	1.30	39.55	1.09	37.6	20.1	17.5	19.9	59	21.1	50	3
A10	8.3–8.5	1.33	40.84	1.03	39.0	21.8	17.2	25.8	59.8	14.4	50	3
A11	7.8–8.0	1.26	41.45	1.13	36.3	19.0	17.3	13.8	55	31.2	50	3
A12	5.4–5.6	1.39	42.94	0.96	38.1	21.3	16.8	9.0	66.3	24.7	50	3
B1	13.7–13.9	1.36	40.06	0.97	42.1	24.7	17.4	8.9	68.7	22.4	85	1
B2	14.5–14.7	1.33	39.91	1.04	38.2	21.9	16.3	5.4	74.5	20.1	85	1
B3	14.5–14.7	1.29	40.5	1.09	39.5	21.9	17.6	5.4	74.5	20.1	85	1
B4	12.5–12.7	1.25	40.26	1.14	35.2	18.7	16.5	31.8	51.2	17.0	85	1
B5	14.1–14.3	1.26	39.59	1.16	39.0	22.1	16.9	6.8	76.3	16.9	85	1
B6	14.3–14.5	1.34	40.42	1.01	36.9	20.4	16.5	10.0	64.7	25.3	85	2
B7	13.1–13.3	1.3	40.14	1.08	34.7	18.0	16.7	15.9	66.3	17.8	85	2
B8	14.8–15.0	1.31	41.05	1.09	31.5	14.5	17.0	2.1	76.4	21.5	85	2
B9	14.7–14.9	1.25	41.27	1.14	39.1	21.9	17.2	5.4	74.5	20.1	85	2
B10	13.8–14.0	1.19	40.6	1.26	41.4	24.7	16.7	8.9	68.7	22.4	85	2
B11	14.5–14.7	1.35	38.87	1.02	42.7	27.6	15.1	1.1	77.8	21.1	85	3
B12	13.1–13.3	1.30	38.43	1.09	35.2	18.0	17.2	15.9	66.3	17.8	85	3
B13	13.5–13.7	1.27	40.21	1.12	41.4	24.7	16.7	8.9	68.7	22.4	85	3
B14	14.9–15.1	1.24	41.7	1.14	40.5	23.6	16.9	7.4	76.8	15.8	85	3
C1	15.4–15.6	1.40	42.41	0.95	38.9	21.8	17.1	30.6	48.5	20.9	120	1
C2	17.1–17.3	1.31	38.42	1.04	39.5	23.3	16.2	5.6	67.3	27.1	120	1
C3	16.1–16.3	1.29	41.38	1.08	39.3	23.0	16.3	7.9	66.3	25.8	120	1
C4	15.3–15.5	1.27	39.33	1.13	41.4	23.6	17.8	7.4	76.8	15.8	120	1
C5	16.2–16.4	1.27	38.31	1.15	39.3	23.0	16.3	7.9	66.3	25.8	120	1
C6	15.7–15.9	1.32	39.87	1.03	39.9	23.0	16.9	7.9	66.3	25.8	120	2
C7	15.1–15.3	1.29	40.24	1.08	39.0	21.9	17.1	4.9	69.3	25.8	120	2
C8	15.8–16.0	1.28	36.93	1.11	38.4	23.0	15.4	7.9	66.3	25.8	120	2
C9	16.5–16.7	1.25	41.66	1.13	34.0	16.8	17.2	11.1	24.4	64.5	120	2
C10	16.3–16.5	1.32	37.17	1.03	32.7	16.8	15.9	11.1	24.4	64.5	120	3
C11	15.2–15.4	1.3	39.24	1.08	38.4	23.6	14.8	7.4	76.8	15.8	120	3
C12	15.6–15.8	1.28	40.92	1.12	40.5	23.0	17.5	7.9	66.3	25.8	120	3
C13	15.7–15.9	1.26	39.82	1.14	39.4	23.0	16.4	7.9	66.3	25.8	120	3
D1	21.1–21.3	1.35	41.43	0.99	37.7	20.2	17.5	3.4	72.2	24.4	150	1
D2	22.3–22.5	1.33	40.92	1.05	37.7	21.2	16.5	3.2	72.2	24.6	150	1
D3	23.1–23.3	1.28	41.59	1.11	40.9	23.8	17.1	3.4	74.9	21.7	150	1
D4	23.8–24.0	1.26	40.92	1.15	41.2	23.7	17.5	3.4	73.2	23.4	150	1
D5	21.6–21.8	1.23	41.66	1.18	39.3	22.3	17.0	3.4	83.4	13.2	150	2
D6	22.5–22.7	1.27	39.45	1.11	39.6	22.5	17.1	5.0	69.5	25.5	150	2
D7	23.3–23.5	1.3	41.11	1.05	39.1	21.8	17.3	4.4	83.4	12.2	150	2
D8	23.8–24.0	1.38	38.23	0.98	37.9	21.5	16.4	8.6	68.2	23.2	150	2
D9	21.9–22.1	1.37	39.83	0.95	40.4	23.8	16.6	3.4	85.2	11.4	150	3
D10	22.9–23.1	1.33	40.14	1.03	38.1	20.3	17.8	15.6	60.9	23.5	150	3
D11	23.4–23.6	1.28	39.82	1.1	37.5	20.4	17.1	14.5	71.7	13.8	150	3
D12	24.6–24.8	1.25	38.92	1.16	38.3	20.9	17.4	3.4	83.4	13.2	150	3

Note: 'bgs' means 'below ground surface,' LL = liquid limit, PL = plastic limit. [a] The grain size of sand particles ranges from 0.075~0.1 mm. [b] The grain size of silt particles ranges from 0.005~0.075 mm. [c] The grain size of clay particles is below 0.005 mm.

The standard dimensions of samples (50 × 100 mm) were prepared and saturated using the vacuum saturation method according to the ASTM D5311-13 [26]. The samples were saturated in the saturation vessel for 10 h. Then, the specimen was placed in the apparatus, and the back pressure saturation was conducted. Back pressure was applied step by step until the value of Skempton's B was greater than 0.97, which can guarantee the saturation to prevent differences in sample preparation from affecting the test results. Then they were placed in the pressure chamber and consolidated according to the corresponding consolidation conditions. When the average strain rate of the soil sample was less than

$1 \times 10^{-3}\%/\text{min}$, consolidation was achieved, followed by the bender element test. The time of consolidation was about 4~5 days.

2.3. Testing Apparatus and Process

The bender element measurement system installed in the GCTS HCA-300 static and dynamic triaxial instrument is used for the test. The bender element includes two piezoelectric ceramic sheets and a central copper stiffening layer. During the test, two bender elements, excitation and receiving, are installed at both ends of the soil sample. The excitation element produces horizontal vibration under the excitation of a specific pulse voltage. In addition, the shear wave reaches the receiving element following the sample propagation. A weak electrical signal is generated at this time, and the oscilloscope can calculate the propagation time (t) of the shear wave. A series of sinusoid signals from 1 to 40 kHz was used as the excitation, and the received signals corresponding to these excitation frequencies were examined to better identify the t. The 10 kHz excitation signal consistently yielded a clear arrival of the shear wave for floodplain soft soils. This is in good agreement with the observation of Yang and Liu [27] and Chen et al. [28]. The height of the soil sample minus the length of the bender element deep into the soil body can be utilized to calculate the shear wave propagation distance and thus determine the shear wave velocity. The equation for calculating the soil shear wave velocity (Vs) using the indoor bender element test is:

$$V_s = L_0/t \tag{1}$$

where $L_0 = L - L_b$, L is the height of the sample; L_b is the length of the bender element deep into the soil; t is the propagation time.

Calculating the shear wave velocity (V_s) and the maximum dynamic shear modulus G_0 requires accurately determining the shear wave's arrival time. Different scholars have proposed various analytical methods for identifying the shear wave arrival time. The widely used methods are the time-domain initial wave, peak-to-peak, and intercorrelation methods. Brigonoli et al. [29], Lee et al. [30], and Zhou et al. [31] indicated that the shear wave velocity propagation time t could be determined easily and accurately using the time-domain initial wave method. In this experiment, a sinusoidal pulse frequency of 10 kHz was applied, and the shear wave always arrived clearly and effectively under the excitation frequency of 10 kHz. The time domain initial wave approach is employed to determine the arrival time of shear wave velocity, i.e., depending on the first turning point A of the received signal as the arrival point of the shear wave. Figure 3 depicts a typical bender element test received signal diagram. After determining $V_{s.}$ for each sample, the following equation is utilized to calculate G_{max} according to elastic theory:

$$G_{max} = \rho V_s^2 \tag{2}$$

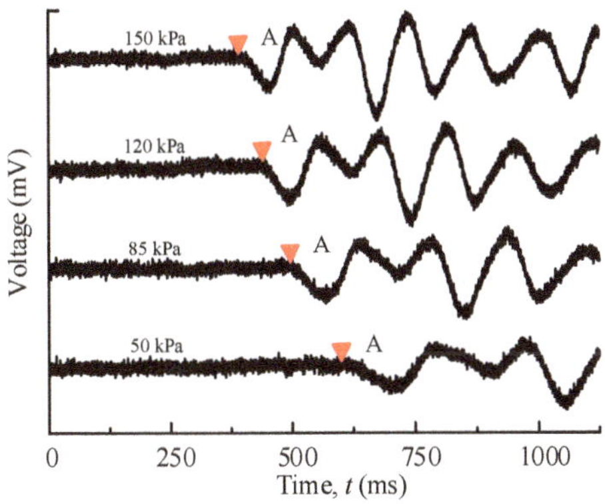

Figure 3. Typical time histories of output signals from bender element tests.

3. Test Results and Analysis

Figure 4 illustrates the relationship between the G_{max} and e for the Yangtze River floodplain's soft soils under different σ_{3c}'. For the given σ_{3c}' and OCR, G_{max} decreases as e increases. This is because the smaller e is, the looser the soil, the weaker the cohesion and cementation between the particles, the more open the pores in the soil, and the more unstable the soil structure, leading to the smaller G_{max}. In addition, under the same σ_{3c}' and e, G_{max} improves with increasing OCR, revealing that the G_{max} is larger, the higher the pre-consolidation pressure. When Figure 4a–d are combined with the given e and OCR, G_{max} rises as σ_{3c}' grows. The increase in σ_{3c}' has a noticeable hoop-tightening effect on the soil, which can reduce the void ratio of the soil and strengthen the bond between soil particles. The greater the effective envelope pressure, the greater the soil sample's resistance to shear deformation, as measured by an increase in G_{max}. In each graph, three red trend lines represent the effect of e on G_{max}, respectively, and Jamiolkowski et al. [32] proposed the equation $F(e)$ to characterize the effect of e on G_{max}.

$$F(e) = e^{-d} \tag{3}$$

where d is the fitting parameter characterizing the effect of increasing e on the degree of G_{max} decay. When σ_{3c}' and OCR are constants, d can be determined using regression analysis, and the corresponding values of d are displayed in Figure 5. It reveals that under the same OCR, the increase in σ_{3c}' does not significantly affect d, which remains nearly constant. As OCR rises, d reduces gradually, with d values of 2.981, 2.523, and 2.055 for OCR = 1, 2, and 3, respectively. This indicates that the decay rate of G_{max} with e is independent of σ_{3c}' but decreases as OCR grows.

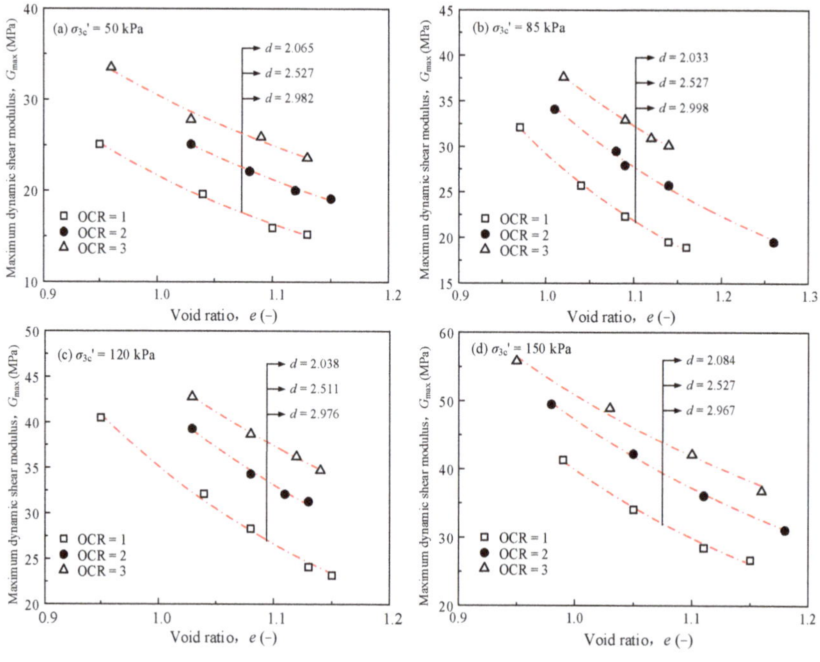

Figure 4. The relationship between G_{max} and e under different σ_{3c}': (**a**) σ_{3c}' = 50 kPa, (**b**) σ_{3c}' = 85 kPa,. (**c**) σ_{3c}' = 120 kPa, (**d**) σ_{3c}' = 150 kPa.

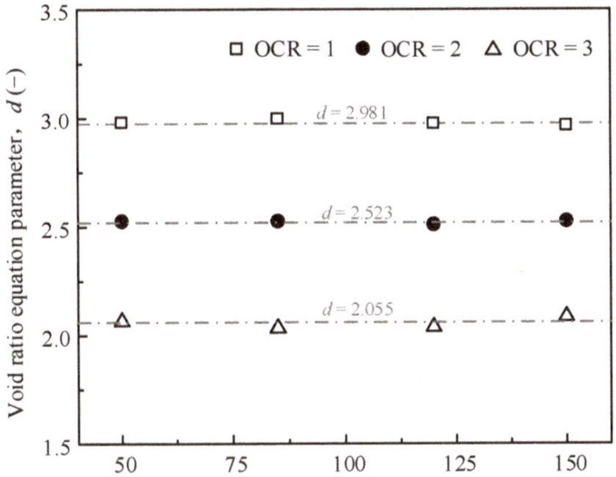

Figure 5. Relationship between d and σ_{3c}' with different OCR.

Figure 6 displays the relationship between the normalized maximum shear modulus $G_{max}/F(e)$ and the normalized initial effective confining pressure σ_{3c}'/P_a, where P_a is the standard atmospheric pressure, assumed to be 100 kPa. For given OCR, $G_{max}/F(e)$ grows as $\sigma3c'/Pa$ improves, but the growth rate gradually decreases, and a power function can describe the relationship between $G_{max}/F(e)$ and σ_{3c}'/P_a:

$$G_{max} = AF(e)(\sigma_{3c}'/P_a)^n \tag{4}$$

where A is a measured fitting parameter whose value corresponds to the $G_{max}/F(e)$ value of the floodplain's soft soil at $\sigma3c'$ = 100 kPa; n is the stress index, describing the influence of σ_{3c}' on $G_{max}/F(e)$. In addition, the $G_{max}/F(e) \sim \sigma_{3c}'/P_a$ relationship curve gradually shift upward as the OCR increases, with the A value gradually increasing from 31.8 to 42.0 MPa and n decreasing from 0.552 to 0.465 as the OCR rises from 1 to 3.

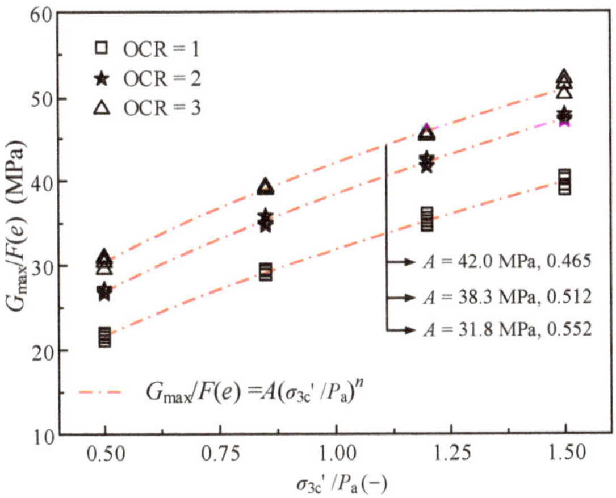

Figure 6. Relationship between $G_{max}/F(e)$ and σ_{3c}'/P_a.

Figures 5 and 6 demonstrate that the OCR significantly impacts the fitted parameters d, A, and n. In contrast, Figure 7 shows that as OCR enhances, d grows linearly, A rises, and n decreases linearly. The following equations give the relationship between d, A, n, and OCR:

$$d = 0.463 \times \text{OCR} + 1.594 \tag{5}$$

$$A = 31.9 \times \mathrm{OCR}^{0.255} \tag{6}$$

$$n = -0.043 \times \mathrm{OCR} + 0.597 \tag{7}$$

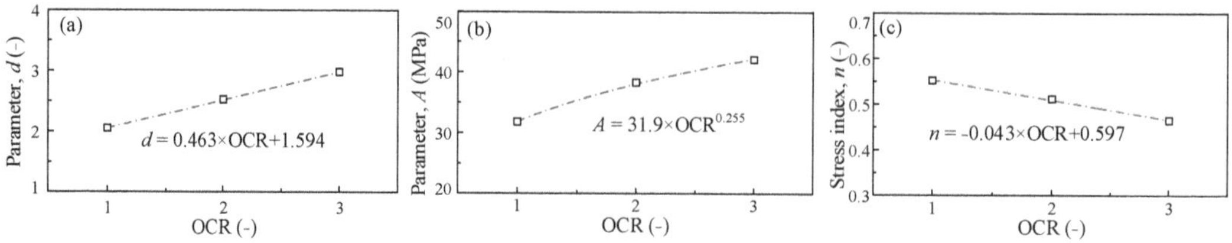

Figure 7. The relationship between parameters d, A, n, and OCR: (**a**) d versus OCR, (**b**) A versus OCR, (**c**) n versus OCR.

The G_{max} prediction equation that integrates and considers OCR, e, and σ_{3c}' can be established by combining Equations (3) and (4).

$$G_{max} = 31.9\mathrm{OCR}^{0.255} \times e^{-0.463\mathrm{OCR}+1.594} \times (\sigma_{3c}'/P_a)^{-0.043\mathrm{OCR}+0.597} \tag{8}$$

Based on this prediction equation, the values of G_{max} of the Yangtze River overconsolidated floodplain's soft soils are predicted by the following step:

Step 1: Determining the effective stress, σ_{3c}', based on the depth of the soil extracted in situ.

Step 2: Identifying the void ratio, e, from laboratory experiments.

Step 3: Calculating the value of the per-consolidation pressure on the compression curve ($e - p$ curve) using the Casagrande method and computing the over-consolidation ratio, OCR.

Step 4: Based on the modified σ_{3c}', e, OCR, estimating G_{max} using Equation (8).

Figure 8 compares the G_{max} tested and predicted values from Equation (8). For Yangtze River floodplain soft soils with various e, σ_{3c}', and OCR, the difference of G_{max} predicted values using the proposed evaluation model is less than 10%. Liang et al. [33] developed a G_{max} prediction model for sand with different values of relative densities, D_r, and the deviation between the predicted and the measured values of G_{max} was within a range of 20%. Zhang et al. [34] established a G_{max} prediction model for marine soils, and the deviation between predicted and measured values of G_{max} was within a range of 15%. Therefore, it is encouraging that the theoretically estimated G_{max} of Yangtze River floodplain soft soils (Equation (8)) coincide well with the experimentally measured G_{max} values. The significant implication of this study is that the predicted Gmax of Yangtze River floodplain soft soils serve as a valuable reference for the site seismic response analysis in the Yangtze delta region.

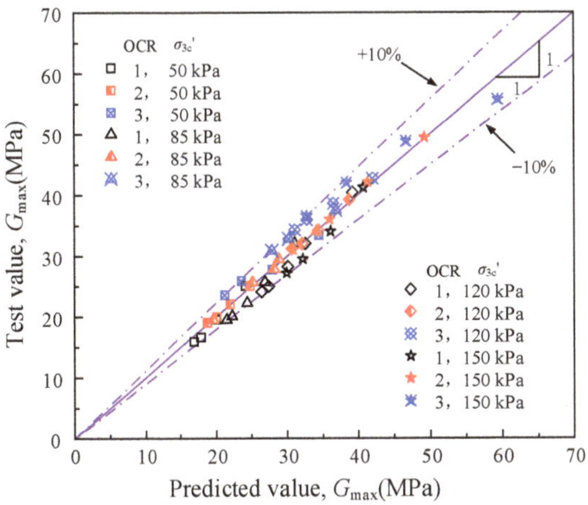

Figure 8. Comparison of G_{max} predicted values and test values.

4. Conclusions

The maximum dynamic shear modulus G_{max} of Yangtze River floodplain soft soil with sand trap structure and horizontal layer texture is examined using a series of bender element tests. The effects of void ratio e, initial effective confining pressure σ_{3c}', and overconsolidated ratio OCR on G_{max} are investigated, and the analysis indicated that G_{max} decreases with e, increasing OCR causes a gradual decrease in the decay rate of G_{max} with e, whereas increasing σ_{3c}' does not affect G_{max} decay rate. Furthermore, the normalized maximum shear modulus $G_{max}/F(e)$ increases as normalized initial effective confining pressure σ_{3c}'/Pa increases. However, its growth rate gradually reduces, and $G_{max}/F(e)$ has a power function relationship with σ_{3c}'/Pa. As the OCR rises, fitting parameters d and A grow while stress index n decreases. d and n exhibit a linear correlation with OCR, whereas A has a power function relationship with OCR. Based on the regression analysis, a G_{max} prediction method is proposed to adequately characterize the Yangtze River floodplain soft soils with various over-consolidation states, initial stress conditions, and densities, with a prediction error of less than 10%.

In the future, the dynamic behavior of soft floodplain soils subjected to cyclic loading will be conducted to investigate the cyclic degradation and establish the cyclic degradation model based on G_{max}. Additionally, microcosmic tests will be performed on floodplain soft soils to explain the dynamic characteristics.

Author Contributions: Methodology, Y.Z., Z.Z. and Q.W.; Validation, X.X.; Investigation, X.X.; Writing—original draft, Y.Z.; Writing—review & editing, Z.Z. and Q.W.; Project administration, Q.W.; Funding acquisition, Q.W. All authors have read and agreed to the published version of the manuscript.

Funding: This research received no external funding.

Institutional Review Board Statement: Not applicable.

Informed Consent Statement: Not applicable.

Data Availability Statement: Data is available on request through the author named Qi Wu.

Conflicts of Interest: The authors declare that there is no conflict of interest regarding the publication of this paper.

References

1. Senetakis, K.; He, H. Dynamic characterization of a biogenic sand with a resonant column of fixed-partly fixed boundary conditions. *Soil Dyn. Earthq. Eng.* **2017**, *95*, 180–187. [CrossRef]
2. An, R.; Kong, L.; Zhang, X.; Li, C. Effects of dry-wet cycles on three-dimensional pore structure and permeability characteristics of granite residual soil using X-ray micro computed tomography. *J. Rock Mech. Geotech. Eng.* **2022**, *14*, 851–860. [CrossRef]
3. Chen, G.X.; Liang, K.; Zhao, K.; Yang, J. Shear modulus and damping ratio of saturated coral sand under generalised cyclic loadings. *Géotechnique* **2022**, 1–18. [CrossRef]
4. Wu, Q.; Wang, Z.; Qin, Y.; Yang, W. Intelligent Model for Dynamic Shear Modulus and Damping Ratio of Undisturbed Marine Clay Based on Back-Propagation Neural Network. *J. Mar. Sci. Eng.* **2023**, *11*, 249. [CrossRef]
5. He, H.; Li, M.N.; Senetakis, K. A note on influence of stress anisotropy on the Poisson's ratio of dry sand. *J. Rock Mech. Geotech. Eng.* **2017**, *9*, 1159–1164. [CrossRef]
6. Morsy, A.M.; Salem, M.A.; Elmamlouk, H.H. Evaluation of dynamic properties of calcareous sands in Egypt at small and medium shear strain ranges. *Soil Dyn. Earthq. Eng.* **2019**, *116*, 692–708. [CrossRef]
7. Subramanian, S.; Khan, Q.; Ku, T. Effect of sand on the stiffness characteristics of cement-stabilized clay. *Constr. Build. Mater.* **2020**, *264*, 120192. [CrossRef]
8. Jafarian, Y.; Javdanian, H.; Haddad, A. Dynamic properties of calcareous and siliceous sands under isotropic and anisotropic stress conditions. *Soils Found.* **2018**, *58*, 172–184. [CrossRef]
9. Yang, S.L.; Ren, Y.B.; Andersen, K.N. Effects of thixotropy and reconsolidation on the undrained shear characteristics of remoulded marine clays. *Ocean Eng.* **2021**, *239*, 109888. [CrossRef]
10. Wu, Q.; Liu, Q.F.; Zhuang, H.Y.; Xu, C.X.; Chen, G.X. Experimental investigation of dynamic shear modulus of saturated marine coral sand. *Ocean Eng.* **2022**, *264*, 112428.
11. Kim, D.S.; Stokoe, K.H. Deformational characteristics of soils at small to medium strains. In Proceedings of the First International Conference on Earthquake Geotechnical Engineering, Tokyo, Japan, 14–16 November 1995; pp. 89–94.
12. Irfan, M.; Cascante, G.; Basu, D.; LeBoeuf, D. Low-strain dynamic characterization of undisturbed Leda clay. *Can. Geotech. J.* **2022**, *59*, 631–643. [CrossRef]
13. Hardin, B.O.; Black, W.L. Vibration modulus of normally consolidated clay. *J. Soil Mech. Found. Div.* **1968**, *92*, 27–42. [CrossRef]
14. Kim, T.C.; Novak, M. Dynamic properties of some cohesive soils of Ontario. *Can. Geotech. J.* **1981**, *18*, 371–389. [CrossRef]
15. Payan, M.; Chenari, R.J. Small strain shear modulus of anisotropically loaded sands. *Soil Dyn. Earthq. Eng.* **2019**, *125*, 105726. [CrossRef]
16. Liu, J.; Otsubo, M.; Kawaguchi, Y.; Kuwano, R. Anisotropy in small-strain shear modulus of granular materials: Effects of particle properties and experimental conditions. *Soils Found.* **2022**, *62*, 101105. [CrossRef]
17. Gu, C.; Wang, J.; Cai, Y.; Guo, L. Influence of cyclic loading history on small strain shear modulus of saturated clays. *Soil Dyn. Earthq. Eng.* **2014**, *66*, 1–12. [CrossRef]
18. Kokusho, T.; Yoshida, Y.; Esashi, Y. Dynamic Properties of Soft Clay for Wide Strain Range. *Soils Found.* **1982**, *22*, 1–18. [CrossRef]
19. Vucetic, M.; Dobry, R. Effect of soil plasticity on cyclic response. *J. Geotech. Geoenviron. Eng.* **1991**, *117*, 89–107. [CrossRef]
20. Santagata, M.; Germaine, J.T.; Ladd, C.C. Factors affecting the initial stiffness of cohesive soils. *J. Geotech. Geoenviron. Eng.* **2005**, *131*, 430–441. [CrossRef]
21. Hoyos, L.R.; Puppala, A.J.; Chainuwat, P. Dynamic Properties of Chemically Stabilized Sulfate Rich Clay. *J. Geotech. Geoenviron. Eng.* **2004**, *130*, 153–162. [CrossRef]
22. Lin, P.; Ni, J.-J.; Garg, A.; Yu, S.-M. Effects of Clay Minerals on Small-Strain Shear Modulus and Damping Ratio of Saturated Clay. *Soil Mech. Found. Eng.* **2020**, *57*, 105–109. [CrossRef]
23. Sadeghzadegan, R.; Naeini, S.A.; Mirzaii, A. Effect of clay content on the small and mid to large strain shear modulus of an unsaturated sand. *Eur. J. Environ. Civ. Eng.* **2020**, *24*, 631–649. [CrossRef]
24. Kantesaria, N.; Sachan, A. Small-strain shear modulus and yielding characteristics of compacted high-plasticity clay. *Géotechnique* **2022**, *72*, 424–437. [CrossRef]
25. *D4318-17E01*; Standard test method for sand content by volume of bentonitic slurries. ASTM International: West Conshohocken, PA, USA, 2017.
26. *D5311-13*; Standard Test Method for Load Controlled Cyclic Triaxial Strength of Soil. ASTM International: West Conshohocken, PA, USA, 2013.
27. Yang, J.; Liu, X. Shear wave velocity and stiffness of sand: The role of non-plastic fines. *Géotechnique* **2016**, *66*, 500–514. [CrossRef]
28. Chen, G.X.; Wu, Q.; Zhao, K.; Shen, Z.F.; Yang, J. A Binary packing material-based procedure for evaluating soil liquefaction triggering during earthquakes. *J. Geotech. Geoenviron. Eng.* **2020**, *146*, 04020040. [CrossRef]
29. Chaney, R.; Demars, K.; Brignoli, E.; Gotti, M.; Stokoe, K. Measurement of Shear Waves in Laboratory Specimens by Means of Piezoelectric Transducers. *Geotech. Test. J.* **1996**, *19*, 384–397. [CrossRef]
30. Lee, J.-S.; Santamarina, J.C. Bender Elements: Performance and Signal Interpretation. *J. Geotech. Geoenviron. Eng.* **2005**, *131*, 1063–1070. [CrossRef]
31. Zhou, Y.-G.; Chen, Y.-M. Laboratory Investigation on Assessing Liquefaction Resistance of Sandy Soils by Shear Wave Velocity. *J. Geotech. Geoenviron. Eng.* **2007**, *133*, 959–972. [CrossRef]

32. Jamiolkowski, M.; Lancellotta, R.; Lopresti, D.C.F. Remarks on the stiffness at small strains of six Italian clays. In Proceedings of the Pre-Failure Deformation of Geomaterials, Sapporo, Japan, 12–14 September 1994; pp. 817–836.
33. Liang, K.; Chen, G.; Du, X.; Xu, C.; Yang, J. A Unified Formula for Small-Strain Shear Modulus of Sandy Soils Based on Extreme Void Ratios. *J. Geotech. Geoenviron. Eng.* **2023**, *149*, 04022127. [CrossRef]
34. Zhang, Y.; Zhao, K.; Peng, Y.; Chen, G. Dynamic shear modulus and damping ratio characteristics of undisturbed marine soils in the Bohai Sea, China. *Earthq. Eng. Vib.* **2022**, *21*, 297–312.

applied sciences

MDPI

Article

Sand Rubber Mixtures under Oedometric Loading: Sand-like vs. Rubber-like Behavior

Pravin Badarayani [1], Bogdan Cazacliu [1], Erdin Ibraim [2], Riccardo Artoni [1] and Patrick Richard [1,*]

[1] Université Gustave Eiffel, MAST, GPEM, F-44344 Bouguenais, France
[2] Department of Civil Engineering, University of Bristol, Bristol BS8 1TR, UK
* Correspondence: patrick.richard@univ-eiffel.fr

Abstract: Each year, the number of scrap tires disposed of in huge piles across the world continuously increases. Consequently, new recycling solutions for these materials have to be proposed. Among them, one possibility consists of shredding tires and mixing the obtained tire chips with sand, which can be used as alternative soils in various geotechnical applications, such as backfilling for retaining structures, slope and highway embankment stabilization, road constructions, soil erosion prevention, and seismic isolation of foundations. Such types of mixtures are highly heterogeneous due to the important difference in elasticity and deformability between the two constituents, which leads to complex mechanical behavior. In this article, the one-dimensional loading/unloading behavior of sand-rubber mixtures is investigated by laboratory strain-controlled experiments performed for different packing densities, particle sizes, rubber contents, and sand/rubber size ratios. After a global analysis of the increase of the packing deformation with the rubber fraction and the stress level, a novel criterion to classify the behavior of the mixture as sand-like or rubber-like was proposed, based on the concavity of the void ratio—log of vertical stress curve. The concavity increased with the stress level and the rubber fraction, up to the limits where the saturation of the voids due to their filling with rubber induces a rubber-like behavior. A simplified phase diagram, limited to the range of this study, is proposed. The one-dimensional confined stiffness and the swelling behavior were also analyzed.

Keywords: sand rubber mixture; granular materials; oedometric compression

1. Introduction

Multistate and multiphase particulate materials may provide a high degree of flexibility in both the design and performance of engineering structures, including the potential for project cost efficiency. This is certainly true for civil engineering materials made from combinations of rigid, liquid, and soft deformable constituents. Soft rubber particles, resulting mainly from recycled waste tires, are used as a partial replacement for granular soils in a large variety of geotechnical applications, such as backfill of retaining walls [1], highway embankments [2–4], subgrade road construction [5,6], ground erosion control, and stabilization of slopes [7,8], as well as seismic foundation isolation [9–12]. While the reduced weight and improved damping properties of sand–rubber mixtures, resulting from the addition of rubber particles, provide a reduction of lateral earth pressures on the retaining walls [1,13] and increased capacity against seismic actions and soil liquefaction occurrence [14–16], the use of rubber as a secondary raw material complements the recovery rate of the scrap tires and reduces the pressure on the landfill disposal solutions.

The engineering properties of sand–rubber composite mixtures have been extensively studied over the past two–three decades in both laboratories and in situ setups, although the latter to a lesser extent. Segregation in the sand–rubber mixtures have been characterized in [17]. In the case of classical laboratory experiments, most of the research has mainly focused on the characterization of the compressibility either in one-dimension or in isotropic loading conditions [18–23], and the monotonic strain-stress-strength behavior of various combinations of soils, rubber types, and particle size dimensions [24–29]. Exploration of the small strain stiffness of sand–rubber composites has also been conducted [19,30–34]. The applicability of the critical state framework to sand–rubber mixtures has been reported in [35,36]. A simple model for predicting the isotropic compression behavior based on soft–rigid interactions at the particle scale has been developed in [37]. Particle-scale mechanisms that underlie the energy dissipation were observed in the sand–rubber mixtures under oedometer testing and have been previously reported [38]. An excellent summary of a range of experimental studies, including the physical characteristics of the mixture materials used as well as the testing conditions and main conclusions, can be found in [35,36]. While, as equally emphasized by the same authors, comparison between different studies remains challenging, the examination of published results shows that the response of the sand–rubber composite mixtures is mainly controlled by the volume fraction of the components and the characteristic geometrical contrasts of the rigid/soft particles through either the aspect ratio (length over width) of fiber-type rubber inclusions, which encourages additional fiber-soil type internal interaction mechanisms [22,26,39,40] or the rubber to sand particle size ratio, as shown by [19,30,41].

Based on a given set of these variables, the behavior of sand–rubber mixtures can vary from sand-like to rubber-like with a transitional state confinement stress-dependent. Large triaxial tests on sand–rubber mixtures and the behavior of mixtures as sand-like (for rubber contents < 34%) and rubber-like (rubber contents > 34%) have been reported in [42]. Further, Mashiri et al. [43] proposed three different behavioral zones, viz. sand-like (rubber contents < 18%), rubber-like (rubber contents > 35%), and sand-rubber-like (18% < rubber contents < 35%). Lee et al. [19] also investigated the small strain and large strain response of sand-rubber mixtures and reported rubber-like (for rubber contents > 60%) and sand-like behaviors (for rubber contents < 30%), whereas for rubber content equal to 40%, it was rubber-like (for low stress) and sand-like (for high stress). While the characterization of these behavioral zones is fundamental for the use of such composites in practical applications and the design of geotechnical systems, their identification is based on qualitative observations of the sand–rubber mixture response.

This article presents the results of strain-controlled one-dimensional loading/unloading experiments performed on sand–rubber mixtures. The influence of different packing densities, particle sizes, rubber contents, and sand/rubber size ratios was evaluated. The one-dimensional confined stiffness and the swelling behavior were more particularly studied. Based on these results, a novel criterion to classify the behavior of the mixture as sand-like or rubber-like is proposed and used to study how this behavior varies with the stress level and the rubber content, in the ranges used within these experiments.

2. Materials and Methods

2.1. Materials, Sample Fabrication, and Experimental Setup

The oedometer tests on the sand and sand–rubber mixtures were conducted on cylindrical, rigidly laterally confined samples of 74 mm diameter and 40 mm height. While Leighton Buzzard fraction A sand [44] and rubber particles obtained from the shredding of used lorry tires [17,37,45] were used as base materials, the sand–rubber mixtures contained specific sand and rubber fraction dimensions. Pretest sample preparation procedures, therefore, included sieving both sand and rubber particles and particle size separation into the following grading sizes: 0.6–0.8 mm (D_{50} = 0.7 mm), 1.25–1.6 mm (D_{50} = 1.4 mm), 1.6–2.0 mm (D_{50} = 1.8 mm), and 2.0–2.24 mm (D_{50} = 2.1 mm) for LB sand, and 1.25–1.6 mm (D_{50} = 1.4 mm) and 1.6–2.0 mm (D_{50} = 1.8 mm) for the rubber, where D_{50} represents the mean grain size of the selected fraction.

All the samples were fabricated in two layers, which were 20 mm each in height, and followed the moist tamping procedure [45–47] (see Figure 1a). The required quantities of dry sand or sand and rubber for each sample layer were, first, manually mixed in a container together with 10% water (by dry weight of solids). Then, the mixture was gently placed in the rigid oedometer cylinder using a spoon and compacted by a rigid steel cylindrical tamper, which had a diameter half that of the sample's diameter [48], until the required target sample void ratio was achieved. Following the completion of fabrication, the sample was placed on the testing configuration and saturated by flushing water through the porous stone cylinder base. Once the saturation was completed, a relatively light steel cylinder top cap was placed, the sample settlement was recorded, and the fabrication void ratio was updated.

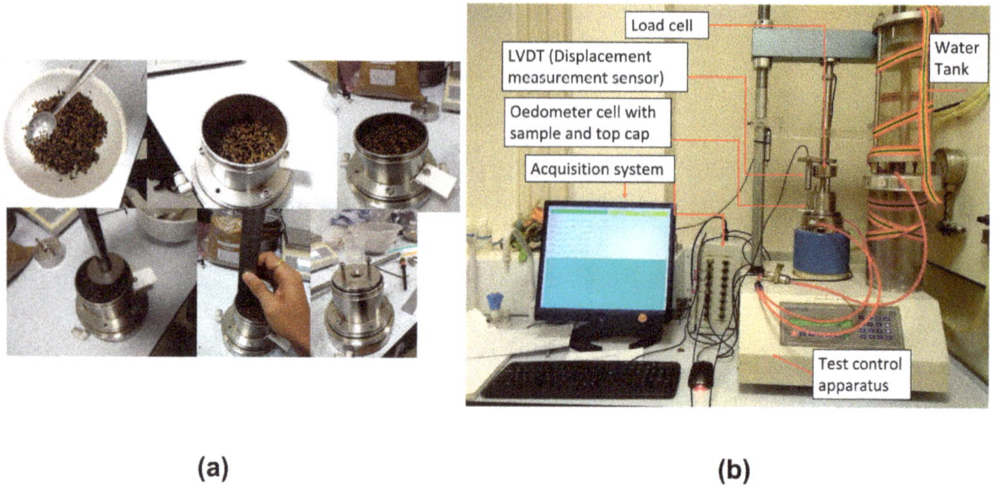

Figure 1. Figure showing (**a**) fabrication of mixtures using the moist tamping method and (**b**) experimental setup for the oedometric compression tests [46].

The one-dimensional testing was conducted using a displacement-controlled electromechanical loading frame (refer to Figure 1b). The lower platen supporting the sample moves upwards with a constant speed set for all tests at 0.5 mm/min, while the sample top cap remains fixed and in contact with the loading ram, incorporating a linear variable differential transformer (LVDT) for vertical displacement measurements, and a 10 kN-load cell (with a linear response throughout the whole measurement range). The one-dimensional compression was stopped once the force approached the load cell limit capacity (corresponding to approximately 2.3 MPa of vertical stress) and the sample was, then, fully unloaded by reversing the lower platen movement at a similar speed. However, at the unloading point, the sample experienced some relaxation due to the backlash in the electromechanical loading system.

2.2. Experimental Program

Several rubber fractions (x_r) were chosen for the tested sand–rubber mixtures: 0%, 10%, 20%, 25%, 30%, 40%, and 50%. The rubber fraction, x_r, is defined as the ratio between the volume of rubber and the total volume of solids (rubber and sand). In the calculation of the rubber fraction, a solid density of 2.65 g/cm^3 of sand and a solid density of 1.04 g/cm^3 of rubber were used.

Packings with two target fabrication void ratios, 0.65 and 0.75, were produced with equal-sized rubber and sand particles with a mean grain size, D_{50}, of 1.8 mm. Samples with a void ratio of 0.65 and made of equal-sized sand and rubber particles with a mean size D_{50} of 1.4 mm were also tested. For two particular rubber fractions, i.e., 10% and 50%, tests were also performed on sand–rubber mixtures at a void ratio of 0.65 by choosing sand/rubber mean particle size ratios of 0.5 and 1.5. The rubber size used for these tests was a D_{50} of 1.4 mm, while the size of the sand was either a D_{50} of 0.7 mm or 2.1 mm.

It is to be noted that the maximum and minimum void ratios of the Leighton Buzzard fraction A sand were 0.83 and 0.55, respectively. These limits are of course different for the present packings, with close grading sizes and significant rubber content. However, no attempt was made to determine these void ratio limits.

Table 1 presents the characteristics of tested samples and the material constituents. The void ratio, e_o, in Table 1 represents the void ratio at the beginning of the one-dimensional compression test. For a given sand–rubber mixture type, some deviation from the target value of the void ratio was recorded and while the gap increased with the rubber content, its extent is regarded as very limited. In the name of the tests, capital letters 'S/R' opposite to small caps letters 's/r' designate samples with a higher sand and rubber particle size, 1.8 mm as opposed to 1.4 mm, while D or L refers to a dense sample condition (0.65 void ratio) or a loose sample condition (0.75 void ratio), respectively. The use of sR-D-10 refers to a dense sample (0.65 void ratio) with 10% rubber content and sand (s) particles smaller than the rubber (R) particles. Similarly, Sr-D-50 refers to a dense sample (0.65 void ratio) with 50% rubber content, yet with sand (S) particles larger than the rubber (r) particles.

Table 1. Summary of the testing series and sample characteristics.

Sand–Rubber Mixture Type	Sand Particle Mean Size, D_{50} (mm)	Rubber Particle Mean Size, D_{50} (mm)	Sand/Rubber Mean Size Ratio	Rubber Fraction, x_r (%)	Sample Void Ratio at the Beginning of Test, e_o
SR-D	1.8	1.8	1.0	0 10 20 25 30 40 50	0.647 0.646 0.644 0.643 0.642 0.641 0.640
SR-L	1.8	1.8	1.0	0 10 20 25 30 40 50	0.748 0.746 0.744 0.744 0.742 0.742 0.740
sr-D	1.4	1.4	1.0	0 10 20 25 30 40 50	0.647 0.645 0.645 0.644 0.643 0.642 0.642
sR-D-10 sR-D-50	0.7	1.4	0.5	10 50	0.646 0.645
Sr-D-10 Sr-D-50	2.1	1.4	1.5	10 50	0.642 0.642

3. Results and Discussion

3.1. General One-Dimensional Response

The vertical strain, ε_v, and stress, σ_v, during an oedometer test are represented in Figure 2 by a conventional semi-logarithmic scale for the different rubber fractions at the two initial densities, SR-D and SR-L tests. As expected, the magnitude of the packing deformation increases with the rubber fraction, while for a given rubber fraction, the deformation of the dense packing is smaller than that of the loose packing.

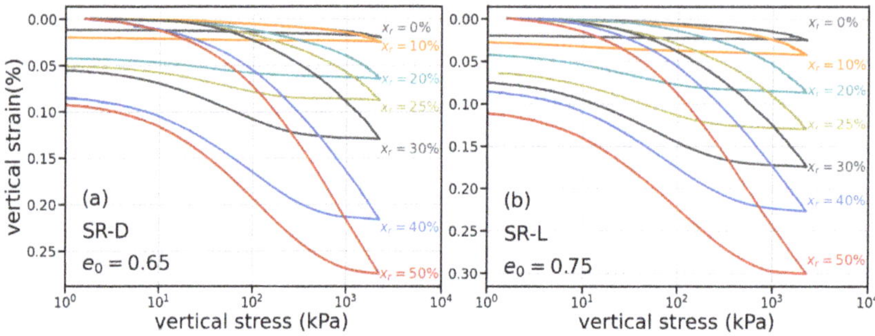

Figure 2. Vertical strain as a function of the vertical stress for samples with 0% to 50% rubber fractions, loading, and unloading during oedometer tests: (**a**) SR-D tests; (**b**) SR-L tests (see Table 1).

The evolutions of the maximum vertical strain ε_{max}, obtained at the maximum applied vertical stress, σ_{max}, of 2.3 MPa, and unrecovered vertical strain, ε_o, after complete unloading with the rubber fraction, x_r, for SR-D, SR-L, and sr-D series tests, and Sr-D-10, Sr-D-50, sR-D-10, and sR-D-50 tests are given in Figure 3. As can be observed, independent of the rubber content, the effect of the particle size for the SR-D and sr-D tests in the measured maximum or unrecovered strain appears negligible (the non-null difference is smaller than 5% of the reference strain). The standard deviations of the strain range differences for the maximum and unrecovered vertical strain at each rubber fraction for these two series are around 0.7% and 0.4%, respectively. For both strain limits, the looser packing of SR-L tests induced, on average, 0.9% more vertical strain compared with the denser SR-D and sr-D samples.

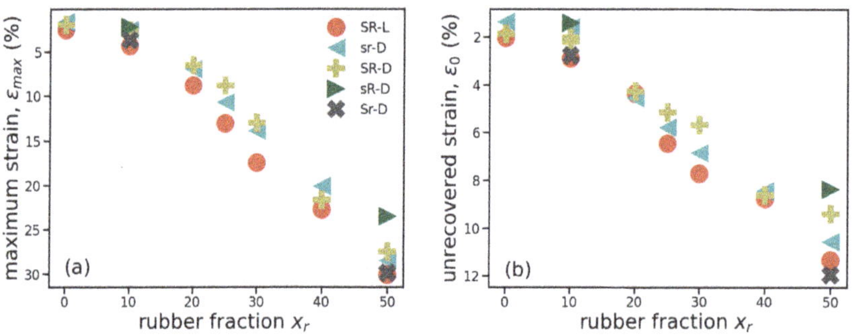

Figure 3. Evolution of (**a**) the maximum vertical strain and (**b**) the unrecovered strain after complete unloading with the rubber fraction for SR-D, SR-L, and sr-D series tests, and Sr-D-10, Sr-D-50, sR-D-10, and sR-D-50 tests (see Table 1).

Figure 3 also plots the corresponding values of the strain limits for tests on samples made by non-equal sand and rubber particle sizes. For the 10% rubber fractions, while the volume of sand is nine times bigger than the volume of rubber, the behavior is dominated by the sand phase and very close to the SR or sr cases. When x_r = 50%, the volume of sand equals the volume of rubber, yet the number of sand particles is about eight times larger than the number of rubber particles when the sand/rubber mean size ratio is 0.5 (sR-D tests), while the number of rubber particles is three times bigger than the sand particles when the sand/rubber mean size ratio is 1.5 (Sr-D tests). As a consequence, the response is more compressible for the latter Sr-D case and less compressible for the former sR-D case, when compared with equal sand and rubber particle sizes tests. A similar observation for the effect of the smaller-sized rubber particles on the one-dimensional compression was also witnessed by [40]. The unrecovered strains follow the same trends, with lower and higher strains for sR and Sr cases.

The loading behavior in one-dimensional testing is generally represented by the evolution of the current void ratio, e, with the vertical stress in the log scale ($\log \sigma_v$). The current void ratio is related to the vertical strain, ε_v, by the equation:

$$e = e_0 - \varepsilon_v(1 + e_0) \quad (1)$$

where e_0 is the initial sample void ratio.

The one-dimensional loading behavior of the pure sand is mostly linear in the e—($\log \sigma_v$) plane, slightly concave for large normal stresses (i.e., negative second derivative, see Figure 2). When rubber is added to the mixture, the stiffness progressively decreases, and the concavity becomes more pronounced as the behavior asymptotically converges to a normal compression line [22]. However, for high rubber fractions and at high vertical stresses (generally larger than 1000 kPa), one can observe a change of curvature in the void ratio—vertical stress semi-logarithmic plane—which becomes slightly convex, see for instance [22,38,41]. This is the effect of the saturation of the voids filled by the rubber grains and, thus, a clear sign of the influence of the rubber grains on the macroscopic behavior of the system. We, firstly, propose here to estimate the mean curvature of the e—($\log \sigma_v$) curve between 1 and 2 MPa vertical stress. This property leads to a method that can be used to determine the behavior or the mixture: "sand-like" if the concavity increases with the rubber content and "rubber-like" if an increase in the rubber content induces decreasing concavity (suggesting, as mentioned above, that the pores are being saturated progressively).

Then, the sign of the following curvature evolution with the rubber fraction:

$$\widetilde{C}(\sigma_v, x_r) = \frac{\partial}{\partial x_r}\frac{\partial^2 e(\sigma_v, x_r)}{\partial (log\sigma_v)2} = \frac{\partial^3 e(\sigma_v, x_r)}{\partial x_r \partial (log\sigma_v)2} \quad (2)$$

indicates whether the mixture is sand (negative $\widetilde{C}$) or rubber type (positive $\widetilde{C}$) at the corresponding stress level. In the following, for practical reasons, the sign is determined by the mean of the curvature over a given vertical stress interval.

Figure 4 shows the evolution of the mean curvature between 1 and 2 MPa of vertical stress with the rubber fraction for the two densities of the mixture and for the two particle sizes. It can be observed that the concavity of the e—($\log \sigma_v$) curve increased with the rubber fraction, up to about 25–30% of rubber, and then decreased. The transition point between sand and rubber-type behaviors seems fairly similar for the two densities and the two particle sizes.

Figure 5 presents the evolution of the mean curvature for five vertical stress intervals. For one rubber content, one point in the figure represents the mean curvature value for the SR-L, SR-D, and sr-D tests. This averaging is justified by the weak dependence of this curvature on the mixture type (see Figure 3). For vertical stresses higher than 600 kPa, the transition point seems to be fairly independent of the stress level. On the contrary, at vertical stresses lower than 600 kPa, the one-dimensional response of the mixture seems to be a sand-like type in all the tested ranges of rubber fractions. It is also interesting to observe that the curvature of the e—($\log \sigma_v$) curve became convex for a rubber fraction of 50%, a phenomenon that clearly reflects a behavior closer to pure rubber packing.

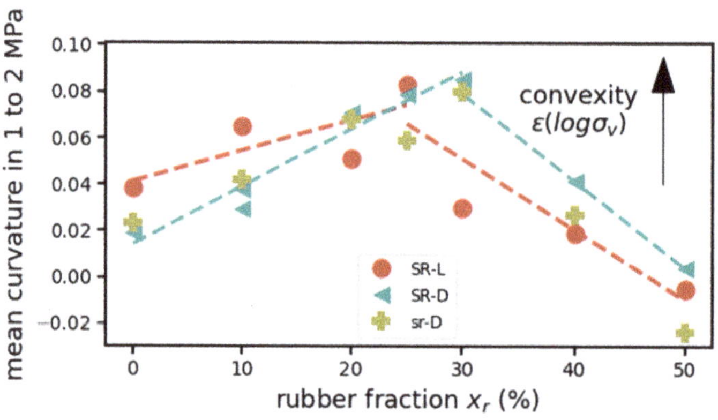

Figure 4. Mean curvature of the e—log (σ_v) evolution between 1 and 2 MPa of vertical stress for equally sized particle mixtures of SR-L, SR-D, and sr-D (see Table 1).

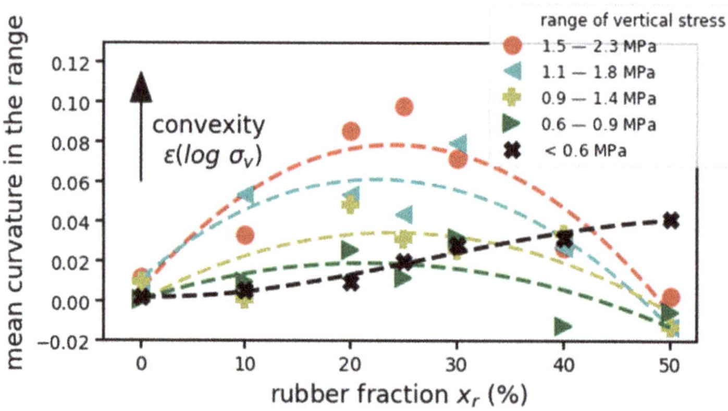

Figure 5. Mean curvature of the e—log σ_v evolution in several ranges of vertical stress; the mean of the values is determined for the equally sized SR-D, SR-L, and sr-D mixtures at the same rubber fraction.

The overall information is schematically presented in the phase diagram (see Figure 6). The rubber-like behavior was observed for vertical stress levels higher than around 600 kPa and a rubber fraction higher than around 25%. In all other situations, up to a rubber fraction of 50%, the behavior (the sign of $\widetilde{C}$) reflected a sand-type mixture.

3.2. One-Dimensional Confined Stiffness

The one-dimensional compression of the sand–rubber mixtures is non-linear and irreversible. The incremental confined stiffness, M, is given by the following relation:

$$M = \delta\sigma_v / \delta\varepsilon_v \tag{3}$$

where $\delta\sigma_v$ is the increment of vertical effective stress and $\delta\varepsilon_v$ is the increment of vertical strain, which is, therefore, not constant during the one-dimensional loading. A power law has been suggested to describe the link between the one-dimensional incremental confined stiffness, M, and the vertical effective stress, σ_v, [49]:

$$M = m\sigma_{ref}\left(\frac{\sigma_v}{\sigma_{ref}}\right)^{\alpha} \tag{4}$$

where σ_{ref} is a reference pressure, introduced for dimensional consistency, taken as 100 kPa, and m and α are material parametersf. The exponent α controls the way in which the incremental stiffness changes with the level of vertical stress. The m, material parameter, (known also as modulus number) is the magnitude of the confined stiffness at the reference pressure.

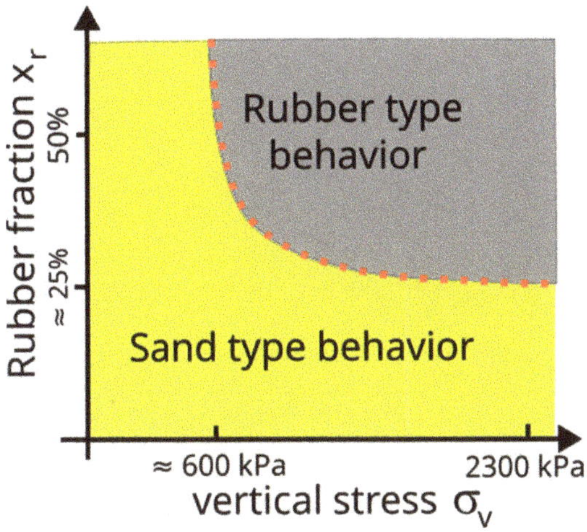

Figure 6. Schematic phase diagram showing the typical behavior of the equally sized sand–rubber mixtures as a function of the stress level and the rubber fraction.

Thus, the confined stiffness was analyzed for vertical stresses ranging from 100 kPa to 500 kPa. This choice was made to avoid the influence of particle crushing, which may occur at higher levels of compression stresses [50]. Post-test particle size analysis showed considerable sand grain crushing.

The parameters m and α can be easily obtained by plotting M/σ_{ref} versus the σ_v/σ_{ref} ratio. Figure 7 reports the corresponding curves for the equally sized sand–rubber particles, SR-D and SR-L test series. The parameter α seems to be independent of the rubber fraction and slightly influenced by the initial void ratio, with averages of 0.69 for SR-D and 0.71 for SR-L. However, the observed difference of α-values between the two packing densities is not statistically relevant given the standard errors, 0.012 and 0.022 respectively. The sr-D test series having smaller particles (D_{50} = 1.4 mm) showed identical trends and similar α values. In all, the observed values of the power index α are close to the results of [30].

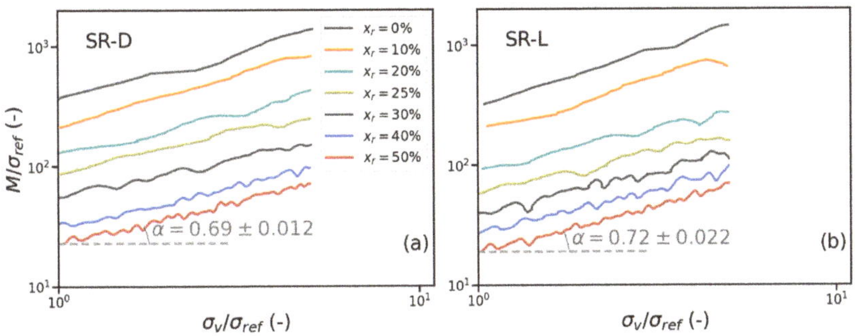

Figure 7. Confined modulus, M, versus vertical stress, σ_v, relation for equally sized particles (D_{50} = 1.8 mm): (**a**) SR-D and (**b**) SR-L for different rubber fractions x_r (low-pass filter was applied on the confined modulus).

The values of the modulus number, m, with the rubber content for all SR-L, SR-D, and sr-D test series are presented in Figure 8. A power law decrease of the modulus number, m, with the rubber fraction is observed for both densities. The ratio between the m-values for loose and dense packings is relatively constant, around 75% (standard deviation of 7%), in the ranges of all rubber fractions.

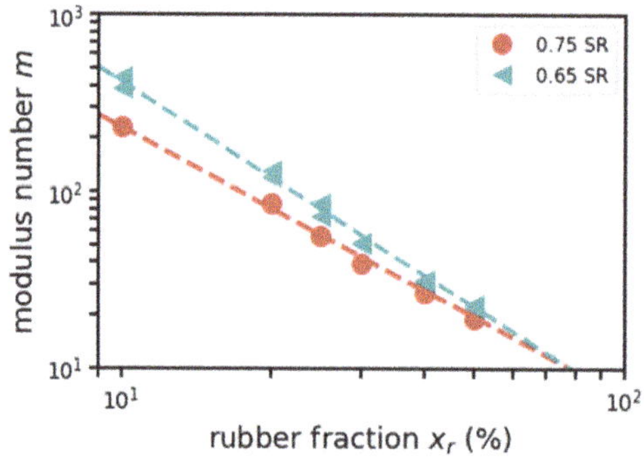

Figure 8. Evolution of modulus number with the rubber fraction for the two packing initial densities (SR-L and SR-D series).

The influence of the particle size ratio on α and m parameters is analyzed in Figure 9. It can be observed that α is not significantly affected by the particle size ratio. The slopes of the one-dimensional confined stiffness versus vertical stress for both sR-D and Sr-D tests are because of the SR-D for both the 10% and 50% rubber contents. However, the influence of the sand–rubber particle size ratio on the modulus number is non-negligible, when the rubber particles are bigger than the sand particles (sR-D-10 and sR-D-50 tests). At these rubber contents, the sR-D tests were also shown to be less compressible than all the other test series, which indicates that stronger sand-to-sand contact networks are dominating in these particular test mixtures. In [40], it was also observed that for mixtures with coarser rubber particles than sand, the behavior of the packings was stiffer, while for the rubber/sand size ratios smaller or equal to unity, the constrained modulus was only slightly increased.

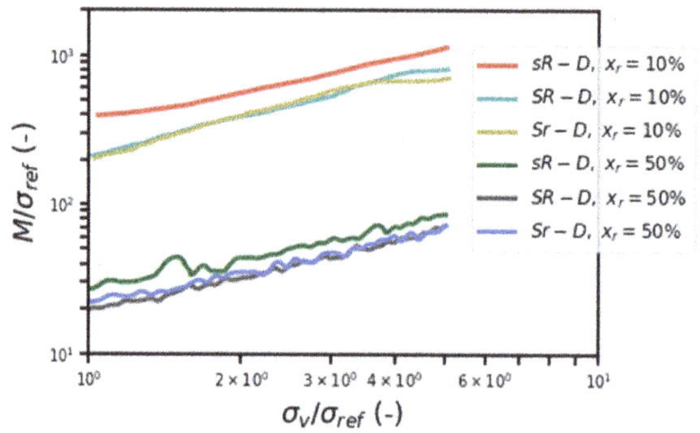

Figure 9. Confined modulus, M, versus vertical stress, σ_v, for the sR-D and Rs-D tests series compared with SR-D at 10% and 50% rubber contents, x_r.

3.3. Swelling Behavior

Typical one-dimensional loading-unloading responses for sand and sand-rubber mixtures are shown in Figure 2. A clear change from the almost rigid unloading for the pure sands to a much more recoverable strain response, rubber content dependent, is observed.

The swelling curves for the sand–rubber mixtures have a distinct S-shape, which was also observed in [19,22,30,35,41]. While the effects of adding rubber could be assessed through the swelling index, C_s, the slope of the straight swelling line in void ratio—log (σ_v) plane, given the high non-linearity of the swelling curves, the swelling index is expressed in an incremental form:

$$C_s = \frac{-\delta e}{\delta(log\sigma_v)} = \frac{(1+e_o)\delta\varepsilon_v}{\delta(log\sigma_v)} \tag{5}$$

The evolution of the incremental swelling index with the vertical stress for the tests presented in Figure 2 is presented in Figure 10. With the exception of pure sand and the sand–rubber mixture with the lowest rubber content of 10%, all the curves reflect the S-shape variation of the swelling curve, with low C_s values at the unloading and the end of test stages, and a steady increase and decrease following the peak value. The value of the peak and the rate of the increase/decrease of the swelling index is highly dependent on the rubber content. The vertical stress value corresponding to the peak swelling index also appears to decrease with the reduction of the rubber content from 100 kPa to 30 kPa.

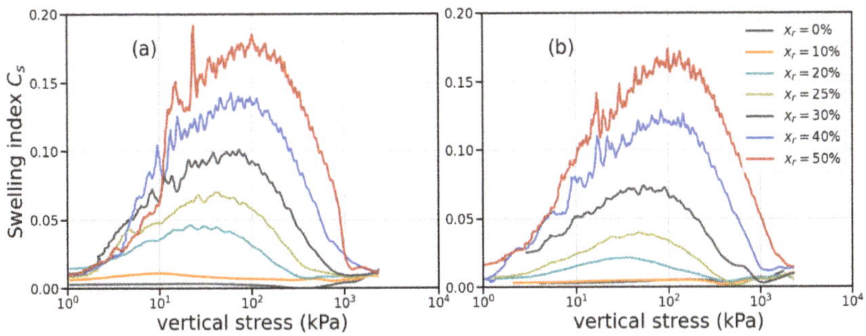

Figure 10. Evolution of the swelling index at different levels of unloading vertical stress for the two densities of the packing and different rubber contents.

4. Conclusions

The mechanical response of the sand–rubber mixtures under one-dimensional loading was studied in this work. Different parameters, such as the initial void ratio, rubber fraction, and sand/rubber size ratio were varied in order to quantify their effect on the mechanical behavior of the mixtures. The different results obtained are presented, briefly, in the following:

- The magnitude of the packing deformation increased as the rubber fraction increased. Further, for a given rubber fraction, the mixture with larger sand particles (Sr-D) was more compressible, while the mixture with smaller sand particles (sR-D) was less compressible than the mixture with equal-sized rubber and sand particles. This trend is in accordance with previous observations by [41]. One reason could be that a greater number of rubber (flexible) particles for the same rubber volume fraction increases the displacement/rotation of all particles, including rigid ones, thus, relaxing the particle jamming.
- The concavity of the loading curve in the e—log (σ_v) plane increased for pure sand packing with an increase in the normal stress. However, with the addition of rubber, the curve was found to become more convex at higher values of stress. The observed curvature change can be attributed to the saturation of the voids due to their filling with rubber, thus, indicating the impact of rubber addition on the macroscopic response of the mixture. An estimation of the mean curvature was performed. The mixtures were categorized into "sand-like" if the concavity increased and "rubber-like" if it decreased with the rubber fraction. This classification is helpful since it gives an objective definition of packing behavior on the basis of vertical stress and rubber fraction.

- The one-dimensional confined stiffness was found to decrease with an increase in the rubber fraction. A power law decrease of the modulus number, *m*, with the rubber fraction was observed for both densities. The ratio between the *m*-values for loose and dense packings was relatively constant around 75% (standard deviation of 7%) on all the range of rubber fractions. The power law index was relatively constant for all tests, with slightly higher values for looser packings. Moreover, for a given rubber fraction, coarser rubber particle packings (sR-D) showed a stiffer response than the corresponding equal-sized particle mixtures, indicating a stronger sand contact force network.

This work strengthens the understanding of the mechanical response of sand–rubber mixtures. Such mixtures can be an interesting solution as a composite soil in numerous geotechnical applications. Thus, the results presented here are a step further to better understanding the behaviors of such mixtures. Numerical studies, using DEM, can be the perspective for such experiments, which will help link the micromechanical interactions with the macroscale responses observed.

Author Contributions: Conceptualization, B.C., E.I., P.R.; methodology, B.C., E.I., P.B.; formal analysis, P.B., B.C.; investigation, P.B., E.I.; resources, P.B., E.I., B.C.; writing—original draft preparation, P.B., B.C.; writing—review and editing, E.I., P.R., R.A.; supervision, P.R., B.C., E.I., R.A. All authors have read and agreed to the published version of the manuscript.

Funding: This research received no external funding.

Institutional Review Board Statement: Not applicable.

Informed Consent Statement: Not applicable.

Data Availability Statement: Data sharing is not applicable.

Conflicts of Interest: The authors declare no conflict of interest.

References

1. Lee, J.H.; Salgado, R.; Bernal, A.; Lovell, C.W. Shredded tires and rubber-sand as lightweight backfill. *J. Geotech. Geoenviron. Eng.* **1999**, *125*, 132–141. [CrossRef]
2. Bosscher, P.J.; Edil, T.B.; Kuraoka, S. Design of highway embankments using tire chips. *J. Geotech. Geoenviron. Eng.* **1997**, *123*, 295–304. [CrossRef]
3. Youwai, S.; Bergado, D.T. Numerical analysis of reinforced wall using rubber tire chips–sand mixtures as backfill material. *Comput. Geotech.* **2004**, *31*, 103–114. [CrossRef]
4. Yoon, S.; Prezzi, M.; Siddiki, N.Z.; Kim, B. Construction of a test embankment using a sand–tire shred mixture as fill material. *Waste Manag.* **2006**, *26*, 1033–1044. [CrossRef] [PubMed]
5. Blumenthal, M.; Zelibor, J.L. Scrap tires used in rubber-modified asphalt pavement and civil engineering applications. In *Utilization of Waste Materials in Civil Engineering Construction*; ASCE: Reston, VA, USA, 1993; pp. 182–192.
6. Humphrey, D.N.; Eaton, R.A. Field performance of tire chips as subgrade insulation for rural roads. In Proceedings of the 6th International Conference on Low-Volume Roads, Minneapolis, MN, USA, 25–29 June 1995; Volume 2, pp. 77–86.
7. Poh, P.S.H.; Broms, B.B. Slope stabilization using old rubber tires and geotextiles. *J. Perform. Constr. Facil.* **1995**, *9*, 76–79. [CrossRef]
8. Garga, V.K.; O'Shaughnessy, V. Tire-reinforced earthfill. Part 1: Construction of a test fill, performance, and retaining wall design. *Can. Geotech. J.* **2000**, *37*, 75–96. [CrossRef]
9. Brunet, S.; De La Llera, J.C.; Kausel, E. Non-linear modeling of seismic isolation systems made of recycled tire-rubber. *Soil Dyn. Earthq. Eng.* **2016**, *85*, 134–145. [CrossRef]
10. Kaneko, T.; Orense, R.P.; Hyodo, M.; Yoshimoto, N. Seismic response characteristics of saturated sand deposits mixed with tire chips. *J. Geotech. Geoenviron. Eng.* **2013**, *139*, 633–643. [CrossRef]
11. Xiong, W.; Yan, M.R.; Li, Y.Z. Geotechnical seismic isolation system—Further experimental study. *Appl. Mech. Mater.* **2014**, *580–583*, 1490–1493. [CrossRef]
12. Tsiavos, A.; Alexander, N.A.; Diambra, A.; Ibraim, E.; Vardanega, P.J.; Gonzalez-Buelga, A.; Sextos, A. A sand-rubber deformable granular layer as a low-cost seismic isolation strategy in developing countries: Experimental investigation. *Soil Dyn. Earthq. Eng.* **2019**, *125*, 105731. [CrossRef]
13. Kaneda, K.; Hazarika, H.; Yamazaki, H. Numerical simulations of earth pressure reduction using tire chips in sand backfill. *J. Appl. Mech.* **2007**, *10*, 467–476. [CrossRef]

14. Hazarika, H.; Kohama, E.; Sugano, T. Underwater shake table tests on waterfront structures protected with tire chips cushion. *J. Geotech. Geoenviron. Eng.* **2008**, *134*, 1706–1719. [CrossRef]
15. Senetakis, K.; Anastasiadis, A.; Pitilakis, K. Dynamic properties of dry sand/rubber (Srm) and gravel/rubber (Grm) mixtures in a wide range of shearing strain amplitudes. *Soil Dyn. Earthq. Eng.* **2012**, *33*, 38–53. [CrossRef]
16. Tsang, H.-H. Seismic isolation by rubber–soil mixtures for developing countries. *Earthq. Eng. Struct. Dyn.* **2008**, *37*, 283–303. [CrossRef]
17. Badarayani, P.R.; Artoni, R.; Cazacliu, B.; Ibraim, E.; Richard, P. Segregation of sand-rubber chips mixtures subject to vertical tapping under confinement. *Powder Technol.* **2021**, *393*, 764–772. [CrossRef]
18. Attom, M.; Khedaywi, T.; Mousa, S.A. The effect of shredded waste tire on the shear strength, swelling and compressibility properties of the clayey soil. *J. Solid Waste Technol. Manag.* **2007**, *33*, 219–227.
19. Lee, J.-S.; Dodds, J.; Santamarina, J.C. Behavior of rigid-soft particle mixtures. *J. Mater. Civ. Eng.* **2007**, *19*, 179–184. [CrossRef]
20. Promputthangkoon, P.; Hyde, A. Compressibility and liquefaction potential of rubber composite soils. In Proceedings of the International Workshop on Scrap Tire Derived Geomaterials-Opportunities and Challenges, Yokosuka, Japan, 23–24 March 2007; pp. 161–170.
21. Trouzine, H.; Bekhiti, M.; Asroun, A. Effects of scrap tyre rubber fibre on swelling behaviour of two clayey soils in Algeria. *Geosynth. Int.* **2012**, *19*, 124–132. [CrossRef]
22. Fu, R.; Coop, M.R.; Li, X.Q. The mechanics of a compressive sand mixed with tyre rubber. *Géotech. Lett.* **2014**, *4*, 238–243. [CrossRef]
23. Benessalah, I.; Arab, A.; Sadek, M.; Bouferra, R. Laboratory study on the compressibility of sand–rubber mixtures under one dimensional consolidation loading conditions. *Granul. Matter* **2019**, *21*, 7. [CrossRef]
24. Pincus, H.; Edil, T.; Bosscher, P. Engineering properties of tire chips and soil mixtures. *Geotech. Test. J.* **1994**, *17*, 453. [CrossRef]
25. Foose, G.J.; Benson, C.H.; Bosscher, P.J. Sand reinforced with shredded waste tires. *J. Geotech. Eng.* **1996**, *122*, 760–767. [CrossRef]
26. Zornberg, J.G.; Cabral, A.R.; Viratjandr, C. Behaviour of tire shred sand mixtures. *Can. Geotech. J.* **2004**, *41*, 227–241. [CrossRef]
27. Asadi, M.; Thoeni, K.; Mahboubi, A. An experimental and numerical study on the compressive behavior of sand-rubber particle mixtures. *Comput. Geotech.* **2018**, *104*, 185–195. [CrossRef]
28. Neaz Sheikh, M.; Mashiri, M.S.; Vinod, J.S.; Tsang, H.-H. Shear and compressibility behavior of sand–tire crumb mixtures. *J. Mater. Civ. Eng.* **2013**, *25*, 1366–1374. [CrossRef]
29. Rouhanifar, S.; Afrazi, M.; Fakhimi, A.; Yazdani, M. Strength and deformation behaviour of sand-rubber mixture. *Int. J. Geotech. Eng.* **2021**, *15*, 1078–1092. [CrossRef]
30. Kim, H.-K.; Santamarina, J.C. Sand–rubber mixtures (Large rubber chips). *Can. Geotech. J.* **2008**, *45*, 1457–1466. [CrossRef]
31. Lee, C.; Truong, Q.H.; Lee, J.-S. Cementation and bond degradation of rubber–sand mixtures. *Can. Geotech. J.* **2010**, *47*, 763–774. [CrossRef]
32. Anastasiadis, A.; Senetakis, K.; Pitilakis, K. Small-strain shear modulus and damping ratio of sand-rubber and gravel-rubber mixtures. *Geotech. Geol. Eng.* **2012**, *30*, 363–382. [CrossRef]
33. Senetakis, K.; Anastasiadis, A. Effects of state of test sample, specimen geometry and sample preparation on dynamic properties of rubber–sand mixtures. *Geosynth. Int.* **2015**, *22*, 301–310. [CrossRef]
34. Tian, Y.; Senetakis, K. Influence of creep on the small-strain stiffness of sand–rubber mixtures. *Géotechnique* **2022**, *72*, 899–910. [CrossRef]
35. Fu, R.; Coop, M.R.; Li, X.Q. Influence of particle type on the mechanics of sand–rubber mixtures. *J. Geotech. Geoenviron. Eng.* **2017**, *143*, 04017059. [CrossRef]
36. Tasalloti, A.; Chiaro, G.; Murali, A.; Banasiak, L. Physical and mechanical properties of granulated rubber mixed with granular soils—A literature review. *Sustainability* **2021**, *13*, 4309. [CrossRef]
37. Platzer, A.; Rouhanifar, S.; Richard, P.; Cazacliu, B.; Ibraim, E. Sand–rubber mixtures undergoing isotropic loading: Derivation and experimental probing of a physical model. *Granul. Matter* **2018**, *20*, 81. [CrossRef]
38. Fonseca, J.; Riaz, A.; Bernal-Sanchez, J.; Barreto, D.; McDougall, J.; Miranda-Manzanares, M.; Marinelli, A.; Dimitriadi, V. Particle–scale interactions and energy dissipation mechanisms in sand–rubber mixtures. *Géotechnique Lett.* **2019**, *9*, 263–268. [CrossRef]
39. David Suits, L.; Sheahan, T.; Ghazavi, M.; Sakhi, M. Optimization of aspect ratio of waste tire shreds in sand-shred mixtures using cbr tests. *Geotech. Test. J.* **2005**, *28*, 12126. [CrossRef]
40. Rao, G.V.; Dutta, R.K. Compressibility and strength behaviour of sand–tyre chip mixtures. *Geotech. Geol. Eng.* **2006**, *24*, 711–724. [CrossRef]
41. Lee, C.; Truong, Q.H.; Lee, W.; Lee, J.-S. Characteristics of rubber-sand particle mixtures according to size ratio. *J. Mater. Civ. Eng.* **2010**, *22*, 323–331. [CrossRef]
42. Gotteland, P.; Lambert, S.; Balachowski, L. Strength characteristics of tyre chips-sand mixtures. *Stud. Geotech. Mech.* **2005**, *27*, 55–66.
43. Mashiri, M.S.; Vinod, J.S.; Sheikh, M.N.; Tsang, H.-H. Shear strength and dilatancy behaviour of sand–tyre chip mixtures. *Soils Found.* **2015**, *55*, 517–528. [CrossRef]
44. Lanzano, G.; Visone, C.; Bilotta, E.; Santucci De Magistris, F. Experimental assessment of the stress–strain behaviour of leighton buzzard sand for the calibration of a constitutive model. *Geotech. Geol. Eng.* **2016**, *34*, 991–1012. [CrossRef]

45. Rouhanifar, S. *Mechanics of Soft-Rigid Soil Mixtures*; University of Bristol: Bristol, UK, 2017.
46. Badarayani, P.; Richard, P.; Cazacliu, B.; Artoni, R.; Ibraim, E. A numerical and experimental study of sand-rubber mixtures subjected to oedometric compression. *E3S Web Conf.* **2019**, *92*, 14010. [CrossRef]
47. Selig, E.; Ladd, R. Preparing test specimens using undercompaction. *Geotech. Test. J.* **1978**, *1*, 16–23. [CrossRef]
48. Ibraim, E.; Diambra, A.; Russell, A.R.; Muir Wood, D. Assessment of laboratory sample preparation for fibre reinforced sands. *Geotext. Geomembr.* **2012**, *34*, 69–79. [CrossRef]
49. Janbu, N. Soil compressibility as determined by oedometer and triaxial tests. In Proceedings of the European Conference on Soil Mechanics and Foundation Engineering (SMFE), Budapest, Hungary, 24–27 September 1963; Volume 1, pp. 19–25.
50. McDowell, G.R.; Bolton, M.D.; Robertson, D. The fractal crushing of granular materials. *J. Mech. Phys. Solids* **1996**, *44*, 2079–2101. [CrossRef]

applied sciences

MDPI

Article

Role of Diatom Microstructure in Determining the Atterberg Limits of Fine-Grained Diatomaceous Soil

Yiqing Xu [1,2], Xianwei Zhang [1,*], Gang Wang [1,2], Xinyu Liu [3] and Aiwu Yang [4]

1 State Key Laboratory of Geomechanics and Geotechnical Engineering, Institute of Rock and Soil Mechanics, Chinese Academy of Sciences, Wuhan 430071, China
2 University of Chinese Academy of Sciences, Beijing 101408, China
3 School of Civil and Hydraulic Engineering, Huazhong University of Science and Technology, Wuhan 430074, China
4 School of Environmental Science and Engineering, Donghua University, Shanghai 200051, China
* Correspondence: xwzhang@whrsm.ac.cn; Tel.: +86-138-8608-4677

Abstract: The presence of diatoms in diatomaceous soil gives it geotechnical properties that are unusual compared with common clays. The most typical physical property of diatomaceous soil is its abnormally high Atterberg limits compared to fine-grained soil without diatoms. For diatomaceous soil, the Atterberg limits are associated with many factors, such as diatom content, diatom crushing degree, etc. In the study reported here, it was ascertained that more diatoms lead to higher plastic and liquid limits. Once the diatoms are crushed, the plastic and liquid limits decrease. The pore fluid salt concentration barely influenced the Atterberg limits of diatomaceous soil. Additionally, the porous diatom microstructure and trimodal pore size distribution of diatomaceous soil were investigated via scanning electron microscopy, transmission electron microscopy, and mercury intrusion porosimetry. The underlying mechanism of abnormally high liquid and plastic limits of diatomaceous soil is revealed as the water stored in the special diatom microstructure. However, water in diatoms has no contribution to plasticity. Also discussed is the applicability of the current soil classification systems for diatomaceous soil. The findings of this study can help for a better understanding of Atterberg limits of diatomaceous soil and provide suggestions for the classification of diatomaceous soil.

Keywords: diatom; diatomaceous soil; Atterberg limits; microstructure; diatom crushing; soil classification

1. Introduction

Lacustrine and marine diatomaceous soil (DS) is found extensively in many countries, such as Mexico [1], Japan [2], Korea [3], Columbia [4], and China [5]. DS contains numerous fossilized diatoms with sizes in a range of several to dozens of micrometers [6] as a result of the accumulation of unicellular diatom cell walls (the diatom cell wall is referred to as frustules) [7]. A diatom frustule comprises mainly amorphous SiO_2 (Opal A) and some feldspar, quartz, pyrite, etc. [8–10]. The geotechnical characteristics of DS are significantly different from those of fine-grained soil without diatom; for example, (i) the porosity of natural DS can be as high as 74% [11], (ii) DS is characterized by a low specific gravity (G_S) of 2.26–2.38, which is lower than that of common clayey soil such as kaolin (G_S = 2.775) and Singapore clay (G_S = 2.770) [12], and (iii) natural DS has a high water content which can exceed 400% [13]. However, although DS has these relatively poor physical properties, it also has a high yield stress of ca. 540 kPa [14] and a high friction angle of 32–43° [15].

The Atterberg limits are among the most important physical properties of fine-grained soil and the plasticity index (I_P) is critical in soil classification. Many studies have investigated the plasticity properties of DS and provided a preliminary understanding of its Atterberg limits [6,12,16–18], showing that the Atterberg limits of DS are higher than the Atterberg limits of fine-grained soil without diatoms. For example, the liquid limit (w_L) and plastic limit (w_p) of artificial DS with a high content of pure diatoms can be as high as 289.7% and 205.5%, respectively [19]; additionally, the w_L value was found to be over 400% for natural DSs from Mexico and Columbia, and their w_p value was ca. 80% [1,4]. Some studies showed that the Atterberg limits of DS are influenced by the mineralogy and fines fraction [12,20]. More diatoms may lead to a steady increase in Atterberg limits and the horizontal rightward proceeding across the A-Line in the plasticity chart [6]. Different diatom species also affect the plasticity properties of the DS; for example, DS contains disk-shaped diatoms that cause higher plastic and liquid limits than that contain tube-shaped diatoms [16].

Despite the current studies, more work is required to better understand the Atterberg limits of DS, i.e., (i) how the NaCl concentration of pore fluid and crushing degree of diatoms influence the Atterberg limits of DS (these two factors should be considered for DS because it conventionally forms in marine environments and contains either completely or partially crushed diatoms under the effect of geological burial), (ii) the microstructural difference between DS and common natural fine-grained deposits without diatom (also the association between the diatom microstructure and the Atterberg limits of DS and the corresponding underlying mechanism), and (iii) the applicability of the Unified Soil Classification System (USCS) for fine-grained DS, because this classification method involves the Atterberg limits and the particle size characteristics.

The study reported herein aimed to investigate the Atterberg limits of DS as affected by diatom content, diatom crushing, and pore fluid ionic concentration. For this purpose, the studied DS was obtained by mixing kaolin with pure diatoms of differing content and crushing degree, and NaCl solution replaced deionized water as the pore fluid for some of the diatomaceous samples. The plastic and liquid limits of this soil were measured by thread-rolling and fall-cone tests, the diatom and DS structures were studied microscopically via scanning electron microscopy (SEM) and transmission electron microscopy (TEM), and the pore characteristics of the DS were revealed via mercury intrusion porosimetry (MIP). Typical pores of the DS were analyzed quantitatively considering the diatom microstructural features. Next, the effects of the diatom content, diatom crushing degrees, and pore-fluid NaCl concentration on w_L, w_p, and I_P were studied, thereby revealing the relationship between the microstructure and the plasticity properties of DS. Also discussed is the applicability of current classification methods (including USCS and pore fluid sensitivity classification methods) for fine-grained soil containing diatoms, based on the present investigations of Atterberg limits and particle size distribution (PSD). The present findings offer an improved understanding of the Atterberg limits of DS and insights into its classification.

2. Materials and Methods

2.1. Sample Preparation

The studied soil was a diatom–kaolin mixture (DKM) comprising Georgia kaolin (RP-2; Active Minerals International, Baltimore, MD, USA) and diatom from Changbai Prefecture, Jilin, China. Figure 1 gives the mineral and chemical compositions of kaolin and diatom as revealed by X-ray diffraction and X-ray energy-dispersive spectroscopy: diatom comprises quartz and opal with no clay minerals, which is consistent with the high contents of Si and O elements; kaolin has a dominant mineral content of kaolinite with traces of illite and quartz, in which the content of illite and quartz are ca. 1.8% and 1.2%, respectively. Table 1 summarizes the physical properties of the diatom and kaolin as determined by ASTM standards. Diatom has lower G_S (2.25), significantly higher specific surface area (179.5 m^2/g), higher activity (A_d) (0.87), and cation exchange capacity (32.3 meq/100 g) compared with kaolin.

DKM samples with varying diatom proportions (0–100%) in terms of dry mass ratio were prepared to investigate the influence of diatom content on the consistency limits of fine-grained DS. To ensure that the water in diatom materials is completely removed before preparing DKM with a certain diatom content, the diatoms and kaolin materials were dried at 105 °C for 24 h. The DKM samples were prepared by thoroughly mixing diatom and kaolin using mechanical stirring. For convenience, each prepared DKM sample was indexed as "diatom content D-kaolin content K", where D refers to the diatoms and K refers to the kaolin; for example, the DKM sample indexed as 40D-60K contained 40% diatoms and 60% kaolin.

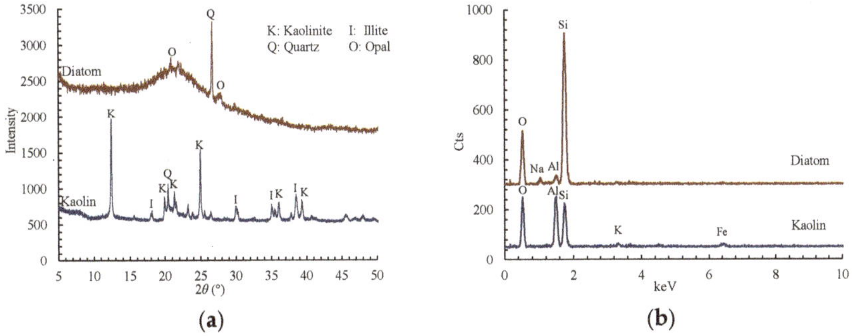

Figure 1. Mineral and chemical compositions of diatoms and kaolin as determined by (**a**) X-ray diffraction and (**b**) X-ray energy-dispersive spectroscopy.

Table 1. Physical properties of pure diatom and kaolin.

Parameter	Value	
	Diatom	**Kaolin**
Specific gravity G_S	2.25	2.65
Activity A_d [1]	0.87	0.26
Average particle diameter d_{50} (μm)	6.80	0.59
Specific surface area (m^2/g) [2]	179.5	28.3
Liquid limit w_L (%)	175	42
Plastic limit w_P (%)	153	25
Plastic index I_P	22	17
USCS classification	MH	CL
pH	8.11	7.26
Cation exchange capacity (meq/100 g)	32.3	9.8

[1] Activity [= ratio of plasticity index to a percentage by weight of clay-size particles]. [2] SSA was determined using the European methylene blue spot method [21].

2.2. Microstructure Observation

SEM and TEM observations were performed on the pure diatom and kaolin and their mixture to explore the microstructure characteristics of DS. The SEM observations involved a Quanta 250 (FEI, Eindhoven, The Netherlands) with an accelerating voltage of 20 kV. The samples were prepared via vacuum freeze-drying to avoid disturbing the microstructure during drying [22]. Samples were observed at the magnifications of 100×, 800×, 2000×, and 5000×, with particular attention paid to the interparticle linkages at the latter two magnifications. The TEM observations involved a field-emission TEM microscope (JEM 2100F; JEOL, Tokyo, Japan) operated at 200 kV. The samples were prepared by dropping a suspension of diatoms and kaolin onto a carbon-coated copper grid; the suspension was obtained by dispersing dry diatom and kaolin particles in deionized water.

Additionally, MIP tests were performed to reveal the pore size distribution of the DKM. DKM slurry was pre-consolidated at 60 kPa, vacuum freeze-dried, and cut into 5 mm × 10 mm × 5 mm pieces to prepare a sample for an MIP test. The MIP tests involved a Quantachrome Pore Master 33, which could detect pores with a diameter in the range of 0.006–300 μm, and they were based on the equation proposed by Washburn [23]:

$$P = -\frac{4\sigma_{Hg}\cos\theta_{Hg}}{d} \tag{1}$$

which describes the correlation between the applied pressure P and the equivalent pore diameter d. Here, θ_{Hg} is the mercury–soil contact angle, and σ_{Hg} is the mercury–soil surface tension; the values used in this study were 141.3° and 0.48 N/m, respectively. Using the pore diameter d_i calculated from this equation, the cumulative pore volume (CV) and the log-incremental pore volume (IV) are calculated as:

$$CV = \sum_i \Delta V_i \quad (2)$$

$$IV = \frac{-\Delta V_i}{\Delta \log d_i} \quad (3)$$

The MIP results are given as cumulative and log-incremental pore volume distribution curves, i.e., CV–d and IV–d curves.

2.3. Details of Tests for Atterberg Limits

In this study, the w_P of the soil was determined using thread-rolling tests following ASTM D4318-17 [24], while the w_L was determined by fall-cone tests, which give less variable experimental results compared to the standard "Casagrande cup" method [25–27]. Note that a thread-rolling test was inappropriate for sample 100D-0K because of the non-plastic nature of pure diatoms (Figure 2); therefore, the method proposed by Vardanega et al. [28] was used to determine w_P for pure diatoms from the correlation between I_P and the flow index of the fall-cone test. Additionally, the PSD curves of the prepared DKM samples were obtained to determine their clay and silt fractions; for this, the hydrometer method [29] was used because the DKM samples contained no coarse particles (>0.075 mm).

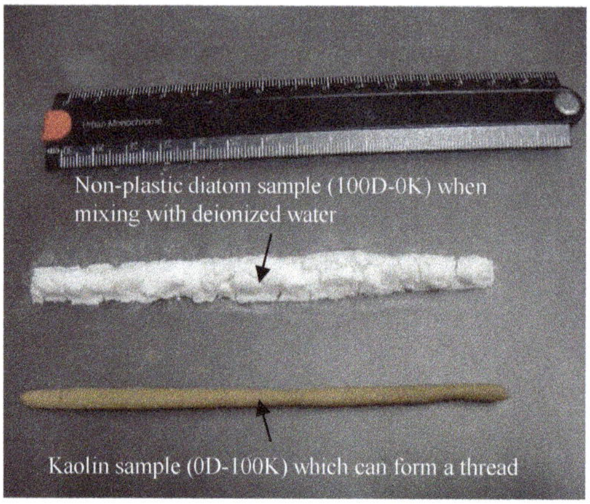

Figure 2. The successful case of kaolin and failed attempt of diatoms to form a thread.

To study the effect of the pore fluid ionic concentration on the Atterberg limits of diatom-rich soil, NaCl solutions with a concentration of 0.01 M, 0.1 M, 1 M, and 2 M were prepared. These NaCl solutions are used to replace deionized water as the pore fluid of DKM. Additionally, kerosene was used as a pore fluid to determine the liquid limit $w_{L,ker}$. The Atterberg limits of these DKMs with their original pore fluids (deionized water) replaced by NaCl solutions and kerosene were measured through thread-rolling and fall-cone tests.

As well as the NaCl concentration of pore fluid, the crushing of diatoms was also considered because the diatoms in natural DS are usually crushed to various degrees. To prepare a DKM sample with crushed diatoms (hereinafter referred to as crushed DKM, e.g., crushed 80D-20K), the ball milling method was performed to obtain crushed diatoms: dry raw diatoms were sent to a planetary ball mill (FRITSCH pulverisette; Shanghai, China) and were mechanically stirred in an agate vessel (500 mL) filled with hundreds of corundum grinding balls (8 mm in diameter) and operated at a rotational speed of 100 rpm; diatoms with various crushing degrees were obtained by changing the milling duration (1 h, 2 h, and 4 h), with 4 h expected to be sufficient to crush diatoms completely. The crushed DKM sample was then prepared by mixing the crushed diatoms as in the case of uncrushed DKM (see Section 2.1). The w_L and w_P values of the crushed DKM samples were also measured using thread-rolling and fall-cone tests and the complete crushed diatoms (with milling duration of 4 h) were observed using SEM to examine their altered microstructure due to diatom crushing.

3. Results

3.1. Microstructure of DKM

3.1.1. Overview of DKM Microstructure

Figure 3 shows the DKM microstructure as revealed by SEM images. Kaolin aggregates are observed in the form of layered kaolinite platelets and the kaolinite platelets are associated with each other via face-to-face, edge-to-face, and edge-to-edge interparticle contacts (Figure 3a). The diatoms are mainly intact disk-shaped ones (*Stephanodiscus* and *Cyclotella*) and their fragments (Figure 3b): the diameters of intact *Cyclotella* and *Stephanodiscus* diatoms are ca. 70 μm and ca. 30 μm, respectively, whereas the size of a smallest diatom fragment is equal to that of a kaolin platelet; also presented in this figure are some tube-shaped diatoms (*Aulacoseira*, with a diameter of ca. 10 μm). Figure 3d–f shows the typical microstructural features of DKM samples. The kaolinite platelets and diatoms gathered together to form the diatom–kaolin aggregate, which has massive pores within it. According to Tanaka and Locat [2], the pore types of DKM include inter-aggregate, intra-aggregate, skeletal, and intra-skeletal: the former two are due to kaolin aggregates, whereas the latter two are due to diatom frustules. For kaolin, the inter-aggregate and intra-aggregate pores of various sizes are identified in Figure 3a. For the diatom, the skeletal pores with a diameter of 0.1–1 μm and an intra-skeletal pore can be seen in the magnified image of a single disk-shaped diatom (Figure 3c). For DKM, all types of pore are found in diatom–kaolin aggregates (Figure 3f); some skeletal pores therein are surrounded or even filled by kaolinite clay particles, which provide a connection between the intact diatoms and diatom fragments. Figure 3g–i shows SEM images of a chain-like skeleton structure of crushed diatoms, which is formed by the 15–25 μm long fragments of diatom frustules. Figure 3g,h shows that a few tiny frustules with diameters of 12–25 μm remain intact. Additionally, the intact diatoms have dominant intra-skeletal and skeletal pores, but these types of pores mostly disappear in the crushed diatoms because the skeletal structure of the diatom has been damaged, leaving the inter-aggregate and intra-aggregate pores to dominate.

TEM images of the kaolin and diatoms are shown in Figure 4. The kaolinite platelets with a size of several hundred nanometers are stacked mainly via face-to-face contacts (Figure 4a,b). Also visible in these images is the tube-like halloysite formed by rolled hydrate kaolinite flakes; the halloysite occasionally appeared along with the kaolinite platelets and interspersed among them. Figure 4c shows a two-dimensional image of the internal structure of a 3.5 μm diameter diatom particle; inter-connective pores are uniformly and hexagonally arrayed on the profile surface of the frustule and the TEM image with magnification of 80,000× (Figure 4d) shows that the diameter of the pores observed in the hexagonal arrangement from Figure 4c is ca. 200 nm.

3.1.2. Pore Characteristics of DKM

The pore size distribution curve of pure kaolin is characterized by a single prominent peak at diameter of 0.2–0.6 μm (Figure 5a), which represents the inter-aggregate pores according to Figure 3. The domain of intra-aggregate pores was not found in the kaolin sample because this type of pore is not abundant, resulting from the dominant face-to-face contact between kaolinites (Figures 3 and 4). For diatom, the dominant domain corresponding to diameters of 0.06–0.2 μm and the subdomain at diameter of 10 μm represent the skeletal and intra-skeletal pores, respectively. Skeletal pores are small in volume but numerous, which is opposite to the case of intra-skeletal pores. The pore size distribution curve of diatoms in the range of d = 0.2–0.6 μm is more like a plateau than a peak. This plateau, as well as the second-highest peak (d = 0.6–2 μm), correspond to the pores in the open fabric of the diatom skeletons (Figure 3b). The pore size distribution of DKM is trimodal and located between those of kaolin and diatoms. For DKM, the domain for d = 0.2–2 μm corresponds to the intra-aggregate pores, inter-aggregate pores, and the pores in the diatom skeletons. The skeletal and intra-skeletal pores in the DKM are reflected by the peaks at diameters of 0.06–0.2 μm and 10 μm, respectively, which is consistent with the case for the diatoms. With increasing diatom content, the pore size distribution curve of the DKM approaches that of the diatoms in shape.

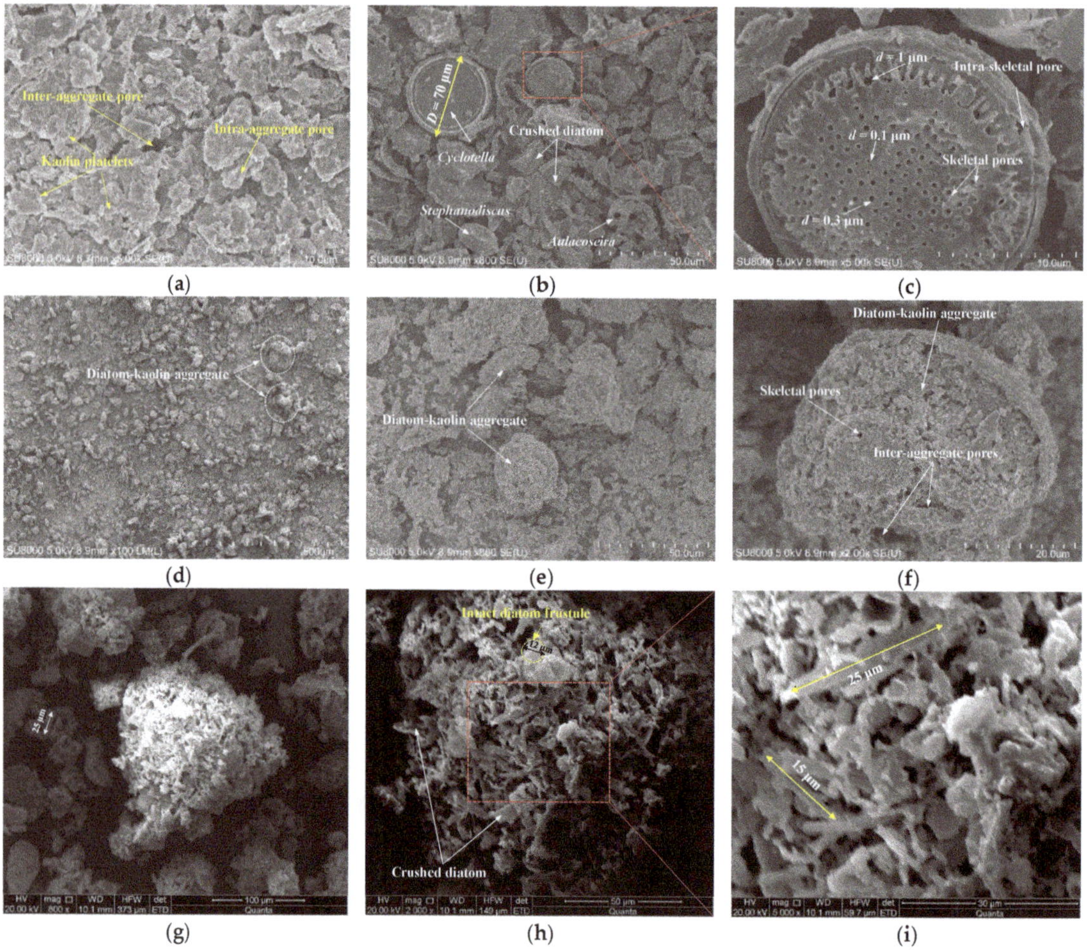

Figure 3. Scanning electron microscopy images of (**a**) kaolin (0D-100K) (**b**,**c**) diatoms (100D-0K), (**d**–**f**) a DKM (60D-40K), and (**g**–**i**) crushed diatoms.

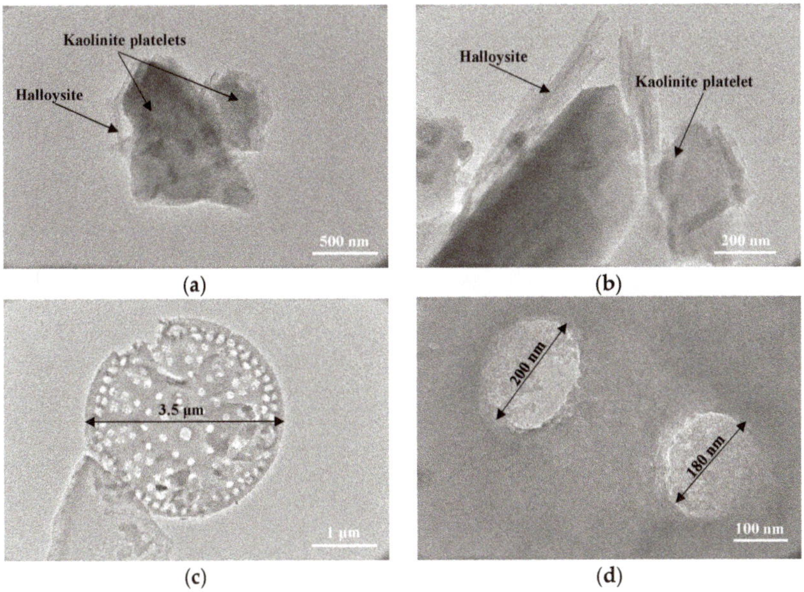

Figure 4. Transmission electron microscopy images of (**a**) kaolin (15,000× magnification); (**b**) kaolin (40,000×); (**c**) diatoms (8000×); and (**d**) diatoms (80,000×).

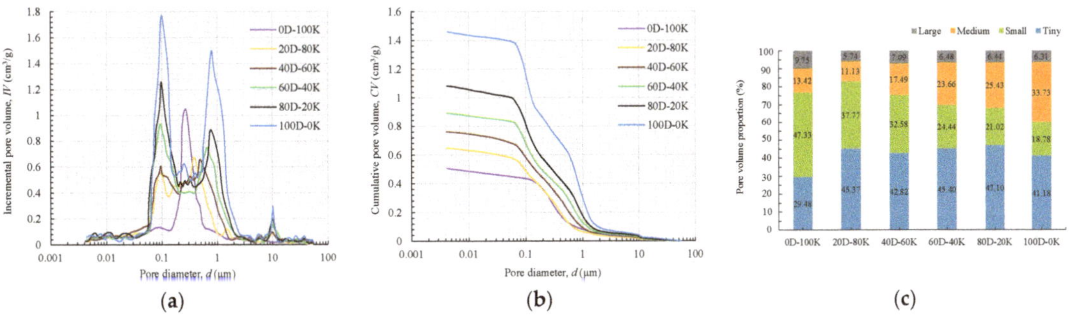

Figure 5. Results of mercury intrusion porosimetry tests on DKM samples: (**a**) *IV–d* curves; (**b**) *CV–d* curves; (**c**) volume proportion of pores with different diameters.

Figure 5b shows the cumulative pore distribution of DKM, which is intermediate to that of diatom and kaolin. The *CV–d* curve of kaolin falls rapidly as *d* increases from 0.2 μm to 0.6 μm, while that of diatom falls in a stepwise manner as *d* increases from 0.06 μm to 2 μm; the *CV–d* curves of both kaolin and diatom for other pore ranges are approximately unchanged. As the diatom proportion increases, the pore volume considerably increases, and the pattern of the *CV–d* curve progressively changes from the kaolin to the diatom.

Based on the *IV–d* curves, the DKM pores are classified as tiny ($d < 0.2$ μm), small ($0.2 < d < 0.6$ μm), medium ($0.6 < d < 2$ μm), and large ($d > 2$ μm) following the study of Zhang et al. [30]. The diatoms have more medium and tiny pores, but far fewer small pores compared to the kaolin (Figure 5c). When increasing the diatom content in DKM, the number of medium pores and small pores increases, which corresponds to the rightward shift of the second dominant peak (Figure 5a). By contrast, the proportion of tiny pores is scarcely affected by the diatom content and is higher than that of the kaolin. From a micro-level perspective (Figure 3a,f), the difference between the pore volume distribution curves of kaolin and DKMs in Figure 5c is due mainly to the skeletal and intra-skeletal pores from the diatom frustules and the varying sizes and shapes of inter-aggregate and intra-aggregate pores in diatom–kaolin aggregates.

3.2. Atterberg Limits of DKM

3.2.1. Effect of Diatom Content

The Atterberg limits of DKM samples with varying diatom contents are presented in Figure 6, along with the values of w_L, w_p, and I_P for different DKMs from previous studies [6,12,31–34]. The more diatoms in the DKM, the higher w_L and w_p are. The Atterberg limits of DKM with a diatom content of 0–60% are lower than DKM with a diatom content of 80–100%; for example, sample 60D-40K has w_L = 70% and w_p = 46%, while sample 80D-20K has the higher values of w_L = 128% and w_p = 98%. However, although w_L and w_p of the DKM change significantly with diatom content, I_P remains relatively stable because the increments in w_L and w_p are approximately the same. Notably, w_L and w_p increase remarkably when the diatom content increases from 60% to 80%. Some previous studies [12,35] suggested that the correlation between increasing w_L and w_p and diatom content is because w_L and w_p are affected by the unique microstructure of diatom frustules.

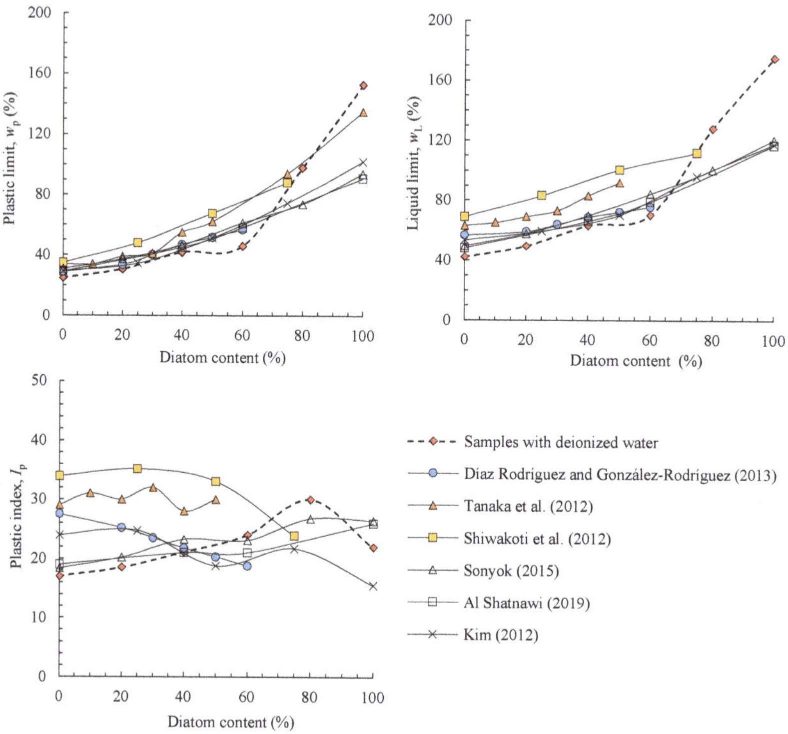

Figure 6. Values of w_p, w_L, and I_P for DKM samples versus diatom content. Data sources: DKMs [6,12,31–34].

The relationships between I_P and w_L for various DKMs are shown in Figure 7 as a Casagrande plasticity chart, along with data points for natural diatomaceous sediments from China [5], Korea [3], Japan [2], and Columbia [4]. The data points in Figure 7 for DKM samples move rightward with increasing diatom content, indicating that the plasticity of a DKM is hardly affected by its diatom content. Additionally, all of the DKM data points fall below the A-line, whereas the data of natural DSs are above the A-line and approach the U-line, as shown by the colored areas in Figure 7. The higher positions of the data points for natural DS compared with those for artificial DKM samples may be due to the existence of high plasticity mineral montmorillonite [1,35]. Figure 8 shows the relationships between I_P and clay content for DKMs and natural DSs, where the slope of the curve represents soil activity A_d. The A_d value increases with the clay fraction and the DKM data points change from the line of A_d = 0.5 to A_d = 1, which is consistent with the findings of Shiwakoti et al. [12].

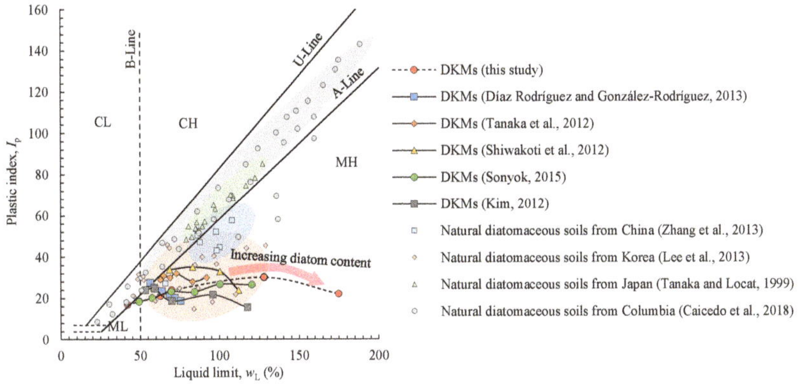

Figure 7. Casagrande plasticity chart for diatomaceous soil (DS). Data sources: DKMs [6,12,31,32,34], Natural diatomaceous soils [2–5].

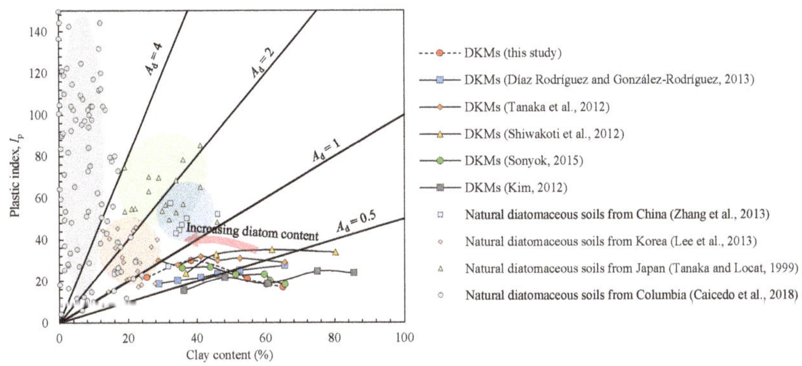

Figure 8. Relationship between I_P and clay fraction for different DKM samples. Data sources: DKMs [6,12,31,32,34], Natural diatomaceous soils [2–5].

3.2.2. Effect of Diatom Crushing

Figure 9 shows how diatom crushing affects the Atterberg limits of the DKM samples. The crushed DKM samples have lower w_L and w_p values than uncrushed ones. In particular, this difference in w_L and w_p between crushed and uncrushed DKMs is more remarkable for samples 80D-20K and 100D-0K. The w_L values of crushed 80D-20K (58.4%) and 100D-0K (62.4%) are much lower than those of uncrushed 80D-20K (128%) and 100D-0K (175%). This pattern of w_L is similar to w_p. The results mean that the high w_L and w_p values for samples 80D-20K and 100D-0K are due mainly to the intact microstructure of the diatoms. As shown in Figure 3h,i, crushed diatom fragments form a chain-like structure that cannot store water, because the water in the crushed diatom can only be stored in the skeletal pores of the fragments and not in the intra-skeletal pores.

Naturally, milling for longer leads to further fragmentation of diatom frustules and lower w_L and w_p (Figure 9) because crushing intact diatoms causes collapsed intra-skeletal pores and reduced water retention capacity. Nevertheless, the I_P values for the DKM samples were scarcely affected by diatom crushing, meaning that the water stored in the intra-skeletal pores of intact diatom frustules contributes little to the plasticity of the DS.

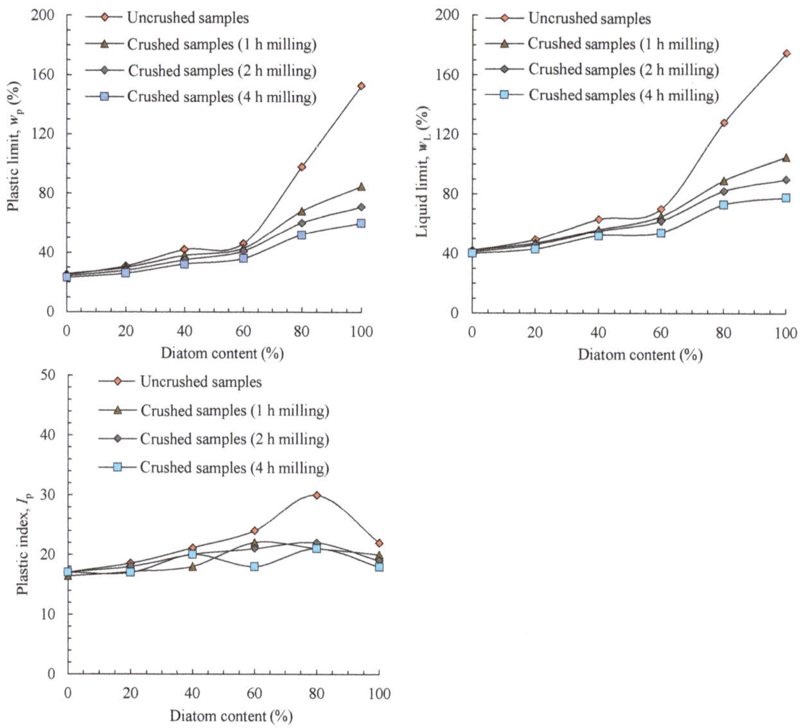

Figure 9. Values of w_P, w_L, and I_P for DKM samples with different diatom contents.

3.2.3. Effect of Pore Fluid Ionic Concentration

Figure 10 shows the test results for the Atterberg limits of the DKM samples with different values of NaCl concentration. For the DKM samples, w_L, w_P, and I_P scarcely change with increasing NaCl concentration; however, for pure kaolin, w_L, and w_P increase initially with increasing NaCl concentration and then decrease when the latter reaches 0.1 mol/L, while I_P tends to decrease with increasing NaCl concentration. Similar conclusions for kaolin were reached by Palomino and Santamarina [36], who reported that NaCl concentration significantly affects the interparticle association of the kaolinite fabric, thereby affecting the Atterberg limits. For kaolin with a high ionic concentration (>0.1 mol/L), face-to-face contacts dominate and the double layer is thin, leading to low w_L and w_P; this effect of NaCl concentration on the kaolinite fabric weakens as more diatoms are added to the kaolin [36]. Although the Atterberg limits of DKM are insensitive to the ionic concentration, they increase apparently with increasing diatom content; this indicates that diatom content has far more influence on the Atterberg limits than does ionic concentration, which is consistent with the findings of Palomino et al. [37].

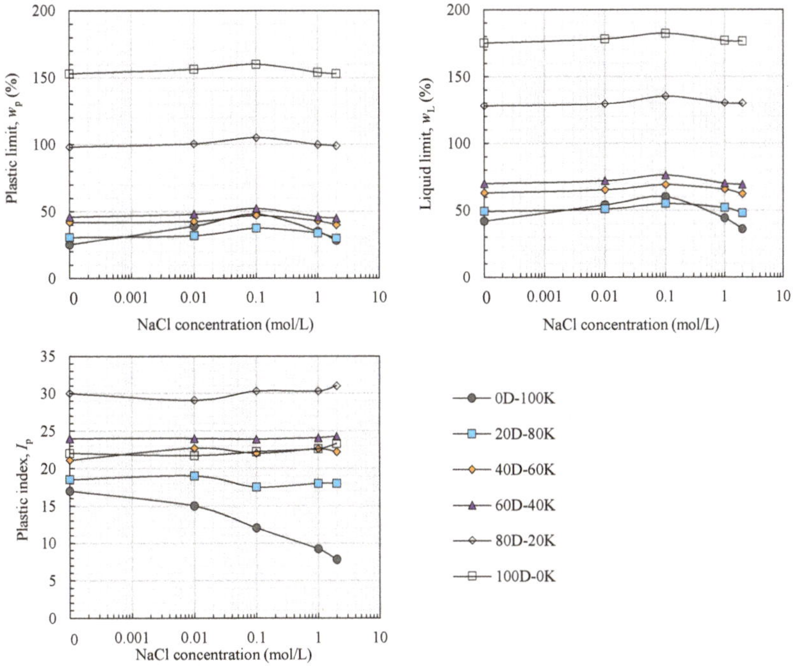

Figure 10. Values of w_P, w_L, and I_P for DKM samples with different NaCl concentrations.

4. Discussion

4.1. Influence of Diatoms on Atterberg Limits from a Microstructural Perspective

The diatom and kaolin structures are shown schematically in Figure 11. A disk-shaped diatom is a spheroid cell with two symmetrical porous shells connected by a girdle band. The shells have a regular and centrosymmetric arrangement of skeletal pores on their surfaces. Inside the spheroid cell is a large inner chamber that stores the biological protoplasm of the diatom; this chamber becomes a typical intra-skeletal pore in a frustule after the diatom dies. Furthermore, Losic et al. [38] showed that the disk-shaped diatom frustule comprises three membranes at the nanoscale—namely the areola, cribium, and cribellum—containing large, small, and tiny nanoholes, respectively. Figure 11b shows the distribution of nanoholes in cribellum and cribium corresponding to a large nanohole in the areola membrane. Notably, the pores in the cribellum were observed in the TEM images (Figure 4c,d) from the vertical direction of a frustule, as shown by the arrow in Figure 11b. As described above, the microstructure of the diatom frustule is porous and water is stored in the skeletal and intra-skeletal pores. These pores are connected through the nanoholes inside the frustule so that the w_L and w_P values and water content of DS are measured high. Figure 11c shows the crystal structure of kaolinite with two tetrahedral and octahedral sheets. The structure of kaolin clays is a layered stacking with limited space inside, and the water in kaolin is adsorbed in the surface of kaolinite platelets in the form of a double layer, which is different from the case of the diatom. However, although the higher consistency limits of diatomaceous samples compared to kaolin caused by the water inside the diatom frustules were measured, the kaolin and diatomaceous samples have similar I_P values. This is because the water inside the diatoms contributes little to the plasticity of the soil. Furthermore, it is observed that the diatoms with $I_P = 22$ are non-plastic, whereas the kaolin with $I_P = 17$ is plastic; therefore, the applicability of I_P as a parameter for soil plasticity evaluation and soil classification for DS should be assessed carefully, especially when the DS contains 20% or more diatom microfossils (by weight).

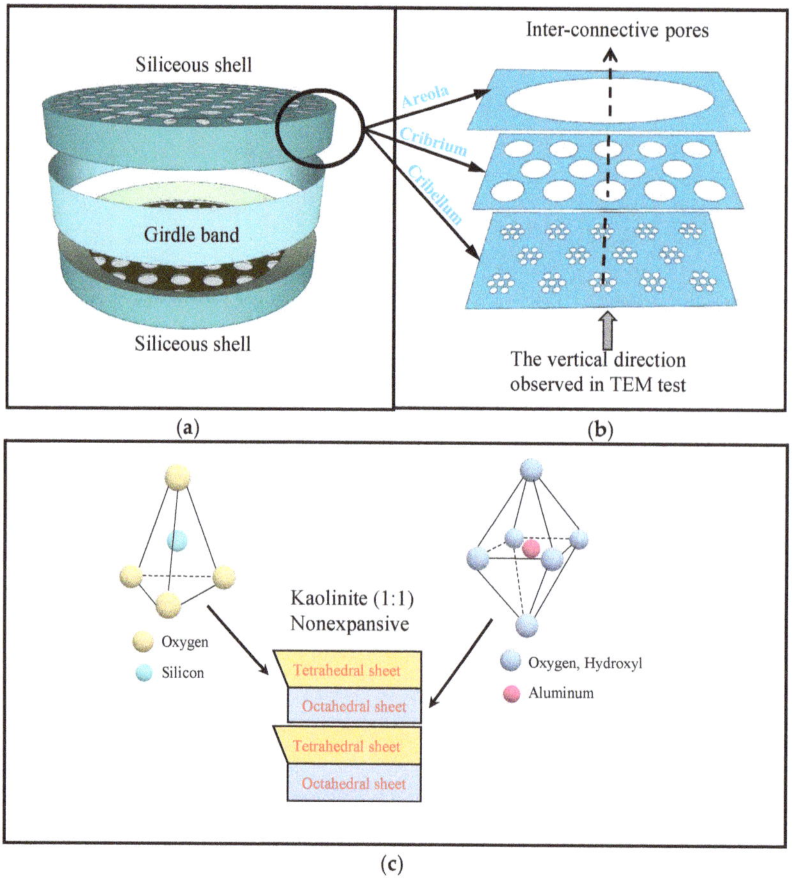

Figure 11. Schematics of (**a**) disk-shaped diatom frustule, (**b**) profile structure of frustule, and (**c**) crystal structure of kaolinite.

Apart from the diatom content, the Atterberg limits of DS are also related to diatom crushing. The crushed DKM samples have lower w_L and w_p values than uncrushed ones. This is because the hollow intra-skeletal and skeletal pores collapsed, thereby reducing water retention capacity, and the higher crushing degree of diatoms leads to the lower Atterberg limits. Moreover, the presence of diatoms affects the response of the Atterberg limits of the soil to the NaCl concentration. The w_L, w_p, and I_P values of the DS hardly changed upon changing the NaCl concentration, whereas those of the kaolin did (Figure 10). For pure kaolin, different values of NaCl concentration lead to various particle associations among kaolinites and further changes in the forces among clay particles [36]. For pure diatoms, the double layer is thin because of the low surface charge density of a frustule and does not change with NaCl concentration [20,39]. The water that significantly affects the consistency limits of the diatoms is stored mainly in the intra-skeletal pores instead of in the double layer. Interestingly, the effects of NaCl concentration on the kaolin sample almost disappear with the addition of diatoms (Figure 10), which may be because of the unique interactions between kaolin and diatom particles of the diatom–kaolin aggregates.

4.2. Classification of Fine-Grained Soil Containing Diatoms

Before classifying the DKM samples, their particle size compositions are obtained from the PSD curves (Figure 12). The diatoms comprise mainly silt, while the kaolin contains more clay than silt. The PSD curves of the DKM samples move downward with increasing diatom content. Higher diatom content in the DKM corresponds to more silt and less clay (Figure 13), and DKM with a diatom content of more than 60% has a dominant composition of silt and is known as silt-dominated soil. Figure 14 shows that the d_{50} value of the DKM increases with increasing diatom content: d_{50} increases by 131% when the diatom proportion changes from 60% to 80%. Similar phenomena have been observed in previous studies [6,32], except for that by Wiemer and Kopf [19], who used clayey silt as a substitute for kaolin to prepare the studied DS.

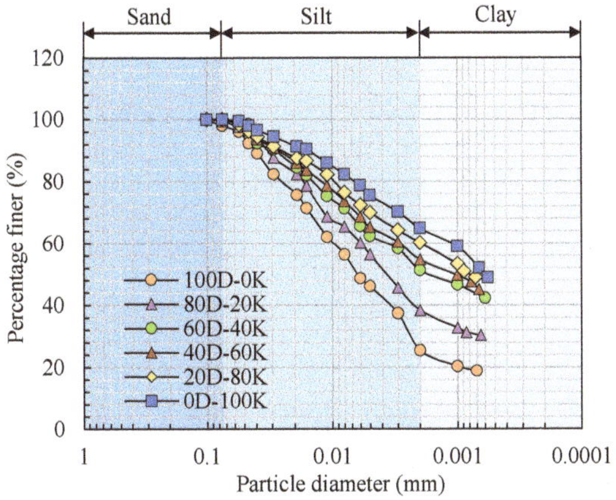

Figure 12. Particle size distribution of DS with different diatom content.

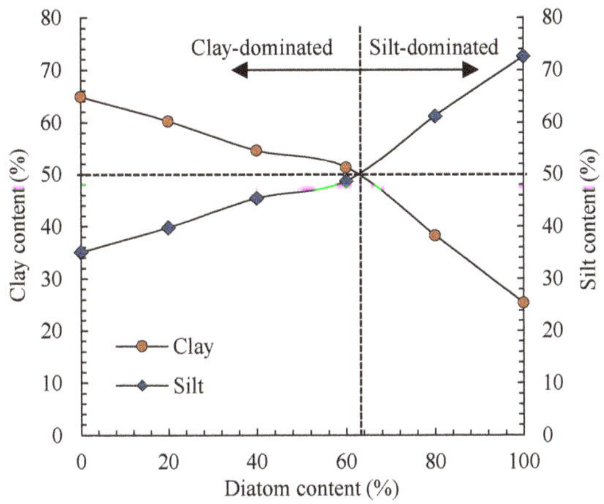

Figure 13. Variations of silt and clay particle contents of DKM samples with diatom content.

Following the USCS (Figure 7), samples 0D-100K, 20D-80K, and the others were classified as low-plasticity clay (CL), low-plasticity silt (ML), and high-plasticity silt (MH), respectively. According to the Casagrande plasticity chart, the rightward shift of the DKM data points with increasing diatom content indicates the change of soil behavior from clayey to silty, and the DKM samples with $w_L > 50\%$ are of high plasticity. Similar tendencies for types of DKM in other studies can be seen in Figure 7. Considering the classification results, the transition of dominant particle size from clay to silt with increasing diatom content for the DKM samples is confirmed by the PSD curves in Figures 12 and 13. However, the results for high-plasticity soil are inappropriate according to the investigation of the microstructure and consistency limits of the DKM samples. For example, sample 100D-0K is non-plastic and should not be classified as soil with high plasticity (Figure 2). Additionally, sample 20D-80K contains dominant clay-sized particles but is misclassified as silt by the USCS. Therefore, the USCS may well result in improper classification of DS because of its unique consistency limits due to the diatom microstructure.

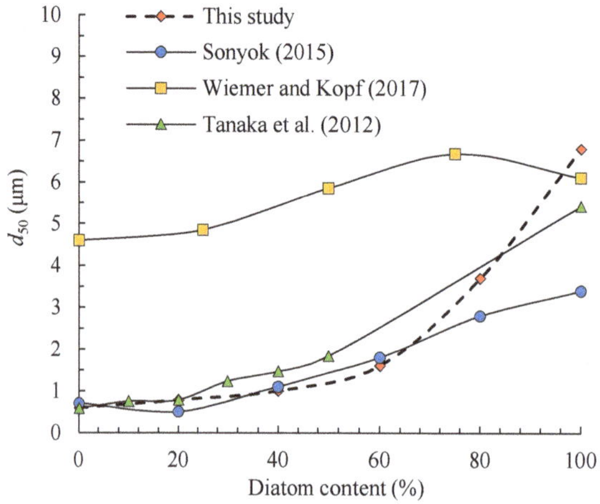

Figure 14. Relationship between average particle diameter d_{50} and diatom content. Data sources: DKMs [6,19,32].

Jang and Santamarina [27] proposed a classification system for fine-grained soil based on pore fluid sensitivity. This classification focuses on the different types of interparticle interaction (i.e., van der Waals and double-layer effects) in fine-grained soil. It uses electrical sensitivity (S_E) to measure the permittivity and conductivity of the soil. S_E is calculated using the fall-cone liquid limits, which are measured with different pore fluids of deionized water, 2-M NaCl solution, and kerosene (i.e., $w_{L,DW}$, $w_{L,brine}$, and $w_{L,ker}$, respectively). Figure 15 shows the classification results according to this system for different types of DKM and the DS studied by Al Shatnawi and Bandini [7], along with those for silica silt, kaolin, and other minerals (i.e., mica, bentonite, and illite). With increasing diatom proportion, the permittivity decreases, and the conductivity increases slightly, which results in decreasing S_E (Figure 15a). As shown in Figure 15b, a DKM with a diatom content of less than 40% is classified as "intermediate plasticity, intermediate S_E", while one with a diatom content of more than 80% is classified as "high plasticity, low S_E". Sample 60D-40K is classified as "high plasticity, intermediate S_E". However, the pore fluid sensitivity classification recognizes pure diatoms as high plasticity soil, although this sample is confirmed to be non-plastic (Figure 2).

Shiwakoti et al. [12] showed that the silt-sized diatom particles in soil are, in effect, like clay particles, leading to a high value of A_d. This is also the case in the present study, and Figure 8 shows that adding diatoms to kaolin results in a higher value of A_d. However, unlike clay particles, diatoms are non-plastic because their frustules are connected through a chain-like skeleton structure (Figure 3i). In this case, the consistency limits of DS are essentially different from those of fine-grained soil without diatoms. Consequently, further investigations are necessary to propose a more reasonable classification system for DS, considering the unique microstructure of diatom frustules.

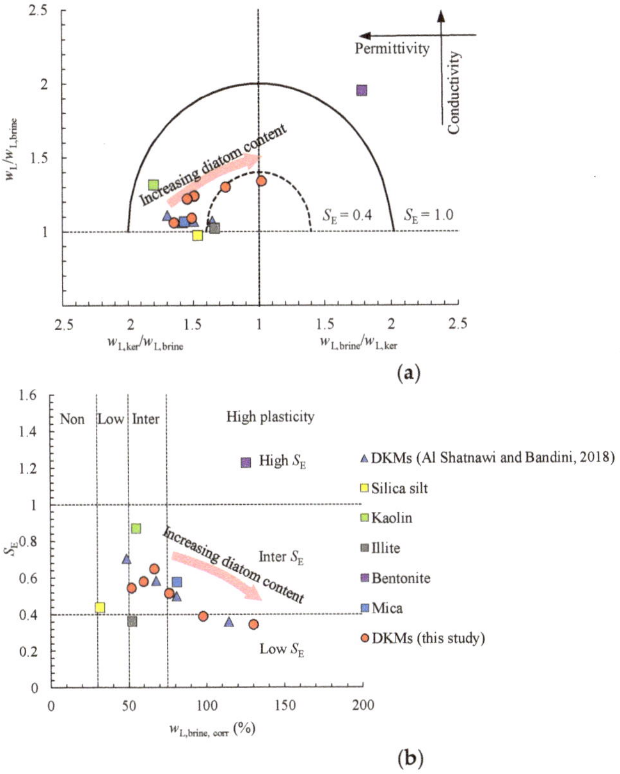

Figure 15. Classification of DS according to pore fluid sensitivity: (**a**) permittivity and conductivity of the samples; (**b**) electrical sensitivity of the samples. Data sources: DKMs [7].

5. Summary and Conclusions

This study investigated how diatom content, diatom crushing degree, and pore fluid ionic concentration influence the consistency limits of DS, along with the underlying microscopic mechanism. The main conclusions are as follows:

Diatom frustules have a porous microstructure with massive skeletal and intra-skeletal pores, which provide space to store water. The crushed diatom fragments form chain-like connective skeletons and have no intra-skeletal pores. The microstructure of a DKM is characterized by diatom–kaolin aggregates, with the pores of the diatom frustules surrounded or filled by clay particles.

The pore size distribution curve of kaolin changes from unimodal to trimodal with increasing diatom content because of the introduction of skeletal and intra-skeletal pores. With increasing diatom content in a DKM, the volume proportion of tiny and medium pores increases, and that of small pores decreases, which is closer to the pore characteristics of diatoms than those of kaolin.

A DKM with a diatom content of over 60% is silt-dominated; otherwise, it is clay-dominated. With an increasing diatom proportion, the clay fraction of a DKM decreases approximately linearly, but its activity increases with the clay fraction. The diatom frustules are active silt-sized particles, in effect, like clay particles in the soil.

The w_L and w_P values increase significantly with increasing diatom content, but I_P remains almost constant. The extremely high w_L and w_P values of DS are associated with the intact diatom frustules: the w_L and w_P values of a crushed DKM are lower than those of an uncrushed one because the collapsed intra-skeletal pores in the former can no longer retain water; however, the water stored in the intra-skeletal pores of diatom frustules contributes little to the plasticity of the soil.

The Atterberg limits of kaolin are affected by NaCl concentration, whereas those of pure diatoms are almost unaffected. Similar to the consistency limits of pure diatoms, the w_L, w_p, and I_P values for a DKM with even a small proportion of diatoms (20%) are insensitive to NaCl concentration.

The plasticity and pore fluid sensitivity charts fail to classify DS accurately, especially when the diatom content is high. Further investigations are needed to establish a more reasonable classification system for DS, considering its unique microstructure and diatom content.

This study considered the effect of geological burial in the marine environment during the forming process of DS by investigating the influences of diatom content, diatom crushing degree, and pore fluid ionic concentration on Atterberg limits. Additionally, the microstructure and pore characteristics of DS were examined via microscopic tests. Through the test results of Atterberg limits and microstructure of DS, this study revealed the micro-level mechanism of the high Atterberg limits of DS. In this study, only disk-shaped diatoms are considered, which means the results of this work may not apply to DS dominating other diatom species. More diatom species and other factors, such as complex mineral components of natural DS, should be considered in further research.

Author Contributions: Conceptualization, X.Z. and X.L.; Methodology, Y.X., X.L. and A.Y.; Validation, Y.X.; Formal analysis, Y.X., G.W. and X.L.; Writing—original draft, Y.X. and X.L.; Writing—review & editing, X.Z. and G.W.; Supervision, X.Z.; Project administration, X.Z.; Funding acquisition, X.Z. All authors have read and agreed to the published version of the manuscript.

Funding: This work was supported by the National Natural Science Foundation of China (Nos. 41972285, 42177148, 41972293), the Youth Innovation Promotion Association CAS (No. 2018363), and Science Fund for Distinguished Young Scholars of Hubei Province (Grant No. 2020CFA103). The APC was funded by the National Natural Science Foundation of China (No. 41972285).

Data Availability Statement: All data that supports the findings of this study are available from the corresponding author upon reasonable request.

Acknowledgments: The author would like to thank Fei Wang of China Railway Design Corporation (Tianjin, China) for his invaluable help in methodology. The financial support of the National Natural Science Foundation of China (Nos. 41972285, 42177148, 41972293), the Youth Innovation Promotion Association CAS (No. 2018363), and Science Fund for Distinguished Young Scholars of Hubei Province (Grant No. 2020CFA103) are gratefully thanked.

Conflicts of Interest: The authors declare no conflict of interest.

Nomenclature

G_S	Specific gravity
DS	Diatomaceous soil
I_P	Plasticity index
d_{50}	Average particle diameter (μm)
w_L	Liquid limit (%)
w_p	Plastic limit (%)
USCS	Unified Soil Classification System
SEM	Scanning electron microscopy
TEM	Transmission electron microscopy
MIP	Mercury intrusion porosimetry
DKM	Diatom-kaolin mixture
A_d	Activity
D	Diatom
K	Kaolin
P	Applied pressure (Pa)
d	Equivalent pore diameter (m)

θ_{Hg}	Mercury–soil contact angle (°)
σ_{Hg}	Mercury–soil surface tension (N/m)
CV	Cumulative pore volume
IV	Log-incremental pore volume
PSD	Particle size distribution
CL	Low-plasticity clay
ML	Low-plasticity silt
MH	High-plasticity silt
$w_{L,ker}$	Liquid limit using kerosene (%)
$w_{L,brine}$	Liquid limit using NaCl concentration (%)
S_E	Electrical sensitivity

References

1. Díaz-Rodríguez, J.A. Characterization and engineering properties of Mexico City lacustrine soils. In *Characterization and Engineering Properties of Natural Soils*; Swets & Zeitlinger: Lisse, The Netherlands, 2003; Volume 1, pp. 725–756.
2. Tanaka, H.; Locat, J. A microstructural investigation of Osaka Bay clay: The impact of microfossils on its mechanical behaviour. *Can. Geotech. J.* **1999**, *36*, 493–508. [CrossRef]
3. Lee, J.S.; Lee, J.Y.; Kim, Y.M.; Lee, C. Stress-dependent and strength properties of gas hydrate-bearing marine sediments from the Ulleung Basin, East Sea, Korea. *Mar. Petrol. Geol.* **2013**, *47*, 66–76. [CrossRef]
4. Caicedo, B.; Mendoza, C.; López, F.; Lizcano, A. Behavior of diatomaceous soil in lacustrine deposits of Bogotá, Colombia. *J. Rock Mech. Geotech. Eng.* **2018**, *10*, 367–379. [CrossRef]
5. Zhang, Y.S.; Guo, C.B.; Yao, X.; Qu, Y.X.; Zhou, N.J. Engineering geological characterization of clayey diatomaceous earth deposits encountered in highway projects in the Tengchong region, Yunnan, China. *Eng. Geol.* **2013**, *167*, 95–104. [CrossRef]
6. Tanaka, M.; Watabe, Y.; Tomita, R.; Kamei, T. Effects of Diatom Microfossils Content on Physical Properties of Clays. In Proceedings of the 22nd International Offshore and Polar Engineering Conference, Rhodes, Greece, 17 June 2012.
7. Al Shatnawi, H.H.; Bandini, P. Limitations of classifications for soils that contain diatom microfossils. In *IFCEE 2018*; American Society of Civil Engineers: Reston, VA, USA, 2018; pp. 123–132.
8. Locat, J.; Tanaka, H. A new class of soils: Fossiliferous soils? In Proceedings of the 15th International Conference on Soil Mechanics and Geotechnical Engineering, Istanbul, Turkey, 28–31 August 2001; pp. 1–3, 2295–2300.
9. Ilia, I.; Stamatakis, M.; Perraki, T. Mineralogy and technical properties of clayey diatomites from north and central Greece. *Open Geosci.* **2009**, *1*, 393–403. [CrossRef]
10. Ghisoli, C.; Caucia, F.; Marinoni, L. XRPD patterns of opals: A brief review and new results from recent studies. *Powder Diffr.* **2010**, *25*, 274–282. [CrossRef]
11. Galán, E.; González, I.; Mayoral, E.; Miras, A. Properties and applications of diatomitic materials from SW Spain. *Appl. Clay Sci.* **1993**, *8*, 1–18. [CrossRef]
12. Shiwakoti, D.R.; Tanaka, H.; Tanaka, M.; Locat, J. Influences of diatom microfossils on engineering properties of soils. *Soils Found.* **2002**, *42*, 1–17. [CrossRef]
13. Díaz-Rodríguez, J.A.; Cruz, R.L.S.; Dávila-Alcocer, V.M.; Vallejo, E.; Girón, P. Physical, chemical, and mineralogical properties of Mexico City sediments: A geotechnical perspective. *Can. Geotech. J.* **1998**, *35*, 600–610. [CrossRef]
14. Perisic, G.A.; Ovalle, C.; Barrios, A. Compressibility and creep of a diatomaceous soil. *Eng. Geol.* **2019**, *258*, 105145. [CrossRef]
15. Díaz-Rodríguez, J.A.; Leroueil, S.; Aleman, J.D. Yielding of Mexico City clay and other natural clays. *J. Geotech. Eng.* **1992**, *118*, 981–995. [CrossRef]
16. Caicedo, B.; Zuluaga, D.; Slebi, C. Effects of micro-features of fossil diatom on the macroscopic behaviour of soils. *Géotechnique Lett.* **2019**, *9*, 322–327. [CrossRef]
17. Zhang, X.W.; Liu, X.Y.; Xu, Y.Q.; Wang, G.; Wang, F. Swelling–shrinkage mechanisms of diatomaceous soil and its geohazard implication. *Eng. Geol.* **2023**, *314*, 107009. [CrossRef]
18. Zhang, X.W.; Liu, X.Y.; Xu, Y.Q.; Wang, G.; Ren, Y.B. Compressibility, permeability and microstructure of fine-grained soils containing diatom microfossils. *Géotechnique* **2022**, 1–50. [CrossRef]
19. Wiemer, G.; Kopf, A. Influence of diatom microfossils on sediment shear strength and slope stability. *Geochem. Geophy. Geosy.* **2017**, *18*, 333–345. [CrossRef]
20. Xu, Y.Q.; Zhang, X.W.; Liu, X.Y.; Wang, G. Alterations of physical properties and microstructure of marine diatomite owing to variation of diatom content. *Mar. Georesour. Geotech.* **2022**, 1–12. [CrossRef]
21. Santamarina, J.C.; Klein, K.A.; Wang, Y.H.; Prencke, E. Specific surface: Determination and relevance. *Can. Geotech. J.* **2002**, *39*, 233–241. [CrossRef]
22. Zhang, X.W.; Liu, X.Y.; Chen, C.; Kong, L.W.; Wang, G. Engineering geology of residual soil derived from mudstone in Zimbabwe. *Eng. Geol.* **2020**, *277*, 105785. [CrossRef]
23. Washburn, E.W. Note on a method of determining the distribution of pore sizes in a porous material. *Proc. Natl. Acad. Sci. USA* **1921**, *7*, 115–116. [CrossRef] [PubMed]

24. *ASTM D4318-17*; Standard Test Methods for Liquid Limit, Plastic Limit, and Plasticity Index of Soils. ASTM International: West Conshohocken, PA, USA, 2017.
25. Casagrande, A. Notes on the design of the liquid limit device. *Géotechnique* **1958**, *8*, 84–91. [CrossRef]
26. *BSI BS EN ISO 1377-2*; Methods of Test for Soils for Civil Engineering Purposes (Classification Tests). BSI: London, UK, 1990.
27. Jang, J.B.; Santamarina, J.C. Fines classification based on sensitivity to pore-fluid chemistry. *J. Geotech. Geoenviron. Eng.* **2016**, *142*, 06015018. [CrossRef]
28. Vardanega, P.J.; Haigh, S.K.; O'Kelly, B.C.; Zhang, X.W.; Liu, X.Y.; Chen, C.; Wang, G. Discussion: Use of fall-cone flow index for soil classification: A new plasticity chart. *Géotechnique* **2022**, 1–23. [CrossRef]
29. *ASTM D7928-21*; Standard Test Method for Particle-Size Distribution (Gradation) of Fine-Grained Soils Using the Sedimentation (Hydrometer) Analysis. ASTM International: West Conshohocken, PA, USA, 2021.
30. Zhang, X.W.; Kong, L.W.; Yang, A.W.; Sayem, H.M. Thixotropic mechanism of clay: A microstructural investigation. *Soils Found.* **2017**, *57*, 23–35. [CrossRef]
31. Díaz-Rodríguez, J.A.; González-Rodríguez, R. Influence of diatom microfossils on soil compressibility. In Proceedings of the 18th International Conference on Soil Mechanics and Geotechnical Engineering, Presses des Ponts, Paris, France, 2–6 September 2013; pp. 325–328.
32. Sonyok, D.R. Effect of Diatoms on Index Properties, Compressibility, Suction and Stiffness of Diatomite-Kaolin Mixtures. Ph.D. Thesis, New Mexico State University, Las Cruces, NM, USA, 2015.
33. Al Shatnawi, H.H. Compressibility, Fabric, and Classification of Fine-Grained Soils Containing Diatom Microfossils. Ph.D. Thesis, New Mexico State University, Las Cruces, NM, USA, 2019.
34. Kim, K.S. Engineering characteristics of diatom modified soil mixture. *J. Kor. Geotech. Soc.* **2012**, *28*, 77–84. [CrossRef]
35. Tanaka, M.; Watabe, Y.; Tomita, R. Effects of crushed diatom microfossils content on physical properties of japanese clays. In Proceedings of the 23rd International Offshore and Polar Engineering Conference, OnePetro, Anchorage, AK, USA, 30 June–5 July 2013.
36. Palomino, A.M.; Santamarina, J.C. Fabric map for kaolinite: Effects of pH and ionic concentration on behavior. *Clay. Clay Miner.* **2005**, *53*, 211–223. [CrossRef]
37. Palomino, A.M.; Kim, S.; Summit, A.; Fratta, D. Impact of diatoms on fabric and chemical stability of diatom–kaolin mixtures. *Appl. Clay Sci.* **2011**, *51*, 287–294. [CrossRef]
38. Losic, D.; Mitchell, J.G.; Voelcker, N.H. Fabrication of gold nanostructures by templating from porous diatom frustules. *New J. Chem.* **2006**, *30*, 908–914. [CrossRef]
39. Lee, J. Elastic, Electromagnetic, Physical, and Mechanical Characterization of Diatomaceous Earth and Offshore Sediments. Ph.D. Thesis, The University of Wisconsin-Madison, Madison, WI, USA, 2020.

applied sciences

Article

Model Test Analysis of Groundwater Level Fluctuations on Karst Cover Deformation Taking the Monolithic Structure of Guilin as an Example

Xuejun Chen [1,2], Xiaotong Gao [1,2], Hui Li [1,2], Mingming Xue [1,2], Xiaohui Gan [1,2] and Yu Song [1,*]

1 College of Civil and Architecture Engineering, Guilin University of Technology, Guilin 541004, China
2 Guangxi Key Laboratory of Geomechanics and Geotechnical Engineering, Guilin 541004, China
* Correspondence: songyu119@glut.edu.cn

Abstract: Engineering practice and real-life cases show that the geological conditions of the Guilin overlying karst site are complex. In particular, the groundwater, which drives the accelerated formation of soil cavities, and the thickness of the overlying soil layer, which affects the speed of the groundwater subsidence process. Therefore, this paper is based on a physical model to evaluate the effects of groundwater level changes caused by different factors on the deformation of karst cover. The model tests are simulated for different cover thicknesses (6 cm, 9 cm, 12 cm, 15 cm, 18 cm) under rainfall and other recharge, cavity supply, and drainage conditions at the same density (1.40 g/cm^3) and initial water content (30%), respectively. The results show that with the increase of rainfall and other recharge time, the basic change trend of different cover thicknesses is that the infiltration curve changes faster at the beginning and slows down at the end, but the thicker the cover, the slower the overall deformation; at a certain rate of cavity recharge and drainage, the thicker the cover, the smaller the deformation caused by the fluctuation of groundwater level. The cavity recharge makes the cover displacement obvious, in the order of 0.304 cm, 0.173 cm, 0.118 cm, 0.068 cm, and 0.056 cm. After the formation of the cavity, the rainfall, other recharge, and the cavity supply and drainage accelerate the destruction and deformation of the soil body and the upward development of the cavity. The research results provide theoretical support for the subsequent prevention and control of karst collapse in covered karst areas, and have certain practical engineering significance.

Keywords: karst collapse; physical model; groundwater level changes; cover deformation

1. Introduction

Karst collapse is a karst-driven geological action and phenomenon in which the rock and soil above a karst cavity are damaged by deformation caused by natural or man-made factors, and a collapse pit (hole) is formed on the ground [1]. Karst collapse is a widespread geohazard problem worldwide, with more than 1/3 of China's land area covered by soluble rock [2]. The widespread distribution and sudden hazard of karst collapse have attracted the great attention of the international community. In the past two decades, the problem of karst collapse has become more and more serious, especially in the present day when roads are opened on every mountain and bridges are built over every river; some underground projects can hardly avoid crossing the karst development area, leading to the accelerated generation of karst collapse [3–5]. This geological disaster problem has seriously hindered local economic construction and social development and has posed a certain threat to the safety of people's lives and properties [6]. Therefore, scholars at home and abroad have paid great attention to karst collapse, and have carried out a lot of scientific research work and achieved fruitful scientific results. At present, the research results on karst collapse at home and abroad can be summarized as follows: collapse causes and collapse mechanisms, influencing factors, stability analysis and prediction, physical model tests and numerical research, monitoring, and prevention [7–11].

Karst collapse in China is mainly influenced by a variety of factors such as topography, stratigraphic lithology, geological formations, changes in groundwater dynamics, and human engineering-economic activities [12,13]. Field tests are limited in many ways, making it difficult to map the characteristics of the overlying karst soil cavities and the corresponding collapse conditions, and even more difficult to determine a series of standard test conditions that are fully compatible with them. Among the many triggering factors affecting karst ground collapse, groundwater is the most important one. The changes in groundwater level, magnitude, and frequency, and its hydrodynamic characteristics are the most active and dominant dynamic factors in the formation of karst collapse [14–18]. The vast majority of karst collapses occur under dramatic changes in groundwater dynamic conditions, and the causes are categorized by domestic scholars as subduction, disintegration, vacuum absorption, and airburst [19].

In the karst collapse of groundwater changes in the collapse mechanism, some scholars have studied many cases of karst ground collapse closely related to the changes in the water table, analysis of the impact of groundwater on karst ground collapse, verification of the "subduction theory, vacuum absorption theory, airburst theory" of the three existing mechanisms [20,21], and it is also pointed out that the types of collapse and the collapse mechanism are different under different geological conditions. The number of karst collapse hazards caused by fluctuating water levels is also increasing, exacerbating the threat and damage to life and property [22]. Therefore, for the collapse induced by the change in groundwater level, the study of the development law of cover deformation due to groundwater fluctuation is particularly important for karst collapse, and it is of great value for the prediction and prevention of karst collapse. However, as karst collapse is characterized by suddenness, concealment, complexity and uncertainty [23–26], a comprehensive and systematic theory of collapse has not yet been established. Other scholars have carried out research work on the causes and influencing factors of karst collapse by taking the final state of karst collapse as the research object [27]. It is difficult to understand the process of karst collapse, which leads to an inability to fully and deeply cognize the mechanism of collapse formation [28], so it is necessary that the formation process and evolution mechanism of karst collapse can be dynamically observed with the help of indoor physical model tests.

At the same time, indoor model tests are also one of the effective methods to verify the existence of the proposed collapse theory [29]. For example, Jin Honghua et al. [30] established a cylindrical soil collapse model and derived the calculation method of stability coefficients for four different working conditions based on the limit equilibrium theory. Sun Jinhui [31] carried out physical model tests on karst collapse with clay overburden and obtained a critical collapse width of 2–3 mm. However, the model device collapse hole was in contact with the atmosphere and in an unpressurized state, which did not match the actual karst collapse opening or collapse pipe in a pressurized state. Li Caihua et al. [32] established a physical model based on the exploration data of a collapse site, and conducted a series of tests mainly for the binary structure (clay in the upper layer and sandy soil in the lower layer). It was concluded that fluctuations in the karst water table were the main cause of the induced karst collapse, but the process of collapse formation could not be quantified. Although the use of physical model tests also verified the validity of some of the collapse theories and reproduced the process of karst collapse, which improved the understanding of the evolution of the karst collapse mechanism, there are still many questions on how to quantify the process of karst collapse, especially how to obtain the extent of cover collapse during the collapse process. Therefore, in this paper, clayey soils are selected as the cover layer, and physical model tests are carried out to simulate the collapse process by vacuum suction and collapse process by positive pressure airburst, and to simulate the cover deformation at the same density and supply and drainage rate for five different cover thicknesses respectively.

Therefore, this paper takes the one-dimensional structural overburden in the karst collapse-prone area of Lingui District, Guilin City as the research object. Based on the previous research results, a physical model of the overburden type is constructed, a typical test model of groundwater level fluctuation is established and model tests are carried out. In this study, the effects of rainfall and other recharge and cavity supply and drainage on the settlement process of the overlying karst cover are simulated. The effect of groundwater level fluctuations on the cover deformation was then analyzed by means of the cover displacement. This study provides a theoretical and technical basis for practical engineering activities in covered karst areas.

2. Geological Setting of the Study Area

2.1. Meteorology and Hydrology

The study area is located in Lingui District, Guilin, at the southern end of the Xianggui Corridor, at an average altitude of 150 m. Meteorological observations show that the annual mean temperature in the study area ranges from 17.0 to 20.0 °C, with cold winters, the average daily temperature in January and February remaining at 5 to 7 °C, and the annual extreme minimum temperature at −5 to 0 °C. The summer and autumn seasons are significantly warmer, with extreme maximum temperatures of 37.0 to 39.0 °C. Temperatures are significantly higher in summer and autumn, with annual maximum temperatures ranging from 37.0 to 39.0 °C. The total annual precipitation is 1266.0~1986.0 mm, and the spatial and temporal distribution of rainfall is uneven, with more precipitation in spring and summer, and severe droughts in autumn and winter, with heavy rainfall mainly concentrated in May–June. The water level in the creek is 154.60 m. The water level in the creek is affected by the seasons and is mainly recharged by rainfall and domestic wastewater from the surrounding area, with a small volume of water. The water table in the study area is basically located at the bedrock surface and is subject to seasonal variations, with a range of 2–3 m.

2.2. Geological Features

The geological drilling data available in the study area shows that the Quaternary strata in the study area are widely distributed, mainly consisting of red clay formed by the Upper Pleistocene Alluvium ($Q_3^{al\text{-}pl}$), the Upper Pleistocene Residual Slope Formation ($Q_3^{dl\text{-}el}$) and the Upper Devonian Rongxian Formation (D_3r). The Upper Pleistocene Alluvium ($Q_3^{al\text{-}pl}$) is a yellow and tawny pebble-gravel sandy clay and chalky clay, composed of gravel, clay and a small amount of fine sand and pebbles, 2–14 m thick; the Upper Pleistocene Residual Slope Formation ($Q_3^{dl\text{-}el}$) is a brick-red and light yellow clay, sandy clay and sandy-clay gravel layer, containing a small amount of quartz particles, 0.1–10 m thick; the Upper Devonian Rongxian Formation (D_3r) is greyish-white, slightly weathered, cryptocrystalline in structure, medium-thick laminated. The main mineral is calcite, with calcite veins developed and tightly cemented, the rock is relatively intact, the core is columnar, the bedrock surface within the site is highly undulating, and the karst fissures are relatively developed (Figure 1).

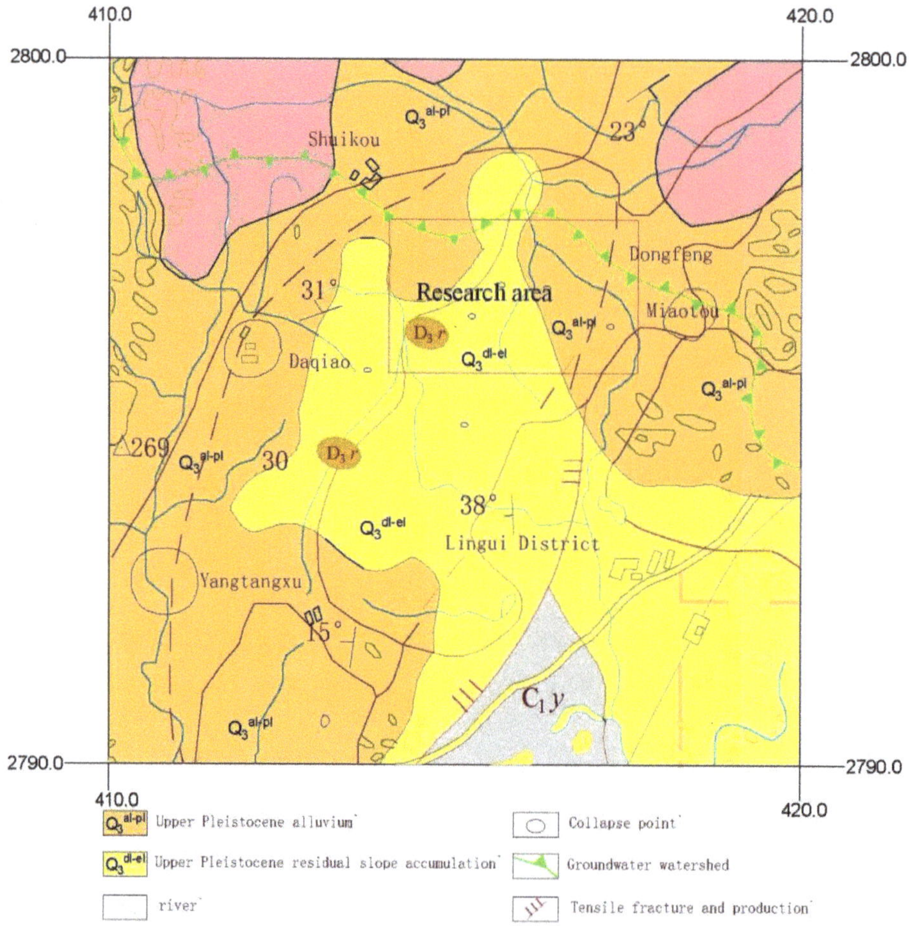

Figure 1. Engineering geological map.

3. Test Materials and Methods

3.1. Generalisation of the Geological Model of the Study Area

Due to the complex geological conditions in the study area, there are many uncertainties. If the karst-induced collapse test were to be carried out directly at the study site, it would be constrained by many factors. It is therefore difficult to carry out in-situ karst collapse tests at the site. In order to obtain the results of the karst collapse test under changing groundwater levels and to carry out the test successfully, a geological model of the geological conditions in the study area needs to be generalised here.

The geological drilling data collected from Lingui District, Guilin City, shows that the overlying karst in Lingui District can be mainly classified into single or double-layer structures. Among the monolayer structures, the overlying soil layer is clayey. In this simulation test study, a specific study area was selected to focus on the monolithic structure (Figure 2).

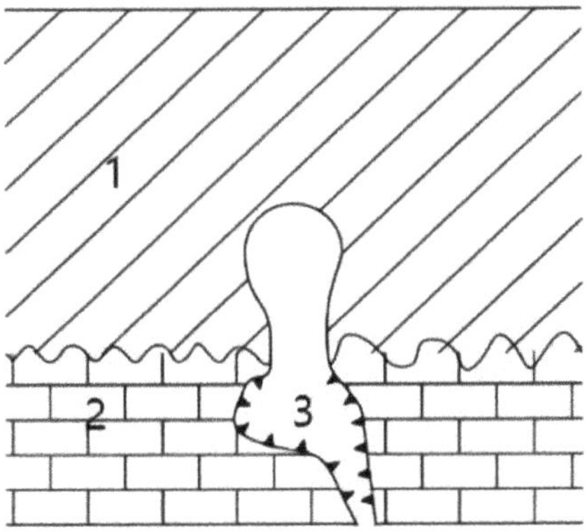

Figure 2. Geological generalization of a single structure. 1—Overburden clayey soil; 2—Limestone; 3—Caves.

3.2. Physical Modelling Principles

When conducting physical model tests, not only should the test requirements and operability be met, but similar issues should also be considered. According to the geological reports available in the study area, it is known that the thickness of the collapse prone overburden in the study area ranges from 0.7 to 5.6 m, and the diameter of the karst openings on the bedrock surface varies from 0.5 to 1.2 m. The geometric similarity ratio was finally determined to be 1:15 by combining the geological, hydrological and tectonic conditions of the study area. 6 cm diameter of the reserved soil hole was set, and the thickness of the overburden layer was 6 cm, 9 cm, 12 cm, 15 cm, and 18 cm. The soil samples of the overburden layer were taken directly from the site during the test, and the moisture content of the test was close to that of the site. However, the density was difficult to achieve in situ, and after repeated indoor tests, the density was finally determined to be 1.4 g/cm^3. In order to make the simulation conditions as close to the site as possible, to take into account the factors involved in karst collapse at the site and to ensure that the soil simulation parameters are basically the same as the original soil at the site, the karst collapse test device was specially tailored. In the course of the test, the test device is subjected to a large load when filling a predetermined thickness of overlying soil. This is why the frame of the device is made of a high hardness aluminum alloy and the side walls are inlaid with 5 mm thick toughened glass. To accommodate the inflow of lateral recharge water during the test, holes were drilled in the glass on the left and right sides, as shown in Figure 3a. To prevent loss of cover soil in the small holes on the left and right sides, towels were taped to the left and right sides before the test began to fill with soil, as shown in Figure 3b. In order to have a clear view of the deformation and damage process inside the cover soil, the karst access opening was set up in a semi-circle at the midpoint of the front side. A special ice block was pre-buried above the passage opening to form a pre-existing soil cavity in the shape of a 1/4 sphere during the test. The physical model consists of four parts: the main model, the rainfall system, the water supply and drainage system, and the monitoring system. The structure of the physical model is shown in Figure 3c.

(a) Lateral weep holes (b) Prevention of loss of cover soil

(c) Overall model drawing

Figure 3. Physical model installations. 1—Special rainfall shower 2—model unit side water tank. 3—overburden soil 4—existing earth cave. 5—dissolving chamber water tank 6—dissolution chamber inlet. 7—dissolution chamber drain 8—drainage monitoring water meters. 9—cameras.

3.2.1. Body of the Model

The main model box has dimensions of L (120 cm) × W (70 cm) × H (100 cm) and is made up of two parts, the upper and lower, as shown in Figure 3. The upper part of this consists of a central soil box (100 cm × 70 cm × 60 cm)) and water tanks (10 cm × 70 cm × 60 cm long) on the left and right sides. The middle soil box is used to simulate the Quaternary soil layer and the water tanks on either side are used to simulate other recharge sources. The lower part is a rectangular (120 cm × 70 cm × 40 cm) water tank, which is used to simulate the changing water level in the solution chamber. The toughened glass at the bottom of the cover soil is set with a semi-circular hole of 3 cm radius to simulate a soil cavity for the karst channel. The karst passages allow the water inside the cavity to communicate with the upper cover soil and thus interact and influence each other when the water level changes.

3.2.2. Rainfall Systems

The rainfall system in this experiment means it is difficult to control the intensity of rainfall accurately, and after reviewing the research results of relevant scholars, it was decided to use the total daily rainfall as the control factor for artificial rainfall [20,32]. The rainfall system is an independent external system, erected on a steel frame above the main model. A special showerhead is used to simulate artificial rainfall, with a switch, timer and water meter at the inlet pipe of the showerhead. The switch and timer control the timing of the rainfall and the water meter controls the daily rainfall amount.

3.2.3. Water Supply and Drainage Systems

The water supply and drainage system are divided into two parts, upper and lower, whose two systems are independent and do not affect each other. The upper left and right tanks are designed to recharge the overburdened soil from the lateral direction, while the lower karst cavity water supply and drainage system are designed to provide water levels for the karst channels. The lower tank is connected to a water pipe connected to the mains as a water supply outlet pipe for the karst cavity and the rate and time of the water supply are controlled by a control switch and water meter. The lower tank is also equipped with a 100 mm diameter water pipe with a switch as a drainage pipe. The switch of the drain is set to close when the water is supplied to the karst cavity to simulate the rise of the water table in the karst area. When the drainage pipe is switched on for drainage, it simulates the fall of the water table (pumping or groundwater runoff) in the karst area and the amount of drainage can be measured by setting a flow meter. Throughout the test, the water supply and drainage can be regulated according to your test requirements.

3.2.4. Monitoring Systems

The deformation of the cover soil and the pressure of the karst cavity are monitored according to the test requirements of this paper. The displacement meter sensor model DMWY-100 is used for the determination of the cover soil deformation, with a compact size suitable for small-scale indoor model tests. The output is highly sensitive and can resolve displacement changes of less than 0.005 mm; has wide range, high accuracy, small drift, and stable performance; stainless steel full bridge housing, which is corrosion resistant and pressure resistant; technical parameters are shown in Table 1. The pressure in the karst cavity is monitored using a levelogger 5 barometric pressure sensor to monitor changes in pressure during supply and drainage. A video camera is used to record the deformation to collapse of the overburden on the karst bedrock, with a focus on capturing the deformation, intensification, and end of the karst collapse caused by rainfall and lateral recharge, water supply, and drainage.

Table 1. Technical parameters of DMWY-100 surface type displacement meter.

Dimensions	**Φ25 × 345**	**Measuring Range (mm)**	**0~100**
Full output (με)	8000	Correction factor (mm/με)	0.0125
Precision	≤0.2%F•S	Error (mm)	≤0.01

Note: F•S is the full-scale output value, around 8000 microstrain.

3.3. *Test Program*

The experiments in this paper focus on the effect of dynamic changes in groundwater levels on the deformation to collapse of overlying karst caps. Therefore, two different hydrodynamic models were developed: (1) the effect of different recharge conditions on the deformation to collapse of the cover soil; (2) the effect of groundwater fluctuation on the deformation to collapse of the cover soil. Physical model tests were also carried out for five different cover thicknesses under two hydrodynamic models, i.e., to quantitatively study the process of deformation to collapse of overlying karst cover. Their indoor physical model test protocols are shown in Table 2.

Table 2. Physical simulation test program for karst collapse.

Test Program	**Control Variables**	**Karst Opening Diameter/cm**	**Stratigraphic Conditions**	**Dry Density/ (g/cm^3)**	**Initial Moisture Content/%**
1	Rainfall and other recharge	6	Red clay Five different thicknesses	1.40	30
2	Dissolution chamber supply and drainage	6	Red clay Five different thicknesses	1.40	30

3.4. Model Filling

Firstly, based on the moisture content w_0 = 5% measured after drying the soil samples taken from the study area, combined with the initial moisture content w_1 = 30% and dry density p_d = 1.4 g/cm^3 determined in the test protocol, and the different cover thicknesses required for each test (6 cm, 9 cm, 12 cm, 15 cm, 18 cm), the mass of soil and the mass of water required for each test were calculated respectively. The calculated mass of soil and the mass of water are then configured to obtain the soil required for the test. Finally, the configured soil samples are filled in batches in the model box. As both scenarios 1 and 2 were the same soil structure in 5 different thicknesses, the filling protocol was the same. For the first 6 cm soil thickness test, the soil sample was divided into two parts and filled in two batches, with one sample filled to 3 cm and the remaining sample filled to 6 cm with surface trimming. Each subsequent test was filled to the required height for each test, following the same filling method as the first 6 cm.

3.5. Different Recharge and Supply and Drainage Controls

3.5.1. Different Supplies

To study the evolution of the cover soil disaster under rainfall conditions, a total rainfall of 255.4 mm was selected as the basis for the artificial rainfall conditions. In order to study the deformation of the parameters during the rainfall cessation period, a 52 min rainfall cessation period was used during the rainfall process. The rainfall starts at 13:30 and is suspended from 17:24 to 18:16. The rainfall stops at 19:30 when the cumulative rainfall reaches 255.4 mm. Throughout the rainfall process, there is a period when rainfall is stopped in order to analyze the effect of rainfall on the trend of cover displacement and deformation. Figure 4 shows the cumulative rainfall time curve taken during one rainfall test.

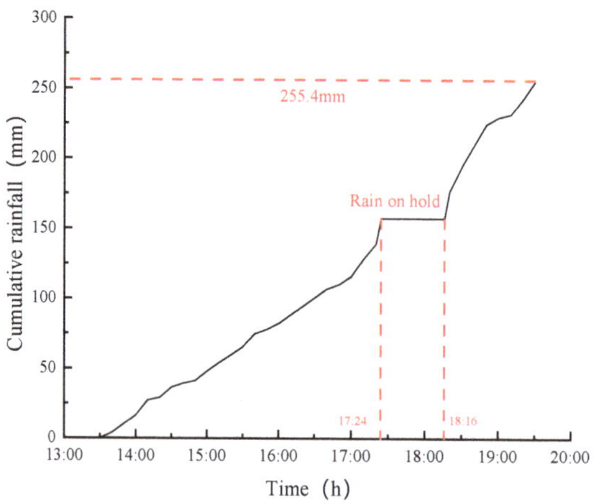

Figure 4. Rainfall vs. time curve.

3.5.2. Dissolution Chamber Supply and Drainage

The water level in the lower karst cavity was 4 cm before the model was filled. After the model was filled, the rainfall device was turned on to allow the covered soil to be fully saturated with water. After the model's seepage field had stabilized, the water level in the karst cavity was slowly raised, i.e., water was supplied at a rate of 1.2 cm/min. When the water supply level is flush with the bottom of the cover layer, the water supply is stopped and the deformation is left to stabilize for 15 min. The water level in the karst cavity is then slowly lowered, i.e., drained at a rate of 1.5 cm/min. When the drainage level reaches 4 cm at the bottom of the cavity, the drainage is stopped and the deformation is stabilized for 15 min. This was repeatedly cycled and the displacement and positive and negative pressure in the cavity were monitored during the deformation to collapse process. A video camera was used to capture any changes or leakage in the karst channel throughout the supply and drainage test.

4. Analysis of Model Test Results

4.1. Effect of Different Recharge on Cover Deformation

The tests show that there is no obvious damage and deformation of the cover soil during the initial rainfall and lateral recharge, only a significant rise in the water level of the cover soil, but scattered soil in the karst channels while the rainfall and lateral recharge infiltrate the soil. With the increase of cumulative recharge and continuous infiltration, the soil above the karst channels gradually forms soil cavities in the process of continuous descent. After the formation of the soil cavities, the continuous infiltration of recharge continues to raise the water level of the cover soil. The infiltration of soil water from the karst channel into the underlying karst cavity intensifies, and when there is a sudden change in the infiltration flow, the infiltrated water carries a small amount of soil particles, but when the flow is stable, the water is clearer. However, after the formation of the soil cavity, the continuous rainfall accelerates the destructive deformation of the soil and accelerates the upward development of the cavity. Taking the first recharge at the beginning of the test as an example, the relationship between the magnitude of displacement of each cover layer by extracting displacement meter data at the end of the test is shown in Figure 5: 6 cm (1.44 mm) > 9 cm (1.13 mm) > 12 cm (0.69 mm) > 15 cm (0.58 mm) > 18 cm (0.45 mm), the data indicates that for this cover type karst collapse, the thicker the cover layer the smaller the deformation displacement of the cover soil. However, the subsequent cumulative recharge will cause the cumulative deformation of the cover to become more and more severe until the collapse eventually occurs.

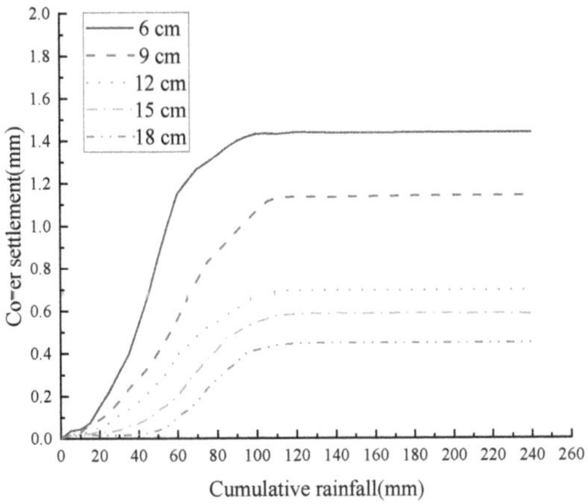

Figure 5. The relationship between rainfall and caprock deformation.

4.2. Effect of Groundwater Level Fluctuations on Cover Deformation

4.2.1. Analysis of the Influence of Water Level Rise on the Dissolution Cavity

The case in which the caprock density is 1.4 g/cm^3 and the average water supply rate is 1.2 cm/min was simulated. That is, when the positive pressure change in the dissolution chamber is the same, for different thicknesses of 6 cm, 9 cm, 12 cm, 15 cm, and 18 cm, the impact test of the karst collapse of the caprock in the karst cavity is shown in Figure 6.

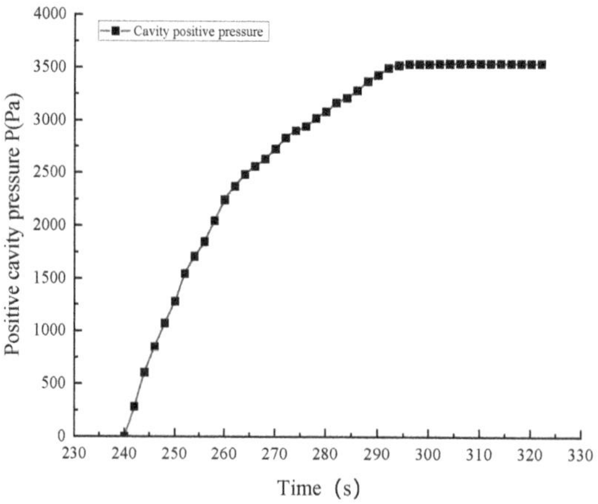

Figure 6. Time-lapse curve of the water supply and positive pressure of the dissolution chamber.

From Figure 6, it can be observed that there was no vacuum negative pressure in the dissolution chamber before 240 s. This is because the water inlet valve was closed and the pressure in the dissolution chamber was consistent with the external atmospheric pressure during this period. After 240 s, the water inlet valve was opened, and the instrument began to supply water, resulting in the positive pressure in the chamber. When the water supply continued, the groundwater level continued to rise, the space in the chamber continued to decrease, and the positive pressure value continued to increase. After 295 s, the positive pressure of the dissolution chamber reached the maximum of 3542 Pa. When the instrument started to supply water, the positive pressure of the dissolution chamber increased at the fastest rate of 375 Pa/s. With the increase of time, the positive pressure change rate gradually decreased. When the time was greater than 295 s, the positive pressure in the dissolution chamber tended to be stable, and the change rate was 0. This is mainly because with an increase in positive pressure in the cavity, the positive pressure acting on the cap layer gradually increases. When the pressure increases to a certain extent, the particles of the cap layer are connected to the outside pores, so that the gas in the cavity can be discharged. When the volume of water supplied is equal to that of the gas discharged from the cavity, the pressure in the cavity remains unchanged.

Figure 7 shows the deformation time curve of the caprock. It can be seen that the deformation-time curve of the caprocks with different thicknesses is similar to the positive pressure curve in the cavity, whereas the time required for the stability of caprock deformation is greater than that of the positive pressure curve. That is, when the positive pressure reaches the maximum, it still affects the deformation of the cover layer soil. After a period, the stress state of the cover layer reaches equilibrium again, and the cover layer soil no longer deforms. Before starting the water supply, the covering layer does not deform, the deformation amount is almost 0, and the structure is relatively stable. When the water supply starts, the covering layer begins to deform slowly. In addition, with a continuous water supply, the amount of deformation gradually increases, the final deformation is unchanged, and the structure tends to be stable.

When the thickness of the covering layer is 6 cm, the covering layer is stable at 376 s, and its deformation is 3.016 mm. When the covering layer thickness is 18 cm, the deformation of the covering layer remains unchanged at 342 s, and its deformation is 0.526 mm. When the caprock density and positive pressure in the cavity are the same, with an increase in caprock thickness, the caprock deformation decreases continuously. Namely, the caprock thickness is inversely proportional to the caprock deformation (Figure 8). In the meantime, when the thickness of the cap layer is larger, the deformation rate in the deformation stage is smaller. When the thickness of the cap layer increases from 6 cm to 18 cm, the deformation rate decreases from 0.132 mm/min to 0.0283 mm/min. This result indicates that the larger the cap layer thickness is, the more difficult it is for the cap layer to deform and fail, and the more stable the structure is.

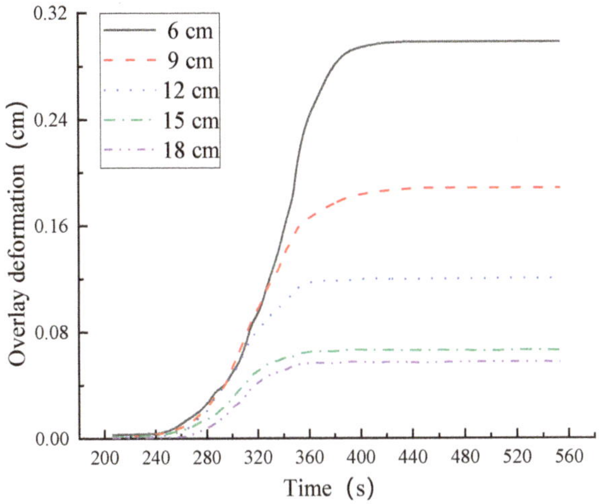

Figure 7. Rising water table cover settlement.

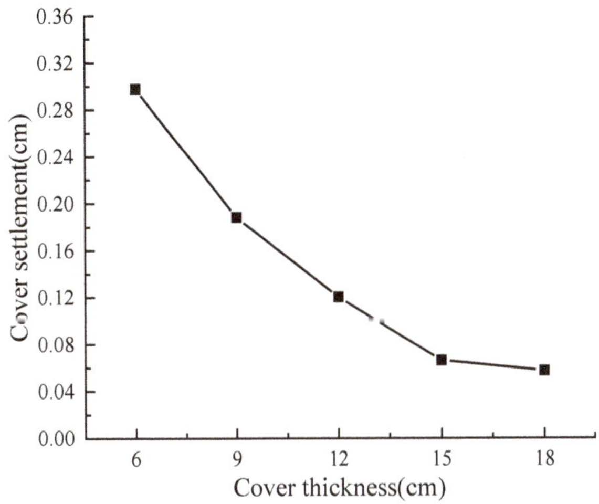

Figure 8. Settlement curve for each cover layer during water level rise.

4.2.2. Analysis of the Results of the Dissolved Cavity Water Level Drop Test

In the simulation test, the density of the cap layer is 1.4 g/cm^3 and the drainage rate is 1.5 cm/min. That is, with the same change of vacuum negative pressure in the dissolution chamber, different thicknesses of 6 cm, 9 cm, 12 cm, 15 cm, and 18 cm are simulated via experiments to study the effect of caprock on karst collapse. The variations of air pressure in the cavity are shown in Figure 9.

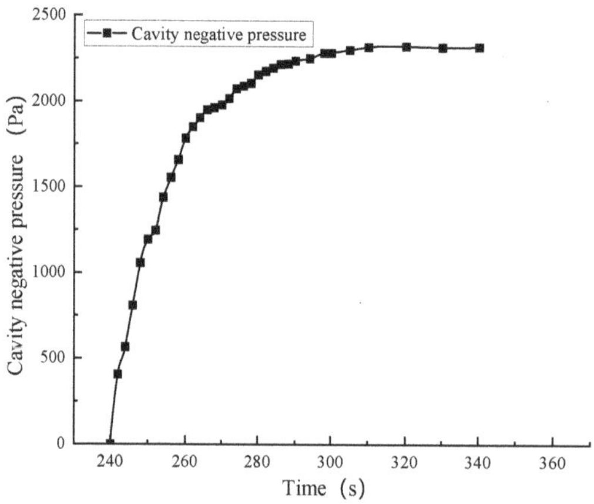

Figure 9. Time-lapse curve of the drainage and cavity pressure.

It can be seen from Figure 9 that before 240 s, there is no vacuum negative pressure in the dissolution chamber in which the pressure is equal to the external atmospheric pressure. At 240 s, the dissolution chamber drainage switch is turned on to start drainage. The continuous process makes the space in the dissolution chamber increase continuously, and the vacuum negative pressure also keeps increasing. When the test goes on for 310 s, the vacuum negative pressure reaches the maximum of 2321 Pa. After drainage, the vacuum negative pressure increases rapidly during 240–270 s. After 270 s, the vacuum negative pressure increases slowly with the increase of time. When the time is greater than 310 s, the negative pressure in the dissolution chamber tends to be stable. Due to the pressure difference between the top and bottom of the cap layer, the vacuum suction force acting on the cap layer gradually increases with the vacuum negative pressure. The external air enters the karst cavity through the connected pores. When the volume of the discharged water is equal to that of the gas entering from the outside, the negative pressure in the karst cavity remains constant.

Figure 10 shows the deformation-time curve of the caprock. Before drainage, the cover layer soil is relatively stable and hardly deforms, and the deformation amount is approximately 0; after the drainage starts, the cover layer soil begins to deform slightly. With the increase in drainage time, the cumulative deformation of the covered soil also increases. As the test continues, the cover soil no longer deforms when other conditions remain unchanged, and finally tends to be stable. When the thickness of the cover layer is 6 cm, the deformation rate in the curve deformation stage is the largest. As the thickness of the cover layer increases, the deformation rate of each curve becomes smaller and smaller, indicating that the deformation and displacement of the cover layer are inversely proportional to its thickness. That is, the thicker the cover layer, the more difficult it is to deform and break. It can be seen from Figure 11 that under the same negative pressure of the cavity, the larger the thickness of the overburden, the smaller the deformation, and the more stable the overburden soil is. the shear strength increases accordingly, and the ability of the cover layer to resist deformation and damage also increases. The deformations of the cover may have a nonlinear relationship with its thickness. With the increase of cover thickness, the change rate of soil deformation gradually slows down.

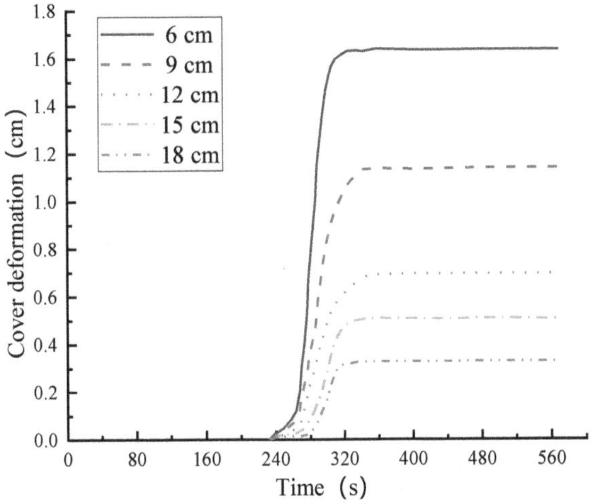

Figure 10. Declining water table cover settlement.

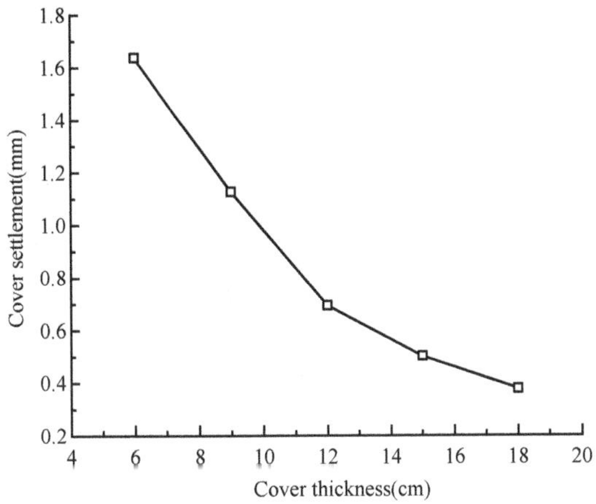

Figure 11. Settlement curves for each cover layer as the water level drops.

4.3. *Analysis of the Collapse Development Process*

The surface water, pore water, and groundwater in the lower part of the soil box are closely linked to form a complete hydrological cycle in the test model installation. The infiltration of surface water can change the water content of the cover soil. When rainfall or other recharge infiltrates, the water content of the cover soil increases, the cohesion and angle of internal friction decrease, the capacity and self-gravity increase, and the shear strength decreases, which ultimately leads to a reduction in collapse resistance. The effect of groundwater on karst ground collapse can be analyzed in many ways, but the magnitude and frequency of groundwater fluctuations are the most important. Based on the development of the soil cavity under the two test scenarios the whole process can be roughly divided into a softening phase of the soil above the cavity, a soil loss-spalling phase and a rapid-integral collapse phase (Figure 12).

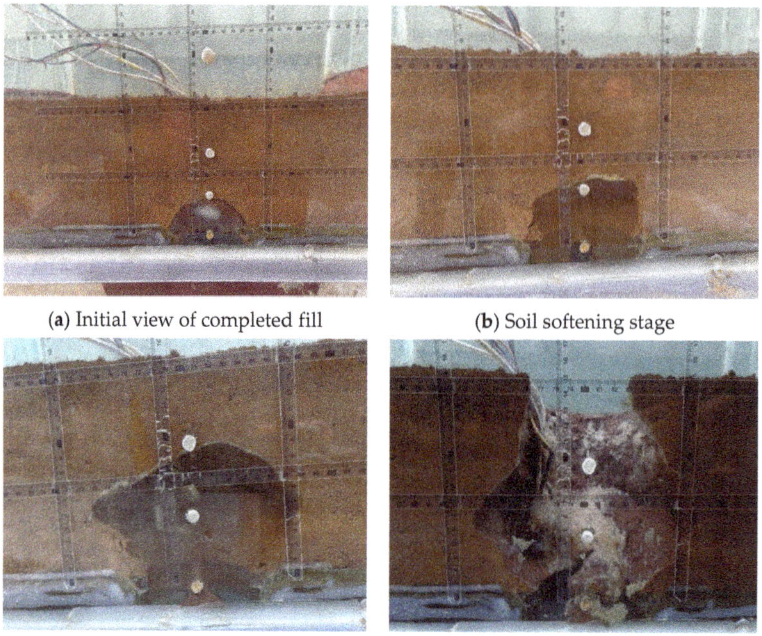

(a) Initial view of completed fill (b) Soil softening stage

(c) Soil erosion–exfoliation phase (d) Rapid overall collapse phase

Figure 12. Schematic diagram of the development of the earth cave collapse.

(1) Soil softening stage: At the beginning of the test, rainfall and lateral seepage recharge the groundwater and a gradient difference is formed between pore water and groundwater, resulting in vertical seepage. As the support for the cover soil above the soil hole is weak, it is susceptible to submerged erosion making the soil particles soften.
(2) Soil loss exfoliation stage: With the increase of rainfall and lateral seepage recharge time, the saturation of the soil increases and the permeability becomes poor. When the water level rises, the gas inside the cavern cannot be discharged and is squeezed to form a high-pressure air mass, which produces tiny fissures in the cover soil. When the water level falls, the negative pressure inside the cavern increases the additional force of downward seepage of pore water in the cover soil, which intensifies the loss and spalling of the soil.
(3) Rapid integral collapse phase: long periods of continuous recharge and fluctuations in the water table resulted in a large loss and spalling of the cover soil above the cavity in the later stages of the test, and the arch structure no longer existed. The cavity was observed earlier in the form of an altar shown in Figure 13, at which point the self-weight of the soil increased with recharge, while the thickness of the soil was still being lost and spalled with fluctuations in the water table. When the collapse develops to the point where the collapse causing force is greater than the collapse resisting force, the soil undergoes shear deformation falling off as a whole, eventually leading to the overall collapse of the surface, as shown in Figure 14.

Figure 13. Cavity altarpiece.

Figure 14. Surface collapse map.

5. Conclusions

In this paper, based on the results of previous studies on overlying karst in Guilin, indoor model tests on karst cover deformation under groundwater level fluctuations were carried out, which mainly led to the following conclusions.

(1) Under the action of rainfall and other recharge, cavity supply, and drainage, the infiltration curve of the soil in the cover layer changes faster at the beginning and slows down gradually at the later stage as the time and intensity of rainfall and other recharge increase. The thicker the cover layer, the slower the overall deformation; when the rate of cavity supply and drainage is certain, the thicker the cover layer, the smaller the deformation produced by the fluctuation of the groundwater level; after the formation of the cavity, rainfall, and other recharge, cavity supply and drainage accelerate the deformation of the soil and the upward development of the cavity.

(2) Under the condition of uniform initial water content and density, the cumulative amount and time of rainfall and other recharge required for the deformation to collapse process increases as the thickness of the cover soil increases, and the number of cycles of cavity drainage required also increases. This indicates that the fluctuation frequency of groundwater level can aggravate the expansion of karst collapse.

(3) The karst cover is affected by a combination of self-gravity, subduction by surface water seepage, suction by water-air change and airburst during the deformation to collapse process. For the same thickness of cover soil, the relationship between the deformation due to different effects is: water level rise is greater than water level fall which is greater than rainfall and lateral recharge; for example, with a cover of 6 cm, water level rise (3.04 mm) > water level fall (1.63 mm) > rainfall and lateral recharge (1.44 mm).

Author Contributions: Conceptualization, X.C., Y.S. and X.G. (Xiaotong Gao); methodology, X.G. (Xiaotong Gao); validation, H.L., X.G. (Xiaohui Gan) and M.X.; investigation, X.G. (Xiaotong Gao); writing—original draft preparation, X.G. (Xiaotong Gao); writing—review and editing, Y.S.; supervision, H.L. and X.G. (Xiaohui Gan); project administration, X.C. All authors have read and agreed to the published version of the manuscript.

Funding: (1) National key research and development program subject "R&D and experimental demonstration of key technologies for water resource regulation in Karst wetlands in the Lijiang River Basin" (2019YFC0507502). (2) The National Natural Science Foundation of China Project "Research on the Collapse Mechanism of Karst Water-soil Coupling in Guilin under Extreme Climate Conditions" (41967037). (3) Guangxi Science and Technology Program Project "Wetland Water Resources Utilization and Water Ecology" (RZ2100000161).

Institutional Review Board Statement: Not applicable.

Informed Consent Statement: Not applicable.

Data Availability Statement: Not applicable.

Conflicts of Interest: The authors declare no conflict of interest.

References

1. Chen, G. *Karst Collapses*; China Railway Publishing House Co., Ltd.: Beijing, China, 1994.
2. He, K.; Guo, D.; Du, W.; Wang, R. The effects of karst collapse on the environments in north China. *Environ. Geol.* **2006**, *52*, 449–455.
3. Yao, C.; Yang, C.; Yang, Q.; Li, Z. Risk Control of Karst Collapse. *J. Phys. Conf. Ser.* **2020**, *1549*. [CrossRef]
4. Chu, X.; Dang, S.; Ding, J. Analysis of karst collapse distribution in Guizhou Province and influential factors. *Yangtze River* **2015**, *46*, 42–44+71.
5. Fidelibus, M.D.; Balacco, G.; Gioia, A.; Iacobellis, V.; Spilotro, G. Mass transport triggered by heavy rainfall: The role of endorheic basins and epikarst in a regional karst aquifer. *Hydrol. Process.* **2017**, *31*, 394–408. [CrossRef]
6. Jiang, X.; Lei, M.; Zhao, H. Review of the advanced monitoring technology of groundwater–air pressure (enclosed potentiometric) for karst collapse studies. *Environ. Earth Sci.* **2019**, *78*, 701. [CrossRef]
7. Hu, J.; Mahdi, M.; Wang, J.; Qin, F.; Zhang, J.; Wu, W.; Han, Y. Karst Collapse Risk Zonation and Evaluation in Wuhan, China Based on Analytic Hierarchy Process, Logistic Regression, and InSAR Angular Distortion Approaches. *Remote Sens.* **2021**, *13*, 5063. [CrossRef]
8. Taheri, K.; Missimer, T.M.; Mohseni, H.; Fidelibus, M.D.; Fathollahy, M.; Taheri, M. Enhancing spatial prediction of sinkhole susceptibility by mixed waters geochemistry evaluation: Application of ROC and GIS. *Environ. Earth Sci.* **2021**, *80*, 470. [CrossRef]
9. Luo, X. On the history of tectonic evolution and karstification in Wuhan. *Carsologica Sin.* **2013**, *32*, 195–202.
10. Kim, Y.J.; Nam, B.H.; Shamet, R.; Soliman, M.; Youn, H. Development of Sinkhole Susceptibility Map of East Central Florida. *Nat. Hazards Rev.* **2020**, *21*, 04020035. [CrossRef]
11. Zhou, W.; Lei, M. Summary editorial for karst hydrogeology: Advances in karst collapse studies. *Environ. Earth Sci.* **2018**, *77*, 803. [CrossRef]
12. Wang, Y.; Chen, W.; Jiang, X.; Guan, Z. Development features and formation mechanisms of karst collapses in the Tailai basin, Shandong Province. *Carsologica Sin.* **2015**, *34*, 495–506.
13. He, K.; Jia, Y.; Wang, B.; Wang, R.; Luo, H. Comprehensive fuzzy evaluation model and evaluation of the karst collapse susceptibility in Zaozhuang Region, China. *Nat. Hazards* **2013**, *68*, 613–629. [CrossRef]
14. Gao, P.; Wang, L. Analysis of collapse mechanism for mantled karst collapse. *Carsologica Sin.* **2017**, *36*, 770–776.
15. Ding, Q. *Weakly Permeable Overburden of Water-Gas Pressure to Collapse Mechanism of the Karst Collapse Physical Model Research*; Chengdu University of Technology: Chengdu, China, 2016.
16. Karimi, H.; Taheri, K. Hazards and mechanism of sinkholes on Kabudar Ahang and Famenin plains of Hamadan, Iran. *Nat. Hazards* **2010**, *55*, 481–499. [CrossRef]
17. Wu, Q.; Zhang, W.; Liu, Y.; Cui, H. Quantifying the process of karst collapse by a physical model. *J. Yangtze River Sci. Res. Inst.* **2018**, *3*, 52–58.
18. Zhang, H.; Xu, G.; Liu, M. Formation environments and mechanisms of multistage paleokarst of Ordovician carbonates in Southern North China Basin. *Sci. Rep.* **2021**, *11*, 819. [CrossRef] [PubMed]
19. Pando, L.; Pulgar, J.A.; Gutiérrez-Claverol, M. A case of man-induced ground subsidence and building settlement related to karstified gypsum (Oviedo, NW Spain). *Environ. Earth Sci.* **2013**, *68*, 507–519. [CrossRef]
20. Xiao, G. *The Forming Mechanism and Forecast of Fatalness about Karst Subsidence in Guilin City*; Jilin University: Changchun, China, 2005.
21. Wei, A.; Li, D.; Zhou, Y.; Deng, Q.; Yan, L. A novel combination approach for karst collapse susceptibility assessment using the analytic hierarchy process, catastrophe, and entropy model. *Nat. Hazards* **2020**, *105*, 405–430. [CrossRef]

22. Shi, S.; Zhang, Q.; Kang, Z. Formation mechanism and development process of karst collapses in Tanluo town of Nanning City. *Carsologica Sin.* **2015**, *34*, 507–514.
23. Vigna, B.; Fiorucci, A.; Banzato, C.; Forti, P.; De Waele, J. Hypogene gypsum karst and sinkhole formation at Moncalvo (Asti, Italy). *Z. Geomorphol. Suppl.* **2010**, *54*, 285. [CrossRef]
24. Shi, H.; Li, Q.; Zhang, Q.; Yu, Y.; Xing, Y.; Yu, K. Mechanism of Shallow Soil Cave-type Karst Collapse Induced by Water Inrush in Underground Engineering Construction. *J. Technol. Sci.* **2020**, *34*, 04019091. [CrossRef]
25. Zhang, H.; Xu, G.; Zhan, H.; Zheng, J.; Wang, M.; Liu, M.; Pan, S.; Wang, N. Formation mechanisms of paleokarst and karst collapse columns of the Middle Cambrian-Lower Ordovician carbonates in Huainan coalfield, Northern China. *J. Hydrol.* **2021**, *601*, 126634. [CrossRef]
26. Liu, H.; Li, L.; Li, Z.; Yu, G. Numerical Modelling of Mining-induced Inrushes from Subjacent Water Conducting Karst Collapse Columns in Northern China. *Mine Water Environ.* **2018**, *37*, 652–662. [CrossRef]
27. Voglar, G.E.; Leštan, D. Solidification/stabilization of metals contaminated industrial soil from former Zn smelter in Celje, Slovenia, using cement as a hydraulic binder. *J. Hazard. Mater.* **2010**, *178*, 926–933. [CrossRef] [PubMed]
28. Wei, Y.; Sun, S. Comprehensive critical mechanical model of covered karst collapse under the effects of positive and negative pressure. *Bull. Eng. Geol. Environ.* **2018**, *77*, 177–190.
29. Lu, Y.; Liu, Q.; Zhang, F. Environmental characteristics of karst in China and their effect on engineering. *Carbonates Evaporites* **2013**, *28*, 251–258. [CrossRef]
30. Jin, H.; Wang, L.; Yang, P.; Song, N. Stability analysis of karst collapse considering joint effect of rainfall infiltration and groundwater leakage. *Yangtze River* **2019**, *9*, 105–112.
31. Sun, J. *Study on Critical Parameter in Cover Karst Collapse by Model Experiment and Numerical Simulation*; Southwest Jiaotong University: Chengdu, China, 2011.
32. Li, C.; Dou, P. Physical model test analysis of the collapse mechanism of karst ground. *Eng. J. Wuhan Univ.* **2021**, *S2*, 239–242.

applied sciences

MDPI

Article

Bearing Capacity of Shallow Foundations on Unsaturated Silty Soils

Mehmet Ugur Yilmazoglu [1,*] and Askin Ozocak [2]

[1] Institute of Natural Sciences, Sakarya University, Serdivan 54050, Turkey
[2] Department of Civil Engineering, Faculty of Engineering, Sakarya University, Serdivan 54050, Turkey
* Correspondence: myilmazoglu@kastamonu.edu.tr

Abstract: In general, the ultimate bearing capacity (UBC) of shallow foundations on unsaturated soils is characterized by the conventional shear strength (SS) parameters in which saturated theories are applied. However, in this case, it is clear that the foundations designed using the obtained values from the saturated cases not be economical. In recent years, procedures have been developed to estimate the UBC of foundations on unsaturated soils, that take into account drained and undrained loading conditions. However, these studies generally concentrate on sandy soils. The validity of the results proposed in the literature should be tested for other soils. Therefore, this paper includes a conventional direct shear box (DSB) test to determine the unsaturated SS of statically compacted silty soil, and a series of model tests were performed to determine the foundation's UBC. In the experimental model setup, the UBC values of different types and sizes of model foundations on silty soil layers with a different soil saturation degrees (SSDs)/matric suctions (MSs) and different void ratio values were measured. In addition, the soil-water characteristic curves (SWCCs) and SS parameters of unsaturated silt were obtained. Using the experimental results, a new equation is proposed for the characterization of the UBC of shallow foundations on unsaturated silty soils. Using this equation, the UBC of unsaturated soils can be determined based on the results of unconfined compressive strength tests (UC) measured on unsaturated soil samples and based on the degree of saturation and the fitting parameter. The results indicate that the measured bearing capacity values obtained via the model footing test, shows a good consistency with those obtained by the proposed equation.

Keywords: unsaturated soil; shallow foundation; silt; matric suction; soil–water characteristic curve

1. Introduction

The design of shallow foundations involves two main objectives: determining the ultimate bearing capacity (UBC) and ensuring compliance with allowable settlement limits. Traditional methods for determining UBC and settlement behavior on saturated soils can be divided into two groups: the Effective Stress Approach [1] (ESA) and the Total Stress Approach [2] (TSA). The ESA requires effective soil parameters and uses c' (effective cohesion) and $\varnothing'$ (effective internal friction angle), whereas the TSA is based on the $\varnothing_u = 0$ analysis and is used for undrained conditions. Until recently, UBC values for footings on unsaturated soil layers were determined based on the assumption of saturated soil [1–3]. However, this assumption is not always accurate and can lead to non-economical solutions. To ensure a realistic design in unsaturated the soil environments, the effect of soil saturation degree (SSD)/matric suction (MS) must be taken into account. MS, which is frequently used as a measure of suction, is a critical characteristic that determines the behavior and mechanical properties of unsaturated soil [4–6]. Therefore, in geotechnical applications, the the MS value must be taken into account when calculating soil strength and volume changes. Capillary tension is the ability of soil to absorb and retain water and is directly related to water content [7].

Research on the behavior of unsaturated soils, which make up a large portion of the Earth's land, has been important in recent years. The majority of shallow foundations are found in areas of unsaturated soil, depending on the depth of the groundwater table. The SS of unsaturated soils is important in the design of shallow foundations because the failure in these foundations is usually caused by exceeding the UBC value [8–11]. Therefore, research on the behavior of unsaturated soils has been prioritized over the last ten years. In the literature, experimental studies have been conducted under both drained and undrained conditions [12–29]. In general, research in this area has aimed to improve understanding of the mechanical behavior of unsaturated soils and to develop appropriate methods for assessing the bearing capacity of unsaturated soils. Studies have focused on factors such as matric suction, soil structure, and soil water content and their effects on the bearing capacity of unsaturated soils. Some studies have also investigated the use of various test methods to assess the bearing capacity of unsaturated soils, such as triaxial tests, direct shear tests, and plate load tests. Additionally, numerical models have been used to simulate the mechanical behavior of unsaturated soils. These studies have shown that the UBC values of foundations, calculated using traditional saturated soil mechanics concepts, are quite far from the UBC values of foundations on unsaturated soil conditions. Studies on the estimation of the UBC of foundations on unsaturated soil layers can be grouped under two main groups, which are similar to traditional soil mechanics groupings. These are the proposed approaches to estimate the SS of unsaturated soils under drained and undrained conditions. There are various structural models and modeling techniques that can be used for both approaches, but these models require a variety of soil parameters that necessitate complex and extensive laboratory tests. On the other hand, there are various numerical modeling techniques presented in the literature that are extended by modifying the ESA and the TSA to take into account the effect of MS [11,12,30–37]. The approaches introduced in the literature are called the Modified Effective Stress Approach (MESA) and the Modified Total Stress Approach (MTSA). These methods, which can be estimated with the test results obtained for unsaturated conditions for the MSEA and the MTSA in the literature, are simple and fast. Additionally, these numerical techniques can be verified by comparing the experimentally measured UBC values with the values obtained through the proposed equations.

Previous studies have shown that the influence of MS on soil strength is not linear [38]. MS is considered to have a non-linear, hydrostatic suction profile. The MS value can change due to changes in the water level or weather conditions, even when the groundwater level is constant [39,40]. This means that foundations on soil layers above the groundwater level may experience larger UBC values than calculations made assuming saturated conditions.

The focus of this research is to use model foundations of various forms and sizes to investigate the UBC of shallow foundations on silty soil. Strength tests and model foundation loading tests on samples with varied void ratios, saturation levels, and the MS values were conducted in order to ascertain the effect of the MS on SS and the UBC values for the unsaturated soil zone.

Estimation of UBC of Foundations on Unsaturated Soils

Several methods have been proposed in the literature for estimating the UBC of foundations on unsaturated soils. These methods mainly involve modifying the equations proposed by Terzaghi [1] and Skempton [2] to take into account the effect of matric suction. Research has shown that foundations on unsaturated soils typically fail through the punching shear failure mode [11,12,15,41]. In the case of cohesive soils, it is assumed the a constant pore water pressure under undrained conditions can provide accurate results. Tang et al. [42] also found that in-place plate bearing tests typically occur under constant water content condition, and that assuming a constant value for the effective stress parameter does not lead to errors in the results. Oh and Vanapalli [18] proposed an equation

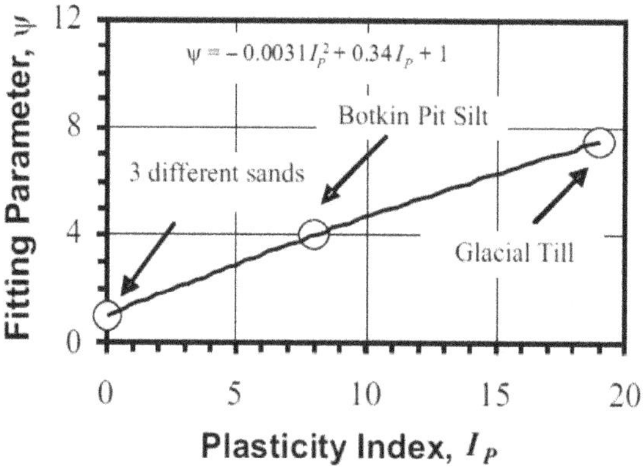

Figure 2. The relationship between Ψ and I_p [14].

In this study, a new equation is proposed in which the UBC of foundations on unsaturated silty soils is determined by adopting the total stress analysis philosophy presented above, with a bearing capacity fitting parameter (S^{Ψ}) based on the SS determined via the UCS tests and the SSD/MS value.

Various models have been proposed in the literature to estimate the UBC of unsaturated soils (see Table 1). Some of these proposed equations directly use the MS value, whereas the other group indirectly uses MS.

Table 1. Bearing capacity models for unsaturated soil.

Authors	UBC Equation	Soil Type
Oloo et al. (1997) [12]	$q_{ult} = \left\{c' + (u_a - u_w)_b tan\varnothing' + [(u_a - u_w) - (u_a - u_w)_b] tan\varnothing^b\right\} N_c + \frac{1}{2} B\gamma N_\gamma$	-
Vanapalli & Mohamed (2007) [14]	$q_{ult} = \{[c' + (u_a - u_w)_b tan\varnothing' - S^{\Psi} tan\varnothing'] + (u_a - u_w)_{AVE} S^{\Psi} tan\varnothing'\} N_c \xi_c + \frac{1}{2} B\gamma N_\gamma \xi_\gamma$	Poorly graded sand
Oh & Vanapalli (2013) [18]	$q_{ult(unsat)} = \left[\frac{q_{u(unsat)}}{2}\right] \times \left[1 + 0.2\frac{B}{L}\right] \times N_{unsat}$	Low plasticity clay
Vahedifard & Robinson (2016) [32]	$S_e = \left(\frac{1}{1+\{-ln(1+\frac{q}{k_s})e^{-\gamma_w \alpha z} - \frac{q}{k_s}\}^n}\right)^{n-1/n}$	-
Tang et al. (2017) [42]	$q_{ult} = [c' + (\chi^s)_{AVE} tan\varnothing'] N_c d_c + q_{ult} q N_q d_q + 0.5\gamma B N_\gamma d_\gamma$	-
Garakani et al. (2020) [37]	$q_{ult} = [c' + c_{AVE}] N_c \xi_c + q N_q \xi_q + 0.5 B\gamma N_\gamma + s(N_s)_{cst} tan\varnothing^b \xi_s$	Poorly graded sand
Zhang et al. (2020) [17]	$q_{ult} = \left[c' + (u_a - u_w)_m tan\varnothing^b\right] N_c^M + \sigma_0 N_q^M \frac{1}{2}(B\gamma - u_a) N_\gamma N_\gamma^M$	Sand

2. Materials and Methods

2.1. Material

Soil samples taken from the city of Adapazarı were used in the experimental studies to examine the change in the UBC of the shallow footings on unsaturated soils. The samples were taken from the Yenigün District of Adapazarı (Turkey), at a depth of 3–4 m. The properties of the silty soil sample used in this study according to ASTM D4318 [49] are given in Table 2. The soil sample was classified as low plasticity silt (ML) according to the USCS. Figure 3 shows the grain size distribution curve of the sample.

Table 2. Physical properties of the soil used in the experimental study.

Property	Value	Symbol and Unit
No 200#	89	FC (%)
Liquid limit	35.8	LL (%)
Plastic limit	25.2	PL (%)
Plasticity index	10.6	IP (%)
Specific gravity	2.692	G_s
Clay ratio	17	C (%)
Silt ratio	72	M (%)
Sand ratio	11	S (%)
Soil class: Low Plasticity Silt (ML)		

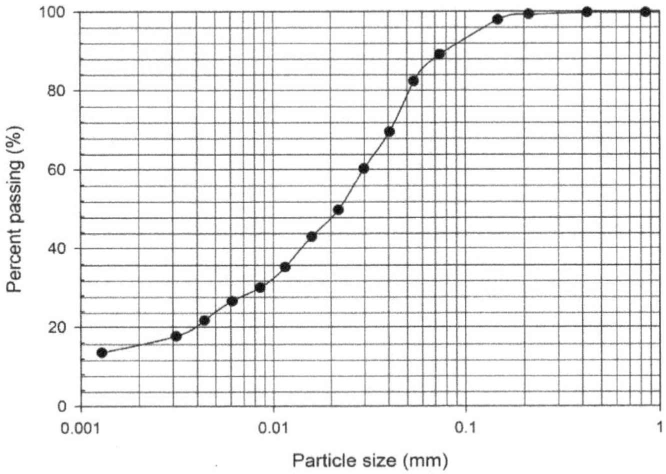

Figure 3. Grain-size distribution of samples used in the study.

The mineralogical composition of the silt used in this study was determined with an X-ray diffraction analysis and is shown in Table 3. It indicates that elements such as calcium oxide and magnesium oxide, which are the primary cementing agents in the soil, are present in significant amounts in the soil sample [50,51]. However, studies in the literature have demonstrated that the cementing bonds that are disrupted in remolded samples do not re-form [50,52]. As a result, when compared to experiments carried out on in situ or undisturbed materials, the cohesiveness of the remolded samples is lower. Furthermore, research suggests that the shear resistance angle is less influenced than the cohesion value [53].

Table 3. Mineralogical properties of the soil sample.

Symbol	Element	Value (%)
SiO_2	Silicon dioxide	40.94
Al_2O_3	Alumina	11.09
Fe_2O_3	Iron oxide	5.36
MgO	Magnesium oxide	4.48
CaO	Quicklime	15.58
K_2O	Potassium oxide	1.72
Na_2O	Sodium oxide	2.09
TiO_2	Titanium oxide	0.82
-	Other	17.92

2.2. Method

In this study, the ultimate bearing capacity (UBC) of shallow foundations on unsaturated soils was investigated. The basic physical properties of silt samples taken from Adapazarı were determined and void ratios were selected for use in experiments. The natural void ratios of silts in Adapazarı were examined [54], and samples with void ratios of 0.65, 0.70, and 0.75 were used in the experiments. The sample SSDs were set at 65%, 70%, 75%, and 85% to account for seasonal groundwater level changes in Adapazarı. The shear strength angle and cohesion values of the silt samples were obtained by using direct shear tests.

The SS parameters of unsaturated soils can be measured using a modified MS-controlled triaxial apparatus or a direct shear box (DSB) test [53,55–57]. However, these methods can be costly and time-consuming, and the results are not as practical as those from the saturated state. A more widely accepted method is to combine the application of the targeted suction magnitude with a DSB test. Studies have shown that this method is consistent with the results obtained using a modified DSB test capable of MS control [53,58–60]. In this study, the shear strength of unsaturated Adapazarı silt was determined using a conventional direct shear test. The samples were prepared at the targeted SSD, and MS values were measured using a pressure plate and filter paper. The shear box test was carried out on a fully automatic direct shearing apparatus with a square shear box 6 × 6 cm in diameter and 2.5 cm in height. During the test, the samples were carefully isolated from air to maintain their saturation degree/suction values. The $\varnothing^b$ values of silt soil were calculated from the test results and MS values, and the UBC was determined using these values.

The relationship between the WC/SSD and MS is important in evaluating the behavior of unsaturated soils. In this study, the MS values of the samples were determined using the filter paper method and pressure plate method to create the SWCCs. The filter paper method, which was proposed in the 1930s, is an economical but slow method for measuring suction, and it can measure total and the MS values up to 30 MPa [61]. This method uses a filter paper whose calibration curve is known and places it in full contact with the sample to determine pore water flow. The measurement period is 6–8 days. After the moisture balance is achieved, the wet and dry weight of the filter paper is weighed on a balance with a sensitivity of 0.0001 g, and the corresponding MS value is found in the calibration curve according to the filter paper water content. The SWCC of the sample is formed by repeating these processes at different WCs from the saturated state to the air-dried state [28,62]. In this study, the MS values were determined with the filter paper method by using the calibration curve recommended for the Whatman No. 42 filter paper in ASTM 5298-92 [63].

The pressure plate method is used to apply suction using the axis translation technique [64]. Air pressure (u_a) is applied to the samples placed on a ceramic with the high AEV in the steel cell coming from the inlet at the top of the cell, and pore water pressure (u_w) is applied trough the inlet connected to the ceramic. This process must be continued until the water flow is balanced [28,65]. When the water at the output reaches equilibrium after the applied suction, the sample weight/volume is taken and the next suction stage is started. The MS (u_a–u_w) values were applied in incremental steps from 5 kPa to 1500 kPa. A 1500F1 pressure plate extractor with a 15-bar capacity was used to conduct the pressure plate experiments.

The UBC of foundations on unsaturated silt soils was examined by model tests. Previous research and experimental studies have shown that the applied loads will have no effect on the boundaries of the sample container if the soil model tank size is 6 or more times higher than the model foundation size [14,66,67]. For this reason, of the soil tank used in the current study was chosen to be greater than 6 times higher than the model foundation dimensions. The model test used a cylindrical model test box with a height of 240 mm and a diameter of 310 mm filled with the soil sample at the desired void ratio and soil saturation degree values. The foundation bearing capacity was investigated with the help of 50 mm diameter circle, 50 × 50 mm square, 42.3 mm circle, and 37.5 × 37.5 square foundation models placed on the model test box. The 42.3 mm circle and 37.5 × 37.5 mm square foundations were selected for further experiments as they have an equal base area.

Sample Preparation

The silty soil taken from the Adapazarı city center was first air-dried in the laboratory environment. The air-dried sample was pulverized by hand. In order to prepare the samples homogeneously at the desired water content, it was sieved through the No.10 (2 mm) sieve with for the maximum grain diameter, both to prevent lumping and to remove organic substances. The sieved soil sample was divided into groups of 2 kg, which were wetted by spraying deionized and deaired water for the desired degree of saturation was reached and mixed with the help of a sample preparation mixer. This process was repeated for each sample of different VR and SSD values. The prepared soil samples were placed in airtight containers and left at room temperature for 24 h for curing. With this procedure, a homogeneous soil-water mixture was obtained with no measurable difference between the desired WCs. After the standard 24 h curing period was completed, three different samples were taken from the different parts of each airtight container to check the WC and it was confirmed that the WCs of the mixtures in all the bags were equal within a ±0.5% evaporation/moisture change margin of error.

The prepared soil samples were compacted at a statically constant velocity in a model tank specifically designed for hydraulic press equipment. A metal plate with a diameter of 305 mm and a thickness of 30 mm was used as the clamping apparatus. While preparing the samples for the model experiments, they were statically compacted into two layers to obtain a homogeneous void ratio. By creating scratches between the layers, adherence losses that could occur between these layers were prevented. For each VR and SSD values, a prepared and cured soil sample equal to the calculated sample weights by using the volume-weight ratio of the soil was statically compacted in the model tank. After the UBC tests were carried out, two specimens were taken out from the compacted sample using 50 mm diameter stainless steel thin-wall tubes for the UC test. With the help of a small cylindrical sample mold that had a certain volume which was obtained from the model test mold, the weight and the WC measurements of the samples taken from different depths, as well as the VR and SSD were checked. In addition, the MS values were measured using the filter paper method on samples with a height of 20 mm and a diameter of 50 mm, which were taken from depths of 0.5B and 1.5B (Figure 4).

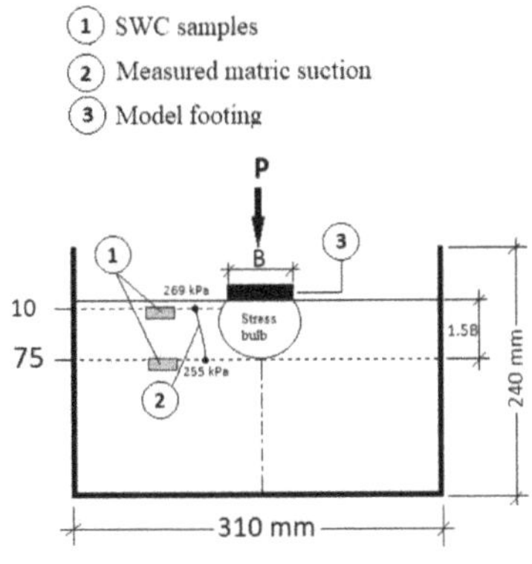

Figure 4. Schema showing the procedure used to determine the MC under the foundation.

The samples for the DSB test were statically compacted, the same as those used in the model footing testing. The samples, prepared with the WC values corresponding to the desired MS values from the SWCC curves, were insulated from the air with silicone grease so that there was no loss in the WC during rapid shearing and no change in the MS value [53,58,68]. The shear rate applied in this study was determined to be 1.15 mm/min after considering other studies in the literature [12,14,53,60]. Experiments on the model load tests were run as displacement-controlled tests. According to earlier research on compacted samples, soil specimens can reach drained conditions when loaded at a strain rate between 0.0102 mm/min and 0.0132 mm/min [46,48]. In the current investigation, undrained loading conditions were simulated using a reasonably quick loading rate of 1.15 mm/min. The DSB tests were conducted in accordance with the ASTM D3080-98 [69] at normal stress levels of 59, 98, and 147 kPa. Due to the fast loading rate applied in the direct shear tests, the undrained shearing conditions were assumed to prevail. In order to confirm the assumption that there is no change in the gravimetric water content (GWC) value of the samples during the DSB test process, the sheared samples were subjected to MS and WC measurements directly after the test.

3. Results and Discussion

After the soil samples were created for the pressure plate tests and filter paper tests, they were statically compressed, as was carried out for the other tests, and the samples were taken with the help of metal rings with a diameter of 50 mm and a height of 20 mm. The samples were fixed by placing porous stones on the samples in order to reach the SSD and were kept in distilled water for 7 days. The samples taken from the model loading tests, on the other hand, were placed immediately on the filter paper to start the equilibrium stage and the MS values of the prepared samples were measured.

The SWCCs of the samples formed at three different void ratios were determined by the pressure plate and filter paper method. In the experimental study, model tests were conducted with soil samples compacted with targeted VRs (0.65, 0.7 and 0.75) and SSDs (i.e., 65%, 70%, 75%, 85%, 100%) considering seasonal groundwater level changes. To reflect this scenario, the samples were statically compressed with targeted VRs at four different SSDs. Since experiments were carried out on samples with three different VRs, the samples were prepared at these VRs for the SWCC experiments. Therefore, SWCC experiments were carried out on three samples (0.65, 0.7 and 0.75 void ratios) and SWCCs have been created according to the GWC. The most ideal curves of the experimental data were created with the Van Genuchten (1980) [70] model, which is one of the SWCC generation models available in the literature, for MS values ranges between 0 and 1,000,000 kPa (Figure 5). Since the SWCC data at high MS range could not be determined with the pressure plate (the suction was determined with the pressure plate up to 800 kPa), it was measured by the filter paper method, and it was again found to be compatible with the curves obtained from the Van Genuchten (1980) [70] model. The AEVs of the samples with 0.65, 0.70 and 0.75 void ratios have been determined as 12.4, 10.2 and 8.0 kPa, respectively.

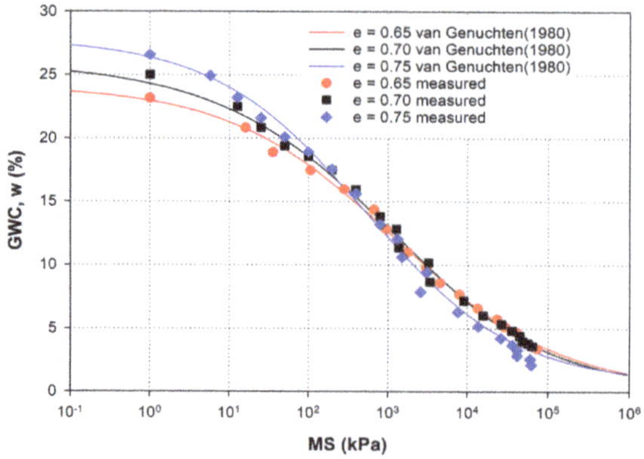

Figure 5. SWCCs of silty soil prepared at different VRs.

Samples with VRs of 0.65, 0.70 and 0.75 and SSDs of 65%, 70%, 75%, 85% and 100% were created using static compression. The samples were saturated under water for 9 days. The SS were recorded as a function of horizontal displacement up to the maximum value. As seen in Figure 6, a clear peak is observed in the stress-strain graphs at each net stress (59, 98 and 147 kPa) at MS values higher than the AEV in the samples with 0.65 VR. No significant peak has been observed for the saturated state. A similar situation is observed for the samples with VRs of 0.70 and 0.75.

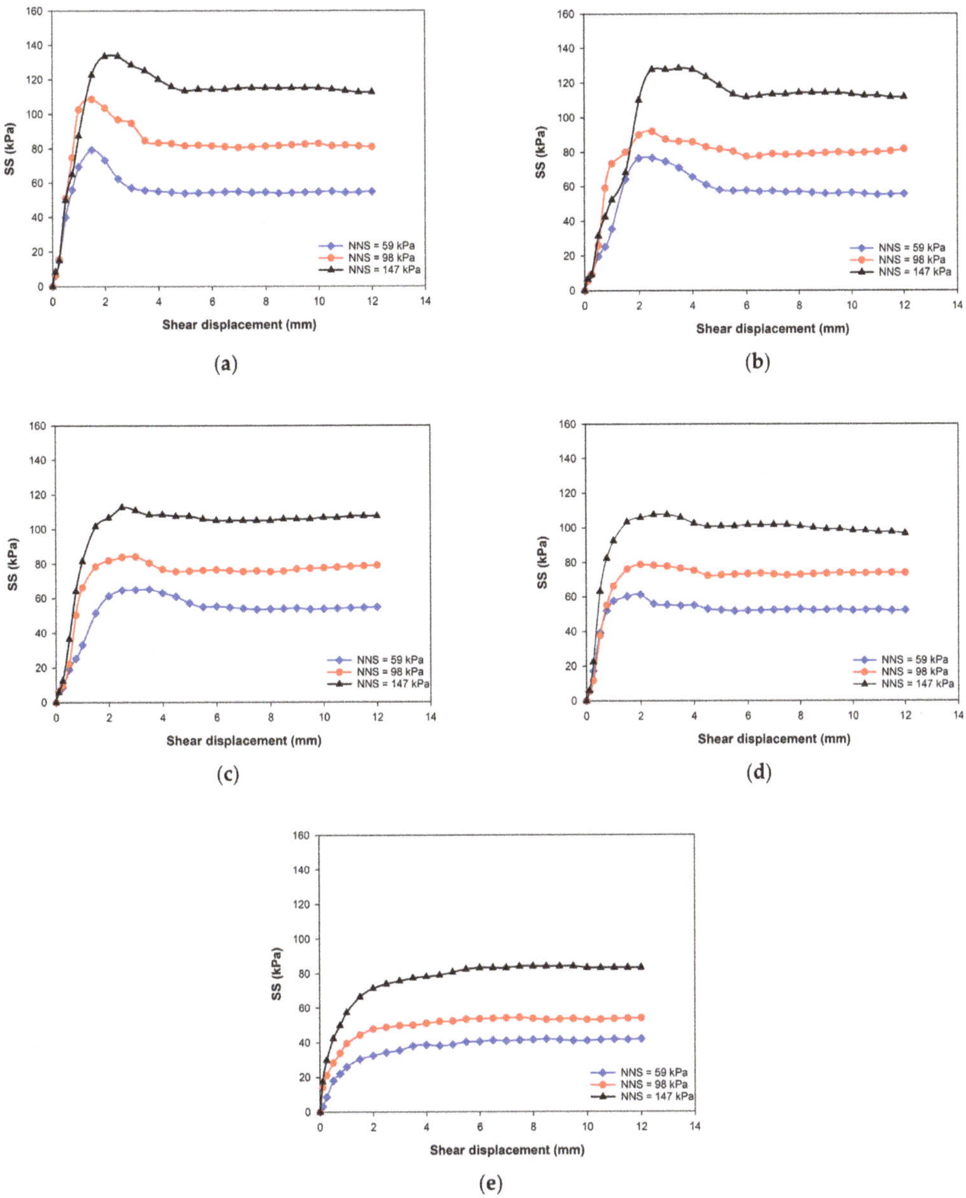

Figure 6. Shear strain curves for silt sample with VR = 0.65 (from shear box tests): (**a**) SSD = 65%; (**b**) SSD = 70%; (**c**) SSD = 75%; (**d**) SSD = 85%; (**e**) SSD = 100%.

In each VR, the MS values taken from the SWCCs and the results of the shear box test performed at 65%, 70%, 75%, 85% and 100% saturation degrees were combined, and the MS—SS net normal stress (NNS) plots were drawn (Figure 7). The $\varnothing^b$ values calculated on these graphs are presented in Table 4. When Table 4 is examined, it is seen that lower $\varnothing^b$ values are calculated for the three different VRs at low NNS. The increase in the VR also increases the $\varnothing^b$ value.

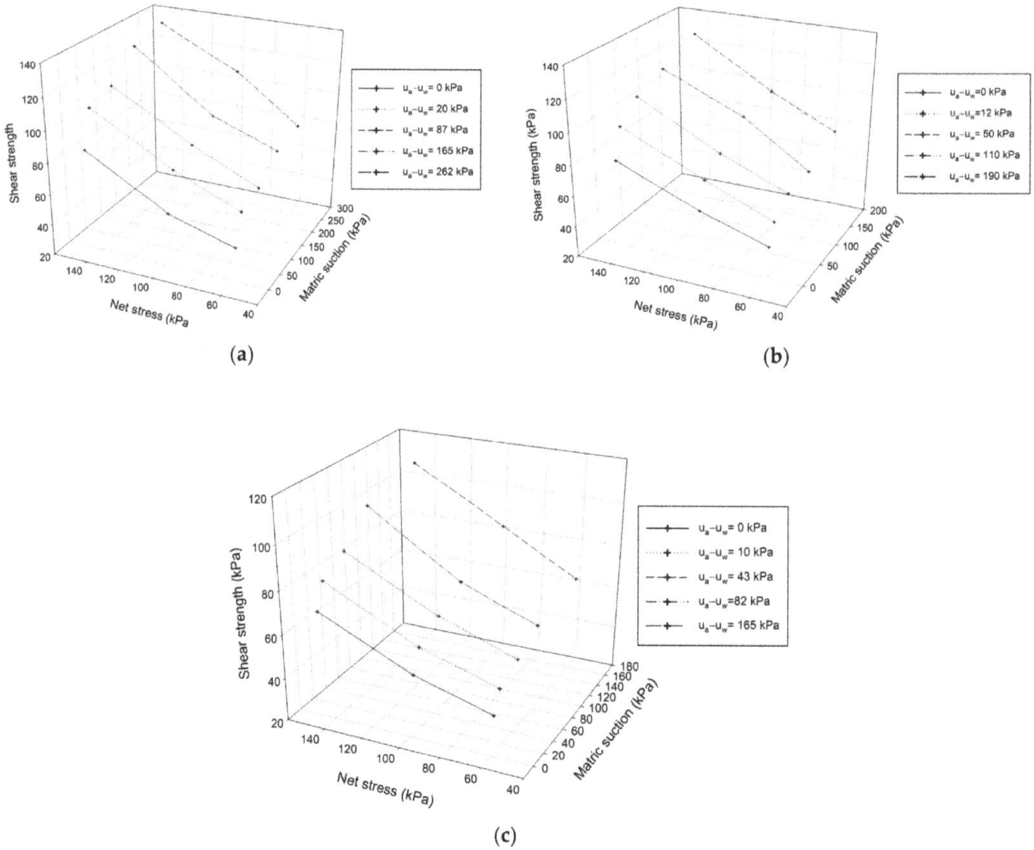

Figure 7. MS-NNS-SS variation for void ratios: (**a**) VR = 0.65; (**b**) VR = 0.70; (**c**) VR = 0.75.

Table 4. Angles showing the slope of the SS increase with respect to the MS.

e	$(\sigma_n - u_a)$ kPa	$\varnothing^b$	$\varnothing^b_{average}$
0.65	59	4.6	6.5
	98	7.6	
	147	7.2	
0.70	59	5.7	7.3
	98	7.1	
	147	9.0	
0.75	59	5.6	8.1
	98	8.8	
	147	10.0	

Figure 8 shows the failure envelopes formed by considering the maximum SS values of the samples. Table 5 lists the SS parameters of the soils obtained from these failure envelopes. With the increase of the SSD, a decreasing trend is observed in the cohesion values and the SS angles of the samples with three different VRs. In the literature, it is a known behavior that a decrease in the SSD/increase in the MS value will cause an increase in the cohesion value [55]. In addition, the observation that the increase in the MS value causes an increase in the angle of SS is compatible with the data of Wen and Yan (2014) [53].

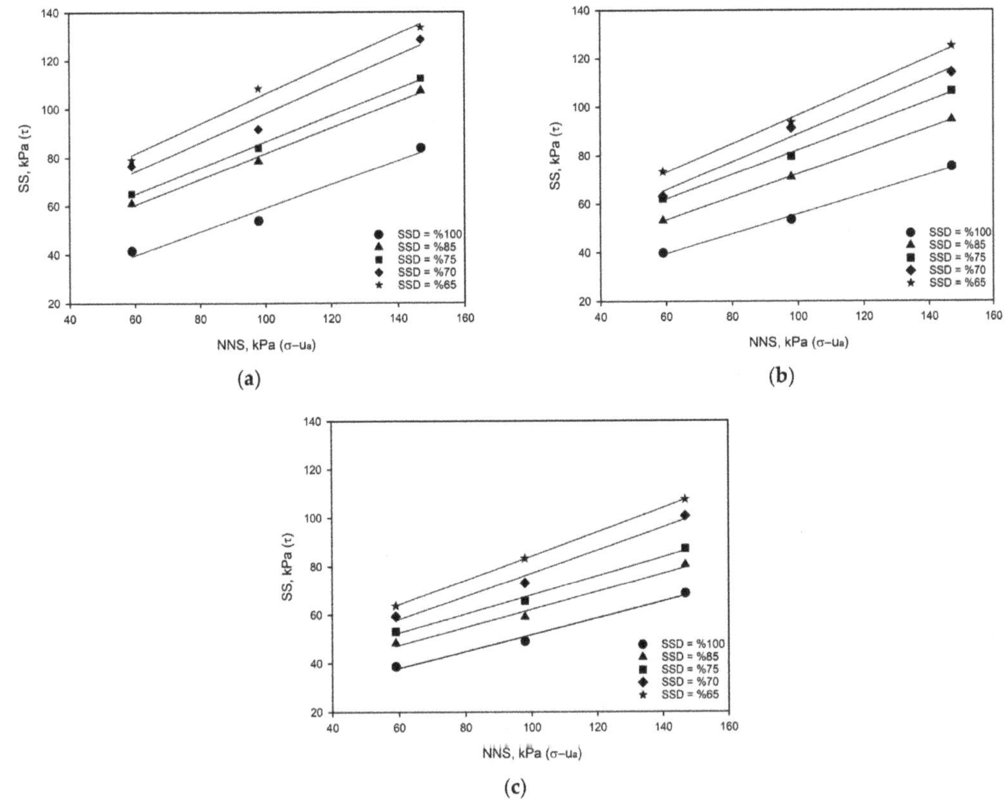

Figure 8. SS-SSD relations under different VR and NNS: (**a**) VR = 0.65; (**b**) VR = 0.70; (**c**) VR = 0.75.

Table 5. SS parameters of samples with different SSD.

e	SS Parameters	SSD, %				
		100	85	75	70	65
0.65	c	5.5	29.1	32.9	38.8	45.1
	∅	27.4	27.8	28.3	30.8	31.5
0.70	c	7.1	31.7	39.1	38.9	45.9
	∅	24.2	22.6	23.8	26.9	27.6
0.75	c	9.6	30.9	35.1	36.8	41.5
	∅	21.1	17.8	18.8	22.4	23.5

Figure 9 shows how the SS of silt soils tested at three different VRs varies with both the NNS and the SSD. The results show that the effect of the SSD on the SS decreases as the NNS value increases. In the tests carried out under NNSs of 59 and 98 kPa, there was a peak/trough fluctuation between the SSDs of 70% and 75% SSDs. This shows that in addition to the MS of the silty soil sample, a different void structure has the strongest or weakest effect on the SS at a certain SSD [53,60].

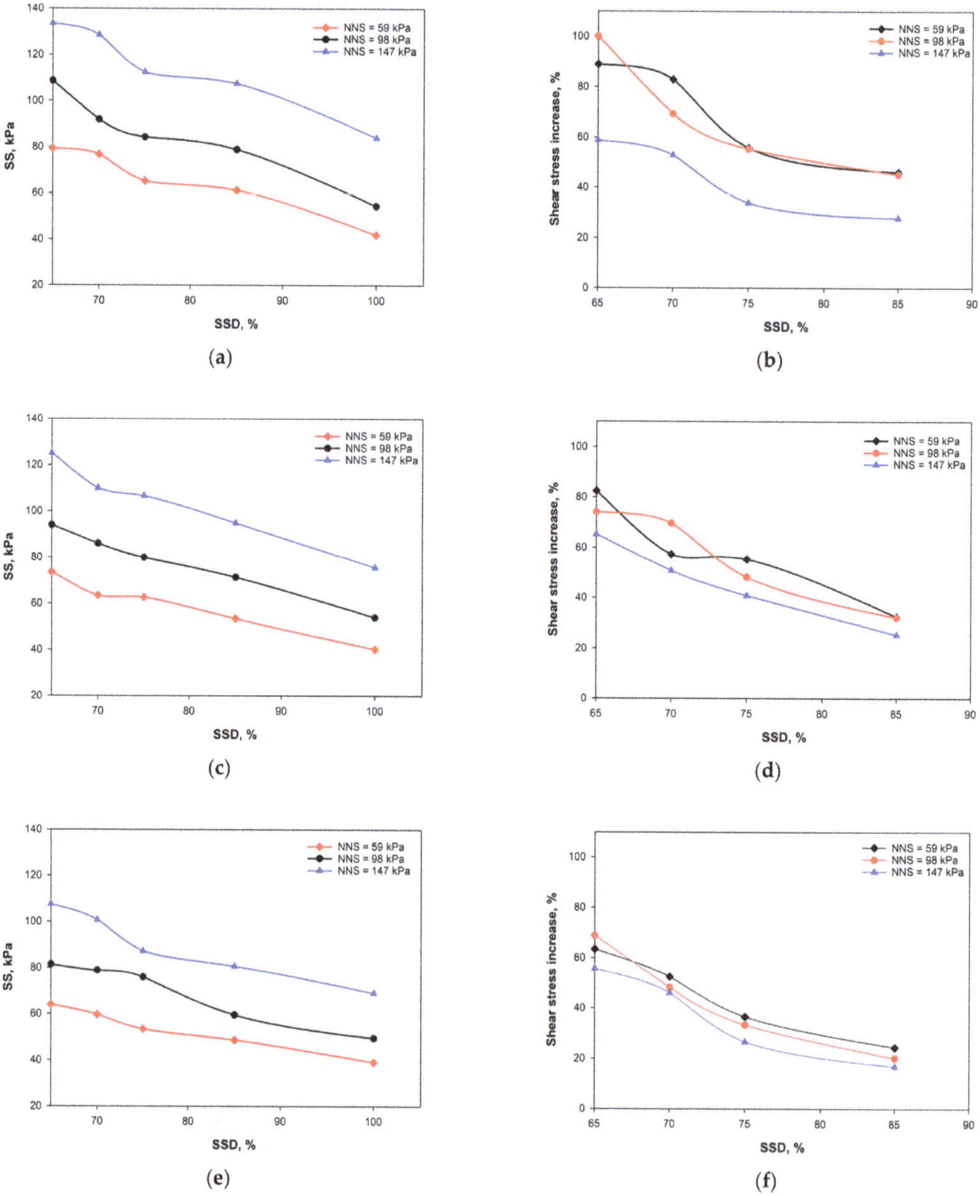

Figure 9. Saturation-shear stress relationships: (**a**) VR = 0.65; (**c**) VR = 0.70; (**e**) VR = 0.75, Saturation-% shear stress increase relationships: (**b**) VR = 0.65; (**d**) VR = 0.70; (**f**) VR = 0.75.

In the current work, model foundations experiments were conducted in a cylindrical tank filled with statically compacted soils to examine the UBC of foundations on unsaturated soils under undrained loading circumstances. The model foundations were loaded relatively quickly (1.15 mm/min) to avoid any changes to the suction value of the sample during the loading process. Square and circular shallow foundation models were loaded up to the peak failure conditions to determine the UBC. The dimensions of the glass fiber reinforced polyester tank used are 310 mm in diameter, height of 250 mm and a thickness of 10.2 mm. Model foundations with the largest foundation width of 50 mm were employed in the study. The dimensions of the model mold were selected large enough (>6D) that stresses on the boundaries of the mold due to the loads applied to the model foundation would have a negligible effect. It was confirmed by the control measurements carried out during the tests that no deformation occurred in the mold (The deformation was checked at three different measurement points on the outer surface of the mold.) during loading. Model foundation loadings were performed by using a triaxial loading frame. The vertical displacements of the model foundations were measured with the deformation gauges, and the loads were measured using a load cell (Figure 10). In both measurements, the data were recorded by taking their values at the predetermined deformations. Tests were terminated until the settlements reached 20 percent of the foundation width.

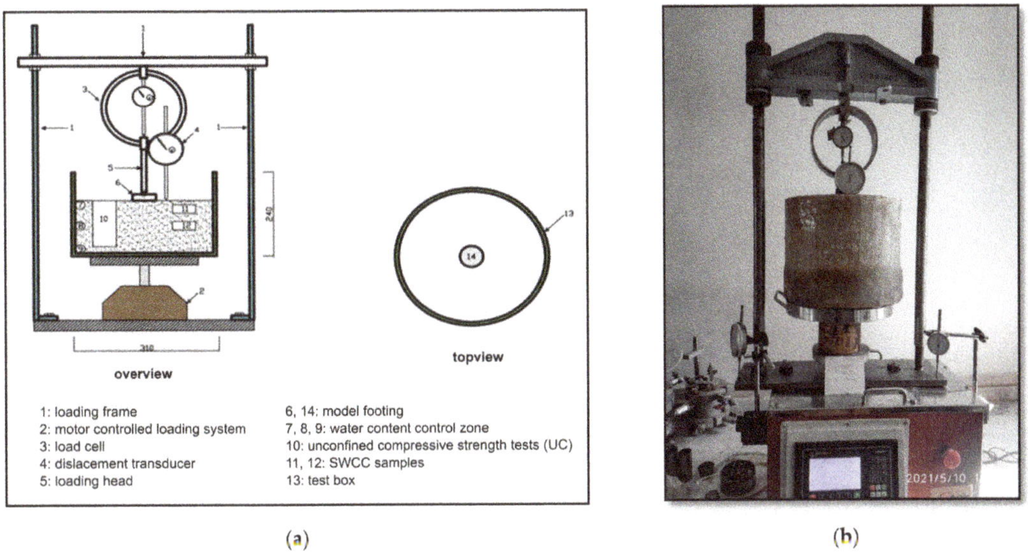

Figure 10. Model test setup: (**a**) schematic; (**b**) experimental picture.

After the UBC tests were performed, two samples were taken from the compacted soils using 50 mm diameter stainless steel thin-walled pipes. A total of 96 UC test were performed on the samples taken. The average values of 12 different scenarios when classified with similar VRs and SSDs are shown in Table 6.

Table 6. Angles showing the slope of the SS increase with respect to the MS.

SSD, %	VR = 0.65	VR = 0.70	VR = 0.75
	q_u (kPa)	q_u (kPa)	q_u (kPa)
65	116.05	71.88	54.31
70	94.68	65.19	37.99
75	75.58	58.89	31.23
85	68.2	36.57	29.31

The model footing test results for both saturated and unsaturated silt sample prepared at the VR of 0.65 are shown in Figure 11. As seen, no clear failure point has been observed in the curves. Therefore, the UBC values have been evaluated by intersecting the tangents of the initial and the final linear part of the curves.

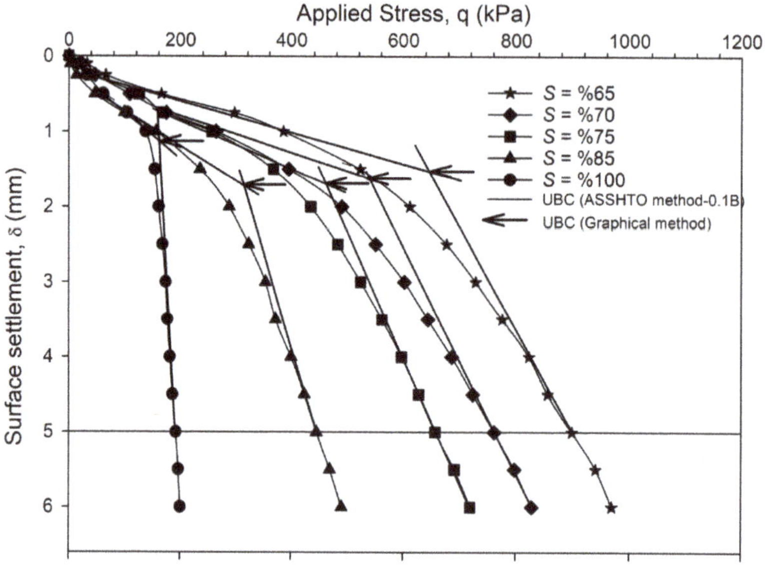

Figure 11. UBC model test results (VR = 0.65 and 50 mm circular model foundation).

The state of the soil and punching shear failure after the model footing tests can be observed from Figure 12. In the punching failure described by Vesic (1973) [44], which is also valid in the current study:

1. A peak is not observed,
2. Except for where the load is applied, no soil heave is visible, and
3. It is defined that it is created by the vertical shear and compression of the soil immediately beneath the footing.

(**a**) (**b**)

Figure 12. Soil settlement after loading test and cracks around model foundation: (**a**) 50 mm × 50 mm square; (**b**) 50 mm circle.

In such cases where the peak is not observed, the UBC of the foundation (q_{ult}) is determined in accordance with ASTM (D1194) [71]. The UBC is defined by ASTM (D1194) [71] as a stress value equal to one tenth (0.1B) of the foundation width. In addition, Steensen-Bach et al. (1987) [72] and Costa et al. (2003) [13] graphically determined the UBC by expanding the elastic and plastic parts of the loading curves profile. In Figure 11, the UBC values were calculated via both the graphical method (with the intersection point of the two linear sections) and the stress value corresponding to the 0.1B settlement value. With the use of the method valid in the literature, the UBC values corresponding to 0.1B have been taken into account in the next evaluations.

In this study, a total stress-based approach was employed. The most crucial part of the paper is introducing a new relationship that can estimate the UBC of foundations on unsaturated silty soils from the UC test and also estimate the UBC fitting parameter (S^{Ψ}) depending on the soil saturation (Equation (8)).

$$q_{ult(unsat)} = \left[\frac{q_{u(unsat)}}{2}\right] \times \left[1 + 0.2\frac{B}{L}\right] \times N_{cs} \times S^{\Psi} \quad (8)$$

where $q_{ult(unsat)}$ is the UBC of the foundation on silty soil; $q_{u(unsat)}$ is the SS of unsaturated soil-based on UC; B is the width of the foundation; L is the length of the foundation; N_{cs} is the UBC; S is the degree of saturation; and Ψ is fitting parameter.

At the end of the experimental studies, the model tests performed on silty soil (72% silt content) showed that the UBC of the foundations was different from that of the foundations on sand or clay soils studied in the literature. As it is known, the behavior of silts is neither exactly similar to clays nor sands, and nor is it the average of the reactions of two neighboring soil types. Silt soils are not under fully undrained conditions as clay soils are, nor are they under drained conditions as sandy soils are. Therefore, a critical part of the unsaturated soil mechanics must be studied to better understand the characterization of the unsaturated silty soils. The results obtained were also plugged into the equations proposed by Oloo et al. (1997) [12] and Vanapalli and Muhammed (2007) [14] for the estimation of a foundation's UBC of unsaturated soils and the results were compared.

Table 7 summarizes the measured the UBC values for all model foundations, including those estimated using the MTSA (i.e., Equations (3) and (8)), and those estimated using MESA (i.e., Equations (5) and (6)). The mean S^{Ψ} values, which were back calculated by using the Equation (8) suggested in this paper, have been found to be −2.89 and −1.33 when the UBC values were determined using 0.1B settlement and graphical methods, respectively. As can be seen from Table 6, although the difference between the calculated UBC values and the measured UBC values is small, the difference between the measured UBC values and the UBC values calculated with Equation (3) [i.e., that estimated using the value $N_{c(unsat)}$ = 5.14 as suggested by Oh and Vanapalli (2013) [18] and the modifications of Skempton (1948) [2] is large.

Equation (6), which is one of the models that estimate the UBC using the MESA, estimated very high UBC values compared to both the graphical method and the UBC values corresponding to the 0.1B settlement. However, Equation (5), which calculates the UBC by using the SS parameters, $\varnothing^{b}$ values, and MS values, gives reliable results for the MS values equivalent to an SSD of 85%, 75% and 70%, according to the UBC values calculated with 0.1B method. In cases where the MS value is high, the model gives very high values according to the test results. The main reason for this is that the model proposed by Oloo et al. (1997) [12] takes into account the increase in SS after high MS values are reached and as the residual WC approaches saturation. It is seen that there are great differences between the UBC values estimated via Equation (5) and the UBC values calculated by using the graphical method on the model test data. In addition, there are significant differences between the UBC results obtained with the test data (and proposed with Equation (8)) and the method produced by Oloo et al., (1997) [12].

Table 7. Model test results and estimated bearing capacity values.

Model Footing	VR	SSD %	WC %	MS kPa	Oloo et al., 1997 (Equation (5))	Vanapalli & Muhammed 2007 (Equation (6))	Oh & Vanapalli 2013 (Equation (3))	0.1B Settlement Method (Exp)	Grap. Method (Exp.)	UBC (Equation (8))	Ψ 0.1B Method	Ψ Grap. Method
50 mm Circle	0.651	64.2	15.69	284.1	1300	1501	348	899	648	826	−2.89	−1.33
	0.647	70.1	16.90	165.96	939	1318	292	762	550	597		
	0.648	74.4	18.00	91.90	712	1121	233	658	470	415		
	0.656	83.8	20.41	21.54	497	767	210	445	300	291		
	0.696	64.3	16.70	229.09	983	934	222	588	385	526	−2.89	−1.33
	0.699	68.6	17.80	141.25	752	857	201	499	345	411		
	0.701	74.5	19.40	63.10	547	679	182	372	280	323		
	0.702	83.2	21.70	15.61	423	521	113	213	140	156		
	0.749	64.7	17.99	157.28	645	587	167	424	320	397	−2.89	−1.33
	0.75	70.5	19.67	82.54	484	522	117	346	210	240		
	0.756	75.5	21.19	28.18	367	426	96	255	180	171		
	0.75	84.4	23.51	14.02	336	413	90	187	110	125		
50 mm × 50 mm Square	0.652	64.8	15.70	284.01	1294	1255	348	905	625	826	−2.89	−1.33
	0.650	69.4	16.80	174.16	958	1136	292	773	560	597		
	0.655	73.9	18.00	91.90	706	946	233	595	410	415		
	0.647	84.1	20.22	23.99	498	567	210	425	290	291		
	0.699	63.7	16.60	241.73	1016	890	222	548	375	526	−2.89	−1.33
	0.701	68.4	17.80	141.25	752	812	201	503	355	411		
	0.706	75.2	19.70	56.67	531	663	182	360	250	323		
	0.706	83.6	21.90	13.28	417	485	113	210	160	156		
	0.751	64.7	18.05	149.05	620	538	167	398	305	397	−2.89	−1.33
	0.749	70.4	19.63	82.54	477	502	117	327	215	240		
	0.748	74.4	21.19	28.18	360	440	96	250	165	171		
	0.751	82.6	23.04	18.38	338	410	90	187	112	125		
42.3 mm Circle	0.651	64.9	15.70	284.01	1294	1255	348	946	630	826	−2.89	−1.33
	0.651	69.3	16.70	157.28	906	1076	292	749	500	597		
	0.644	74.9	18.10	87.10	692	922	233	486	345	415		
	0.652	83.1	20.09	28.18	511	699	210	406	245	291		
	0.698	64.0	16.70	229.09	983	870	222	561	365	526	−2.89	−1.33
	0.693	69.4	17.90	133.86	733	795	201	494	345	411		
	0.698	73.9	19.20	70.25	566	703	182	347	200	323		
	0.706	84.1	22.10	11.93	413	478	113	201	120	156		
	0.754	64.7	18.13	141.25	604	530	167	444	335	397	−2.89	−1.33
	0.751	70.7	19.68	82.54	477	502	117	367	220	240		
	0.748	73.8	20.84	50.89	409	480	96	252	160	171		
	0.749	84.7	23.60	12.59	326	392	90	176	105	125		
37.5 mm × 37.5 mm Square	0.644	64.9	15.72	279.46	1285	1244	348	887	570	826	−2.89	−1.33
	0.657	68.2	16.50	194.98	1026	1211	292	778	510	597		
	0.648	74.3	17.90	92.29	712	948	233	460	330	415		
	0.665	84.4	20.23	23.99	503	664	210	335	225	291		
	0.701	63.6	16.60	241.73	1014	890	222	519	360	526	−2.89	−1.33
	0.699	68.5	17.90	133.86	731	796	201	460	320	411		
	0.699	74.3	19.30	68.25	559	697	182	314	180	323		
	0.703	84.6	22.10	11.93	412	534	113	184	100	156		
	0.747	63.8	17.70	165.96	655	554	167	477	345	397	−2.89	−1.33
	0.753	70.6	19.58	82.54	475	502	117	335	210	240		
	0.747	74.0	20.53	56.67	419	491	96	253	155	171		
	0.769	84.7	24.48	9.12	317	408	90	161	108	125		

Figure 13 compares the estimated ultimate bearing capacity obtained with Equation (8) and the measured the ultimate bearing capacity obtained through the model footing tests. When these graphs are examined, it is seen that the UBC, which is sometimes measured with model tests and sometimes estimated using the proposed equation, is slightly higher than the estimate value, but it is understood that these differences are within a small margin of error. It is also understood that the UBC measured at an SSD of 70% and 75% is slightly higher than the estimated value. This situation is similar to the undulating behavior of the SS both mentioned above and presented in the literature [53]. Considering the contribution of the MS values to the SS, although the MS value of the sample with an SSD of 70% is lower than the MS value of the sample with an SSD of 60%, its contribution to the SS is higher. It also appears that the size of the model foundations (i.e., 50 mm × 50 mm square; 50 mm circle; 42.3 mm circle; and 37.5 mm × 37.5 mm square) rather than the foundation shape has a limited effect on the UBC under MS.

The variation in the UBC measured according to the MS values corresponding to the SSD of 65%, 70%, 75%, 85% and 100% at three different VRs with a 50 mm diameter circular model foundation is shown in Figure 14. The results show a significant increase in UBC due to the contribution of MS in the range of the MS values corresponding to the decreasing SSD. In addition, as the VR increases, the contribution of the MS to the UBC begins to be relatively low compared to the smaller MS values.

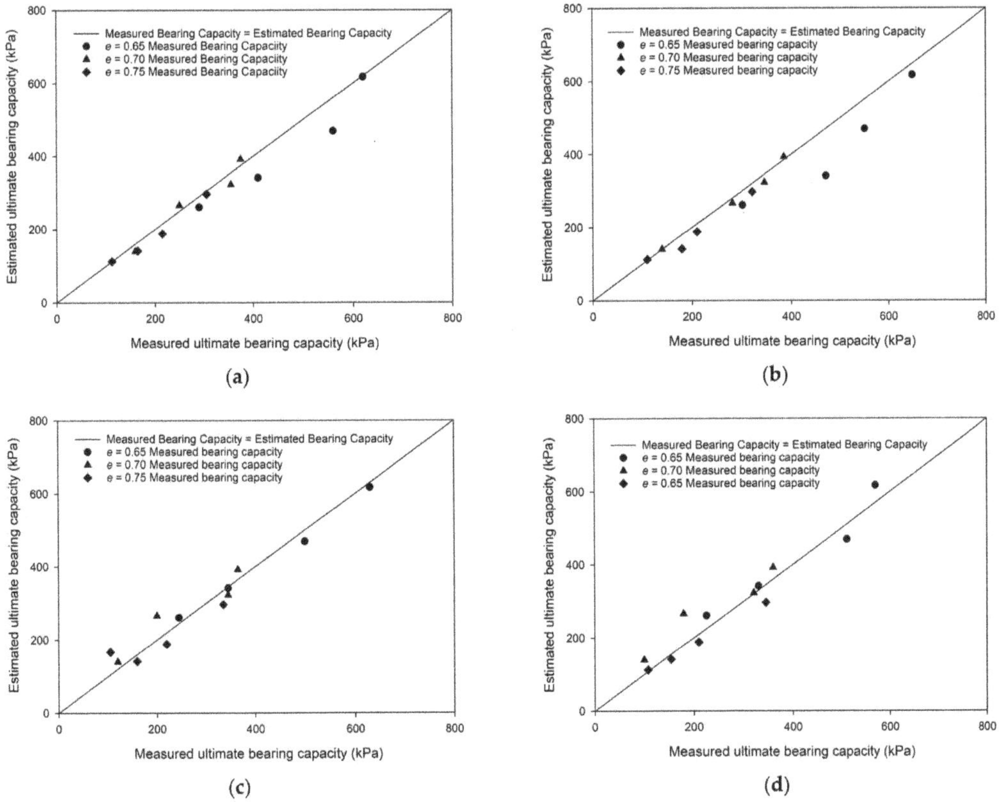

Figure 13. Comparison of the bearing capacity values found using the proposed Equation (8) and those found using the 0.1B technique: (**a**) 50 mm × 50 mm square; (**b**) 50 mm circle; (**c**) 42.3 mm circle; (**d**) 37.5 mm × 37.5 mm square.

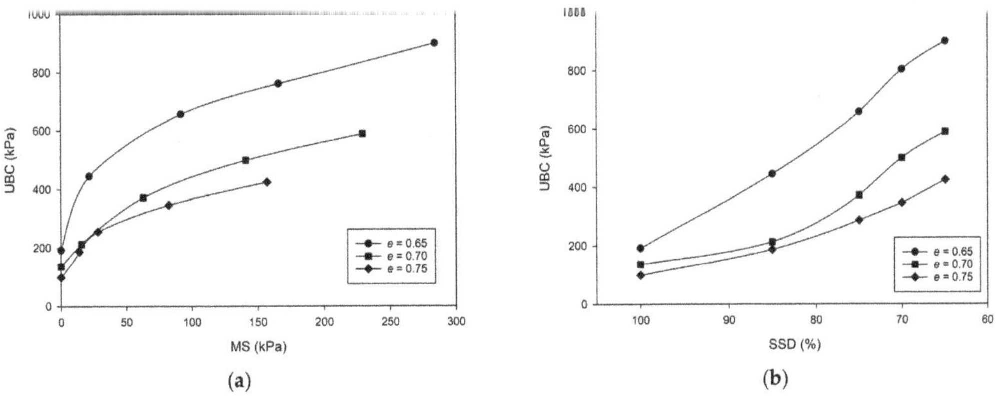

Figure 14. Graphs of 50 mm circle UBC: (**a**) MS (kPa); (**b**) SSD %.

4. Conclusions

In this study, an equation was proposed to determine the UBC of shallow foundations on unsaturated silty soil layers. In an experimental model setup created for this purpose, the UBC values of different types and sizes of model footings on silty soil layers with a different SSD/MS and VR values were measured. In addition, the SWCCs and SS parameters of unsaturated silt were determined. The conclusions obtained as a result of these experimental studies and analyses can be listed as follows:

1. Although the SS parameters of the unsaturated silty soil sample changes depending on the MS and the VR values, the failure behavior (punching failure) remains the same.
2. An increase in cohesion values in accordance with the literature was observed due to the increase in MS for all silt samples with different VRs. However, an increase was observed in the SS angle value due to the MS magnitude.
3. It was observed in the literature that the values calculated with the equations for the UBC of shallow foundations on unsaturated soils do not comply with the results of the model tests carried out on silty soils. Within the scope of this study, a new equation that gives suitable results for shallow foundations on unsaturated silty soils is proposed (Equation (8)).
4. It was determined that the foundation's UBC value for unsaturated silty soils can be estimated by using the unconfined compression test result and the fitting parameter depending on the SSD.
5. To evaluate foundations on unsaturated silty soils, the UBC can be estimated by considering the 0.1B method (ASSHTO method) by using the proposed equation in the current paper and by taking the fitting parameter (Ψ) as -2.89. If the graphical method is taken into account, the estimation can be made by taking the fitting parameter as -1.33 due to the generated relationship.
6. As the foundation size becomes smaller, the values of the foundation's UBC values measured in the model tests and calculated from the proposed equation become closer to each other. In addition, it was observed that the model foundation dimensions, and the variation in the foundation type causes changes to the measured UBC value.
7. In future studies, UBC tests can be performed on silt soils with different plasticity indexes and a relationship can be derived between the plasticity index and the fitting parameter (Ψ).

Author Contributions: Writing—original draft, M.U.Y.; conceptualization, A.O.; supervision, A.O.; investigation, M.U.Y.; methodology, M.U.Y. and A.O.; data curation, M.U.Y. and A.O.; writing—review and editing, A.O. and M.U.Y. All authors have read and agreed to the published version of the manuscript.

Funding: This research received no external funding.

Institutional Review Board Statement: Not applicable.

Informed Consent Statement: Not applicable.

Data Availability Statement: Not applicable.

Acknowledgments: We thank Sedat Sert, Ertan Bol, and Kadir Kocaman (Sakarya University) for comments and suggestions.

Conflicts of Interest: The authors declare no conflict of interest.

References

1. Terzaghi, K. *Theoretical Soil Mechanics*; John Wiley & Sons, Inc.: New York, NY, USA, 1943.
2. Skempton, A.W. The φ 0 Analysis of Stability and Its Theoretical Basis. In Proceedings of the 2nd International Conference on Soil Mechanics and Foundation Engineering (Rotterdam), Rotterdam, The Netherlands, 21–30 June 1948; pp. 72–77.
3. Meyerhof, G.G. The Ultimate Bearing Capacity of Foudations. *Geotechnique* **1951**, *2*, 301–332. [CrossRef]
4. Houston, S.L.; Houston, W.N.; Wagner, A.-M. Laboratory Filter Paper Suction Measurements. *Geotech. Test. J.* **1994**, *17*, 185–194.

5. Guan, Y.; Fredlund, D.G. Use of the Tensile Strength of Water for the Direct Measurement of High Soil Suction. *Can. Geotech. J.* **1997**, *34*, 604–614. [CrossRef]
6. Houston, S.L.; Perez-Garcia, N.; Houston, W.N. Shear Strength and Shear-Induced Volume Change Behavior of Unsaturated Soils from a Triaxial Test Program. *J. Geotech. Geoenviron. Eng.* **2008**, *134*, 1619–1632. [CrossRef]
7. Bulut, R.; Wray, W.K. Free Energy of Water-Suction-in Filter Papers. *Geotech. Test. J.* **2005**, *28*, 355–364.
8. Likos, W.J.; Lu, N. Hysteresis of Capillary Stress in Unsaturated Granular Soil. *J. Eng. Mech.* **2004**, *130*, 646–655. [CrossRef]
9. Aldaood, A. Impact of Fine Materials on the Saturated and Unsaturated Behavior of Silty Sand Soil. *Ain Shams Eng. J.* **2020**, *11*, 717–725. [CrossRef]
10. Zhang, C.; Chen, X.; Fan, W.; Zhao, J. A New Unified Failure Criterion for Unsaturated Soils. *Environ. Earth Sci.* **2015**, *74*, 3345–3356. [CrossRef]
11. Yan, Q.; Zhao, J.; Zhang, C.; Wang, J. Ultimate Bearing Capacity of Strip Foundations in Unsaturated Soils Considering the Intermediate Principal Stress Effect. *Adv. Civ. Eng.* **2020**, *2020*, 8854552. [CrossRef]
12. Oloo, S.Y.; Fredlund, D.G.; Gan, J.K. Bearing Capacity of Unpaved Roads. *Can. Geotech. J.* **1997**, *34*, 398–407. [CrossRef]
13. Costa, Y.D.; Cintra, J.C.; Zornberg, J.G. Influence of Matric Suction on the Results of Plate Load Tests Performed on a Lateritic Soil Deposit. *Geotech. Test. J.* **2003**, *26*, 219–227.
14. Vanapalli, S.K.; Mohamed, F.M. Bearing Capacity of Model Footings in Unsaturated Soils. In *Experimental Unsaturated Soil Mechanics*; Springer: Berlin/Heidelberg, Germany, 2007; pp. 483–493.
15. Rojas, J.C.; Salinas, L.M.; Sejas, C. Plate-Load Tests on an Unsaturated Lean Clay. In *Experimental Unsaturated Soil Mechanics*; Springer: Berlin/Heidelberg, Germany, 2007; pp. 445–452.
16. Zhang, F.; Bao, X.; Ikariya, T. A New Model for Unsaturated Soil Using Skeleton Stress and Degree of Saturation as State Variables. *Soils Found.* **2010**, *50*, 355–370. [CrossRef]
17. Zhang, C.; Yan, Q.; Zhao, J.; Wang, J. Formulation of Ultimate Bearing Capacity for Strip Foundations Based on the Meyerhof Theory and Unsaturated Soil Mechanics. *Comput. Geotech.* **2020**, *126*, 103734. [CrossRef]
18. Oh, W.T.; Vanapalli, S.K. Interpretation of the Bearing Capacity of Unsaturated Fine-Grained Soil Using the Modified Effective and the Modified Total Stress Approaches. *Int. J. Geomech.* **2013**, *13*, 769–778. [CrossRef]
19. Ibrahim, K.M.H.I. Effect of Percentage of Low Plastic Fines on the Unsaturated Shear Strength of Compacted Gravel Soil. *Ain Shams Eng. J.* **2015**, *6*, 413–419. [CrossRef]
20. Lotfizadeh, M.R.; Kamalian, M. Estimating Bearing Capacity of Strip Footings over Two-Layered Sandy Soils Using the Characteristic Lines Method. *Int. J. Civ. Eng.* **2016**, *14*, 107–116. [CrossRef]
21. Maghvan, S.V.; Imam, R.; McCartney, J.S. Relative Density Effects on the Bearing Capacity of Unsaturated Sand. *Soils Found.* **2019**, *59*, 1280–1291. [CrossRef]
22. Chen, T.; Xiao, S. An Upper Bound Solution to Undrained Bearing Capacity of Rigid Strip Footings Near Slopes. *Int. J. Civ. Eng.* **2020**, *18*, 475–485. [CrossRef]
23. Chen, W.; Xia, W.; Zhang, S.; Wang, E. Study on the Influence of Groundwater Variation on the Bearing Capacity of Sandy Shallow Foundation. *Appl. Sci.* **2023**, *13*, 473. [CrossRef]
24. Silveira, I.A.; Rodrigues, R.A. Collapsible Behavior of Lateritic Soil Due to Compacting Conditions. *Int. J. Civ. Eng.* **2020**, *18*, 1157–1166. [CrossRef]
25. Liu, Y.; Vanapalli, S.K. Mechanical Behavior of a Floating Model Pile in Unsaturated Expansive Soil Associated with Water Infiltration: Laboratory Investigations and Numerical Simulations. *Soils Found.* **2021**, *61*, 929–943. [CrossRef]
26. Tan, M.; Cheng, X.; Vanapalli, S. Simple Approaches for the Design of Shallow and Deep Foundations for Unsaturated Soils I: Theoretical and Experimental Studies. *Indian Geotech. J.* **2021**, *51*, 97–114. [CrossRef]
27. Tan, M.; Vanapalli, S.K. Foundation Bearing Capacity Estimation on Unsaturated Soil Slope under Transient Flow Condition Using Slip Line Method. *Comput. Geotech.* **2022**, *148*, 104804. [CrossRef]
28. Vanapalli, S.K.; Salinas, L.M.; Avila, D.; Karube, D. Suction and Storage Characteristics of Unsaturated Soils. In *Unsaturated Soils-Volume 3: Proceedings of the 3rd International Conference on Unsaturated Soils, UNSAT 2002, 10–13 March 2002, Recife, Brazil*; CRC Press: Boca Raton, FL, USA, 2021; p. 1045.
29. Cheng, Y.-Y.; Gao, X.-G.; Liu, T.-H.; Li, L.-X.; Du, W.; Hamad, A.; Wang, J.-P. Effect of Water Content on Strength of Alluvial Silt in The Lower Yellow River. *Water* **2022**, *14*, 3231. [CrossRef]
30. Fredlund, D.G.; Morgenstern, N.R. Stress State Variables for Unsaturated Soils. *J. Geotech. Eng. Div.* **1977**, *103*, 447–466. [CrossRef]
31. Oh, W.T.; Vanapalli, S.K. Modelling the Applied Vertical Stress and Settlement Relationship of Shallow Foundations in Saturated and Unsaturated Sands. *Can. Geotech. J.* **2011**, *48*, 425–438. [CrossRef]
32. Vahedifard, F.; Robinson, J.D. Unified Method for Estimating the Ultimate Bearing Capacity of Shallow Foundations in Variably Saturated Soils under Steady Flow. *J. Geotech. Geoenviron. Eng.* **2016**, *142*, 04015095. [CrossRef]
33. Vo, T.; Russell, A.R. Bearing Capacity of Strip Footings on Unsaturated Soils by the Slip Line Theory. *Comput. Geotech.* **2016**, *74*, 122–131. [CrossRef]
34. Oh, W.T.; Vanapalli, S.K. Modeling the Stress versus Settlement Behavior of Shallow Foundations in Unsaturated Cohesive Soils Extending the Modified Total Stress Approach. *Soils Found.* **2018**, *58*, 382–397. [CrossRef]
35. Ghasemzadeh, H.; Akbari, F. Determining the Bearing Capacity Factor Due to Nonlinear Matric Suction Distribution in the Soil. *Can. J. Soil Sci.* **2019**, *99*, 434–446. [CrossRef]

36. Ghasemzadeh, H.; Akbari, F. Investigation of Soil Active Wedge Angle with Linear Matric Suction Distribution below the Footing. *Int. J. Civ. Eng.* **2020**, *18*, 161–168. [CrossRef]
37. Akbari Garakani, A.; Sadeghi, H.; Saheb, S.; Lamei, A. Bearing Capacity of Shallow Foundations on Unsaturated Soils: Analytical Approach with 3D Numerical Simulations and Experimental Validations. *Int. J. Geomech.* **2020**, *20*, 04019181. [CrossRef]
38. Lu, N.; Godt, J.W.; Wu, D.T. A Closed-Form Equation for Effective Stress in Unsaturated Soil. *Water Resour. Res.* **2010**, *46*, W05515. [CrossRef]
39. Griffiths, D.V.; Lu, N. Unsaturated Slope Stability Analysis with Steady Infiltration or Evaporation Using Elasto-Plastic Finite Elements. *Int. J. Numer. Anal. Methods Geomech.* **2005**, *29*, 249–267. [CrossRef]
40. Zhang, L.L.; Fredlund, D.G.; Fredlund, M.D.; Wilson, G.W. Modeling the Unsaturated Soil Zone in Slope Stability Analysis. *Can. Geotech. J.* **2014**, *51*, 1384–1398. [CrossRef]
41. Consoli, N.C.; Schnaid, F.; Milititsky, J. Interpretation of Plate Load Tests on Residual Soil Site. *J. Geotech. Geoenviron. Eng.* **1998**, *124*, 857–867. [CrossRef]
42. Tang, Y.; Taiebat, H.A.; Russell, A.R. Bearing Capacity of Shallow Foundations in Unsaturated Soil Considering Hydraulic Hysteresis and Three Drainage Conditions. *Int. J. Geomech.* **2017**, *17*, 04016142. [CrossRef]
43. Meyerhof, G.G. Some Recent Research on the Bearing Capacity of Foundations. *Can. Geotech. J.* **1963**, *1*, 16–26. [CrossRef]
44. Vesić, A.S. Analysis of Ultimate Loads of Shallow Foundations. *J. Soil Mech. Found. Div.* **1973**, *99*, 45–73. [CrossRef]
45. Kumbhojkar, A.S. Numerical Evaluation of Terzaghi's N γ. *J. Geotech. Eng.* **1993**, *119*, 598–607. [CrossRef]
46. Gan, J.K.M.; Fredlund, D.G.; Rahardjo, H. Determination of the Shear Strength Parameters of an Unsaturated Soil Using the Direct Shear Test. *Can. Geotech. J.* **1988**, *25*, 500–510. [CrossRef]
47. Escario, V.; Saez, J. The Shear Strength of Partly Saturated Soils. *Geotechnique* **1986**, *36*, 453–456. [CrossRef]
48. Vanapalli, S.K.; Fredlund, D.G.; Pufahl, D.E. The Relationship between the Soil-Water Characteristic Curve and the Unsaturated Shear Strength of a Compacted Glacial Till. *Geotech. Test. J.* **1996**, *19*, 259–268.
49. *ASTM D4318-17e1*; Standard Test Methods for Liquid Limit, Plastic Limit, and Plasticity Index of Soils. ASTM International: West Conshohocken, PA, USA, 2010.
50. Kruse, G.A.; Dijkstra, T.A.; Schokking, F. Effects of Soil Structure on Soil Behaviour: Illustrated with Loess, Glacially Loaded Clay and Simulated Flaser Bedding Examples. *Eng. Geol.* **2007**, *91*, 34–45. [CrossRef]
51. Ltifi, M.; Abichou, T.; Tisot, J.P. Effects of Soil Aging on Mechanical and Hydraulic Properties of a Silty Soil. *Geotech. Geol. Eng.* **2014**, *32*, 1101–1108. [CrossRef]
52. Dijkstra, T.A.; Rogers, C.D.F.; Smalley, I.J.; Derbyshire, E.; Li, Y.J.; Meng, X.M. The Loess of North-Central China: Geotechnical Properties and Their Relation to Slope Stability. *Eng. Geol.* **1994**, *36*, 153–171. [CrossRef]
53. Wen, B.-P.; Yan, Y.-J. Influence of Structure on Shear Characteristics of the Unsaturated Loess in Lanzhou, China. *Eng. Geol.* **2014**, *168*, 46–58. [CrossRef]
54. Önalp, A.; Arel, E.; Bol, E.; Özocak, A.; Sert, S. *The Assessment of Liquefaction Potential with Dissipation Method in Piezocone Penetration Testing (CPTU)*; Scientific and Technological Research Council of Turkey: Ankara, Türkiye, 2007.
55. Fredlund, D.G.; Rahardjo, H. *Soil Mechanics for Unsaturated Soils*; John Wiley & Sons: Hoboken, NJ, USA, 1993.
56. Hu, Z.Q.; Shen, Z.J.; Xie, D.Y. Research on Structural Behavior of Unsaturated Loess. *Chin. J. Rock Mech. Eng.* **2000**, *19*, 775–779.
57. Maleki, M.; Bayat, M. Experimental Evaluation of Mechanical Behavior of Unsaturated Silty Sand under Constant Water Content Condition. *Eng. Geol.* **2012**, *141*, 45–56. [CrossRef]
58. Oloo, S.Y.; Fredlund, D.G. A Method for Determination of φ b for Statically Compacted Soils. *Can. Geotech. J.* **1996**, *33*, 272–280. [CrossRef]
59. Vanapalli, S.K.; Fredlund, D.G. Comparison of Different Procedures to Predict Unsaturated Soil Shear Strength. *Geotech. Spec. Publ.* **2000**, *99*, 195–209.
60. Vanapalli, S.K.; Lane, J.J. A Simple Technique for Determining the Shear Strength of Finegrained Unsaturated Soils Using the Conventional Direct Shear Apparatus. In Proceedings of the Second Canadian Specialty Conference on Computer Applications in Geotechnique, Winnipeg, Canada; Citeseer. 2002; pp. 245–253.
61. Gardner, R. A Method of Measuring the Capillary Tension of Soil Moisture over a Wide Moisture Range. *Soil Sci.* **1937**, *43*, 277–284. [CrossRef]
62. Qian, J.S.; Lu, H. Effect of Compaction Degree on Soil-Water Characteristic Curve of Chongming Clay. In *Applied Mechanics and Materials*; Trans Tech Publications Ltd.: Wollerau, Switzerland, 2011; Volume 90, pp. 701–706.
63. *ASTM D5298-16*; Standard Test Method for Measurement of Soil Potential (Suction) Using Filter Paper. ASTM International: West Conshohocken, PA, USA, 2016.
64. Hilf, J.W. *An Investigation of Pore-Water Pressure in Compacted Cohesive Soils*; University of Colorado at Boulder: Boulder, CO, USA, 1956.
65. Cresswell, H.P.; Green, T.W.; McKenzie, N.J. The Adequacy of Pressure Plate Apparatus for Determining Soil Water Retention. *Soil Sci. Soc. Am. J.* **2008**, *72*, 41–49. [CrossRef]
66. Adams, M.T.; Collin, J.G. Large Model Spread Footing Load Tests on Geosynthetic Reinforced Soil Foundations. *J. Geotech. Geoenviron. Eng.* **1997**, *123*, 66–72. [CrossRef]
67. Oh, W.T.; Garga, V.K.; Vanapalli, S.K. Shear Strength Characteristics of Statically Compacted Unsaturated Kaolin. *Can. Geotech. J.* **2008**, *45*, 910–922. [CrossRef]

68. Oloo, S.Y. *A Bearing Capacity Approach to the Design of Low-Volume Traffic Roads*; University of Saskatchewan: Saskatoon, SK, Canada, 1999.
69. *ASTM D3080-04*; Standard Test Method for Direct Shear Test of Soils under Consolidated Drained Conditions. ASTM International: West Conshohocken, PA, USA, 1999.
70. Van Genuchten, M.T. A Closed-Form Equation for Predicting the Hydraulic Conductivity of Unsaturated Soils. *Soil Sci. Soc. Am. J.* **1980**, *44*, 892–898. [CrossRef]
71. *ASTM D1194-94*; Standard Test Method for Bearing Capacity of Soil for Static Load and Spread Footings. ASTM International: West Conshohocken, PA, USA, 2003.
72. Steensen-Bach, J.O.; Foged, N.; Steenfelt, J.S. Capillary Induced Stresses—fact or Fiction? In Proceedings of the 9th European Conference on Soil Mechanics and foundation Engineering, Dublin, Ireland, 31 August–3 September 1987; pp. 83–89.

applied sciences

Review

Diatomaceous Soils and Advances in Geotechnical Engineering—Part I

Daniel Zuluaga-Astudillo [1,2,*], Juan Carlos Ruge [1], Javier Camacho-Tauta [1], Oscar Reyes-Ortiz [1] and Bernardo Caicedo-Hormaza [3]

1 Programa de Doctorado en Ingeniería, Universidad Militar Nueva Granada, Bogotá 110111, Colombia
2 Programa de Diseño y Construcción de Vías y Aeropistas, Escuela de Ingenieros Militares, Ejército de Colombia, Bogotá 111611, Colombia
3 Departamento de Ingeniería Civil y Ambiental, Universidad de los Andes, Bogotá 110311, Colombia
* Correspondence: est.daniel.zuluaga@unimilitar.edu.co or daniel.zuluaga@esing.edu.co

Abstract: Diatoms are microscopic algae with a skeleton called a frustule, formed chiefly of silica, and are found in almost all aquatic environments and climatic conditions. Diatomaceous soils (DSs) originate from frustule sedimentation. In civil works (design and construction), the uncommon values obtained from DSs are not completely understood. There needs to be more knowledge about the strength and compressibility of DSs. The stability of these deposits is still being determined. Definitions of substances such as diatoms, diatomaceous soils, diatomaceous earth, diatomaceous oozes, frustules, and diatomite need to be clarified. This document references construction processes that face problems such as differential settlements, pile rebounds, and irregular pore pressures due to frustules. This review analyzes multiple sets of results regarding the grain size distribution, specific gravity, consistency, plasticity, compressibility, and shear strength of DSs. It is concluded that the particle size distribution of DSs generally classifies them as silts. Particles are modified by the imposition of stresses (frustule breakage), which impacts compressibility. Microfossils take up stresses, restrict strains, and cause sudden increases in compressibility when their yield stress is exceeded. Currently, their strain mechanisms need to be better understood. The Gs decreases with increasing frustule content, given the high porosity of the skeletons. The intraparticle pores of the frustules explain the high liquid limit (LL) of DSs. DSs can have high shear strengths and large yield surfaces due to the "interlocking" phenomenon and the interparticle contacts' high frictional component caused by their rough surface and high silica content.

Keywords: diatom frustules; diatomaceous soil concept; fossiliferous soils; diatomite; shear resistance; compressibility; foundation problems; index properties

1. Introduction

The design and construction phases of foundations or fills are usually complex when the development occurs on soft soils, and complications can even arise during the works' service periods, particularly due to excessive differential settlements and the inaccuracy of the models used in the calculation of bearing capacity [1]. In the practical exercise of civil engineering, the values associated with the uncommon responses obtained from DSs are not considered [2]. These include, for example, high values of the effective friction angle (ϕ') for the design of foundations and for slope stability. This unusual behavior is not only evidenced in construction processes, but also in laboratory tests [3].

The above is due to a lack of knowledge or sufficient documentary, experimental, or construction background evidence to support any types of decision making [4]. Consequently, the spectrum of possibilities regarding the development of a stable construction on this type of deposit is uncertain, leading to the potential for loss of lives when buildings, bridges, embankments, slopes, or other types of works are executed.

Sometimes, this situation leads to the reprocessing of research and consulting projects, the reconstruction of works, or the need for emergency attention because of the excessive settlement in backfill structures due to secondary consolidation [4], the rejection of piles at shallow driving depths [5], susceptibility to geologic hazards in road projects [6], and liquefaction failures even when index properties state otherwise [7], to name a few. Although the effects of microfossils in marine or lacustrine sediments have been documented, the records are limited to reports on variations in aspects such as cohesion or friction angle. They do not delve into an explanation of differential behavior or consider its origin [8].

An example of this is the secondary consolidation of diatom lacustrine materials from the "La Pine, Deschutes" Basin, where significant settlements were triggered after the installation of the foundation embankment as part of the project called "US 97 Wickiup Junction" (OR, USA), which was intended to eliminate a railroad crossing. This was due to the strain and crushing of skeletal particles, resulting from the imposition of new loads [4]. In 2017, measurements of the settlements taken during the winter months were consistent with those projected in the designs. However, further monitoring five weeks later proved to be quite unusual, to such an extent that it led to the suspension of the work. A new exploration sequence complemented with scanning electron microscopy (SEM) was carried out, finding a significant amount of diatoms (50–80%) in an excellent state of preservation. This discovery was contrary to what was reported in the geological maps of the area. The study concluded that significant long-term secondary consolidation had probably been initiated in the deeper diatom soils. Diatomaceous soils in a saturated-unconsolidated condition have poorly understood properties.

In Bogota (Colombia), there have been cases of buildings in service for 30 to 50 years with optimum performance throughout. Nevertheless, after this period, they began to present excessive differential settlements that exceeded the permissible limits (5 cm isolated footing and 10 cm raft and mat foundations). This phenomenon was previously associated with secondary consolidation processes [9]. More than 60% of the city is built on lacustrine deposits that are soft and highly plastic. SEM images from samples taken at a 20 m depth reflected an open, flocculated soil structure with a considerable amount of diatom frustules of multiple species, sizes, and types of mineralogy. This characteristic is associated with the environment having a low depositional rate (1 m per 3440 years) [9]. The high diatom content compared to other deposits may be explained by the high friction angle, void ratio, and compressibility coefficient. The analysis of the settlements via secondary consolidation is indispensable in designing geotechnical structures supported on lacustrine deposits containing diatoms [9].

The documented buildings have recorded high values for the compression index (Cc), recompression index (Cs), and void ratio. Therefore, the expected number of settlements is high. The settlement criteria include the factors that govern the determination of the available bearing capacity in the compressible soils of Bogota [9].

On the other hand, Ref. [10] studied the diatomaceous deposit of Klamath County, Bonanza, Oregon (USA), which is located in the zone of influence of the bridge replacement project known as "OR140 Buck Creek Bridge". The diatomaceous materials were characterized via field (SPT and CPT) and laboratory tests. The profile indicated that the diatomaceous soil was about 4 m below the surface, and a thickness of 1 m was recorded. Oedometric testing of the "Buck Creek" soil indicated high overconsolidation ratios ranging from 10 to 40. These soils exhibited an unusually high shear strength due to "interlocking" phenomena between the grains, which exhibited considerable irregularity in their shape. In the project, an unexpected rebound phenomenon occurred during the driving of piles designed with a closed toe. This rejection occurred at depths of less than 5 m.

Consequently, the construction process had to be modified to open-toe pile driving. The piezometric data indicated that negative pore pressures were generated during the rebound of the closed-toe piles. In situ testing indicated changes in material stiffness and dilatant tendencies near the refusal depth. Dissipation tests conducted near that point indicate a positive induced pore pressure, equivalent to 100 times the hydrostatic pressure, which dissipates rapidly (in less than one h), implying a high hydraulic conductivity. The latter is a phenomenon contradictory to the negative pore pressures mentioned above, which is why [5] insists on the need for further research. The criteria for pile rebound in DS differ from those previously reported in the literature. The authors state that their study is perhaps the first documented case of negative pore pressures being generated during pile rejection.

Another case refers to the Cooper Marl sector (Charleston, SC, USA), where rejections have also been recorded in the installation of piles, and whose soils share characteristics with DS in terms of engineering and physical properties. In this area, considerable pore pressures have been observed during pile driving. Rejection has been reported especially in large-diameter piles. This situation is generally managed in practice by alternate installation—i.e., when the rebound of one pile is recorded, proceed with the next one and when it reaches rejection, return to the initial one [5].

Ref. [11] provides the results of the characterization of Bogota soils, which are classified as very soft with a high diatom content, similar to those of Mexico City and the Sea of Japan. At depths greater than 10 m, extreme values have been obtained for some geotechnical properties, such as an LL of 400%, moisture content of 200%, void ratio close to 5, and compressibility coefficient of approximately 5.

The usual geotechnical correlations and requirements for the design of foundations need to be revised for soils that present diatomaceous particles in concentrations higher than 40% by weight. From this magnitude, some soils register considerable increases in the liquid limit (LL) and the void ratio, particularly at low stress levels of 12.5 kPa to 100 kPa [12].

The properties of fine soils (silts and clays) are mainly defined by their behavior and not necessarily by their particle size [13]. The mechanical behavior and index properties of soils that record considerable fossil contents cannot be explained by applying the typical patterns associated with organic or inorganic soils. Some of the systems proposed in geotechnical engineering for soil classification—for example, those based on plasticity (Casagrande methods)—are inadequate for correctly categorizing this type of sediments [8,14–16]. Multiple authors [17–20] have warned that the results obtained from DSs are unexpected, inconsistent with classical soil mechanics theories, or present problematic or uncertain engineering properties [19,21–23]. The abundance of fossils in a soil matrix can modify its physical, chemical, and geotechnical properties to such an extent that it differs from conventional results [8].

Ref. [24] reported large diatomite sedimentary deposits in Oita (Japan), where natural moisture contents are high (>141%) but simultaneously present high stiffness. The water-holding capacity is proposed to be associated with the fossil internal volumes and the pores that characterize the frustules. The authors of [11] associate the unusual behavior of these soils to the empty skeletons of the frustules, their irregular patterns, and their abrasive surfaces. However, there is no clear evidence to relate or quantify this aspect, either from the number of pores per unit area, size, or the irregularity of their surface [17,25].

In the Colombian scenario, the lack of information regarding DS is such that the analysis of "soils with special characteristics" is restricted to the study of phenomena such as expansivity, collapsibility, dispersivity, the presence of vegetation, and the presence of water bodies [26], ignoring the effects of secondary consolidation, potential absorption, and high shear strength angles linked to the high liquid limits, usual in DS. Therefore, the characterization of these deposits should at least consider analysis by microscopy [13,27] in order to recognize the presence of fossils, their concentration, and their state of preservation and, thus, to understand the incidence of these variables on the projects' performance. Few studies have attempted to investigate the relationship between mechanical properties and the microstructure of microfossil deposits [24]. Tests such as SEM, X-ray diffraction (XRD), or Porosimetry allow one to obtain a better understanding of the soil, which cannot be conceived by regular mechanical tests [13,28].

In order to characterize and quantify the volumetric proportion of diatomaceous particles and relate them to the physical properties of soils, Ref. [15] extracted samples from central and southern Oregon, based on which they made SEM observations and applied quantitative stereology procedures using manual particle selection. The accuracy of the method was validated by applying it to artificial samples dosed by weight (75% diatomaceous earth—25% kaolin). Some sources that have tried to quantify the fossil content in soil samples have applied indirect methods, for example by assessing the amorphous silica content through dissolution or X-ray diffraction. However, these methods do not provide information on the shape or porosity of the particles, factors that are quickly released by SEM observation [15] (see Appendix A).

There are reports of places where considerable DS deposits have been identified [7,29]—for example, in the Equatorial Pacific [30], Mexico [31,32], Japan [33,34], the USA [4,35], Chile [18], China [23], Colombia [12,19,36,37], Poland [38], as well as places with the presence of living diatoms, which leads us to infer that the underlying strata present fossil sedimentation. In Colombia, more than 73 places with microscopic algae are recorded [39]. It is essential to increase characterization and mechanical evaluation studies in all places that have a history of volcanic activity, such as the Pacific fire belt, since the presence of DS is enhanced there [18,31,32,40,41].

The grain size distribution of DSs generally classifies them as fine soils, mainly within the silt range (0.002 mm < SD < 0.075 mm). The presence of diatoms (frustules) leads to an increase in the liquid limit (LL) [32,40] and plasticity index (PI) [12], simultaneously with an increase in the effective friction angle (ϕ') [18,42]. This contradictory behavior has not been fully understood, nor has it been associated with the diatom species present in the studied samples.

Some researchers have attempted to explain the high values of shear strength in diatomaceous deposits, in terms of ϕ', through an assumption of interlocking between soil particles, of which some are fragments of frustules [2]. However, in this scenario, the correspondence between increases in resistance, the concentration of fossils, the state of deterioration (fracturing) of the pieces, the representative forms of the species, and the compositional condition (single or multispecies) is not precisely defined. The relationship between diatoms and soil particles has not been wholly characterized [32]. Ref. [8] suggests that a way to evaluate the concentration of diatoms in a given soil should be found so that it can be related to the geotechnical behavior of the latter.

It is concluded that the state-of-the-art regarding the variation in the physical properties and mechanical response of soil deposits with the presence of diatom fossils is poorly developed [15,18]. Published results are generally limited to the report of friction angle (ϕ) values and consistency limit records [12,32,37]. Ref. [8] state that many Japanese soils do not follow the patterns that regularly describe soil behavior because they contain significant amounts of diatoms, while [15] posit that diatomaceous soils significantly affect the engineering behavior of the medium, even when they are found in tiny proportions (20%). Although the presence of fossils in soils has been widely assessed from a geological point of view and their geotechnical performance has been observed, explanations for unexpected or non-typical responses have not yet been given. Considering the presence of DS in different parts of the world and its unique behaviors, its influence should be recognized and it should be considered as a distinct soil type in the classification of marine sediments [43]. A new classification system for "fossiliferous soils" is considered for inclusion in addition to the existing classification of organic and inorganic soils [8,44].

Few documentary sources elucidate the influence of species on DS behavior. When they do, are limited to the description of immediate details such as particle size variation and frustule content [14], but do not delve into subjects such as mechanical response or dynamic effects. The authors of [44] compared the geotechnical behavior between samples with the presence of centric and pennate particles. The latter, contained in a diatomaceous ooze, is associated with an open and less dense fabric, high void and water content, and high porosity. This fabric collapses rapidly with increasing vertical stresses. Diatomaceous oozes with mostly pennate particles have a more remarkable ability to reproduce in situ stress conditions when subjected to one-dimensional consolidation tests. This type of diatomaceous ooze is prone to generate high pore pressures under seismic stresses and, therefore, they are cataloged as factors that generate submarine landslides [44]. On the other hand, the disk shape of the centric particles is related to a denser, naturally oriented fabric, which results in a more resistant structure. The effect of diatom species on geotechnical responses is a relevant topic for future research [44].

Diatom-rich sediments show different behaviors when evaluated in the laboratory and analyzed in their natural condition. It has also been concluded that these soils somehow support the lithostatic and tectonic forces that are slowly transmitted to them. However, the reasons for these behaviors and the discrepancies between laboratory and geological evidence remain incompletely understood [44].

Given their unique characteristics such as high water absorption [45], high LL, and high ϕ', there is a great potential to use diatomaceous earth in the development of geotechnical and geo-environmental engineering projects, as would be the case of mechanically competent hydraulic fills [20,23], in addition to soil stabilization or improvement processes, or even in the redefinition of design parameters for slope cuts. The study of the mechanical properties of DSs is of great importance to optimize engineering designs in offshore structures [18] and to enhance the understanding of submarine landslides [7]. Other potential direct beneficiaries of DSs are builders of oil production complexes [46]. There is particular interest in the study of this type of soil in the mining and energy industries. In recent years, this has been of particular relevance in coastal areas, as is the case of Mejillones Bay in Chile [18]. Complementary applications in civil engineering even propose the use of DS in the production of masonry elements [47].

2. Evolution of Concepts Surrounding Diatomaceous Soils

It has been recognized that there is the terminology used when mentioning soils with the presence of diatom fossils is imprecise and arbitrary. In several cases, they are called "diatomaceous soil" when the deposits have been formed from the frustules and not from the algae themselves (diatoms). Likewise, the use of terms such as "diatomites" or "diatomaceous oozes" is evidenced, without specifying the technical particularities that lead to such denominations. Therefore, it is necessary to delve into the methodological proposals that some authors have projected for the correct categorization of this type of fossiliferous soil.

Most diatomaceous soils (DSs) and diatomites originated from sedimentation processes [48]. During the process of stratified accumulation, the frustules can be subject to deterioration phenomena, such as total or partial dissolution [49]; fracturing by friction due to transport (mostly in aqueous media); crushing by excessive loads transmitted by upper strata; and proximity to geological fault zones, among others.

Diatoms are microscopic, unicellular, eukaryotic, protist, photosynthetic algae, with a siliceous skeleton called a frustule [38,50], and are found in almost all fresh and saline aquatic environments [29,48] and even on soils and wet surfaces, either planktonic (floating) or benthic (adhering to substrates) [51], and in all climatic conditions [52]. They are extremely sensitive to physical and chemical changes in such environments, particularly in turbidity, temperature, light, pH, nutrients, and salinity [53]. Typical sizes vary between 20 and 200 μm in diameter or length, depending on the species. Some cases reach 2 mm. These can be recognized as solitary or in colonies and are joined by filaments or chains [54]. They are primarily found in lacustrine and low-energy marine environments [15].

The cell wall of diatoms (frustules) is of special interest, being composed chiefly of silica [55], and gives rise to nanometer structures [56]; see Figure 1. Species evolution creates distinctive patterns in the perforations of the frustules, which are used for visual classification [57]. The physical characteristics of the frustules depend on the biotic and environmental conditions of the depositional periods, especially concerning the nature of the algae, the depth of the aquatic sedimentation media, and the intensity of the light regulating the photosynthetic processes [48].

Figure 1. Representation of recent and fossilized freshwater and saltwater diatoms [52].

Frustules are impregnations of polymerized silicic acid, which is the same material as quartz. There are two types of diatoms: pennate, with bilateral symmetry and predominantly found in freshwater, and central, with radial symmetry and generally found in marine areas [8]; see Figures 2 and 3. As shown in Figure 4, for a specific type of diatom, two overlapping pieces are identified. There is an external theca or epitheca and an internal theca or hypotheca [57], which are made up of 95% silica.

2.1. Diatoms, Diatomaceous Soils, or Diatomaceous Earth

The "ornamentations" or characteristic patterns of the frustules are due to the differential arrangement or fixation of silica in the cell wall. These are notable for having a definite organization and are helpful for classification. The main ornamentations include striae, pores, areoles, spines, raphe, and nodules [58]. From these patterns is possible to differentiate for example, centric and pennate species [59].

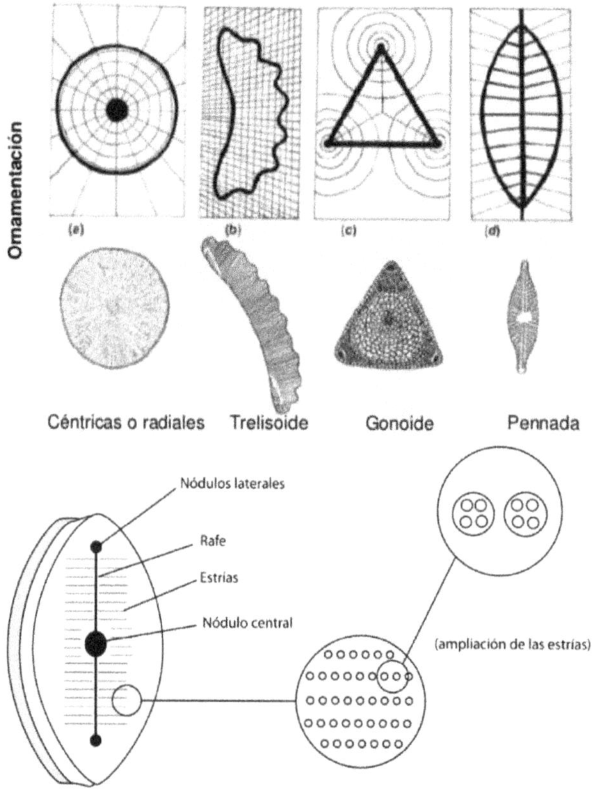

Figure 2. Characteristic ornamentations and components of some frustules [58]. Above: (**a**) centric or radial species (**b**) "trelisoide" species (**c**) "gonoide" species (**d**) pennate species. Below: typical distribution of pores and striae for a raphid species.

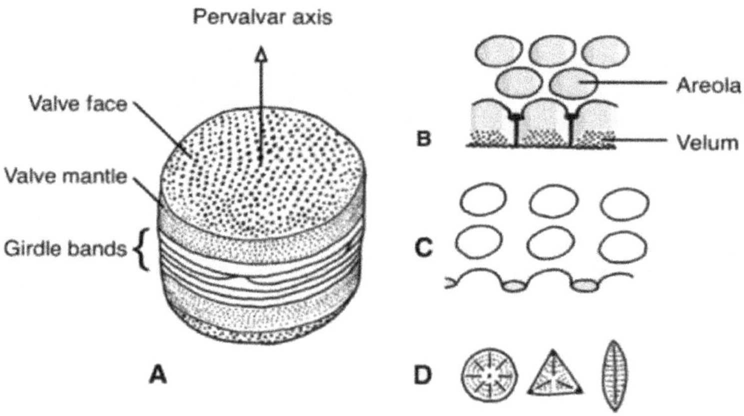

Figure 3. Morphology and classification of diatoms [50]. (**A**) Basic morphology of a diatom frustule. (**B**) Cross section of a typical double-layered pseudoloculated frustule. (**C**) Cross section of a single silica layer. (**D**) Three basic shapes of frustules: centric circular, centric multipolar (bipolar, triangular, star-shaped), and lateral pennate.

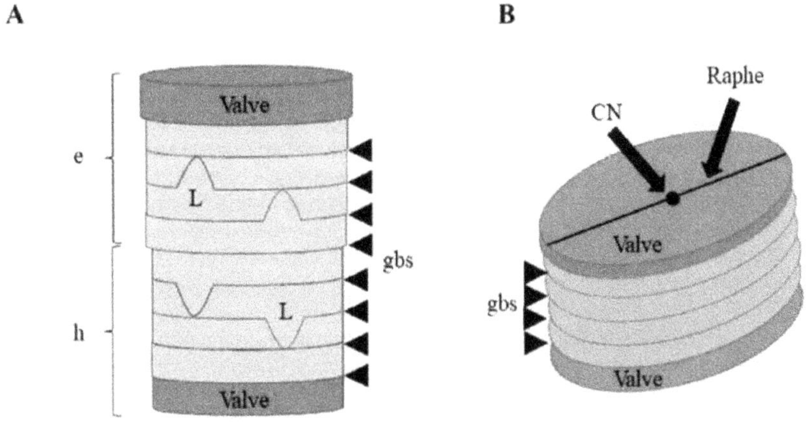

Figure 4. Representation of centric and pennate species frustule [59] cited by [60]. (**A**) The dark gray discs represent the valves. The higher and larger ones are the epitheca (e); the lower and smaller ones are the hypotheca (h). Girdle bands (gbs) are indicated by arrows. Ligules (L) are bell-shaped structures within the girdle bands. (**B**) Diagram of a pennate diatom exposing the central node (CN) and raphe on the upper valve.

To date, more than 200 genera of diatoms are known, and it is estimated that there are about 100,000 extinct species; see Figure 1. The concentration of silica is mostly influenced by the species and not so much by environmental factors [60,61]. More than 10,000 species of living diatoms are recognized [8], and some authors specify a number between 12,000 and 60,000 [51], or even 100,000, of which between 1400 and 1800 are marine planktonic [50]. These magnitudes indicate a broad uncertainty and a sizeable genetic repertoire [52], as well as a great capacity for adaptation and evolution. The fossil record indicates that the global diversity of diatoms increased in the Cenozoic period (65 million years–today), especially in the Neogene (23.3–5.33 million years) [50]. The fossil record indicates that they evolved from the Cretaceous; fossil diatom deposits from the end of this period are known as diatomaceous earth [58].

Some authors define diatoms as soil deposits formed from fossilized algae which have settled in seas and lakes and are currently obtained by mining. Diatoms can be found whole or as the remains of opalized algal cells incorporated within clay, organic, siltstone, or quartz matrices [61]. The deposits result from the accumulation of amorphous silica in oceans or freshwater bodies from the walls (frustules) of diatoms or dead algae [62]. After their death and organic decomposition, the frustules are deposited on the bottom of oceans or lakes, forming DSs [42,47,48].

Ref. [63] refers to several authors and explains that, according to Murray and Renard, the first use of the word diatomite dates back to the 19th century from deep water deposits called diatom ooze. They also refer to Conger, who described "pure diatomaceous earth" as a material that reached a concentration between 95 and 98% silica. They expand on the work of Terzaghi, who defined diatom ooze as a loose, unconsolidated sediment containing mainly diatoms. In the non-scientific colloquial literature, the term diatomaceous earth is used as a name for ground diatomite and diatom ooze, creating ambiguity in the use of diatomaceous earth as a definition. For this reason, Ref. [63] exclude "diatomaceous earth" from a proposed new classification system, as the term should be intended for use in the scientific literature and not in the public domain.

Unconsolidated sediments containing diatom frustules in a proportion more significant than 80% of the total weight shall be termed diatomaceous ooze. Sediments with fractions between 50% and 80% of frustules, whose remaining component is clay and silt, will be termed clayey or silty diatomaceous ooze. When the fraction of frustules is less than

of lakes, the proliferation of diatoms, and the formation of diatomites. Thirteen diatomite occurrences have been identified: eight in the municipality of Siachoque, three in Tunja, and two in Chivatá [64].

On a regional scale in Mexico, the most important deposits are located in the state of Jalisco, where 99% of the exploitation occurs. The main identified uses of diatomite are as a filter, decolorizer, insulator, absorbent, fertilizer, insecticide, filler material, and source of silica [58]. As a filter, it is applied to clarify and purify liquids in chemical, metallurgical, food, pharmaceutical, beverage, and petroleum processes. Due to its porous structure and inert chemical composition, diatomite retains dissolved, suspended, or colloidal particles from the filtered liquid without modifying the physical-chemical characteristics of the final product [64].

The presence of frustules gives diatomites different physical properties (low density and large specific surface area) from those associated with claystones or sandstones. In addition, the mechanisms through which these properties are imposed have yet to be fully understood, particularly from a microscopic perspective. Existing research has provided a partial understanding of the behavior of diatomites, establishing that their index properties are less important than their engineering properties. How the presence of frustules affects index properties has not yet been fully elucidated [14].

2.3. *Diatom Frustules*

Diatom frustules are a clear example of micro- and nano-structured materials available in nature, derived from biomineralization processes. They have four main functions in a diatom: to generate mechanical stability, to separate and filter nutrients from viruses, to regulate the rate of sinking, and to regulate the amount of light entering the cell [68]. They have multiple pores or areoles and are elongated, spherical, or disc-shaped, with variable sizes between 10 and 200 μm [29]. Frustules are amorphous silica nanospheres deposited in geometric [50] and dimensional patterns specific to each species [68].

The morphological characteristics of diatoms and their frustules should be analyzed on two scales. The macro corresponds to the shell's total shape, and the nano is related to the pore networks and the thickness of the walls. These constructive principles aim to generate self-supporting structures from the minimum silica consumption. The frustules must be able to withstand mechanical loads, not only external but also internal, due to the effects of turgor and reproductive processes [52].

The morphological and surface characteristics of the frustules, including size, pores, and stiffness, are utterly different from those of other inorganic particles whose diameters vary between 30 and 100 μm. Ref. [38] report diatom frustules extracted in Jawornik (Poland) with primarily cylindrical and centric (discoid) shapes, radii between 50 and 60 μm, and a highly porous surface. Pore sizes range from 100 to 200 nm. No traces of amorphous or dissolved silica were identified within the pores. Conventional geotechnical methods have not been proven to be sufficient to complete the characterization of DSs [69]. Ref. [14] suggest establishing a classification for diatomites based on their engineering properties and the particular biological structure of the frustules.

The effect of the frustules on the water retention properties of DS and claystones has yet to be fully recorded in the literature. The content of the fossils affects both the fabric and the microstructure of the soils containing them and hence the retention properties [17]. In kaolinitic soils artificially dosed with fossils, the suction values and geometry of the "Water Retention Curve" (WRC) present important changes as the number of fossils is increased. The WRC is restricted with the addition of diatoms; likewise, the range of mobilized suction becomes smaller [17].

Silica uptake in diatoms is mainly related to aerobic respiration processes and less so to photosynthesis [60,70]. Silica in aquatic environments is mainly called silicic acid or Si(OH)4 [71]. When acid concentrations are low, cells activate silica transporters that facilitate fixation [60,72]. The extent of the silicification of frustules depends on the rate of silica fixation, either by availability in the environment or by the effectiveness and genomic characteristics of its transporters [73,74]. Growth depends on factors such as salinity, aeration, pH, and amount of light (natural or artificial). However, there are records of species that can grow in the dark [75].

In metabolic terms, the frustules' silica can be considered an impermeable and solid excrement. However, evolutionary adaptations have led it to be considered a dominant element in the physiology and morphogenesis of cells [52]. The existence of two porous layers in the structure of diatoms helps to reduce friction between them and the ambient water and increases the ability to resist liquid pressure. Depending on the shape and species, the frustules can act as self-repairing and self-lubricating media [76].

2.4. Taphonomic Processes of Diatomaceous Soils

Taphonomy studies the processes in which organic remains pass from the biosphere to the lithosphere, considering the moment of death of an organism or the detachment of its parts. Such a change occurs through decomposition, burial, or preservation in the form of fossils or other types of biostable material. Only a tiny fraction of the organisms that inhabited the planet are preserved as fossils, while organic remains are relatively abundant. Taphonomy is essential for interpreting the meaning of the fossil record [77].

The microfossils or micro-remnants present in DSs, particularly in Japanese marine soils, are usually less than 1 mm in diameter. Of these, both plant and animal origins are reported. Their evolution is due to diagenetic changes technically defined as taphonomic processes [8].

Taphonomic processes alter diatom assemblages in sedimentary deposits depending on the climate, available space, time, and species. Silica is an element concentrated in the photic zone of most marine environments, leading to a higher productivity rate of diatoms. This element is necessary for the growth of frustules and generates ballast for subsequent sedimentation [49]. Beyond the above, very little is known about the evolution of silicification processes in diatom frustules [50]. High oxygen contents and low silica concentrations in a diatom indicate that the diatom developed in a planktonic community in deep freshwater environments. Conversely, high silica concentrations suggest that diatoms developed in benthic environments and shallow lakes [66].

The skeletons of diatoms and radiolarians are composed of amorphous silica (Opal), which is very soluble in seawater. Hence, the preservation of frustules is only possible for larger species. Less than 5% of diatom skeletons reach the seafloor without being dissolved. Larger proportions could accumulate if sedimentary basins were shallow or if the sedimentation rate were higher [8].

Diatoms have not been eliminated in the last few tens of millions of years because of the multilayered porous structures of their frustules, which present unique mechanical properties, such as a high elastic modulus (22.4 GPa), which gives it restorative properties under the imposition of loads and strains [76,78].

3. Review of Progress in Geotechnical Lines

Some significant properties of DSs are their high porosity, high sorption potential [6,33,79], high initial void ratio, and low density [18,19]. High liquid limit values have been identified in DS from Mexico and Japan [32,40]. Void contents between 80 and 90% [65,80] and specific surface areas between 40 and 350 m^2/g have also been reported [32,81]. The permeability coefficients of undisturbed DS have been recorded in magnitude of 10^{-6} cm/s what implies an impervious layer [82]. However, there are few documentary sources in which physical properties are analyzed or identified as a function of frustule concentration, species, type, or state of deterioration such as "crushing" [7].

Investigations have been carried out on the shear strength of kaolin and DS mixtures evaluated in undrained conditions [12] in order to determine the influence of the diatom fossil content on the resistant properties of this type of mixture, particularly regarding the angle of internal friction ϕ. "For diatom additions between 40% and 60%, the increase is significant, 25.3% and 41%, respectively, at an overconsolidation ratio OCR = 1 and 33% and 45%, respectively, for an OCR = 2" [32].

The microscopic skeletons of algae and dead plankton in soils and sediments lead to the measurement of mechanical characteristics in magnitudes that contradict the behaviors defined in classical geotechnics [20] and this represents a lack of experience in their engineering properties [21]. The fluid retention potential and other properties are a function of the geometry and arrangement of the frustules and pores, which are dependent on the species and its homogeneity (single or multispecies condition) [25]. Diatom microfossils affect soil properties such as water content, Atterberg limits, and density because of the sizeable intraparticle porosity of the frustules [29].

Soils containing diatoms have improved connection between microstructures, but due to the presence of expansive minerals such as illite or montmorillonite they may present expansivity problems [6,33]. Similarly, they can retain high water contents inside their frustules [18,22] and simultaneously register high friction angle and shear strength values [12,18,22].

The results presented by [40] indicate significant increases in the permeability and compressibility coefficients of soils with diatom microfossil contents. Ref. [7] concluded that diatom microfossils in marine sediments contribute to increased slope stability under static and cyclic loading conditions because diatoms confer increased resistance for any loading mode. Such an increase is a product of particle entanglement and surface roughness.

Additionally, several studies have been conducted to investigate the water retention curve of diatomites [17] and the cumulative effect of these on consistency limits. These increases are attributed to the enormous water-holding capacity of the internal structure of the frustules [32,40].

The interest in microfossil deposits has reached the point of proposing modeling exercises to predict or correlate their behaviors with physical properties and mechanical responses. Ref. [67] evaluated the effects on the behavior of diatomaceous rocks under the "strain-softening" condition as a function of time. This aspect compares the results obtained in normally consolidated specimens and those obtained from the numerical simulation of undrained triaxial tests. Ref. [83] analyzed the effect of the intermediate principal stress on the strength of diatomaceous rocks under complex stress states. For this purpose, numerical simulations and triaxial tests in CU and CD conditions were compared using 3D finite difference codes.

3.1. Granulometric Distribution in Diatomaceous Soils

The size of diatomaceous fossils is one of the leading research elements, since the expected mechanical responses are initially associated with the forces governing the micro and nano worlds (capillarity forces and electrical forces of attraction and repulsion) [31]. From the size of the frustules, elements of interest are recognized, such as pore magnitude and density, effective water storage volumes, colonies or agglomerations [84], and surface irregularity patterns.

DS investigations are usually based on physical characterization tests. Most DSs are classified as silts, which, depending on their geological formation process, are accompanied by relatively small portions of sand (>0.075 mm) or clay (<2 μm). However, for the case of clay fractions, it is worth highlighting that these may be composed of fragments of frustules and not necessarily of phyllosilicates or alumina sheets (clays), such as kaolinite, illite, montmorillonite, and smectite [28].

Ref. [33] reports that, in the DSs of Japan, fossils comprise more than 50% of the observable surfaces, either with complete or fragmented pieces. Dimensions are variable, but some diatom skeletons are as long as 100 μm. This magnitude influences the pore size distribution.

The particle size distribution leads to DSs not being classified as clays but rather as silts. It is worth noting that the particle size is modified by the imposition of stresses (frustule breakage), which has an impact on the compressibility of the DSs, leading, for example, to the recording of high Cc values [18]. Figure 6 shows a granulometric comparison of DS from different localities.

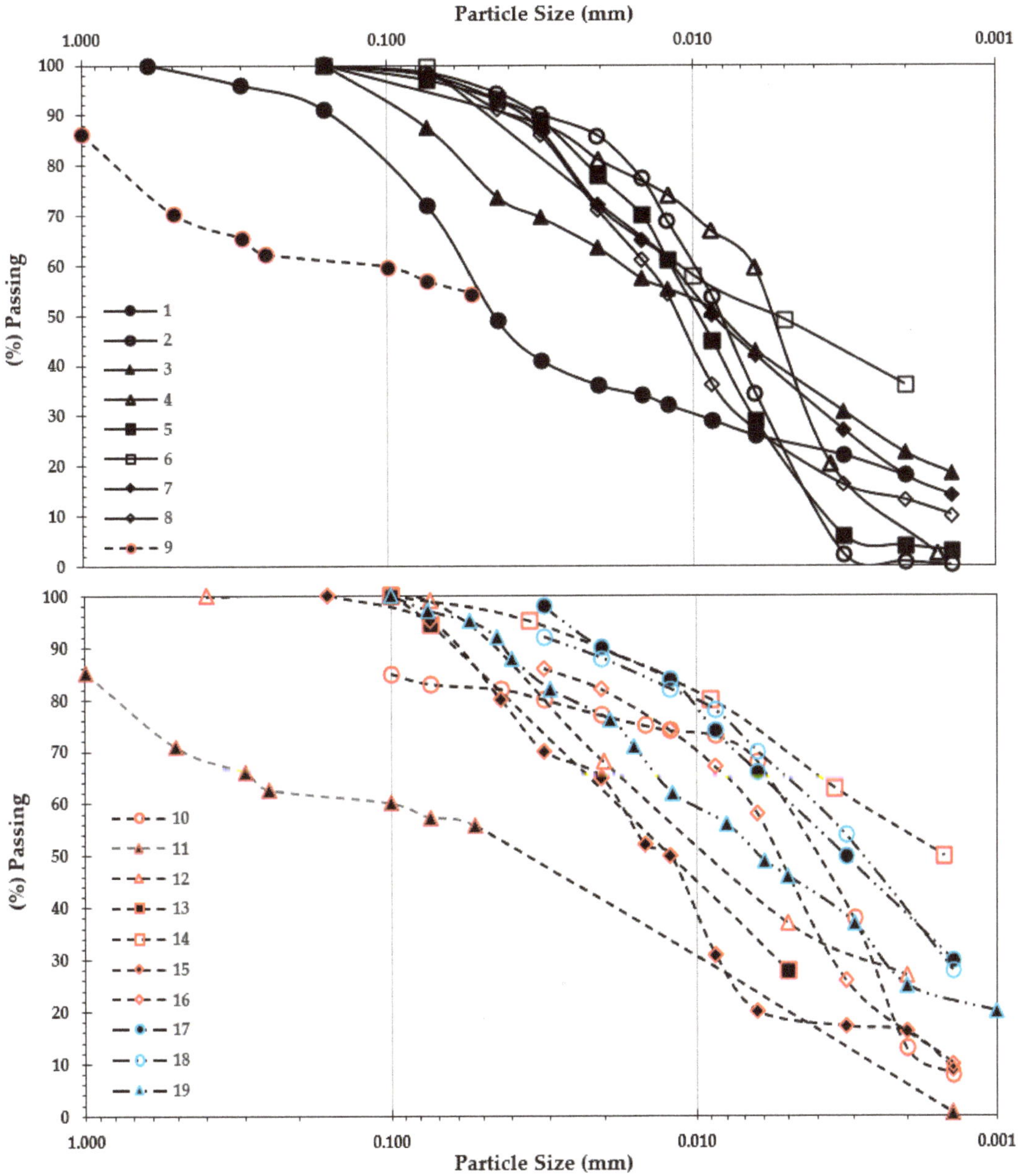

Figure 6. *Cont.*

Reference	Origin	Source
1	Chile	Arenaldi et al., 2019
2	Mexico	Caicedo et al., 2019
3	Colombia	Caicedo et al., 2019
4	USA	Zuluaga et al., 2021
5	Mexico	López, 2009
6	China	Zhang et al., 2013
7	Japan	Tanaka & Local, 1999
8	Japan	Shiwakoti et al., 2002
9	Turkey	Aksakal et al., 2012
10	USA	Palomino et al., 2011
11	China	Qu & Zhao, 2016
12	EE.UU.	Wang et al., 2021 Yazdani et al., 2021
13	Japan	Hong et al., 2006
14	Japan	Tanaka et al., 2001
15	Japan	Locat & Tanaka, 2001
16	Colombia	González, 2020
17	Colombia	González, 2020
18	Colombia	González, 2020
19	China	Xu et al., 2022

Figure 6. Background on particle size distribution in SD. Ref. [18] (Arenaldi et al., 2019), Ref. [12] (Caicedo et al., 2019), Ref. [25] (Zuluaga et al., 2021), Ref. [56] (López, 2009), Ref. [6] (Zhang et al., 2013), Ref. [33] (Tanaka & Local, 1999), Ref. [40] (Shiwakoti et al., 2002), Ref. [65] (Aksakal et al., 2012), Ref. [20] (Palomino et al., 2011), Ref. [23] (Qu & Zhao, 2016), Refs. [5,10] Wang et al., 2021 Yazdani et al., 2021, Ref. [24] (Hong et al., 2006), Ref. [8] (Locat & Tanaka, 2001), Ref. [85] (González, 2020), Ref. [14] (Xu et al., 2022).

3.2. *Specific Gravity of Solids (Gs) in Diatomaceous Soils*

In artificial kaolin–DS mixtures, it has been concluded that the Gs decreases as the content of frustules increases, given the high porosity of the diatomaceous skeletons and the volume they occupy, which is considerably larger than that of kaolin [56]. This trend is present in DSs of different origins, whether single or multi-species [25]; see Figure 7. Other references of Gs in DS are presented in Table 1.

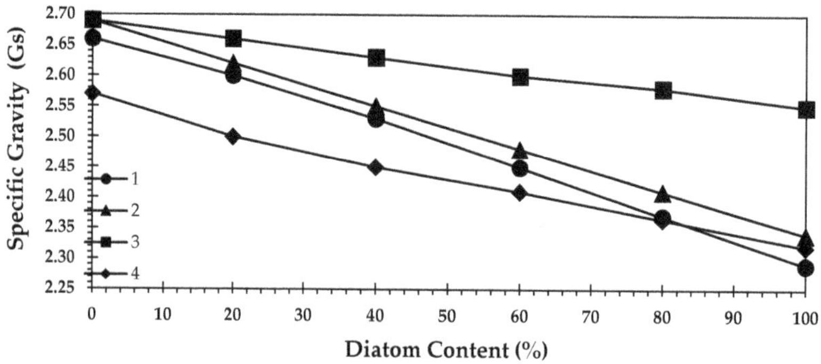

Reference	Origin	Type	Source
1	USA	Multispecies	Zuluaga et al., 2021
2	Mexico	Multispecies	Zuluaga et al., 2021
3	Colombia	Multispecies	Zuluaga et al., 2021
4	Mexico	Multispecies	López, 2009

Figure 7. Variation in the specific gravity of fine soils as a function of diatom content. Ref. [25] (Zuluaga et al., 2021), Ref. [56] (López, 2009).

Table 1. Specific gravity values in SD according to origin and type.

Origin	Type	Gs	Source
Mexico	Not defined	2.72–2.80	[41]
Chile	Centric monospecies	2.63	[22]
Indonesia	Not defined	1.87–2.00	[55]
Japan	Not defined	2.40–2.70	[56]
Mexico	Centric monospecies	2.32–2.35	[56]
México	Not defined	2.6	[79]
USA	Monospecies	2.08	[20]
USA	Multispecies	2.29	[25]
Mexico	Monospecies	2.34	[25]
Colombia	Monospecies	2.55	[25]
USA	Multispecies	2.20–2.57	[5,10]
Japan (Ishikawa)	Not defined	2.183	[86]
Japan	Centric monospecies—Cyclotella	2.13	[24]
Not defined	Diatomaceous Lutite	2.183	[67]
Colombia (Dosquebradas)	Monospecies	2.51	[85]
Colombia (Siachoque)	Monospecies	2.45	[85]
Colombia (Chivatá)	Monospecies	2.43	[85]
USA (Oregon Chiloquin and La Pine)	Not defined	2.2	[15]

Adapted from [25].

3.3. Consistency and Plasticity in Diatomaceous Soils

The high liquid limit (LL) values of soils or soil mixtures containing diatomaceous microfossils can be explained by the large intraparticle pores of the frustules, in which a large amount of water is stored. This intraparticle fluid is in addition to the pore space and does not affect particle–particle interactions [29]. In addition to the dosage, other properties affect index geotechnical parameters, such as the shape of the frustule, its porosity, and its preservation (whole or broken particles). Additionally, in the case of artificial mixtures, the mineralogy of the matrix, which can be kaolin type, will be influential [15].

Figures 8 and 9 show an increase in the consistency limits of different DSs as the fossil content increases, an aspect that has also been highlighted by [15]. These effects are particularly evident for the LL [19], in which similar positive slopes are observed for DSs of different origins. The Osaka sedimentary deposits in Japan [8] show a similar increase in their consistency limits, proportional to the microfossil content, while the plasticity index (PI) remains more or less constant [15]. Thus, it can be inferred that the index properties of DS samples alter as a function of the microfossil concentration. This aspect is due to the geometry and porosity of the frustules, which enhance the amount of water stored in the soil [8,17,18].

The high LL values can be explained by the frustules' high storage and absorption potential [6] on two scales. The first is in the total volume generated by the fossil (micro—dependent on the shape), while the second is in the volumes of the distinctive pores of the frustules (nano—dependent on the biological processes of the species) [43]. It is then posited that, although diatomaceous soil reports a high absorption level, water is encapsulated and does not affect the connectivity between the frustules or their fragments.

Figure 10 shows the classification of some soils using the plasticity chart according to the concentration of fossils (by weight). In some Mexican and Colombian monospecies samples [12], the increase in the PI as a function of LL can be observed under a condition parallel to A-line. Conversely, almost perpendicular behavior concerning the A-line is evident in the monospecies samples from Japan [40] and Mexico [56]. The North American multispecies sample [25], although not entirely parallel to the A-line, retains an upward proportional behavior between PI and LL.

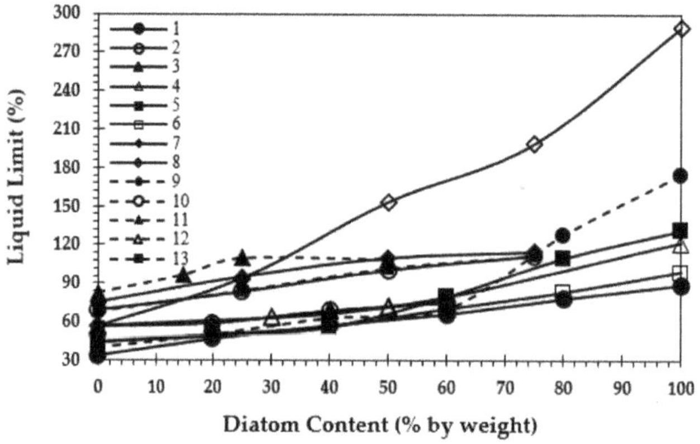

Reference	Origin	Reference
1	USA	Zuluaga et al., 2021
2	Japan	Shiwakoti et al., 2002
3	Mexico	López, 2009
4	USA	Palomino et al., 2011
5	Mexico	Caicedo et al., 2019
6	Colombia	Caicedo et al., 2019
7	Japan	Tanaka et al., 2001
8	Germany	Wiemer et al., 2017
9	China	Xu et al., 2022
10	Mexico	Díaz, 2011
11	Japan	Shiwakoti et al., 2002
12	Mexico	Díaz & González, 2013
13	Japan	Tanaka & Local, 1999

Figure 8. Variation in the liquid limit of fine soils as a function of diatom content. Ref. [25] (Zuluaga et al., 2021), Ref. [40] (Shiwakoti et al., 2002), Ref. [56] (López, 2009), Ref. [20] (Palomino et al., 2011), Ref. [12] (Caicedo et al., 2019), Ref. [87] (Tanaka, 2001), Ref. [44] (Wiemer et al., 2017), Ref. [14] (Xu et al., 2022), Ref. [32] (Díaz, 2011), Ref. [79] (Díaz & González, 2013), Ref. [33] (Tanaka & Local, 1999).

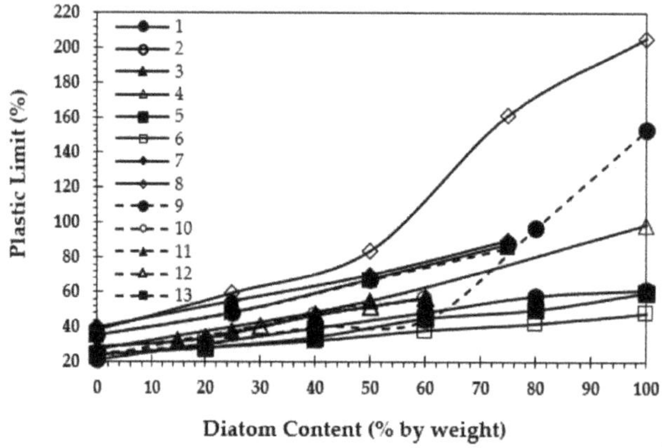

Reference	Origin	Reference
1	USA	Zuluaga et al., 2021
2	Japan	Shiwakoti et al., 2002
3	Mexico	López, 2009
4	USA	Palomino et al., 2011
5	Mexico	Caicedo et al., 2019
6	Colombia	Caicedo et al., 2019
7	Japan	Tanaka et al., 2001
8	Germany	Wiemer et al., 2017
9	China	Xu et al., 2022
10	Mexico	Díaz, 2011
11	Japan	Shiwakoti et al., 2002
12	Mexico	Díaz & González, 2013
13	Japan	Tanaka & Local, 1999

Figure 9. Variation in the plastic limit of fine soils as a function of diatom content. Ref. [25] (Zuluaga et al., 2021), Ref. [40] (Shiwakoti et al., 2002), Ref. [56] (López, 2009), Ref. [20] (Palomino et al., 2011), Ref. [12] (Caicedo et al., 2019), Ref. [87] (Tanaka, 2001), Ref. [44] (Wiemer et al., 2017), Ref. [14] (Xu et al., 2022), Ref. [32] (Díaz, 2011), Ref. [79] (Díaz & González, 2013), Ref. [33] (Tanaka & Local, 1999).

In classical soil mechanics, a higher value of PI in the soil is associated with lower shear strength. However, this phenomenon is the opposite in DSs; due to the record of increments of PI because of the water storage potential, they also tend to register a higher value of friction angle as the fossil content increases [18].

Ref. [14] proposes that the Casagrande Chart is suitable for classifying kaolin-type or diatomaceous-type soils when these have a total concentration—i.e., 100% for either of them. However, they clarify that, for mixtures of these two inputs, such as 20% diatomaceous and 80% kaolin, classification as silt below the "A-line" is inappropriate. These same authors [14] disagree with some values of high plasticity indexes reported for artificial mixtures (diatomite–kaolin) with diatomaceous earth contents higher than 60% due to the more meaningful presence of frustules. However, this leads to a greater water storage capacity, which does not necessarily imply an increase in plasticity. The increases in LL and PL are almost the same; therefore, there should be no variation in PI.

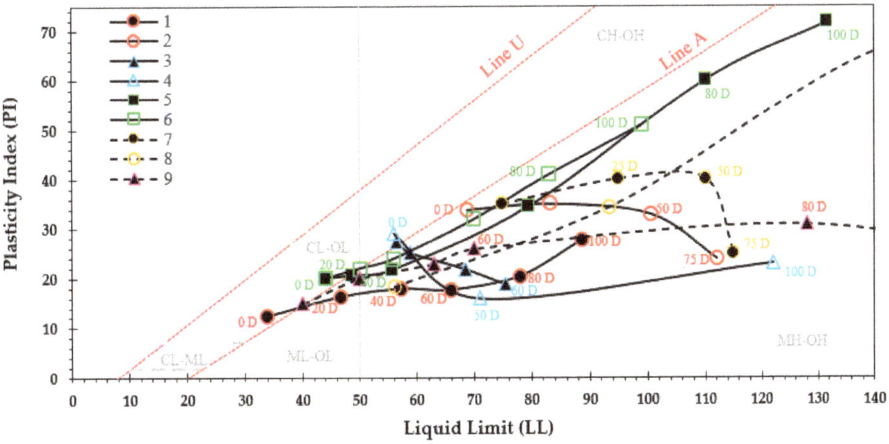

Reference	Origin	Source
1	USA	Zuluaga et al., 2021
2	Japan	Shiwakoti et. al., 2002
3	Mexico	López, 2009
4	USA	Palomino et al., 2011
5	Mexico	Caicedo et al., 2019
6	Colombia	Caicedo et al., 2019
7	Japan	Tanaka et al., 2001
8	Germany	Wiemer et al., 2017
9	China	Xu et al., 2022

Figure 10. Classification of diatomaceous soils of multiple origin according to the plasticity chart. Ref. [25] (Zuluaga et al., 2021), Ref. [40] (Shiwakoti et al., 2002), Ref. [56] (López, 2009), Ref. [20] (Palomino et al., 2011), Ref. [12] (Caicedo et al., 2019), Ref. [87] (Tanaka, 2001), Ref. [44] (Wiemer et al., 2017), Ref. [14] (Xu et al., 2022).

Figure 11 presents the LL vs. PI ratio in DS of multiple origins. The values show the classification below the graphical limits of A-line, which classifies them mostly as high plasticity silts (MH). This representation is given for soils with diatom contents equal to or greater than 50%. From Figure 12 and for the PL vs. PI ratio, no pattern of variability is identified. However, PL values ranging between 30% and 100% are mainly associated with PI records lower than 80%.

The water-holding capacity of diatoms considerably influences the index properties, which are calculated from the total amount of water in the soil. However, Ref. [33] states that the water resting inside the intraskeletal pores has little influence on the index properties—i.e., any relationship based on water proportion or void content for soils that contain microfossils should be redefined in order to avoid calculation and classification errors.

Water held within skeletal and intraskeletal pores is retained by suction. With increasing levels of stress, microfossils will fracture, and thus the number of intraskeletal pores will be drastically reduced. At this point, the overall effects of the index properties will have vanished [33]. If small changes in diatom geometry are evident at a given geostatic stress, the amount of water retained in the frustules will be nearly constant. Representative variation will be evident with depth [8].

Determining the liquid and plastic limits in DS through the traditional Casagrande methods presents certain complications, given the non-plasticity of this type of deposit. This parameter has been calculated indirectly through the relationship between the cone fall method and the flow rate [12,14].

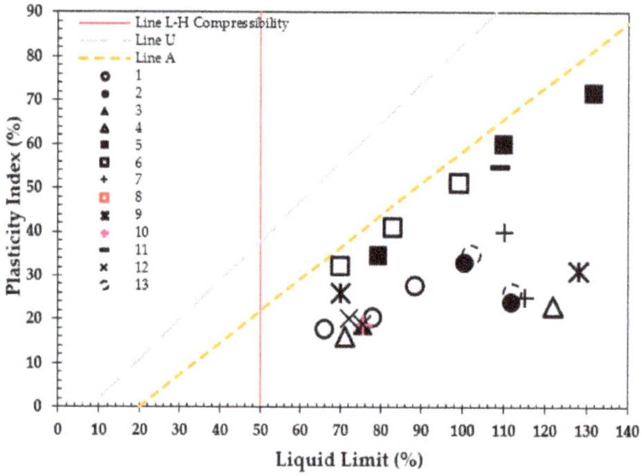

Reference	Origin	Reference
1	USA	Zuluaga et al., 2021
2	Japan	Shiwakoti et al., 2002
3	Mexico	López, 2009
4	USA	Palomino et al., 2011
5	Mexico	Caicedo et al., 2019
6	Colombia	Caicedo et al., 2019
7	Japan	Tanaka et al., 2001
8	Germany	Wiemer et al., 2017
9	China	Xu et al., 2022
10	Mexico	Díaz, 2011
11	Japan	Shiwakoti et al., 2002
12	Mexico	Díaz & González, 2013
13	Japan	Tanaka & Local, 1999

Figure 11. Relationship between LL and plasticity in DSs of different origin—concentrations higher than 50%. Ref. [25] (Zuluaga et al., 2021), Ref. [40] (Shiwakoti et al., 2002), Ref. [56] (López, 2009), Ref. [20] (Palomino et al., 2011), Ref. [12] (Caicedo et al., 2019), Ref. [87] (Tanaka, 2001), Ref. [44] (Wiemer et al., 2017), Ref. [14] (Xu et al., 2022), Ref. [32] (Díaz, 2011), Ref. [79] (Díaz & González, 2013), Ref. [33] (Tanaka & Local, 1999).

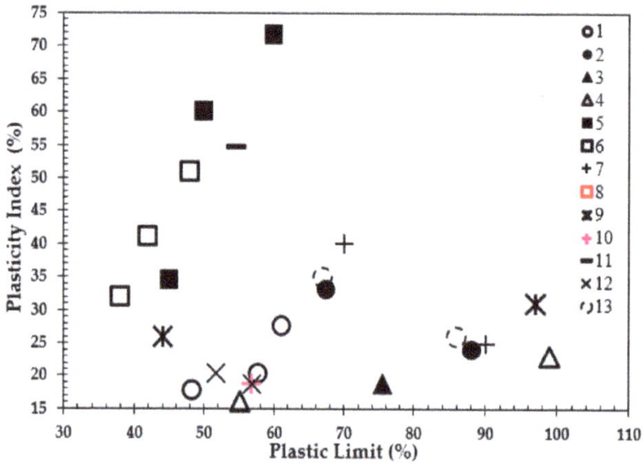

Reference	Origin	Reference
1	USA	Zuluaga et al., 2021
2	Japan	Shiwakoti et al., 2002
3	Mexico	López, 2009
4	USA	Palomino et al., 2011
5	Mexico	Caicedo et al., 2019
6	Colombia	Caicedo et al., 2019
7	Japan	Tanaka et al., 2001
8	Germany	Wiemer et al., 2017
9	China	Xu et al., 2022
10	Mexico	Díaz, 2011
11	Japan	Shiwakoti et al., 2002
12	Mexico	Díaz & González, 2013
13	Japan	Tanaka & Local, 1999

Figure 12. Relationship between plastic limit and plasticity in DSs of different origin. Ref. [25] (Zuluaga et al., 2021), Ref. [40] (Shiwakoti et al., 2002), Ref. [56] (López, 2009), Ref. [20] (Palomino et al., 2011), Ref. [12] (Caicedo et al., 2019), Ref. [87] (Tanaka, 2001), Ref. [44] (Wiemer et al., 2017), Ref. [14] (Xu et al., 2022), Ref. [32] (Díaz, 2011), Ref. [79] (Díaz & González, 2013), Ref. [33] (Tanaka & Local, 1999).

Activity in soils, defined as the ratio PI/clay content (particles less than <2 µm), can also be modified by the presence of microfossils. Such is the case for the Ariake clay soils (Japan), which have high diatom contents and report activity values of between 1.0 and 2.0, while other clays with low fossil contents—for example, Singapore and Bangkok clays—report ranges of 0.5–0.8 and 0.9–1.4, respectively. These differences can be attributed to differences in the composition and the presence of microfossils [87]. Certain diatomaceous deposits have a low bulk density and high moisture content due to the hollow structure of the frustules where water is stored, leading to altered index properties [69].

The outer pores of the frustules are connected to the interior of the diatoms and provide plenty of space for water storage. Consequently, LL and PL theoretically increase in the same proportion as the content of frustules. It is worth clarifying that for different types of diatoms, the consistency limits do not increase strictly in parallel [14]. Fractured diatom frustules are a cause of decreased PI. Likewise, this indicates that the diatom species present can also affect the limits, given the different shapes and microstructures possible; this is the why the PI in the Coscinodiscus species is higher than that associated with the Aulacoseira species.

With the increase in diatom content, the shrinkage limit also increases, and the volumetric shrinkage decreases. This increase can be explained by the effect of water absorption on the interparticle pores, which leads to reduced desiccation cracking. Shrinkage tests indicate a strong water retention capacity associated with the frustules [14]. A record of the consistency limits and plasticity index for DS of different origins is presented in Table 2.

Table 2. Consistency limits and plasticity index in DSs according to origin and type.

Origin	Sample Details	LL	PL	PI	Source
USA	Multispecies	Casagrande 114–142 Cone 83–170	Rolls Method 59–98 Cone 65–124		[5,10]
Thailand	Monospecies	56	50	6	[88]
Armenia	Sample depth 40 m	109–120	88–93	21–88	[89]
	Sample depth 76 m	120–134	89–95	32–39	
Japan	—	55–180		28–115	[87]
Japan	Centric species—Sample depth 383 m—19% clay content	117	42	75	[8]
Not defined	Diatomaceous lutite	172	94	78	[67]
Colombia (Dosquebradas)	Monospecies	140	121	19	[85]
Colombia (Siachoque)	Monospecies	118	76	42	[85]
Colombia (Chivatá)	Monospecies	132	87	45	[85]
USA, Florida	Monospecies	120	94	26	[29]
Japan (Ishikawa)	Soft rock (diatoms, clay, volcanic ash)	172.7	94.7	78	[86]
China (Changbai)	-	175	153	22	[14]
USA (Oregon Chiloquin)	Sample depth 8–33 m diatomaceous earth content 97%–71%	110–136	73–100	35–37	[15]
USA (Oregon La Pine)	Sample depth 37–46 m diatom content 47%–46% diatoms	75–95	50–55	22–45	[15]

For the lacustrine deposit near the city of Bogotá, the content of particles with sizes smaller than 2 µm fluctuates between 0% and 25% (samples taken at depths between 2 m and 25 m). However, according to the size distribution curves, this would be classified as silt. Nevertheless, when categorizing the soil from LL and PI, it would be classified as clay according to the Casagrande chart. This contradictory situation can be explained based on the presence of diatoms. Significant water contents can be harbored in their pores [11].

3.4. *Artificial Soils and Control of Variables in Diatomaceous Soils*

Kaolin is a clay in which the mineral known as kaolinite predominates; it is white, inert to chemical agents, odorless, electrically insulating, moldable, easy to extrude, resistant to high temperatures, non-toxic, and non-abrasive. Kaolinite comes mainly from the feldspars and micas of granitic rocks, developing in regions where precipitation is high and there is adequate drainage to ensure the washing of cations. Kaolinite is one of the most common clay minerals found in sedimentary and residual soils [90]. The plasticity and mechanical strength of kaolin are associated with the degree of packing of its particles and will be greater the finer its particle size distribution is. These properties are usually related to the content of the fraction smaller than 2 µm [91].

Kaolinite has a lower plasticity than other clay minerals such as illite. In contrast to montmorillonite and upon wetting, the bonds between kaolinite particles are strong. It is not susceptible to expansion and has a comparatively low deformability and high shear strength. The reasons why kaolin is more stable in terms of its chemical behavior is because it has a low specific surface area and its cation exchange capacity (CEC) is lower than that of other minerals [90]. Mineralogical studies indicate that kaolin-type soils do not present expansion, since their components present low Atterberg limits and low plasticity compared to other types of clays due to their high contents of quartz [28]. Some characterization results obtained on kaolin samples are Gs: 2.68; LL: 62.4%; PL: 28.7%; and PI: 33.7% [28]. Comparatively, Ref. [92] reports Gs: 2.65; LL: 28.2%; PL: 16.5%; and PI: 11.7%.

Kaolin is a type of clay usually used for the elaboration of artificial soil samples [14], with researchers often dosing it gravimetrically or volumetrically together with other components or independently in order to assess aspects such as its mineralogy, porosimetry, and microstructure [28]. The elaboration of artificial soils from kaolin has even been utilized to evaluate models of materials subjected to geotechnical centrifuge tests due to its versatility in preparation, handling, and consistency. In addition, its mechanical properties mean that its response to load stresses does not differ from the behavior recorded in natural samples [92,93].

Although kaolin-based artificial soils are not a perfect substitute for diatomite, their use has made it possible to identify the impact of each component of the mixture on its physical, chemical, and mechanical properties.

Changes in the geotechnical characteristics of the soils due to the presence of diatom microfossils were examined in [40] using artificially prepared mixtures. The clay used for the mixtures was kaolin, which is commercially available in Japan and has been used as a model for making mixtures with diatoms and crushed sand for comparative purposes. Mechanical mixing considered the application of distilled water. The mixing proportions used for of diatoms in the total mixture were 0, 25, 50, 50, 75, and 100%. The results indicated that the presence of microfossils alters the soils' index properties and other engineering properties. Refs. [33,87] show that the marine clays of Japan contain significant proportions of diatoms. The effects of these fossils have been assessed through tests carried out on mixtures of diatomites and kaolin [40]. One of the conclusions of these studies was that incorporating diatoms into kaolin matrices or clays such as Singapore clay leads to a reduction in soil density and a significant increase in ϕ' values [87]. Similarly, in artificial mixtures of kaolin and diatomite dosed by weight (0, 25, 50, 50, 75, and 100%), the authors of [8] managed to establish that LL and PL increase almost in the same proportion as the diatomite content increases.

In one study, oedometric tests were conducted on pre-consolidated specimens of artificial kaolin–diatomite samples dosed by weight (40%) and on samples conformed only by fractured diatomite [29]. The results indicated that vertical strains derived from mechanisms other than primary consolidation in soils with diatomaceous microfossils depend on the level of effective vertical stress applied. In order to evaluate the influence of diatomaceous microfossils on the cyclic behavior of the soil, the researchers in [56] performed a series of pre-cyclic monotonic, cyclic, and post-cyclic monotonic tests on artificial kaolin–diatomite mixtures under simple shear conditions at a constant volume. The proportions of diatomite used were 0, 20, 40, 60, and 100% by dry weight. The tests were carried out under normally consolidated (OCR = 1) and preconsolidated (OCR = 2) conditions. The percentage of water added to the mixtures was equal to their LL. The results indicated that the cyclic behavior of the kaolin–diatomite mixtures depends on the stress ratio used, the level of strain reached, the OCR, and the diatomite content.

The kaolin used by [56] is commercially distributed in Mexico; its kaolinite content was determined from X-ray diffraction and reported a value of 97%. The Gs and specific surface area were 2.57 and 22 m^2/g, respectively. About 65% of the particles that compose it are smaller than 2 μm in size, and its pH value is close to 5.8. This kaolin has an LL of 56% and a PI of 28% [56].

Ref. [32] evaluated artificial diatomite–kaolinite mixtures by simple shear under undrained conditions to determine the fossils' influence on their strength. The mixtures had proportions of dry weights of diatomite of 0, 20, 40, 60, and 100%. It was concluded that the strain needed to reach the maximum shear stress and friction angle and the Atterberg limit increases as the diatomite content becomes higher. The shrinkage of the mixtures and the PI decreases as the content of frustules increases.

Variations in fabric, chemical stability, water holding capacity, and silica dissolution in mixtures of kaolin and diatomaceous earth were investigated by [20]. The clay used in this research was untreated kaolin washed with sodium chloride solutions to change the surface electrical charge. The kaolin particles were dish-shaped and smaller than the diatomaceous earth particles. The Gs of kaolin was 2.6, and its specific surface area was calculated to be 40.4 (m^2/g) [20]. While the diatoms presented internal porosity, the kaolin particles did not register any opening. It was concluded that the presence of diatoms significantly impacts the general behavior of soil mixtures due to the variation in the concentration of ionic pore fluid and the conformation of a more rigid skeleton that reduces the tendency of the soil to strain. The presence of diatoms in the mixture caused a decrease in coagulation in the presence of salt, increased LL and PL, increased water holding capacity, and reduced solubility in electrolytic solutions [20].

Experimental programs based on oedometric tests were developed by [79] using artificially prepared mixtures of diatoms and kaolin in order to evaluate the influence of microfossils on soil compressibility. The results showed that the compressibility of kaolin increases with the addition of diatomite, and, consequently, the Cc values for mixtures with diatomite are higher than those obtained for pure kaolin. In this investigation, dry kaolinite powder and diatomite were mixed gravimetrically then homogenized, and distilled water was added until the LL was reached. No segregation was observed during the process.

An experimental program was developed to determine the influence of microalgae on kaolinitic soils in one study [17]. Artificial samples with different contents of fossilized microalgae were prepared. The clayey base material used for the investigation was kaolin; some of the characteristics of this were LL: 63%; PL: 30%; and Gs: 2.65. The proportions of fossils were 5, 10, 20, and 40% by weight. It was concluded that the hydraulic properties of kaolin increased as the proportion of micro-fossilized skeleton related to higher diatom content increased.

In order to identify the relationship between the microscopic and macroscopic characteristics of DSs, the authors of [12] prepared artificial mixtures with two species of diatoms, kaolin and drinking water. Some of the characteristics associated with the kaolin used were density: 2.69 g/cm^3; LL: 44%; PL: 24%; fraction <2 μm: 29.1%; and d50: 10.5 μm. From microscopic observations, it was determined that the kaolin platelets were not small enough to fill the nanopores characteristic of diatom frustules. Based on the mixtures, oedometric and triaxial chamber shear strength tests were developed. This study demonstrated the effects of the microscopic forms of the frustules present on the macroscopic responses of the soils.

For the diatom species Coscinodiscus (discoid shape), an increase in the friction angle was reported, reaching values up to 51° for a fossil content of 60%. Meanwhile, for the same proportion of diatoms for the species Aulacoseira (cylindrical shapes), the friction angle reached was 34°. The differences in the friction angle of these mixtures have a clear link with the shear strength of the pure diatoms, evaluated through their angle of repose. The cylindrical shape of Aulacoseira reduces the interaction between the frustules and allows them to roll, while the high angularity and interlocking of the Coscinodiscus increase the friction between particles [12].

The influence of two diatom species and their contents on the peak and residual shear strength in artificial mixtures of kaolin and DS were investigated in one study [37]. The kaolin used was untreated and commercially available in Bogota (Colombia). A Gs of 2.69 was reported. According to scanning electron microscope images, kaolin is composed of overlapping laminar groups and does not feature pores in its internal structure. The applied distilled water content corresponded to LL. It was concluded that both diatom species increased the peak and residual strength of the soil, showing behaviors similar to those seen in dense sand. Ref. [10] refers to the studies of [33,40], in which soil mixtures were prepared with a proportion of 20% diatom particles and 80% kaolin. This dosage led to LL and PL values higher than those reported in 100% kaolin soil being obtained. It is important to clarify that the ability to replicate in situ conditions using artificial samples and to develop undrained shear strength and one-dimensional compression tests is diminished at higher diatom contents [44].

Research conducted using diatomaceous soils and kaolin-type clay matrices, added by weight or volume, has been conducted. In this regard, certain limitations have been identified when comparing the results obtained for volumetric ratios with those for gravimetric order. This difference is derived from the fact that diatomite particles present lower intraparticle porosities (porous surfaces and empty centers) and specific gravities than silica crystals and kaolin [15].

3.5. Compressibility in Diatomaceous Soils

Significant magnitudes varying between 80 and 90% have been reported regarding the void content present in DS or diatomites [65,80]. Diatomaceous deposits are composed of silica frustules with rough surfaces and harbor pores, which confer a low soil density and high sorption potential [18]. Some compressibility studies on DS have been carried out in undisturbed samples recovered from deposits [18] or artificial mixtures [5,23] that generally feature a fine clayey matrix of kaolin or bentonite type [41], in which the content of frustules can be controlled [12,32].

DSs exhibit a higher compressibility range than fine soils with a similar geotechnical classification [18]. DSs have large intraskeletal porosities and perfect skeletal pore networks. As consolidation develops, the clay matrix accompanying the fossils reorganizes around the fossils, almost as if an arching phenomenon is applied [8]. The presence of fossils within a saturated, fine-grained matrix leads to a higher soil compressibility. However, the strain mechanisms involved still need to be better understood [29].

Oedometric tests conducted in DS have shown values of the Cc/Cs ratio (normal compression index/recovery index) greater than 30, a considerable magnitude, as fine soils record values ranging from 3 to 7 [18]. The Cc values obtained in DS in undisturbed conditions (Cc: 4–8) or even remolded conditions (Cc: 2.1) are higher than those of fine clayey soil with an acceptable configuration and fabric (Cc: 0.44–1.78 Louisville Clay—Canada); this is due to the breakdown of the diatomaceous skeletons and rearrangement [40].

With the ample capacity of microfossils to harbor water within their rigid structures, it is not unexpected that these soils might report high compression index values, even at considerable depths. Diatomaceous oozes have been studied in the North Sea and found to have Cc values as high as 5.0 [8].

In artificial samples made up of 100% diatomaceous earth mined in Luneburg Heath (Germany), it was observed that the compression index (Cc) and the secondary compression index (Ca) increase along with the diatom content and vertical stress. This soil reports a constant increase in Cc, reaching values of 2.9 for vertical stresses of 1600 kPa [44].

Differential behaviors have been recorded for submarine sediment samples classified as diatomaceous ooze extracted from the South Sandwich Trench (British Territory located at the southern end of the Atlantic Ocean bordering Argentina to the west), in which peak values of Cc >6 have been reported for vertical stresses near 20 kPa, followed by a reduction to 3.5, a magnitude associated with vertical stresses of 400 kPa. Ca values of approximately 0.065 have been recorded in 100% diatomaceous earth samples for vertical stresses of 1600 kPa. For diatomaceous oozes, the Ca value has been found to be close to 0.1 for vertical stresses of 200 kPa [44].

For soil samples obtained at a depth of 250 m from around Kansai Airport (Japan), abundant in diatoms, Cc values of 4.7 have been obtained. In this case, an alternation between normally consolidated (NC) and overconsolidated (OC) layers was identified. The observed behavior found by the authors [8] indicated that the OC layers presented a higher or stronger content of microfossils. In contrast, the NC layers reported either a low concentration or a high deterioration of frustules, which was generated once the yield stress was overcome due to the effect of deposition. Therefore, the Cc values expected for soils in the OC condition are higher than those that might be expected for the NC condition.

Marine sediments with diatom microfossils are systematically associated with a decrease in consolidation state, going from "highly overconsolidated" in the upper 100 m below sea level to the "normally consolidated" condition below that depth. It is proposed that the wide permeability at low levels of vertical stress allows for rapid drainage and the dissipation of excess pore pressure, which translates into a high degree of apparent overconsolidation. To this must be added the effects of particle locking and cementation, which also contribute to the overconsolidation condition described. Low levels of consolidation below 100 m are related to depositional sequences that have never experienced a dry condition or perceived a state of effective vertical stress. Situations like this generate contradictions between geologic evidence and laboratory data [44].

Microscopic observations carried out after compressional processes show high contents of frustule destruction, a phenomenon that could explain the wide ranges of compressibility recorded [18,40].

Figure 13 presents the compressibility contrast (log σv vs. e) of a series of DSs of different origins. For most species, the void ratio (e) variation ranges from 0.5 to 3.7. Two particular cases have been found for samples of Japanese origin (Centric monospecies) [40] and German origin (Remolded multispecies 100% DS) [44], which record initial values of e greater than 6.5. The North American multispecies report a maximum e of 1.9 for a concentration of 100% DS. For the case of the Colombian and Mexican monospecies samples with 60% DS content, maximum values of 1.87 and 3.67 were obtained, respectively [25].

In DS or samples with a high presence of microfossils, axial strains are simultaneously associated with pore pressure dissipation and particle breakage. However, it is challenging to quantify the individual contribution of each mechanism. Suppose that pore pressure within the sample is not measured in an oedometric test. In that case, it is not feasible to determine whether a specific part of the compression is associated with primary consolidation. In DS, particle breakage is expected to be enhanced when a particular level of stress is exceeded, which will depend on the stress history and diatom content. Cc values increase at higher fossil contents, and Cs values are relatively small [29].

The presence of diatom frustules in a soil mass can derive in delayed compressibility (no volume reduction since the skeletons assume the stresses, restricting strains) or in a sudden increase in compressibility once the yield stress of the microfossils is exceeded [69].

From the results of characterization and compressibility in diatomaceous soils in Bogota, it was possible to identify that high values of compressibility coefficient and void ratio are related to high diatomaceous contents and high LL. In this case, the ratio between the compressibility coefficient and the recompression coefficient is 7.6 [11].

For the case of Oita diatomites (Japan), the pore size distribution of unaltered and unconsolidated samples is practically identical to that recorded in hydrostatically consolidated samples with stresses of 500 kPa. This characteristic indicates that the structure remains almost invariant in the pre-yield state (<2100 kPa), explaining why the strain and strength of this highly structured soil are independent of the stress level when in the pre-yield state [24].

3.6. Shear Strength of Diatomaceous Soils

DS can reach high shear strengths and large yield surfaces [18]. This is understood to occur because of the phenomenon called "interlocking" [5] and the high frictional component in the contact zones between particles that present rough surfaces and a high silica composition [35,40]. The high ϕ' values of Japanese marine clays are due to the composition of clay minerals and the existence of granular materials, that include abundant diatom microfossils [87]. The variation in the arrangement of soil particles and the presence of microfossil fragments generate differences in the behavior of stress paths [87]. The frictional behavior of some diatomaceous deposits is mainly influenced by the size and shape of their particles [69].

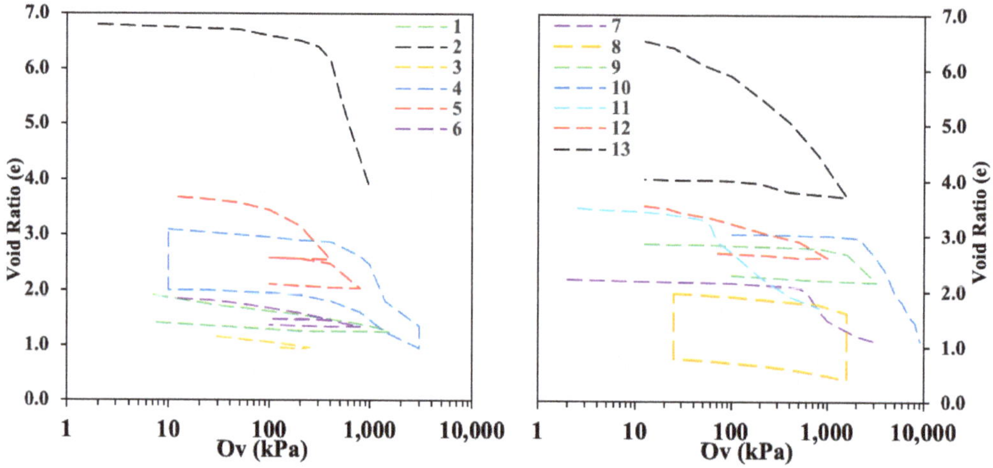

Reference	Origin	Observation	Source
1	USA	Multispecies _ concentration 100% diatoms	Zuluaga et al., 2021
2	Japan	Centric monospecies _ undisturbed sample	Shiwakoti et al., 2002
3	Mexico	Centric monospecies _ concentration 100% diatoms	López, 2009
4	Chile	Centric monospecies	Arenaldi et al., 2019
5	Mexico	Monospecies Coscinodiscus Centralis _ concentration 100% diatoms	Zuluaga et al., 2021
6	Colombia	Monospecies Aulacoseira Granulata _ concentration 100% diatoms	Zuluaga et al., 2021
7	Japan	Centric monospecies _ undisturbed sample _ depth: 72 m	Tanaka & Local, 1999
8	Colombia	Undisturbed sample _ lacustrine sediment _ 20 < LL < 100	Caicedo et al., 2018
9	USA	Multispecies _ undisturbed sample	Yazdani et al., 2021 J. Wang et al., 2021
10	Japan	Monospecies _ undisturbed sample	Hong et al., 2006
11	Japan	Undisturbed sample _ depth: 10 m	Tanaka et al., 2001
12	USA	Preconsolidated sample	Al Shatnawi & Bandini, 2019
13	Germany	Multispecies _ remolded sample _ concentration 100% diatoms	Wiemer et al., 2017

Figure 13. Compressibility in SD of different origin. Adapted from [25]. Ref. [25] (Zuluaga et al., 2021), Ref. [40] (Shiwakoti et al., 2002), Ref. [56] (López, 2009), Ref. [18] (Arenaldi et al., 2019), Ref. [33] (Tanaka & Local, 1999), Ref. [19] (Caicedo et al., 2018), Ref. [5] (Yazdani et al., 2021), Ref. [10] (Wang et al., 2021), Ref. [24] (Hong et al., 2006), Ref. [87] (Tanaka et al., 2001), Ref. [29] (Shatnawi & Bandini, 2019), Ref. [44] (Wiemer et al., 2017).

An increase in diatomaceous content leads to an increase in the friction angle. In triaxial tests carried out in soils of Bogota in undrained conditions, the friction angle was recognized through the increase in the LL and organic matter content, a situation contradictory to that reported in the classical literature [11].

Classical soil mechanics leads to the assumption that the effective friction angle ϕ' of soil tends to decrease when the PI of soils is increased. However, in deposits containing diatoms, this projection is not fulfilled. Mexico City soils, although clayey, report ϕ' values of 40°, even when the PI is close to 300%. The presence of microfossils (diatoms or others) can alter the resistant parameters, particularly the friction angle [8].

According to [32], the behavior of stress–strain curves depends on the content of diatoms. From this research, some of the properties of the volcanic lacustrine soil of Mexico City may be explained. The relationship between the internal friction angle ϕ' and the diatomite content is remarkable. The values of ϕ' for normal stress of 260 kPa were found to vary between 27° and 37°. For a normal stress of 130 kPa, the values ranged between 25° and 38°.

From the evaluation of artificial mixtures of diatomite with Singapore clay and kaolin, it was possible detect an increase in ϕ' as the fossil content increased, particularly from 25%. At this dosage, the ϕ' of the mixture with kaolin rose from 24° to 34°, and for the case of Singapore clay, the change was from 22° to 30°. The above was without significant changes in the PI for concentrations lower than 75% [8].

The shear strength of DSs from Luneburg Heath, Lower Saxony, in Germany, was studied by the authors of [7] using direct shear tests under drained conditions on weight-dosed artificial samples (kaolin + DS), varying normal stresses between 200 kPa and 600 kPa. Increases in the effective friction angle ϕ' were observed going from 20° to 32° when the DS content increased from 0% to 100% for a normal stress of 400 kPa. The most significant gain in shear strength was observed for the 25% DS content.

Some investigations have assessed the undrained shear strength (Su) in DS using Vane tests. This equipment is suitable for use in saturated fine soils in soft and stiff conditions. Particle sizes distinguish DSs in the range of silts and fine sands, which leads to partial drainage during shear tests. Therefore, measurements conducted with this technique may present erroneous values for Su [44].

Ref. [6] investigated a type of DS which has been found in road projects in the Tengchong region of southwest China's Yunnan province. This soil was characterized by various methods such as sieve analysis, chemical analysis, XRD and electron microscopy analysis, shrinkage testing, uniaxial unconfined compression testing, triaxial shear testing, and direct shearing. The results revealed that "Tengchong diatomaceous clay" is an unusual deposit with both expansive clay soil and soft clay rock characteristics. Ref. [6] explains that diatoms improve the connection between microstructures and transform the mechanical properties of soil. Due to its significant number of expansive materials, Tengchong diatomaceous soil is an expansive soft rock prone to engineering problems and geological hazards.

According to [2], the presence of diatoms increases the Atterberg limits and compressibility and allows high pore and low effective confining pressures during shear tests. Ref. [2] explains that all the organic soils used in their research presented high values of ϕ' under triaxial tests in CU conditions. They also mention that the strain mechanisms of the microstructural elements can help to explain the unusual behavior of this type of soil. The theory proposes the contributions of subhorizontal planes and non-organic microstructural elements (microfossil skeletons) to the high values of friction angles. However, this study does not discriminate based on the state of preservation of the frustules nor based on their concentration, shape, or species.

With the above, Ref. [2] explains the reason behind the results obtained for Dutch organic soils when diatoms are present (extremely high effective resistance). The authors point out that these values are not as expected and that this phenomenon has yet to be fully explained. Even though the values obtained are high, they are not applied in practice.

Figure 14 exposes the results presented by [25] regarding the ϕ' of multiple DSs of varied origin as a function of concentration; the parameter increase for any species or origin is highlighted. Table 3 presents references regarding values of effective friction angle ϕ' obtained from triaxial tests. These data are complemented by other characteristics of the deposits, the artificial soils with the presence of fossils, or the tests themselves, highlighting aspects such as the type of fill, the depth of sampling, the type of triaxial test, the shear condition evaluated, and the range of applied stresses, among others. The values of ϕ' recorded in the literature fluctuate between 22° and 65°, highlighting values higher than 60° obtained in deposits with a high organic content.

Other authors [37] evaluated the shear strength of DS of two origins (Mexican and Colombian), in peak and residual conditions, by using the "Annular Shear test" or "Ring Shear test." Including both species in fine kaolin matrices increased the peak and residual strength, with these behaving similarly to dense sand. Regarding the friction angle, as the diatom content increased, the peak and residual conditions also increased. The Mexican sample at a 100% concentration reported friction angles greater than 35° in the peak condition and greater than 19° in the residual condition.

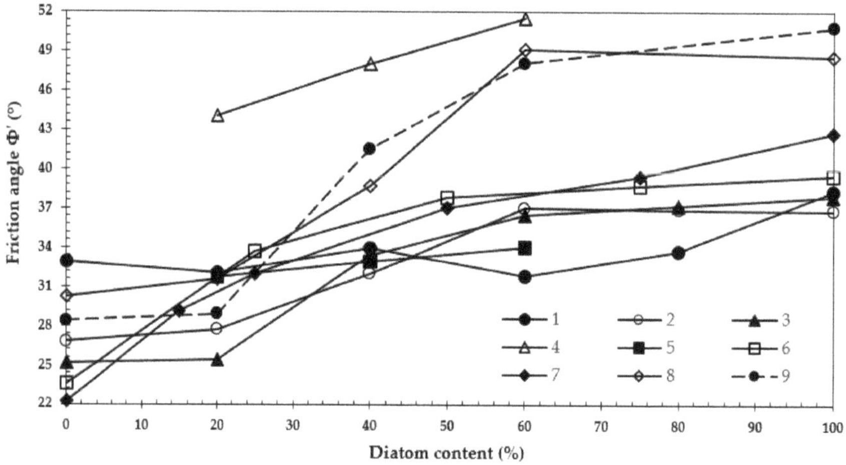

Reference	Origin	Observation	Source
1	USA	Multispecies	Zuluaga et al., 2021
2	Mexico	OCR=1	Díaz, 2011
3	Mexico	OCR=2	Díaz, 2011
4	Mexico	Coscinodiscus Centralis monospecies	Caicedo et al., 2019
5	Colombia	Aulacoseira Granulata monospecies	Caicedo et al., 2019
6	Japan	Centric monospecies + kaolin base	Shiwakoti et al., 2002
7	Japan	Centric monospecies + Singapore clay base	Shiwakoti et al., 2002
8	Mexico	Centric monospecies OCR=1	López, 2009
9	Mexico	Centric monospecies OCR=2	López, 2009

Figure 14. Friction angle records for DSs of different origin. Adapted from [25]. Ref. [25] (Zuluaga et al., 2021), Ref. [32] (Díaz, 2011), Ref. [12] (Caicedo et al., 2019), Ref. [40] (Shiwakoti et al., 2002), Ref. [56] (López, 2009).

Table 3. ϕ' values reported from triaxial tests in soils with the presence of diatoms.

Origin	Soil	Φ′	Recorded stress	Author	Comments
Monterrey California —USA	Compacted diatomaceous soil fill (1980)	44°	σ′3 13.8–27.6 kPa.	[35]	Diatoms have rough surfaces (bumps and indentations) that can increase their frictional resistance. Although the soil fill has a low dry density, the diatoms interlock, resulting in a high effective friction angle. Unconfined submerged triaxial test.
Ariake—Japan	—	a.65° b.47° c.46–57° d.45°	—	[87]	a. Peak condition—depth 3–12 m b. Peak condition—depth 12–16 m c. Residual condition—depth 3–12 m d. Residual condition—depth 12–16 m Unconsolidated undrained triaxial tests (UU)

Table 3. *Cont.*

Origin	Soil	Φ′	Recorded stress	Author	Comments
Oita—Japan	Undisturbed sample. Depth 3 m. Species Cyclotella	30.7°	σ′3 50–6500 kPa	[24]	Consolidated undrained triaxial tests. From the graphical representation between confining stress and undrained shear strength, an inflection point is identified at 1550 kPa. When the confining stresses are lower than this transition point, the undrained shear strength is independent of the confining stress and is more related to the structural strength of the diatomite. When the confining stress is greater than the transition stress, the undrained shear strength increases linearly with the confining stress.
Oostvaardersplasen Nature Reserve—Lelystad, The Netherlands.	Organic Silt	38°–56°	σ′3 19–121 kPa	[2]	Subhorizontal laminae and other non-organic microstructural elements (microfossil skeleton) are reflected in the high values of ϕ′ of organic soils. The deformation mechanisms of microstructural elements explain the unusual geotechnical properties. Despite the dispersion in index properties, organic soils exhibit extremely high values of ϕ ′ during triaxial CU compressions. These values are associated with low effective confining pressure.
	Organic Clay	38°–46°	σ′3 41–83 kPa		
	Peat	63°–65°	σ′3 40–79 kPa		
Yunnan Province—China	Undisturbed samples. Natural water content. Shear testing speed 0.08 mm/min.	22.7°–26.1°	σ′3 100–400 kPa	[6]	Shear test results show a strong microstructural connection in the Tengchong diatomaceous clay soil, which is different from that in common soft clayey rocks. It exhibits elastic deformation under confining pressures between 100 kPa and 250 kPa. However, under a pressure of 400 kPa, obvious plastic deformation is observed, indicating that the internal microstructure is destroyed. Consolidated drained triaxial test.
Bogotá—Colombia	Lacustrine Deposit	22°–47°	p′ 50–500 kPa	[19]	The friction angle increases as the liquid limit increases (linear relationship) and can reach values as high as 47°. The high ϕ ′ can be explained by the presence of diatoms. Consolidated undrained test.
Boyacá—Colombia. Quintana Roo—Mexico	Reconstituted artificial sample. (kaolin + DS). Species Aulacoseira Granulata. Coscinodiscus Centralis.	31°–34° (Col) 44°–51° (Mex)	σ′3 100–400 kPa	[12]	The differences in ϕ ′ of the soil mixtures (kaolin + DS) have a clear link with the shear strength of pure diatoms, evaluated through their angle of repose. These differences are certainly related to their shapes. The cylindrical condition of Aulacoseira (Colombian origin) reduces the interaction between particles and allows the particles to roll, while the high angularity and interlocking of Coscinodiscus (Mexican origin) increases the friction between particles. Consolidated undrained test.
Bahía de Mejillones—Chile	Undisturbed soil samples. Extracted from 3 m depth.	29°	σ′3 50–2000 kPa	[22]	The soil behaves slightly dilatant at confining pressures below 100 kPa and highly contractive at pressures above 1000 kPa. The resistance can be controlled by the microstructure of the soil, influenced by the rearrangement and breakage of diatoms, creating a transition to a frictional behavior. Soil heterogeneity due to diatom layering implies a high variability in triaxial shear tests at confining pressures above 500 kPa. Compared to conventional fine soils with similar PI, the cyclic behavior of diatomaceous soil is more brittle at low stresses, but degradation is greater at high stresses. Isotropically consolidated undrained tests.
Klamath Falls, Oregon—USA	Undisturbed samples taken at depths up to 27 m.	40°–51°	-	[10]	The Buck Creek deposit exhibited unusually high shear strength, considering the elevated porosity, plasticity, and compressibility records. This is attributed to entanglement between diatom particles. The undrained shear strength from the effective friction angle is found to be high, but lower than that calculated using SPT-type procedures. Isotropically consolidated undrained tests.

Table 3. *Cont.*

Origin	Soil	Φ′	Recorded stress	Author	Comments
Buck Creek Bridge, Bonanza, Oregon—USA	Undisturbed samples	26.5°	-	[5]	Laboratory and in situ test results revealed that the diatom deposit at Buck Creek Bridge is stiff and has high shear strength. Elastic displacements are very high for the case of closed-toe piles driving (very low resistance is observed during driving), which conflicts with the measured high stiffness and strength of the soil. Isotropically consolidated undrained triaxial test.

Under the shear ring method, Ref. [89] evaluated samples of unaltered diatomaceous clay extracted from the Shambsk sector in Armenia, applying normal pressures of 0.5, 1.0, and 1.5 kg/cm^2 and shear stresses of 0.50, 0.45, 0.40, and 0.35 kg/cm^2. The latter was kept constant for up to 200 days and in conditions close to the degree of saturation. For all cases, the angle of internal friction varied between 27° and 33°. It could be concluded that Shambsk diatomaceous clays do not show expansion potential.

The interest in the mechanical properties of DSs and diatomites, particularly their shear strength, has led to the study of variables such as the effects of intermediate principal stresses on the response of "soft rocks" (<20 MPa uniaxial compressive strength) under complex stress states (σ1 > σ2 > σ3). Thus, Ref. [83] developed a series of numerical simulations of triaxial tests in drained (CD) and undrained (CU) consolidated conditions. From the simulations of the triaxial test conducted in the (CD) condition in soft diatomaceous rocks, it was concluded that the peak (qp) and residual (qr) strengths increase by about 30% when the effect of the intermediate principal stress is considered. The resistances (qp) and (qr) are lower in the σ2 = σ1 condition with respect to the σ2 = σ3 condition.

Additionally, Ref. [83] assessed the response of a strong rock core with intermediate layers of soft diatomaceous rock by applying two-dimensional finite difference simulations. From this, it was determined that the rock pillar's strengths (qp) and (qr) are significantly influenced by the effect of the intermediate principal stress, especially when the intermediate diatomaceous rock layer approaches verticality. Tests carried out on "diatomaceous shale" samples have shown that, after the peak strength, the deviatoric stress decreases with increasing axial strain. The stress ratio n (q/p′) increases during the strain-softening process because of the excess pore pressure generated by a negative dilatancy phenomenon. The effective stress paths in undrained triaxial tests lie within the yield surface of the Cam Clay model. It is not possible to clearly distinguish the elastic behavior of plastics. It is then recognized that diatomaceous shales behave as typically overconsolidated clay.

The compressive strength of diatomaceous sediments in unconfined undrained conditions is four times lower (Su = 0.1) than the unconfined, undrained shear strength determined in triaxial tests (Su = 0.4). If a soil sample is considered close to an unconfined situation, the results obtained in the laboratory will underestimate the actual properties of this type of deposit [44].

As the content of diatom frustules increases within an artificial mixture (diatomite–kaolin), the mixture's viscosity tends to decrease drastically, since the clay particles provide less cohesion and the frustules generate a weak interaction. However, at high shear rates (rheometer), the breakage and rearrangement of the frustules increase the viscosity, since the interaction between the particles is enhanced. Samples with low diatom contents (20%) report a higher shear strength than a 100% kaolin sample. This attribute indicates that a small amount of frustules reinforces the structural cohesion provided by the clay matrix [14].

4. Discussion

Diatomaceous soils (DSs) constitute an emerging line of research; their presence in multiple places, the possibility of finding them in marine or lacustrine environments (current or desiccated), and their variable thicknesses and depths entail geotechnical evaluation that goes beyond typical characterization procedures.

The primary time frame of study has been identified since 1992. A second literary production peak has evolved since 2012. Some authors stand out as referents for consultation (Tanaka, Locat, Day, Liao), since they generated the first indications regarding the lines of recognition of the DS and their research proposals are still valid. In the current scenario, some results have deepened the understanding of this behavior, particularly in coastal deposits (Wiemer, Arenaldi).

The lines of study in DS can be categorized into two large blocks. The oldest concentrates on particle size distribution, limits of consistency, oedometric tests, and shear resistance tests performed under static conditions. The most recent line covers shear tests conducted under dynamic conditions, analyses with microscopy, and modeling. To date, many authors recognize the need to deepen research on aspects such as secondary consolidation due to the breaking of frustules, the phenomenon of particle interlocking during the application of shear stresses, the elastic properties of microfossils and their resistance before breaking (determination of the yield point), and the generation of pore pressure during construction processes, among others.

Diatom fossils are recorded in multiple types of soils, such as those with a high organic content, clayey matrices, and sandy silt media. The effective friction angles obtained in DS through triaxial tests fluctuate between 22° to 65° depending on the frustule content and other variables. The shear resistance has been evaluated in unaltered, reconstituted, and artificial samples, with some authors noting that the behavior can be different depending on the origin and treatment of the samples.

The higher the content of frustules is, the higher the number of voids will be, given the inter and intraskeletal poral structures. This situation is associated with a broad absorption or encapsulation capacity for water, which directly affects the determination of the consistency limits. The frustules present strain resistance at the micrometric level when stress imposition exists. When the frustules' yield limit is exceeded, DS presents volumetric variations. These are associated with processes such as secondary consolidation and collapse.

Understanding the DS requires recognizing the environmental, biological, taphonomic, and geological conditions surrounding the generation and deposition of the frustules. This literary review has allowed us to identify the imprecise use of terms relating to DSs, since their physical characteristics (silica content, porosities, geostatic stresses, and formation temperatures) are not considered when categorizing them. Suppose that geotechnical behavior uncertainty is added to the above. In that case, it becomes necessary to establish a technical framework for classifying fossiliferous soils independent of the one currently applied to organic and inorganic soils.

The characterization and evaluation of DS are mostly restricted to a place of exploration; that is to say, no contrast is generated with the results of other sources. Therefore, the conceptualization of fossiliferous soils is limited, considering that the fossil's shape, content, and state of conservation must affect all the study parameters.

Author Contributions: Conceptualization, D.Z.-A. and J.C.R.; Methodology, D.Z.-A. and J.C.R.; Formal analysis, D.Z.-A., J.C.R., J.C.-T., O.R.-O. and B.C.-H.; Investigation, D.Z.-A.; Resources, D.Z.-A., J.C.R. and O.R.-O.; Writing—original draft preparation, D.Z.-A., J.C.R., J.C.-T., O.R.-O. and B.C.-H.; Writing—review and editing, D.Z.-A., J.C.R., J.C.-T., O.R.-O. and B.C.-H.; Supervision, J.C.R., J.C.-T., O.R.-O. and B.C.-H.; Funding acquisition, D.Z.-A., J.C.R., J.C.-T. and O.R.-O. All authors have read and agreed to the published version of the manuscript.

Funding: This research received no external funding.

Institutional Review Board Statement: Doctoral monitoring committee, Universidad Militar Nueva Granada, Doctoral Program in Engineering. Centro de Investigaciones, Escuela de Ingenieros Militares, Ejército Nacional de Colombia.

Informed Consent Statement: Not applicable.

Data Availability Statement: Not applicable.

Conflicts of Interest: The authors declare no conflict of interest.

Appendix A

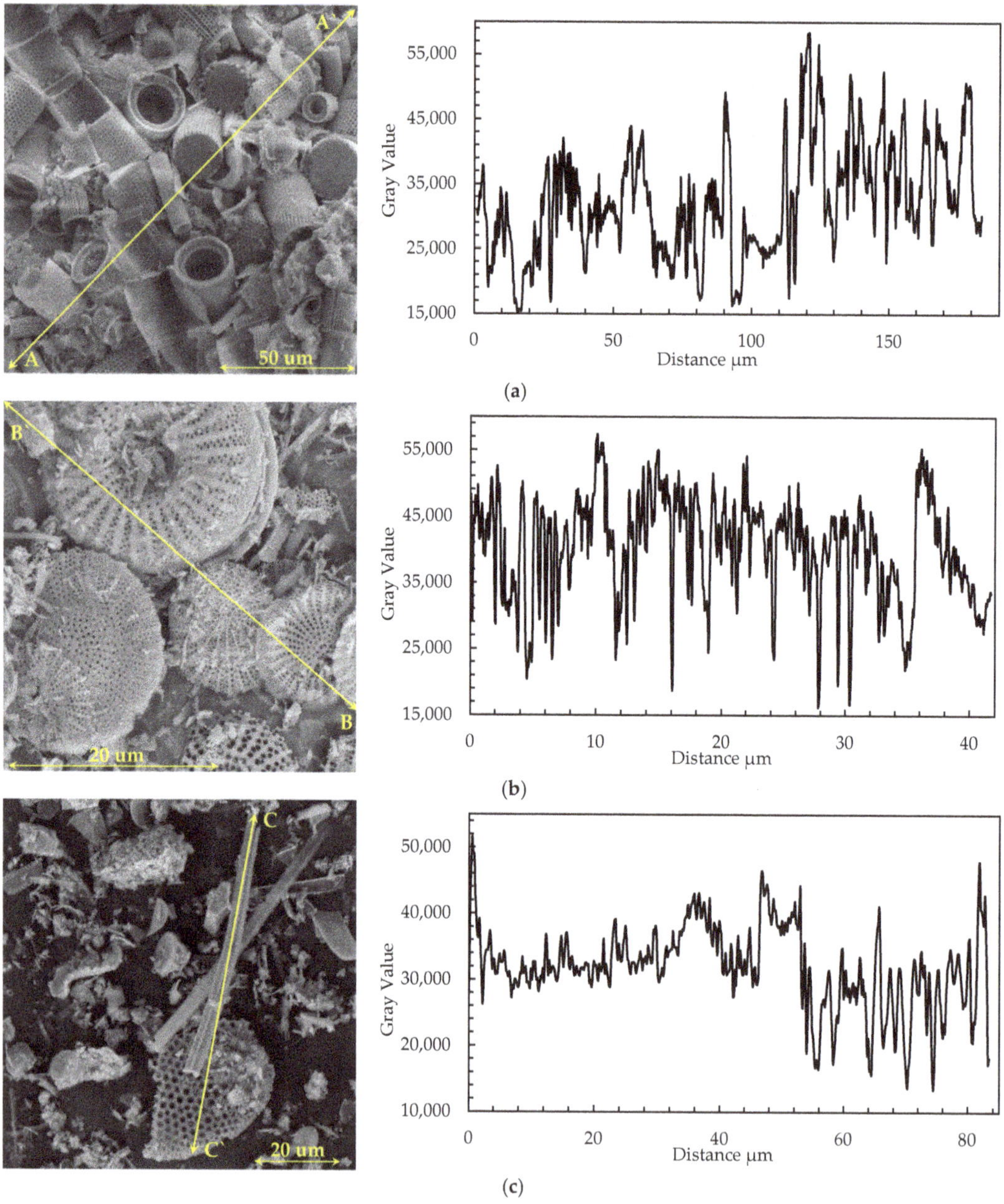

Figure A1. Diatom frustules samples and surface roughness plot. (**a**) Colombian origin soil—centric monospecies Aulacoseira granulata—average diameter 11.51 um. (**b**) Mexican origin soil—centric monospecies Cyclostephanos tholiformis—average diameter 11.44 um. (**c**) Peruvian origin soil—multispecies Thalassiothrix, Thalassionema, Coscinodiscus, Diatomella, and Tabularia. A-A′, B-B′, and C-C′ define the superior projection of the axes on top of samples for their corresponding surface roughness plot.

References

1. Zuluaga-Astudillo, D.A.; Slebi-Acevedo, C.J.; Ruge, J.C.; Olarte, M.C. Physical Modeling in Geotechnical Centrifuge of Foundations Supported on Diatomaceous Soils. *Indian Geotech. J.* **2022**, 1–10. [CrossRef]
2. Cheng, X.; Ngan-Tillard, D.; Haan, E.D. The causes of the high friction angle of Dutch organic soils. *Eng. Geol.* **2007**, *93*, 31–44. [CrossRef]
3. Sonyok, D.R. *Effect of Diatoms on Index Properties, Compressibility, Suction, and Stiffness of Diatomite-Kaolin Mixtures*; New Mexico State University: Las Cruces, NM, USA, 2015.
4. Oregon Department of Transportation. *Geotechnical Design Process, Findings and Recommendations Key # 09679 US97 @ Wickiup Jct. (La Pine) Sec project*; Oregon Department of Transportation: Bend, OR, USA, 2017.
5. Yazdani, E.; Wang, J.; Evans, T.M. Case study of a driven pile foundation in diatomaceous soil. II: Pile installation, dynamic analysis, and pore pressure generation. *J. Rock Mech. Geotech. Eng.* **2021**, *13*, 446–456. [CrossRef]
6. Zhang, Y.; Guo, C.; Yao, X.; Qu, Y.; Zhou, N. Engineering geological characterization of clayey diatomaceous earth deposits encountered in highway projects in the Tengchong region, Yunnan, China. *Eng. Geol.* **2013**, *167*, 95–104. [CrossRef]
7. Wiemer, G.; Kopf, A. Influence of diatom microfossils on sediment shear strength and slope stability. *Geochem. Geophys. Geosystems* **2017**, *18*, 333–345. [CrossRef]
8. Locat, J.; Tanaka, H. A new class of soils: Fossiliferous soils? In Proceedings of the Fifteenth International Conference on Soil Mechanics and Geotechnical Engineering, Istanbul, Turkey, 27–31 August 2001; Volumes 1–3, pp. 2295–2300.
9. Mendoza, C.; Caicedo, B.; Duque, J. Technical report on the compression, structure, and creep behaviors of lacustrine soil deposits in Bogotá, Colombia. *Soils Found.* **2022**, *62*, 101215. [CrossRef]
10. Wang, J.; Yazdani, E.; Evans, T.M. Case study of a driven pile foundation in diatomaceous soil. I: Site characterization and engineering properties. *J. Rock Mech. Geotech. Eng.* **2021**, *13*, 431–445. [CrossRef]
11. Mendoza, C.; Caicedo, B.; Lopez-Caballero, F. Geotechnical behavior of Bogotá lacustrine soil through its geological history. In Proceedings of the 17th European Conference on Soil Mechanics and Geotechnical Engineering, ECSMGE, Reykjavik, Iceland, 1–6 September 2019; pp. 1–8. [CrossRef]
12. Caicedo, B.; Zuluaga, D.; Slebi, C. Effects of micro-features of fossil diatom on the macroscopic behaviour of soils. *Géotechnique Lett.* **2019**, *9*, 322–327. [CrossRef]
13. Ural, N. The significance of scanning electron microscopy (SEM) analysis on the microstructure of improved clay: An overview. *Open Geosci.* **2021**, *13*, 197–218. [CrossRef]
14. Xu, Y.; Zhang, X.; Liu, X.; Wang, G. Alterations of physical properties and microstructure of marine diatomite owing to variation of diatom content. *Mar. Georesources Geotechnol.* **2022**, 1–12. [CrossRef]
15. Covarrubias, A.; Wang, J.; Moug, D.; Evans, M.; Walter, A. "Geo-Congress 2022 American Society of Civil Engineers", in Relating the Proportion of Diatom Particles to the Physical Properties of Natural Diatomaceous Soil. 2022, pp. 479–489. Available online: https://ascelibrary.org/doi/abs/10.1061/9780784484036.047 (accessed on 19 December 2022).
16. Matthew Evans, T.; Moug, D. Diatomaceous soils: A less than cromulent engineering material. In *Geotechnics for Sustainable Infrastructure Development*; Springer: Singapore, 2020; pp. 709–716.
17. Ruge, J.C.; Castro-Rincón, W.; Camacho-Tauta, J.F.; Molina-Gómez, F.A. Dependencia de las propiedades de retención de agua en suelos caoliníticos con contenidos de microalgas fosilizadas adicionadas artificialmente. In *Geotechnical Engineering in the XXI Century: Lessons Learned and Future Challenges*; IOS Press: Amsterdam, The Netherlands, 2019; pp. 780–787.
18. Perisic, G.A.; Ovalle, C.; Barrios, A. Compressibility and creep of a diatomaceous soil. *Eng. Geol.* **2019**, *258*, 105145. [CrossRef]
19. Caicedo, B.; Mendoza, C.; López, F.; Lizcano, A. Behavior of diatomaceous soil in lacustrine deposits of Bogotá, Colombia. *J. Rock Mech. Geotech. Eng.* **2018**, *10*, 367–379. [CrossRef]
20. Palomino, A.M.; Kim, S.; Summitt, A.; Fratta, D. Impact of diatoms on fabric and chemical stability of diatom–kaolin mixtures. *Appl. Clay Sci.* **2011**, *51*, 287–294. [CrossRef]
21. Su, Q.; Deng, Z.; Wang, X.; Jia, W.; Niu, Y. Experimental Study of the Dynamic Characteristics of a New Antidrainage Subgrade Structure for High-Speed Railways in Diatomaceous Earth Areas. *Materials.* **2022**, *15*, 532. [CrossRef]
22. Ovalle, C.; Arenaldi-Perisic, G. Mechanical behaviour of undisturbed diatomaceous soil. *Mar. Georesources Geotechnol.* **2020**, *39*, 623–630. [CrossRef]
23. Qu, J.-L.; Zhao, D.-X. Comparative research on tillable properties of diatomite-improved soils in the Yangtze River Delta region, China. *Sci. Total Environ.* **2016**, *568*, 480–488. [CrossRef]
24. Hong, Z.; Tateishi, Y.; Han, J. Experimental Study of Macro- and Microbehavior of Natural Diatomite. *J. Geotech. Geoenvironmental Eng.* **2006**, *132*, 603–610. [CrossRef]
25. A Zuluaga, D.; Sabogal, D.; A Buenaventura, C.; Slebi, C.J. Physical and mechanical behavior of fine soil according to the content of multispecies diatoms. *J. Phys. Conf. Ser.* **2021**, *2118*, 012011. [CrossRef]
26. M. de A. V. y D. T. Comisión Asesora Permanente para el Régimen de Construcciones Sismo Resistente, "Reglamento Colom-Biano de Construcción Sismo Resistente _ NSR 10 _ Título H", en Reglamento Colombiano de Construcción Sismo Resistente. 2010; p. 4. Available online: https://www.idrd.gov.co/sites/default/files/documentos/Construcciones/8titulo-h-nsr-100.pdf (accessed on 19 December 2022).
27. Igoe, D.; Gavin, K. Characterization of the Blessington sand geotechnical test site. *AIMS Geosci.* **2019**, *5*, 145–162. [CrossRef]

28. Ruge, J.C.; Palacios, X.; Molina-Gómez, F.; Suárez, J.P.R. Mineralogical, microstructural and porosimetry analysis in three different clayey soils. *J. Phys. Conf. Ser.* **2019**, *1388*, 012014. [CrossRef]
29. Shatnawi, H.H.A.; Bandini, P. Oedometric Behavior of a Diatom-Kaolin Mixture. In *Geo-Congress 2019: Geotechnical Materials, Modeling, and Testing*; American Society of Civil Engineers: Reston, VA, USA, 2019; pp. 673–681.
30. MacKillop, A.K.; Moran, K.; Jarrett, K.; Farrell, J.; Murray, D. Consolidation Properties of Equatorial Pacific Ocean Sediments and Their Relationship to Stress History and Offsets in the Leg 138 Composite Depth Sections. In Proceedings of the Integrated Ocean Drilling Program Scientific Results, International Ocean Discovery Program. 1995, Volume 138, pp. 357–369. Available online: https://scholars.unh.edu/ccom_affil/18/ (accessed on 19 December 2022). [CrossRef]
31. Santamarina, J. Energy Geotechnology Enabling New Insights Soil Behavior, GeoCongress 2014 Karl Terzaghi Lecture, GeoInstitute, Atlanta, GA, US. 2014. Available online: https://australiangeomechanics.org/meetings/2014-terzaghi-lecture-energy-geotechnology-enabling-new-insights-into-soil-behaviour/ (accessed on 19 December 2022).
32. Díaz-Rodríguez, J.A. Comportamiento monotónico de suelos diatomáceos. In Revista Investigación de Desastres Naturales, Accidentes e Infraestructura Civil; Universidad de Puerto Rico, Mayagüez, Puerto Rico. 2011, Volume 12, pp. 27–34. Available online: https://core.ac.uk/download/pdf/296528413.pdf (accessed on 19 December 2022).
33. Tanaka, H.; Locat, J. A microstructural investigation of Osaka Bay clay: The impact of microfossils on its mechanical behaviour. *Can. Geotech. J.* **1999**, *36*, 493–508. [CrossRef]
34. Holler, P.R. Consolidation characteristics and permeabilities of sediments from the Japan Sea (Sites 798 and 799). *Proc. Ocean. Drill. Program Sci. Results* **1992**, 1123–1133. [CrossRef]
35. Day, R.W. Engineering Properties of Diatomaceous Fill. *J. Geotech. Eng.* **1995**, *121*, 908–910. [CrossRef]
36. Caicedo, B.; Mendoza, C.; Lizcano, A.; Lopez-Caballero, F. Some contributions to mechanical behaviors of lacustrine deposit in Bogotá, Colombia. *J. Rock Mech. Geotech. Eng.* **2019**, *11*, 837–849. [CrossRef]
37. Slebi-Acevedo, C.; Zuluaga-Astudillo, D.; Ruge, J.; Castro-Fresno, D. Influence of the Diatomite Specie on the Peak and Residual Shear Strength of the Fine-Grained Soil. *Appl. Sci.* **2021**, *11*, 1352. [CrossRef]
38. Lutyński, M.; Sakiewicz, P.; Lutyńska, S. Characterization of Diatomaceous Earth and Halloysite Resources of Poland. *Minerals* **2019**, *9*, 670. [CrossRef]
39. Montoya-Moreno, Y.; Sala, S.; Vouilloud, A.; Aguirre, N.; Plata-Díaz, Y. Lista de las diatomeas de ambientes continentales de Colombia. *Biota Colomb.* **2013**, *14*, 13–78. [CrossRef]
40. Polidori, E. Relationship Between the Atterberg Limits and Clay Content. *Soils Found.* **2007**, *47*, 887–896. [CrossRef]
41. León-Ordoñez, L.A.; Mosquera-Falla, A.A. *Comparación del Comportamiento Físico-mecánico de las Arcillas Bentoníticas Provenientes de Bugalagrande, Valle del Cauca, Colombia con Adición de Diatomeas de la Ciudad de México, México*; Universidad Católica De Colombia: Bogotá, Colombia, 2018.
42. Arenaldi, G.; Ovalle, C.; Barrios, A. Compresibilidad y propiedades dinámicas de suelos diatomáceos de Mejillones. *Obras Proy.* **2019**, *25*, 6–14. [CrossRef]
43. Rajasekaran, G. Influence of microfossils and pyrites on the behaviour of oceanbed sediments. *Ocean Eng.* **2005**, *33*, 517–529. [CrossRef]
44. Wiemer, G.; Dziadek, R.; Kopf, A. The enigmatic consolidation of diatomaceous sediment. *Mar. Geol.* **2017**, *385*, 173–184. [CrossRef]
45. Maringue, J.; Yañez, G.; Sáez, E.; Podestá, L.; Figueroa, R.; Estay, N.P.; Lira, E. Dynamic characterization of the Mejillones Basin in northern Chile, using combined geophysical field measurements. *Eng. Geol.* **2018**, *233*, 238–254. [CrossRef]
46. Fazlutdinova, A.I.; Sukhanova, N.V. Composition of soil diatoms in zones of impact from oil production complexes. *Russ. J. Ecol.* **2014**, *45*, 188–193. [CrossRef]
47. Mejía, J.M.; de Gutiérrez, R.M.; Montes, C. Rice husk ash and spent diatomaceous earth as a source of silica to fabricate a geopolymeric binary binder. *J. Clean. Prod.* **2016**, *118*, 133–139. [CrossRef]
48. Flower, R.J. *Diatomites: Their Formation, Distribution, and Uses*, 2nd ed.; Elsevier: Amsterdam, The Netherlands, 2013.
49. Warnock, J.P.; Scherer, R.P. Diatom species abundance and morphologically-based dissolution proxies in coastal Southern Ocean assemblages. *Cont. Shelf Res.* **2015**, *102*, 1–8. [CrossRef]
50. Finkel, Z.; Kotrc, B. Silica Use Through Time: Macroevolutionary Change in the Morphology of the Diatom Fustule. *Geomicrobiol. J.* **2010**, *27*, 596–608. [CrossRef]
51. Gebeshuber, I.C. Biotribology inspires new technologies. *Nano Today* **2007**, *2*, 30–37. [CrossRef]
52. Sterrenburg, F.A.; Gordon, R.; Tiffany, M.A.; Nagy, S.S. Diatoms: Living in a Constructal Environment. In *Algae and Cyanobacteria in Extreme Environments*, 1st ed.; Seckbach, J., Ed.; Springer: Dordrecht, The Netherlands, 2007; pp. 141–172.
53. Ponce, J.F.; Fernández, M. Diatom Analysis. In *Climatic and Environmental History of Isla de los Estados, Argentina*; Springer: Dordrecht, The Netherlands, 2014; pp. 87–104.
54. U. College London y M. Olney, "Microfossil Image Rocevery and Circulation for Learning and Education", Micropalaeontology Unit. 2002. Available online: https://www.ucl.ac.uk/GeolSci/micropal/diatom.html (accessed on 15 November 2021).
55. Hasan, M.; Saidi, T.; Muyasir, A.; Alkhaly, Y.R.; Muslimsyah, M. Characteristic of calcined diatomaceous earth from Aceh Besar District—Indonesia as cementitious binder. *IOP Conf. Series Mater. Sci. Eng.* **2020**, *933*, 012021. [CrossRef]
56. López, J. *Comportamiento Cíclico de Suelos Diatomáceos*; Universidad Nacional Autónoma de México: México City, México, 2009.

57. Lozano-Duque, Y.; Vidal, L.A.; Navas, G.R. Listado de diatomeas (Bacillariophyta) registradas para el Mar Caribe colombiano. *Boletín De Investig. Mar. Y Costeras-INVEMAR* **2010**, *39*, 83–116. [CrossRef]
58. Zepeda, C. *Bacillariophyta Diatomeas Guía Didáctica*; Universidad Autónoma del Estado de México: México City, México, 2019; Volume 1, pp. 1–50.
59. Hildebrand, M. Diatoms, Biomineralization Processes, and Genomics. *Chem. Rev.* **2008**, *108*, 4855–4874. [CrossRef]
60. Sardo, A.; Orefice, I.; Balzano, S.; Barra, L.; Romano, G. Mini-Review: Potential of Diatom-Derived Silica for Biomedical Applications. *Appl. Sci.* **2021**, *11*, 4533. [CrossRef]
61. Agropuli, "Ficha técnica 'TIERRA DE DIATOMEAS—AGROPULÍ—TDDA Silicio Asimilable: Dióxido de Silicio (SiO_2) Amorfo. Enmienda de uso Agrícola'". Bogotá. 2021, pp. 1–13. Available online: http://www.agropuli.com.co/index.html#tierradediatomeas (accessed on 19 December 2022).
62. Antonides, B.L.E. *Diatomites*; US Geological Survey Publications: Reston VA, USA, 1997.
63. Zahajská, P.; Opfergelt, S.; Fritz, S.C.; Stadmark, J.; Conley, D.J. What is diatomite? *Quat. Res.* **2020**, *96*, 48–52. [CrossRef]
64. Manosalva, S.; Naranjo, W. Diatomitas. In *Geología, Caracterización y Potencial Uso Industrial*, 1st ed.; UPTC Publicaciones: Boyacá, Colombia, 2009.
65. Aksakal, E.L.; Angin, I.; Oztas, T. Effects of diatomite on soil physical properties. *Catena* **2012**, *88*, 1–5. [CrossRef]
66. Avramenko, A.; Cherepanova, M.; Pushkar', V.; Yarusova, S. Diatom characteristics of the Far East siliceous organogenic deposits. *Russ. Geol. Geophys.* **2015**, *56*, 947–958. [CrossRef]
67. Liao, H.; Ning, C.; Akaishi, M.; Zhou, L. Effects of the time-dependent behavior on strain softening of diatomaceous soft rock. *Met. Mater.* **1998**, *4*, 1093–1096. [CrossRef]
68. Rogato, A.; De Tommasi, E. Physical, Chemical, and Genetic Techniques for Diatom Frustule Modification: Applications in Nanotechnology. *Appl. Sci.* **2020**, *10*, 8738. [CrossRef]
69. Wiemer, G.; Moernaut, J.; Stark, N.; Kempf, P.; De Batist, M.; Pino, M.; Urrutia, R.; de Guevara, B.L.; Strasser, M.; Kopf, A. The role of sediment composition and behavior under dynamic loading conditions on slope failure initiation: A study of a subaqueous landslide in earthquake-prone South-Central Chile. *Geol. Rundsch.* **2015**, *104*, 1439–1457. [CrossRef]
70. Lewin, J.C. Silicon Metabolism in Diatoms. *J. Gen. Physiol.* **1955**, *39*, 1–10. [CrossRef]
71. Annenkov, V.V.; Danilovtseva, E.N.; Pal'Shin, V.A.; Verkhozina, O.N.; Zelinskiy, S.N.; Krishnan, U.M. Silicic acid condensation under the influence of water-soluble polymers: From biology to new materials. *RSC Adv.* **2017**, *7*, 20995–21027. [CrossRef]
72. Thamatrakoln, K.; Hildebrand, M. Silicon Uptake in Diatoms Revisited: A Model for Saturable and Nonsaturable Uptake Kinetics and the Role of Silicon Transporters. *Plant Physiol.* **2007**, *146*, 1397–1407. [CrossRef] [PubMed]
73. Hildebrand, M.; Dahlin, K.; Volcani, B.E. Characterization of a silicon transporter gene family in Cylindrotheca fusiformis: Sequences, expression analysis, and identification of homologs in other diatoms. *Mol. Genet. Genom.* **1998**, *260*, 480–486. [CrossRef]
74. Durkin, C.A.; Koester, J.A.; Bender, S.J.; Armbrust, E.V. The evolution of silicon transporters in diatoms. *J. Phycol.* **2016**, *52*, 716–731. [CrossRef] [PubMed]
75. Pahl, S.L.; Lewis, D.M.; Chen, F.; King, K.D. Heterotrophic growth and nutritional aspects of the diatom Cyclotella cryptica (Bacillariophyceae): Effect of some environmental factors. *J. Biosci. Bioeng.* **2010**, *109*, 235–239. [CrossRef]
76. Meng, F.; Gao, G.; Jia, Z. Study on Tribological Mechanism for Multi-layer Porous Structure of Diatom Frustule. *Microb. Ecol.* **2014**, *69*, 45–58. [CrossRef]
77. Behrensmeyer, A. Taphonomy. In *Encyclopedia of Geology*, 2nd ed.; Elsevier: Amsterdam, The Netherlands, 2021; pp. 12–22. [CrossRef]
78. Hamm, C.E.; Merkel, R.; Springer, O.; Jurkojc, P.; Maier, C.; Prechtel, K.; Smetacek, V. Architecture and material properties of diatom shells provide effective mechanical protection. *Nature* **2003**, *421*, 841–843. [CrossRef] [PubMed]
79. Díaz-Rodríguez, J.A.; González-Rodríguez, R. Influence of diatom microfossils on soil compressibility. In Proceedings of the 18th International Conference on Soil Mechanics and Geotechnical Engineering, Paris, France, 2–6 September 2013; pp. 325–328.
80. Khraisheh, M.A.; Al-Degs, Y.S.; Mcminn, W.A. Remediation of wastewater containing heavy metals using raw and modified diatomite. *Chem. Eng. J.* **2004**, *99*, 177–184. [CrossRef]
81. Wu, J.; Yang, Y.; Lin, J. Advanced tertiary treatment of municipal wastewater using raw and modified diatomite. *J. Hazard. Mater.* **2005**, *127*, 196–203. [CrossRef]
82. Deng, Z.; Wang, W.; Yan, T.; Xie, K.; Li, Y.; Liu, Y.; Su, Q. In Situ Experimental Study of Natural Diatomaceous Earth Slopes under Alternating Dry and Wet Conditions. *Water* **2022**, *14*, 831. [CrossRef]
83. Ma, Z.-Y.; Liao, H.-J.; Dang, F.-N. Effect of intermediate principal stress on strength of soft rock under complex stress states. *J. Cent. South Univ.* **2014**, *21*, 1583–1593. [CrossRef]
84. Dobrosielska, M.; Dobrucka, R.; Gloc, M.; Brząkalski, D.; Szymański, M.; Kurzydłowski, K.; Przekop, R. A New Method of Diatomaceous Earth Fractionation—A Bio-Raw Material Source for Epoxy-Based Composites. *Materials* **2021**, *14*, 1663. [CrossRef]
85. González, J. Capacidad de Absorción de las Arcillas Diatoméicas y su Aplicación en la Industria de Hidrocarburos. In *Encuentro Internacional de Educación en Ingeniería*; Asociación Colombiana de Facultades de Ingeniería ACOFI: Bogotá, Colombia, 2020.
86. Kang, X.; Liao, H. Bounding surface plasticity model for jointed soft rocks considering overconsolidation and structural decay. *Comput. Geotech.* **2019**, *108*, 295–307. [CrossRef]

87. Tanaka, H.; Locat, J.; Shibuya, S.; Soon, T.T.; Shiwakoti, D.R. Characterization of Singapore, Bangkok, and Ariake clays. *Can. Geotech. J.* **2001**, *38*, 378–400. [CrossRef]
88. Pimraksa, K.; Chindaprasirt, P. Lightweight bricks made of diatomaceous earth, lime and gypsum. *Ceram. Int.* **2009**, *35*, 471–478. [CrossRef]
89. Petrosyan, G.M. Manifestation of rock pressure in diatomaceous clays. *Hydrotech. Constr.* **1973**, *7*, 248–251. [CrossRef]
90. Nava, O.; Díaz, J. *Influencia de las Diatomeas en la Resistencia al Corte del Caolín*; Universidad Nacional Autónoma De México: Mexico City, Mexico, 2007.
91. Bartolomé, J.F. El Caolín: Composición, estructura, génesis y aplicaciones. *Boletín Soc. Española Cerámica Y Vidr.* **1997**, *36*, 7–20.
92. Jaime, M. *Dos Aplicaciones de la Modelación en Centrífuga Para Suelos Arcillosos: Consolidación Natural de Mezclas y Mecánica de Fracturas en Taludes Verticales*; Proyecto de grado Universidad de los Andes: Bogotá, Colombia, 2002; pp. 1–90.
93. Garzón, L. *Modelación en Centrífuga de Túneles Poco Profundos Con Variación Espacial del Duelo*; Universidad de los Andes: Bogotá, Colombia, 2010.

applied sciences

MDPI

Article

Dynamic Characteristics of Post-Cyclic Saturated Loess

Qian Wang [1,2,3,*], Yan Wang [2], Wenguo Ma [4] and Dongwang Tao [1]

1 Institute of Engineering Mechanics, China Earthquake Administration, Harbin 150001, China
2 Key Laboratory of Loess Earthquake Engineering, China Earthquake Administration & Gansu Province, Lanzhou 730000, China
3 School of Civil Engineering, Qinghai University, Xining 810000, China
4 School of Physics and Electronic–Electrical Engineering, Ningxia University, Yinchuan 750000, China
* Correspondence: wangqian@gsdzj.gov.cn

Abstract: Damage to soil structure caused by strong earthquakes is one of the main reasons for post-earthquake geohazard development. To investigate the nonlinear dynamic behaviors of the post-earthquake loess, a pre-shock reconsolidation test was designed to simulate the process of the loess undergoing earthquake and post-earthquake reconsolidation in a natural state. Furthermore, dynamic triaxial tests of the specimens before and after pre-shock action and consolidation stabilization with different over-consolidation ratios (OCR) were conducted to investigate the variety of the dynamic modulus and damping ratio of the saturated loess. The influence of pre-shock and reconsolidation on the dynamic behaviors was determined. Moreover, the mechanism of the changes after pre-shock and consolidation was discussed by combining the microstructure test results of soil samples before and after pre-shock and reconsolidation. The results suggest that the kinetic stiffness of the pre-shock saturated loess decreases significantly under the same consolidation conditions. The growth of the damping ratio-dynamic strain curve increases, and the deformation potential of the loess has a remarkable growth. With an increase in OCR, the dynamic elastic modulus after pre-shock increases continuously; however, the damping ratio decreases significantly. The dynamic stiffness increases and the deformation potential weakens significantly. The strong earthquake leads to the weakening of interparticle cementation, pore penetration, and structural reorganization in the local area, causes connecting of the macropores, and produces microfractures in the soil, which makes a significant decrease in the dynamic shear modulus ratio and an increase in the damping ratio of the loess, leading to the enhancement of soil dynamic nonlinearity and the attenuation of the dynamic strength. Moreover, the compaction effect of reconsolidation on the soil increases the interparticle friction and heals some microfractures, which leads to an increase in soil stiffness. This makes the maximum dynamic shear modulus and the maximum dynamic shear stress amplitude of the post-cyclic saturated loess perform at the same level compared with the natural loess without shock when the OCR equals three. However, the dynamic shear modulus and the damping ratio of the post-cyclic saturated loess are close to the natural loess when the OCR equals two.

Keywords: pre-shock; dynamic modulus; reconsolidation; damping ratio; post-seismic effect

1. Introduction

As a problematic soil, loess is extensively distributed in Central and Eastern Asia, especially in Northern China. The Loess Plateau mainly includes the provinces such as Gansu, Shaanxi, Henan, Qinghai, and Shanxi, as well as the Autonomous Region of Ningxia and Inner Mongolia. The thickness of the loess layer has obvious regional differences, which vary from several meters to more than 500 m, including loess plateau, loess beam, loess mount, loess terraces, and a variety of geomorphic units [1,2]. The particle composition of the loess is mainly silt content, which is commonly 50–80% [1]. The clay content and sand content have obvious regional differences [3]. The loess region in China is mostly an arid area. The loess is distributed with a large number of pores due to its formation; that is, from the eolian sediment and erosional-depositional by rain-wash [4,5]. A significant portion of these pores are visible to the eyes and consist of multiple particles lapped together, while the cement between the particles is only a small amount of soluble salt and mucilage, which makes these pores exist naturally in a sub-stable state [6,7]. This state causes the loess to have engineering properties such as high compressibility, strong water sensitivity, and high seismic sensitivity.

In Northern China, there are large-scale seismic belt zones including the Haiyuan seismic belt zone, the Fenhe-Weihe fault zone, and the Western Qinling fault zone [8]. The neotectonic movements were strong and the seismic activities were frequent [9]. More than 570 strong earthquakes from 78 B. C., including many huge earthquakes, such as the Huaxian earthquake in 1556, the Haiyuan earthquake in 1920, the Hongtong earthquake in 1303, and the Tianshui earthquake in 1654. More than 1.4 million deaths are blamed, most of which are caused by seismic geohazards including landslides and mud flows caused by loess liquefaction (Figures 1 and 2) [1,10–13].

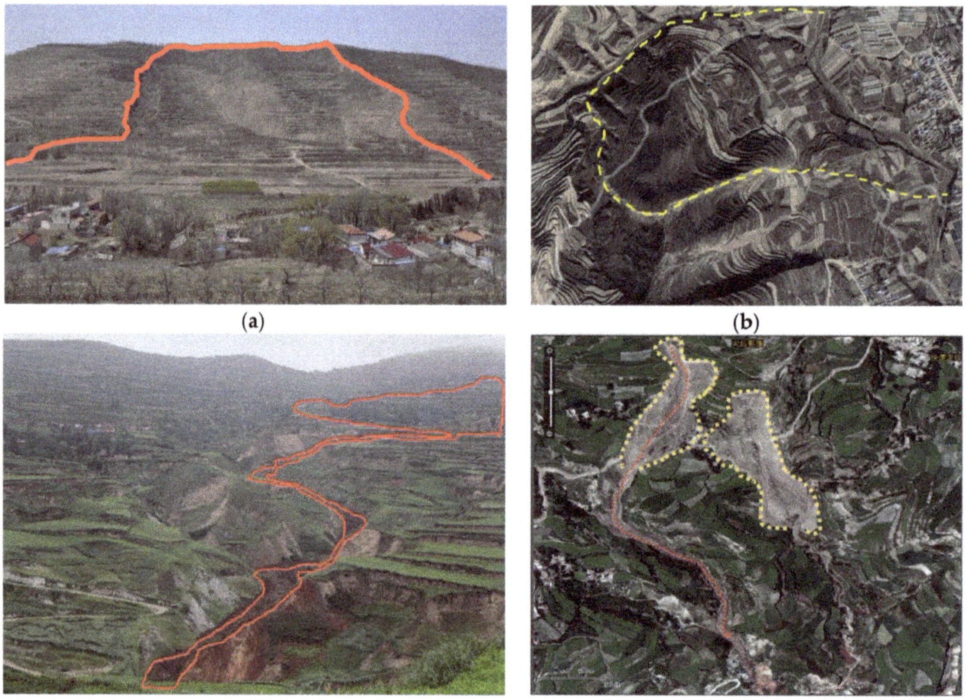

(a) (b) (c) (d)

Figure 1. Earthquake-induced geological disasters in the Loess Plateau: (**a**) the Sunjiagou landslide triggered by the Haiyuan earthquake in 1920, (**b**) satellite imagery of the Sunjiagou landslide and the surrounding areas from Google Earth, (**c**) the Yongguang landslide and mudflow triggered by the Minxian-Zhangxian earthquake in 2013, and (**d**) satellite imagery of the Yongguang landslide and slide path of the mudflow. Note the red line in (**a**,**c**), and the yellow line in (**b**,**d**) are the boundary of the landslides, the red dotted line in (**d**) is the slide path of the mudflow.

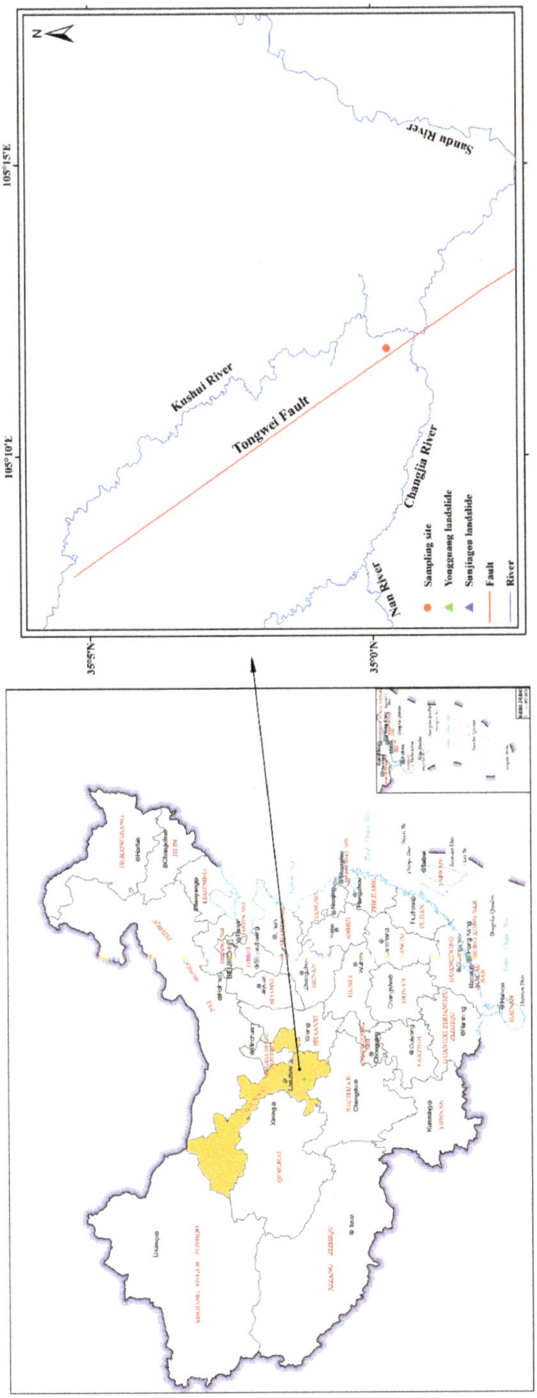

Figure 2. Location of the sampling site.

The natural loess has a sub-stable structure with trellis pores, which are easy to damage under heavy rainfall, seismic action, and human activities such as agricultural irrigation and slope excavation [6,14,15], leading to serious geological hazards in the loess region. However, the co-seismic effect and post-seismic effect of strong earthquakes are one of the key causes of structural variations, as well as the geological hazards of the loess. After the earthquake, the damaged structure of the loess will be re-consolidated and become relatively stable [16]. Due to the structural changes caused by the destruction-consolidation process, the mechanical behaviors of reconsolidated loess aftershocks are significantly different from the behaviors of the original loess in the natural state. Therefore, the loess is loosened and cracked under seismic action, and the structural changes occur by the earthquake and post-earthquake reconsolidation [15–17]. It contributes to the development of geological hazards in the meizoseismal areas, underlying a huge hidden danger to engineering construction [18,19].

The relationship between the post-seismic effects of strong earthquakes and landslides has played an important role in earthquake-induced landslide hazards and prevention. Following the Sichuan *M* 8.0 earthquake that occurred in 2008, the behaviors of the mechanical property of rocks and soils after strong earthquakes and their effects on slope stability have received more extensive attention from scholars [20–24]. Due to the significant seismic sensitivity of loess, its sub-stable structure can be damaged under seismic action with lower intensity [1,25]. In addition, the damage characteristics of the soil are significantly contrasted when the seismic intensity is different, which mainly occurs in the deformation section of the ground caused by the rearrangement of the particles during an earthquake. Furthermore, in the case of reconsolidation under external forces and complex environmental conditions, the reconsolidation conditions also have an appreciable influence on the physical and mechanical qualities of the soil [17,26–28]. However, these studies mainly focus on the characteristics of static parameters and liquefaction of the post-seismic loess [17,26–32]. Considering the background of strong seismic activity in the loess distribution areas in China, studies on the kinetic behaviors of post-earthquake loess are both necessary and significant in analyzing the slope stability under earthquake effects.

The nonlinear dynamic parameters including the dynamic elastic (shear) modulus (E_d and G_d) and the damping ratio (D) are the key factors for analyzing the seismic hazard of the site, which is also an important parameter evaluation of dynamic stability of the slope. The recent studies on nonlinear dynamic properties of the loess mainly include the distribution characteristics and influencing factors of regional nonlinear dynamic properties [1,3,33–39]. However, the studies on nonlinear dynamic properties of reconsolidated loess are lacking. In this paper, the physical process of pre-shock and reconsolidation is simulated in the laboratory. The nonlinear dynamic behaviors of the loess under the effect of pre-shock and reconsolidation are drawn. The impact of the OCR on the dynamic parameters of the loess aftershock is analyzed. The results will greatly aid in interpretation of the post-seismic effects on loess and the stability of loess slopes.

2. Methodology

2.1. Test Materials

The loess for the laboratory experiments is taken from a loess slope in Lijiadian Town, Tongwei County. The sampling site is in the gully region in Central–South Gansu Province. It is also a concentrated distribution area of loess landslide in the affected areas of the Tianshui Earthquake with a magnitude of 8.0 in 1644, the Tongwei earthquake with a magnitude of 7.5 in 1718, and the Haiyuan earthquake with a magnitude of 8.5 in 1920 [1]. The sampling site is located on a platform of the slope, which is the trailing edge of an earthquake-induced landslide. Figure 2 shows the location of the sampling site. The Tongwei fault, which is not only a buried deep-rooted fault but also the seismogenic fault of the *M* 7.5 earthquake that occurred in 1718, is close to the sampling site [40,41].

An exploration well is first dug manually in a flat area on the platform during the process of sampling. Then, the undisturbed soil samples are manually cut with a small shovel and knife from the shaft wall in the exploration well at a depth of 5.0–5.2 m. The samples are carefully taken out of the exploratory well and then processed into cylindrical samples with a diameter of approximately 100 mm and height of 140 to 160 mm by manual cutting. To protect against moisture loss, the cylindrical samples are wrapped with plastic film and taped tightly. Finally, the samples are placed in a box in order, and soft material was placed between the samples to prevent breakage during transportation.

The undisturbed loess samples are processed with the cutting plate in the laboratory. The samples are carefully cut artificially until it being machined into a cylinder of 50 mm in diameter. Then, the cylindrical specimens are carefully removed and placed in a metal mold with a height of 100 mm, and the excess soil at the end is trimmed off. Note the specimens are weighed and measured for diameter and height until it is prepared for use. Figure 3 shows the collection and preparation process of the loess specimen.

Figure 3. Process of sample collection and specimen preparation: (**a**) sampling in the exploration well, (**b**) the clod sample, (**c**) cutting the samples in the field, (**d**) cylinder samples wrapped with plastic cling film and taped, (**e**) samples placed in boxes, (**f**) removed plastic film, (**g**) preparing process of the specimens, (**h**) cylinder specimen after cutting in radial, and (**i**) the specimens with plastic bags in a box.

The loose soil cut from the cutting plate was taken and a series of tests including particle fraction, specific gravity, water content, and liquid–plastic limit are conducted, respectively. The test methods followed the Chinese Code GB50123-2019, as well as the ASTM Code including ASTM D4318-05, ASTM D854-06, and ASTM D422-07 [42–45]. The results are represented in Figure 4 and Table 1. According to the Casagrande plasticity chart, the soil used in this study is CL.

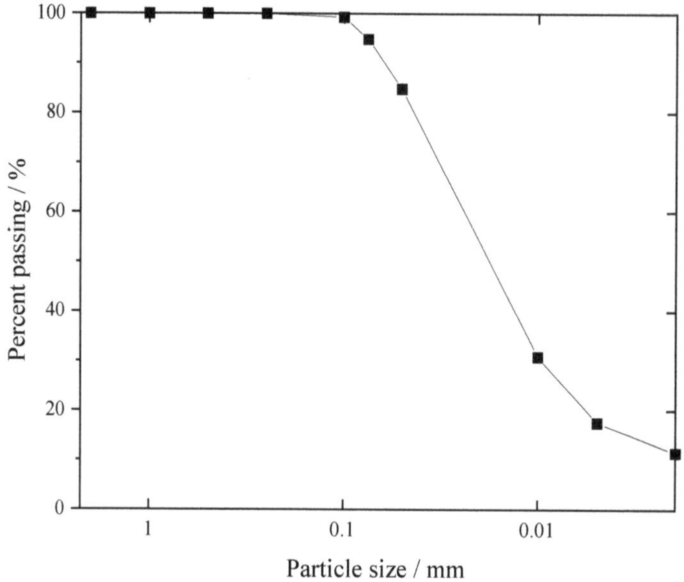

Figure 4. Particle size distribution curve of the loess.

Table 1. Physical properties of the tested loess.

Parameters	Value
Density	1.42 g/cm^3
Moisture	9.95%
Specific gravity	2.71
Plastic limit	16.81%
Liquid limit	26.83%
Plastic index	10.02

2.2. Testing Instruments

The test instruments used in this study mainly include the WF-12440 hollow cylinder apparatus (WF-HCA, Figure 5a) and the scanning electron microscope typed KYKY-2800B (KYKY-SEM, Figure 5b).

The WF-HCA was used to conduct pre-shock tests and dynamic triaxial tests. The instrument, manufactured by Wykeham Farrance Ltd. in the UK, uses an air-to-water converter to apply peripheral pressure and a pneumatic actuator to apply axial tensile and torsional forces, allowing for precise static and dynamic control to carry out vibration triaxial tests and cyclic torsional shear tests under complex conditions. The main modules of the equipment include a pressure chamber module, air–water conversion module, measurement module, control module, and data collection module. A 48-channel data acquisition instrument connected to a computer can realize the implementation of multi-channel data acquisition. The maximum longitudinal displacement is ±25 mm; the maximum longitudinal force is ±10 kN. The measurement limits of the pore water pressure and back pressure are both 1000 kPa. The sample platform is detachable, which can meet the requirements of the triaxial test of cylinder samples with different diameters. The measurement accuracy of the sensors is shown as follows: the force sensor is 0.1 N, the pressure (stress) sensor is 0.1 kPa, and the strain sensor is 0.001% [3,25,30].

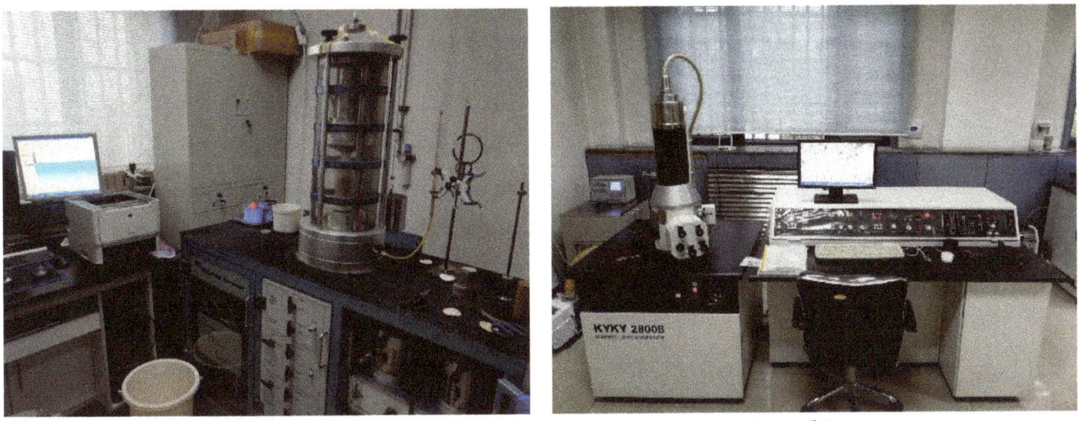

(**a**) (**b**)

Figure 5. Test instruments: (**a**) the WF-12440 dynamic triaxial torsional shear test system, and (**b**) the KYKY-2800B scanning electron microscope.

The KYKY-SEM is used to determine the micrograph of the structure of the loess samples. The instrument is manufactured by KYKY Technology Co., Ltd., Beijing, China. The ion sputtering instrument accompanying the equipment can realize the gold plating treatment of the non-conductive sample surface. The main technical parameters of the equipment include: resolution of 4.5 nm, magnification between 5 and 250,000, acceleration voltage from 0.1 kV to 30 kV, and maximum specimen size of 60 mm in diameter. The sample stage of this instrument can realize displacement in three directions, the XZ plane deflection (T), and the XY plane rotation (R) where, the limit in the X direction, the Y direction, the Z direction, the deflection of T, and rotation in R are 50 mm, 50 mm, 25 mm, $\pm 45^{\circ}$, and 360°, respectively [12,30].

2.3. Test Methods

The test methods of the dynamic triaxial tests (including pre-shock tests) follow the Chinese Code GB50123-2019 and the ASTM Code ASTM D3999-2003 [42,46]. The process of pre-shock is simulated using the WF-HCA. Firstly, latex film is wrapped around the sample with a film bearing cylinder. Then, a permeable stone and a filter are put successively on the sample platform of the WF-HCA. The specimen coated with latex film is carefully placed on the filter paper, and another filter and permeable stone are put on top of the specimen. Make sure the second permeable stone just touches the top seat by adjusting the height of the sample platform. The two ends of the latex film are respectively set on the sample platform and the top seat and tied tightly with rubber rings to prevent water leakage. The chamber is filled with water and the surrounding pressure is simulated by the air–water conversion system. All sensors need to be zeroed out. Before pre-shock tests, the specimen should be pre-consolidated to simulate the original stress state. The axial pressure is determined based on the natural density and sampling depth of the samples. A noninotropic consolidation mode was adopted for the specimens, and the coefficient of cell pressure was 0.59 [1]. The standard for sample consolidation stability is determined when the axial strain within 1 h is no more than 0.01%. After the sample is stabilized, a sinusoidal load with a frequency of 1Hz is conducted on the sample in axial for cyclic shearing. The amplitude of the load is calculated based on Equations (1) and (2) [25,30,47] where σ_d is the dynamic stress; τ_d is the dynamic shear stress; a_{max}/g is the seismic coefficient, indicating the ratio between the peak ground acceleration (PGA) and the gravitational acceleration.

k_d is the reduction coefficient of depth, which can be calculated according to Equation (3); γ is the unit weight of the soil; ds is the depth of the soil sample.

$$\sigma_d = 2\tau_d \tag{1}$$

$$\tau_d = 0.65\frac{a_{max}}{g}\cdot k_d\cdot\gamma\cdot d_s \tag{2}$$

$$k_d = 1 - 0.00765d_s \tag{3}$$

To make sure the damage of the pre-shock action on the soil was more obvious and to increase the comparability of the test results, the seismic coefficient $a_{\max}/g$ = 0.4 was taken in this study by considering the historical strong earthquakes and the seismic ground motion parameter zonation results in the study area [48,49]; it is indicated that there exists strong earthquake action with a simulated earthquake intensity of IX degree. Therefore, the maximum cyclic number for the pre-shock test is 30 [50].

The pre-shock specimens are removed from the platform carefully. Plastic bags are placed on the specimens with the wrapped latex film, and then the bags are tied tightly and placed in a dry dish for 24 h before the dynamic triaxial tests. After that, the specimens are placed again in the chamber of the WF-HCA in the same way as the pre-shock tests. Then, the specimens are saturated on the WF-HCA using a backpressure saturation method. A small cell pressure (approximately 20 kPa) is loaded on the specimen, and then the degassed water is drained into the specimen from bottom to top to fully discharge the air inside it. After the air in the specimen is fully discharged, a series of confining pressure and back pressure is alternately uploaded to the specimen step by step, and the degassed water is drained after each level of the back pressure. Furthermore, the B value is measured after the next cell pressure upload to determine the saturation degree of the specimens. The specimens are determined to have fully saturated when the B value reaches 0.95. An isotropic consolidation method is used to consolidate the specimens, and the axial pressure is calculated based on the saturation density and sampling depth of the specimens [3]. The cell pressure coefficient is 1. For the purpose of studying the impact of OCR on the nonlinear dynamic parameters of the post-seismic saturated loess, the OCR is set as 1, 2, 3, and 4. Note that the consolidation conditions of specimens with each OCR are simulated based on the determined axial pressure. The criteria for consolidation stability of the specimens are the same as those for the pre-shock tests. Considering the non-uniformity of the original loess specimens, and to ensure the data do not have the problem of comparability, multiple continuous sinusoidal loads with more than 10 different amplitudes and a frequency of 1 Hz are exerted on the same specimen. In addition, the number of vibrations under each level of loading was 10 to guarantee the stability of the cyclic load. Moreover, the variation of the σ_d and the dynamic strain, ε_d, are collected in real-time. A schematic diagram of the load path of the pre-shock reconsolidation cyclic shear process is shown in Figure 6.

The micro-structure photos of loess before and after pre-shock and reconsolidation are obtained using the KYKY-SEM. Before the test, the loess specimens are dried naturally. Small pieces of soil samples are taken manually. Sandpaper is used to polish the samples into thin sheets of approximately 1mm thickness. A knife is used to carefully trim the sheet into a circular shape with a diameter of 10mm. The floating soil on the surface of the specimens is blown away using an aurilave, and the specimens are bonded onto the sample holder using conductive tape. Then, the surface of the specimens is coated with gold using the ion sputtering instrument. To ensure the gold is uniformly sprayed on the surface, the duration time is no less than 90 s. Finally, the specimen is taken out with the sample holder and placed in the specimen table of the KYKY-SEM. During the tests, the acceleration voltage is 21 kV and the magnification is 500X.

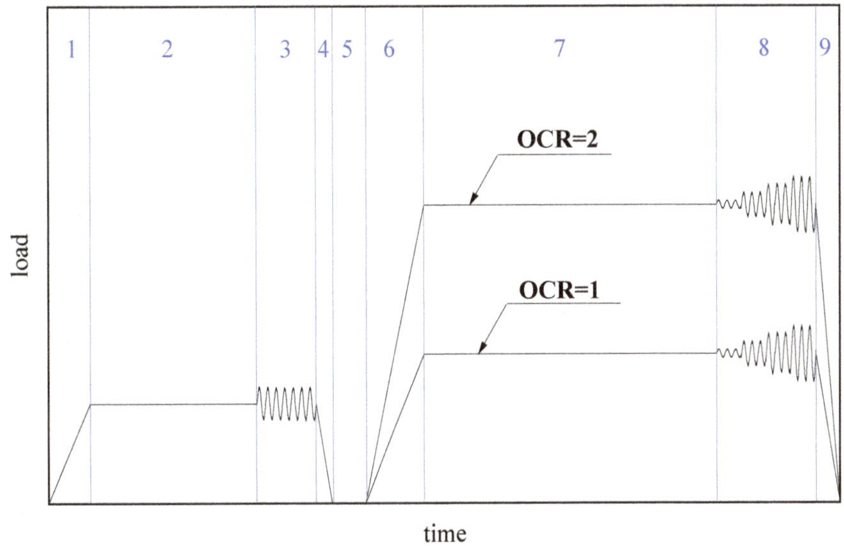

Figure 6. Schematic diagram of the load path of the pre-shock reconsolidation cyclic shear process: stage 1: load for pre-shock, stage 2: consolidation for pre-shock, stage 3: pre-shock, stage 4: unload after pre-shock, stage 5: stabilization and saturation of the specimens, stage 6: load for cyclic shear, stage 7: consolidation for cyclic shear, stage 8: cyclic shear, and stage 9: unload after cyclic shear.

3. Results

3.1. Dynamic Elastic and Shear Modulus Characteristics of the Saturated Loess before and after Pre-Seismic Reconsolidation (PSR)

Based on the dynamic triaxial tests, the $\sigma_d - \varepsilon_d$ curves of the saturated loess before and after the PSR process are drawn in Figure 7. As shown in Figure 7, the ε_d of the saturated loess before and after the PSR process shows a nonlinear growth with the increase in σ_d. The pre-shock action has a significant effect on the $\sigma_d - \varepsilon_d$ curves of saturated loess. Compared with the saturated loess without pre-shock, the σ_d of the saturated loess grows slowly with the increase in ε_d after pre-shock, indicating that the pre-shock caused a nonlinear enhancement of the loess. Furthermore, the $\sigma_d - \varepsilon_d$ curves after pre-shock are significantly influenced by the effect of reconsolidation. The σ_d at the same level of ε_d grows monotonically with the increase in the OCR, indicating that the increase in the OCR attenuates the kinetic nonlinearity of the loess. When the OCR reaches three, the kinetic nonlinearity of the pre-shock saturated loess is weakened compared with that of the original loess before the pre-shock.

The E_d is defined as the ratio of σ_d and ε_d:

$$E_d = \frac{\sigma_d}{\varepsilon_d} \tag{4}$$

Figure 8 shows the E_d versus ε_d of the saturated loess under different PSR processes. The E_d of the saturated loess under different conditions reduces with an increase in the ε_d, nonlinearly. After pre-shock, the E_d of the saturated loess decreases significantly at the same level as the ε_d. However, with an increase in the OCR, the E_d of the saturated loess increases at the same level as the ε_d in an upward trend. When the OCR reaches three, the E_d of the loess after the PSR process is larger.

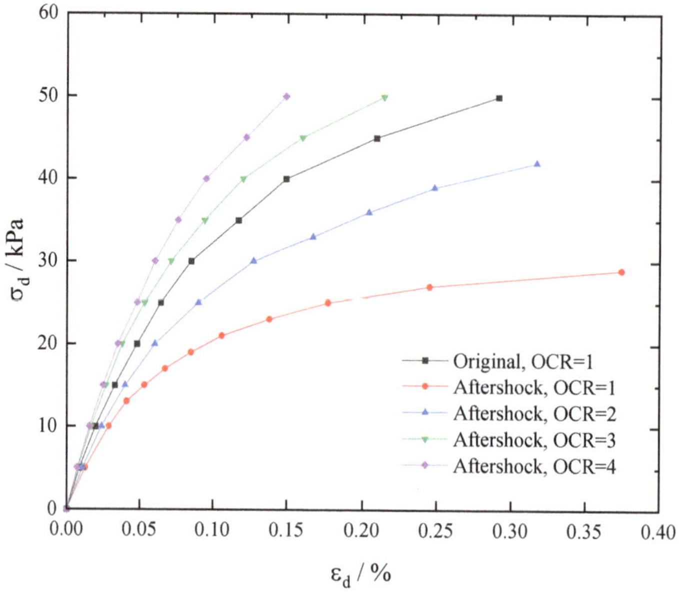

Figure 7. Dynamic stress: dynamic strain curves of the saturated loess before and after pre-shock and reconsolidation.

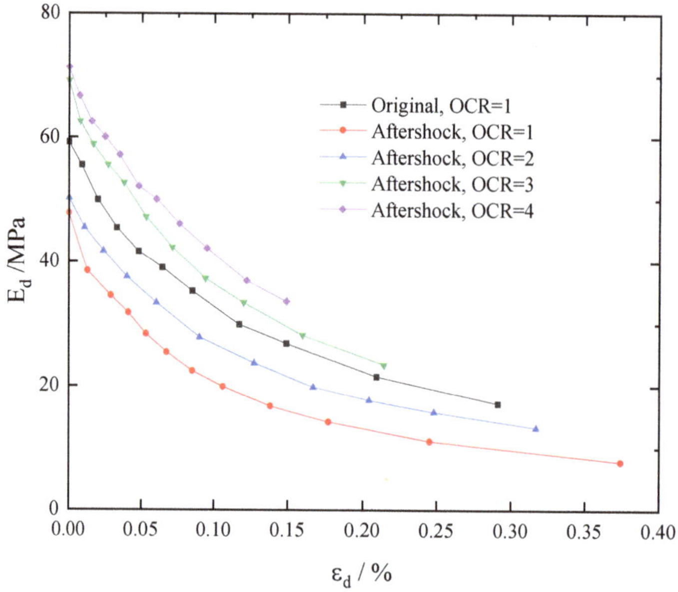

Figure 8. Dynamic elastic modulus: dynamic strain curves of the saturated loess under different reconsolidation conditions.

The relationship between E_d and ε_d of the loess in different moisture contents are consistent with the Hardin–Drnevich (HD) model [1,3,35,51]. The test results in Figure 7 are normalized and fitted nonlinearly using the HD model in the dynamic shear strain (γ_d) range of 10^{-6} to 10^{-2}. The G_d and the γ_d are obtained using Equations (5) and (6), respectively.

$$G_d = \frac{E_d}{2(1+\mu)} \quad (5)$$

$$\gamma_d = (1+\mu)\varepsilon_d \quad (6)$$

where μ is the Poisson's ratio of the loess. On the basis of Figure 7, the G_d, γ_d, and the inverse of the G_d ($1/G_d$) are calculated. Note that the relationship between the $1/G_d$ and the γ_d could be gained based on the test results, as follows in Equation (7), which is used to obtain the Hardin–Drnevich model parameters a and b for the saturated loess under different test conditions.

$$\frac{1}{G_d} = a + b\gamma_d \quad (7)$$

In accordance with the Equation (7), the reference shear strain $\gamma_{dr} = a/b$, so the dynamic shear modulus ratio G_d/G_{dmax} follows as [3,52]:

$$\frac{G_d}{G_{dmax}} = \frac{1}{1+\gamma_d/\gamma_{dr}} \quad (8)$$

The normalized curves of the G_d/G_{dmax} versus the γ_d of the saturated loess before and after the PSR process are calculated using Equation (8) and shown in Figure 9. According to Figure 9, the G_d/G_{dmax} of the saturated loess reduces nonlinearly with the growth of γ_d. Under the same consolidation conditions, the $G_d/G_{dmax} - \gamma_d$ curves of the saturated loess before and after pre-shock show obvious differences. Before the pre-shock, the G_d/G_{dmax} of the saturated loess decreased from 1 to 0.94 in the small shear strain stage ($10^{-6} < \gamma_d < 10^{-4}$). However, in the large shear strain stage ($10^{-4} < \gamma_d < 10^{-2}$), the G_d/G_{dmax} are 0.76, 0.61, 0.24, and 0.13 at the characteristic shear strains of 0.0005, 0.001, 0.005, and 0.01, respectively. After the pre-shock, the G_d/G_{dmax} of the saturated loess reduced to 0.9 with increasing the γ_d in the small shear strain stage. With the large shear strain stage, the G_d/G_{dmax} are 0.76, 0.61, 0.24, and 0.13 at the characteristic shear strains of 0.0005, 0.001, 0.005, and 0.01, respectively. The G_d/G_{dmax} of the saturated loess decays to 0.90 at the small shear strain stage and changes to 0.66, 0.49, 0.14, and 0.09 at the large shear strain stage with characteristic strains of 0.0005, 0.001, 0.005 and 0.01, respectively. The G_d/G_{dmax} of the saturated loess decreased obviously with the increase in the γ_d after the pre-seismic action, indicating that the impact of pre-shock on the kinetic stiffness of the saturated loess is more significant. Note that the difference between the G_d/G_{dmax} of the saturated loess before and after pre-seismic action is more obvious when the γ_d range is 0.0001–0.005; however, the difference presents a decreasing tendency when the γ_d range is 0.005–0.01.

The relationship of the G_d/G_{dmax} versus the γ_d of the after or pre-shock saturated loess under different reconsolidation conditions is shown in Figure 9. As seen in Figure 9, after the OCR reaches two, the G_d/G_{dmax} of the saturated loess reduces to 0.94 at the small shear strain stage, and the G_d/G_{dmax} is 0.74, 0.59, 0.23, and 0.13 at the large shear strain stage with the characteristic strains of 0.0005, 0.001, 0.005, and 0.01, respectively. This result is similar to the G_d/G_{dmax} of the original saturated loess at the same γ_d. Furthermore, the decay of the G_d/G_{dmax} is obviously slower than that of the saturated loess with an OCR of one. When the OCR is three or four, the decay trend of G_d/G_{dmax} of the saturated loess is the same as that of the OCR of two. When the OCR is four, the decay trend of the curves in the small-strain stage is similar to that of the original saturated loess. However, in the large shear strain stage, the G_d/G_{dmax} corresponds to the four characteristic shear strains of 0.78, 0.64, 0.26, and 0.15, respectively, which are slightly stiffer than other groups.

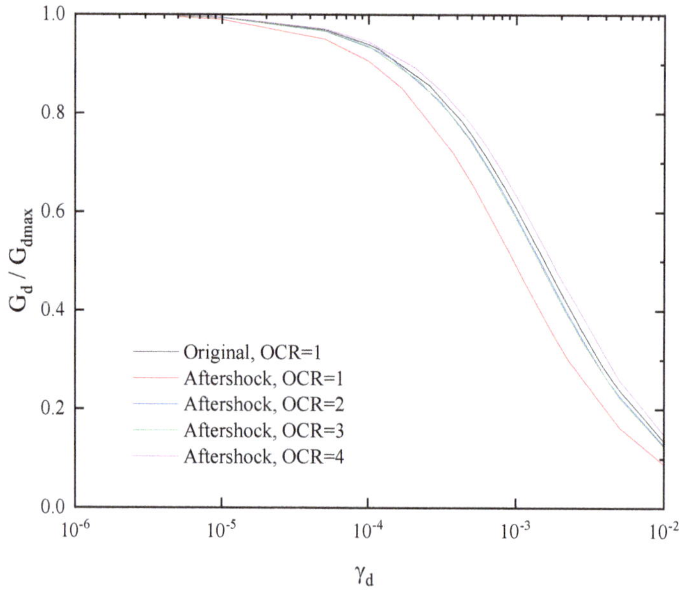

Figure 9. Curves of the dynamic shear modulus ratio versus dynamic shear strain of the saturated loess under different test conditions.

3.2. Damping Ratio of Saturated Loess before and after the PSR Process

The D of the soil is reflected as the hysteresis of the $\sigma_d - \varepsilon_d$ or $\tau_d - \gamma_d$ relationship hysteresis circle performance under the periodic dynamic load. It is caused by the energy consumed by frictional action during soil deformation, reflecting the energy loss resulting in the inner resistance of the soil under cyclic load. Thus, the D is one of the essential dynamic properties of soil and an essential parameter for the analysis of ground motion. The D can be obtained by the $\sigma_d - \varepsilon_d$ or $\tau_d - \gamma_d$ hysteresis curve method, as seen in Figure 10. The D of soil is obtained by calculating the area ΔW and area W of ΔAOB of the area enclosed by the hysteresis curve, and then calculated in accordance with the Equation (9).

$$D = \frac{1}{4\pi}\frac{\Delta W}{W} \tag{9}$$

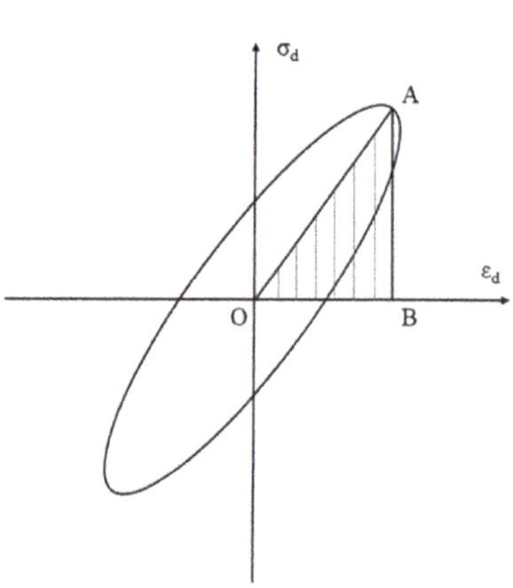

Figure 10. Calculation method of the damping ratio.

Based on Figure 10 and Equation (9), the D of the saturated loess is determined. The curves of the D versus the dynamic strain are drawn, as shown in Figure 11. According to the figure, the D of the saturated loess under different test conditions has a tendency to grow with an increase in the ε_d, and the $D - \varepsilon_d$ curve is steep when the ε_d is small; the curves gradually tend to flatten after a certain dynamic strain. The D of the saturated loess after pre-shock tends to increase more obviously with an increase in the ε_d. The inflection point of the curve corresponds to a larger dynamic strain. Combined with the definition of the D, the energy loss caused by frictional action during the soil deformation after the pre-shock is enhanced. The deformation hysteresis effect is more obvious, and its deformation stability tends to be weakened under the dynamic load. In addition, Figure 11 also shows the $D - \varepsilon_d$ curves of the post-seismic saturated loess with a different OCR. The impact of the OCR on the trend of the $D - \varepsilon_d$ curves is more obvious. The ε_d corresponding to the inflection point of the $D - \varepsilon_d$ curves decreases continuously with an increase in the OCR. When the OCR reaches two, the D of the post-seismic saturated loess at the same level of ε_d is lower than that of the original saturated loess, indicating that the deformation capability of the loess under cyclic load tends to be weakened after the over-consolidation effect.

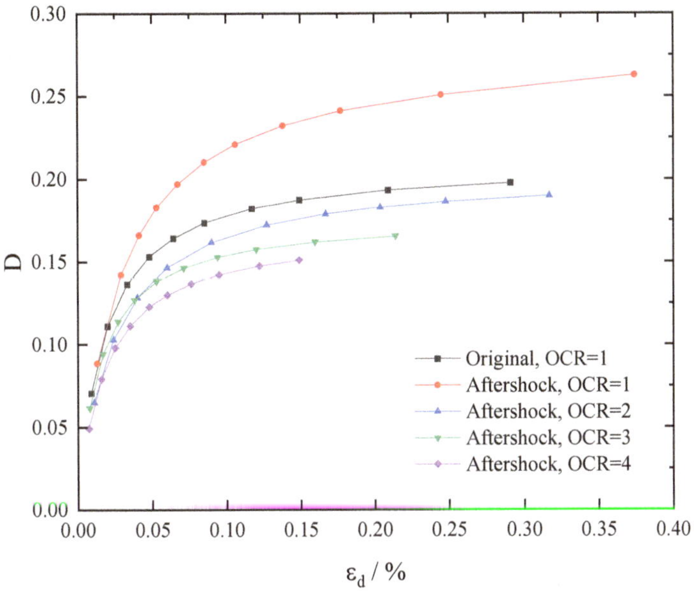

Figure 11. Curves of the damping ratio versus the dynamic strain of the saturated loess under different test conditions.

The ratio of the D to the γ_d, γ_d/D, and the γ_d, has a linear relationship as shown in Equation (10):

$$\frac{\gamma_d}{D} = A + B\gamma_d \tag{10}$$

where A and B are the fitting parameters. Equation (9) could be adapted as:

$$D = \frac{\gamma_d}{A + B\gamma_d} \tag{11}$$

Based on Figure 11, the normalized curves of the D versus the γ_d of the saturated loess before and after the PSR process are calculated, as shown in Figure 12. According to Figure 12, the $D - \gamma_d$ curves of the saturated loess under different test conditions show a characteristic of moderate growth—sharp growth—moderate growth: the shape of the curve is like a capital "S". Under the same consolidation conditions, the growth trends of the $D - \gamma_d$ curves before and after pre-shock are different. However, the difference is mainly manifested in the large shear strain stage ($\gamma_d > 2 \times 10^{-4}$). When the γ_d is 0.0005, 0.001, 0.005, and 0.01, the D of the pre-seismic saturated loess increases by 15.76%, 20.24%, 29.89%, and 31.48%, respectively, compared with that of the original saturated loess. This indicates that the pre-seismic action has a more obvious influence on the D as the γ_d continues to increase in the large shear strain stage. According to Figure 12, there is some influence of the over-consolidation effect on the various characteristics of the $D - \gamma_d$ curves of the saturated loess after pre-shock. In general, with an increase in the OCR, the growth tendency of the curve flattens out. However, the effect of reconsolidation is manifested throughout the shear strain range, and this effect tends to increase in the large shear strain stage compared to the small shear strain stage. When the $\gamma_d < 5 \times 10^{-4}$, the D of the post-seismic saturated loess with an OCR of three is larger than that of the post-earthquake saturated loess with an OCR of two in the same level of the γ_d. When the characteristic γ_d is at 0.0005, 0.0001, 0.0005, 0.001, 0.005, and 0.01, the D of the saturated loess with an OCR of four is reduced by 14.58%, 19.98%, 40.67%, 49.45%, 60.76%, and 62.62%, respectively, compared with that of post-seismic saturated loess with an OCR of one, which is 26.62% less than that of the saturated loess before the PSR process. The D decreased by 26.62%, 26.16%, 24.76%, 24.29%, 23.77%, and 23.69% than those of the original saturated loess, indicating that the D of the saturated loess decreased significantly and the deformation stability increased obviously after the OCR reached to four.

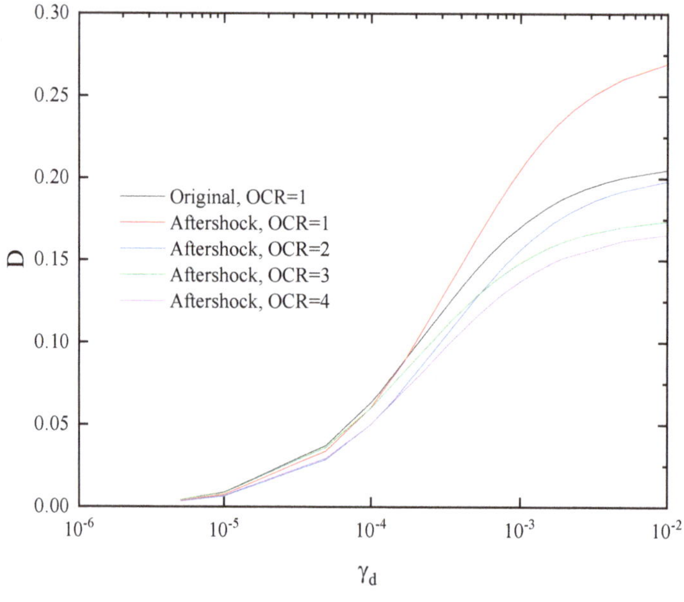

Figure 12. Normalized curves of the damping ratio versus the dynamic shear strain of the saturated loess under different test conditions.

4. Discussion

The experimental results of dynamic characteristics of post-earthquake saturated loess under different reconsolidation conditions show that the pre-seismic action will lead to the kinetic stiffness of the saturated loess having an obvious attenuation and the D of the loess increasing significantly. The dynamic stiffness of the saturated loess after the PSR process increases and the D decreases with an increase in over-consolidation ratio, and there is a nonlinear relationship between G_d/G_{dmax}, D, and the OCR. Considering the demand for parameter selection in dynamic response analysis of slope sites, there is a tremendous need to study the effect of pre-seismic and reconsolidation on nonlinear kinetic characteristic parameters of saturated loess. Furthermore, the structural change of soil is the essential reason for the difference in soil mechanics. The layout of soil particles and the content and distribution of pores in soil have a decisive effect on the mechanical behaviors of soil. Therefore, it is helpful to understand the influence mechanism of strong earthquakes and post-earthquake reconsolidation on the nonlinear dynamic characteristics of saturated loess by analyzing the microstructure characteristics of saturated loess under different PSR conditions.

4.1. Influence Law of Strong Earthquake and Post-Earthquake Reconsolidation on Nonlinear Dynamic Characteristics of the Saturated Loess

According to Figures 9 and 12, the relationships of G_d/G_{dmax} versus OCR and D versus OCR of the saturated loess at different levels of γ_d are drawn, respectively. To accurately describe the variation laws of nonlinear kinetic characteristic parameters with OCR at different levels of the γ_d, the ratio of G_d/G_{dmax} of saturated loess after pre-seismic to that of the original saturated loess at the same level of the γ_d is defined as G_{u}, and the ratio of D of saturated loess after pre-seismic to that of the original saturated loess at the same level of the γ_d is defined as D_{u}. Figure 13a shows the relationship between G_{u} versus OCR, and Figure 13b shows the relationship between D_{u} versus OCR. According to Figure 13a, with the increase in the OCR, G_{u} shows an increasing trend generally. In the small shear strain stage, with the increase in the OCR, the variation of G_{u} is not immediately apparent. In the large shear strain stage, when the OCR increases from one to two, G_{u} increases sharply; when it is between two and three, the change rate of G_{u} is small; when it increases from three to four, the growth trend of G_{u} tends to be obvious, but the growth rate is slower than that of it at one to two. According to Figure 13b, there are significant differences in the changes of D_{u}−OCR curves at different strain stages. In the small shear strain stage, D_{u} shows a decreasing–increasing–increasing tendency with the increase in OCR; however, the change in degree is small in general. The influence of the PSR on D is insignificant at this stage. At the large shear strain stage, D_{u} decreases monotonously with an increase in the OCR, and its decrease in degrees gradually slows down, indicating that the PSR process has a significant effect on D. When the OCR reaches two, the D_{u} of saturated loess after pre-seismic is close to one, which shows its D reaches the level before pre-seismic.

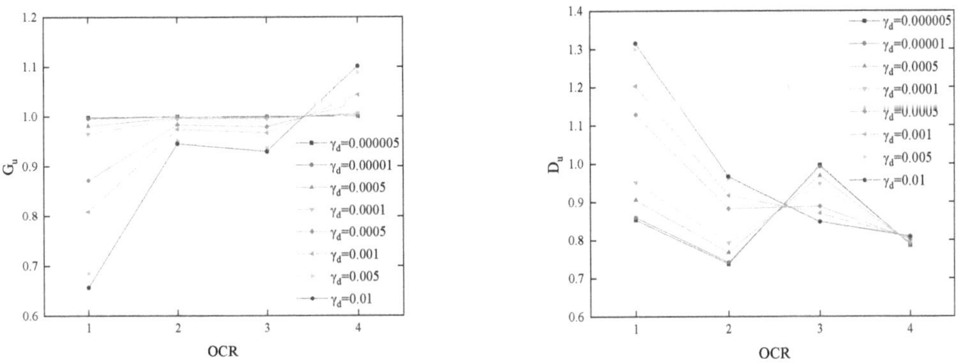

Figure 13. Influence of the OCR on the dynamic shear modulus ratio and damping ratio: (**a**) the relationship between G_{u} and OCR and (**b**) the relationship between D_{u} and OCR.

To further illustrate the influence of the PSR process on the nonlinear kinetic properties of the saturated loess, the variation laws of the maximum dynamic shear modulus G_{dmax}, the maximum dynamic shear stress amplitude τ_{dult} and the maximum damping ratio D_{max}

with pre-seismic and reconsolidation ratio is analyzed. According to Equation (7), the G_{dmax} is as follows:

$$G_{dmax} = \frac{1}{a} \tag{12}$$

Based on the definition of G_d, there are:

$$\tau_d = \frac{\gamma_d}{a + b\gamma_d} \tag{13}$$

According to Equations (7) and (13), there are:

$$\tau_{dult} = \lim_{\gamma_d \to \infty} \frac{\tau_d}{\gamma_d} = \frac{1}{b} \tag{14}$$

According to Equation (10):

$$B = \frac{1}{D} - \frac{A}{\gamma_d} \tag{15}$$

When $\gamma_d \to \infty$,

$$D = D = \frac{1}{B_{max}} \tag{16}$$

The G_{dmax}, $\tau_{dult,}$ and $D_{\max}$ are calculated according to Equations (12), (14) and (16), respectively, and their relationship curves with the OCR were drawn, as shown in Figure 14. In Figure 14, the changes of G_{dmax}, $\tau_{dult,}$ and $D_{\max}$ before and after pre-seismic are given when OCR = 1. After pre-seismic, G_{dmax} decreases significantly and $D_{\max}$ increases significantly, which is consistent with the results in Figures 8 and 11. Furthermore, τ_{dult} decreases significantly after pre-seismic, and it is only half of that before pre-seismic. Considering that τ_{dult} represents the possible extreme value of dynamic shear stress in the $\tau_d - \gamma_d$ curve, it can reflect the dynamic strength of the soil. Therefore, the pre-seismic effect also leads to a significant attenuation of the strength of saturated loess, which is the same as those studies on the seismic subsidence of unsaturated loess in the literature [35].

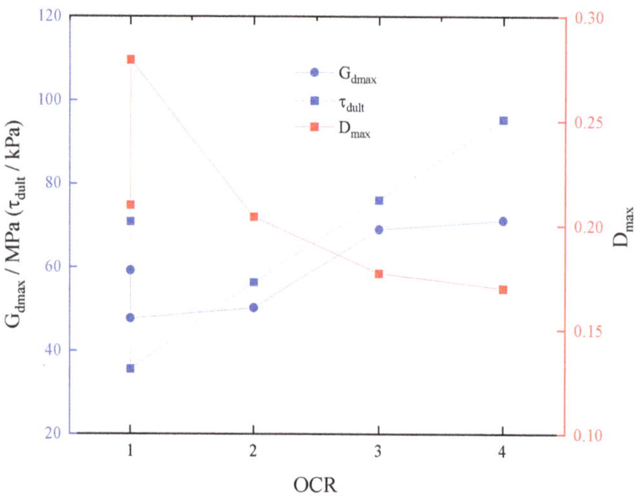

Figure 14. Influence of OCR on G_{dmax}, $\tau_{dult,}$ and $D_{\max}$.

Figure 14 also shows the effect of reconsolidation on G_{dmax}, $\tau_{dult,}$ and $D_{\max}$ of post-earthquake saturated loess. From the graph, it can be seen that G_{dmax} has a trend of moderate growth–sharp growth–moderate growth with an increase in the OCR; when $2 \leq \text{OCR} \leq 3$, the growth is faster. In addition, there is a positive linear correlation between τ_{dult} and OCR. $D_{\max}$ decreases nonlinearly and its increasing trend decreases with an increase in the OCR. Combined with Figure 13a, it should be noted that the G_{dmax} increases rapidly at $2 \leq$ OCR ≤ 3, while G_u hardly increases at this stage, and even decreases slightly at the large shear strain stage. Considering the definition of G_d/G_{dmax}, the variation tendency of G_u is mainly affected by the rapid growth of G_{dmax}. In other words, when $2 \leq$ OCR ≤ 3, reconsolidation still has a significant influence on the dynamic stiffness of saturated loess after an earthquake.

4.2. Influence Mechanism of Strong Earthquake and Post-Earthquake Reconsolidation on Nonlinear Dynamic Characteristics of Saturated Loess

The variation of microstructure of soil under different conditions is the fundamental cause of the change in its mechanical behaviors [6,18,53]. To illustrate the influencing mechanism of a strong earthquake and post-earthquake reconsolidation on the nonlinear kinetic characteristics of saturated loess, the microstructure photos of each group of samples before cyclic shear were obtained based on the SEM test, as shown in Figure 15. Among them, Figure 15a shows the original saturated loess under the OCR of 1, Figure 15b shows the aftershock loess with OCR = 1, Figure 15c shows the aftershock loess with OCR = 2, Figure 15d shows the aftershock loess with OCR = 3, and Figure 15e shows the aftershock loess with OCR = 4. As seen in Figure 15a, before pre-seismic, the boundaries of particles are clear and the particles are arranged in a mosaic, whereas the contact among grains is surface contact. There are many more trellis pores, a relatively uniform distribution, and an obvious overhead pore structure. However, the structure of the soil changed significantly after pre-seismic and over-consolidation. According to Figure 15b, when the soil is normally consolidated after the pre-seismic, the particle boundary is still obvious; however, the contact among grains occurs more as point contact and boundary contact. The connection between the particles is not as tight as before. There are still plenty of pores; however, the trellis pore is no longer obvious due to the deformation of pre-shock. Furthermore, the uniformity of the pore distribution becomes better. Note that the most conspicuous change between Figure 15a,b is a large number of pores connected in tiny cracks in Figure 15b, resulting from the loosening and rearrangement of the gains during vibration. According to Figure 15c,d, after OCR reached two, the particle boundary became blurred, resulting in the consolidation and compaction of soil, and the number of medium and large pores reduced significantly. Moreover, the small gaps formed by pore connection after the cyclic shear were also healed locally under the action of over-consolidated pressure. After the OCR reaches three, only a small number of medium and large pores is distributed in the loess. The compaction effect of soil particles is more obvious. According to Figure 15e, when the OCR reaches four, the boundary of soil particles is blurred and the particles of different sizes are gathered together. The particles are not easy to distinguish and the porosity content is significantly reduced.

Based on the analysis and discussion shown above, combined with the results shown in Figures 8, 9 and 11–14, it is illustrated that the stiffness of the loess has an obvious loss, resulting in the changes of particle arrangement and the loosening of connection among particles after pre-seismic load. Meanwhile, the penetration of medium and large pores caused by cyclic load makes the accumulation and transmission of pore water pressure much easier, which can lead to an obvious reduction in the effective stress as well as a significant enhancement of the viscosity of the soil. The reconsolidation effect on the soil after pre-seismic makes the loose soil gradually compact. With an increase in the OCR, the consolidation pressure continues to increase, and the compaction effect on the loess is gradually increased. Since the pores of the soil are obviously reduced and the connected pores result in pre-shock being gradually blocked under the compaction effect, the factual contacting area between the particles increases and the connection becomes much closer. Thus, the stiffness of the loess is significantly increased, the viscosity tends to decrease, and the deformation potential of the loess gradually tends to decrease [54].

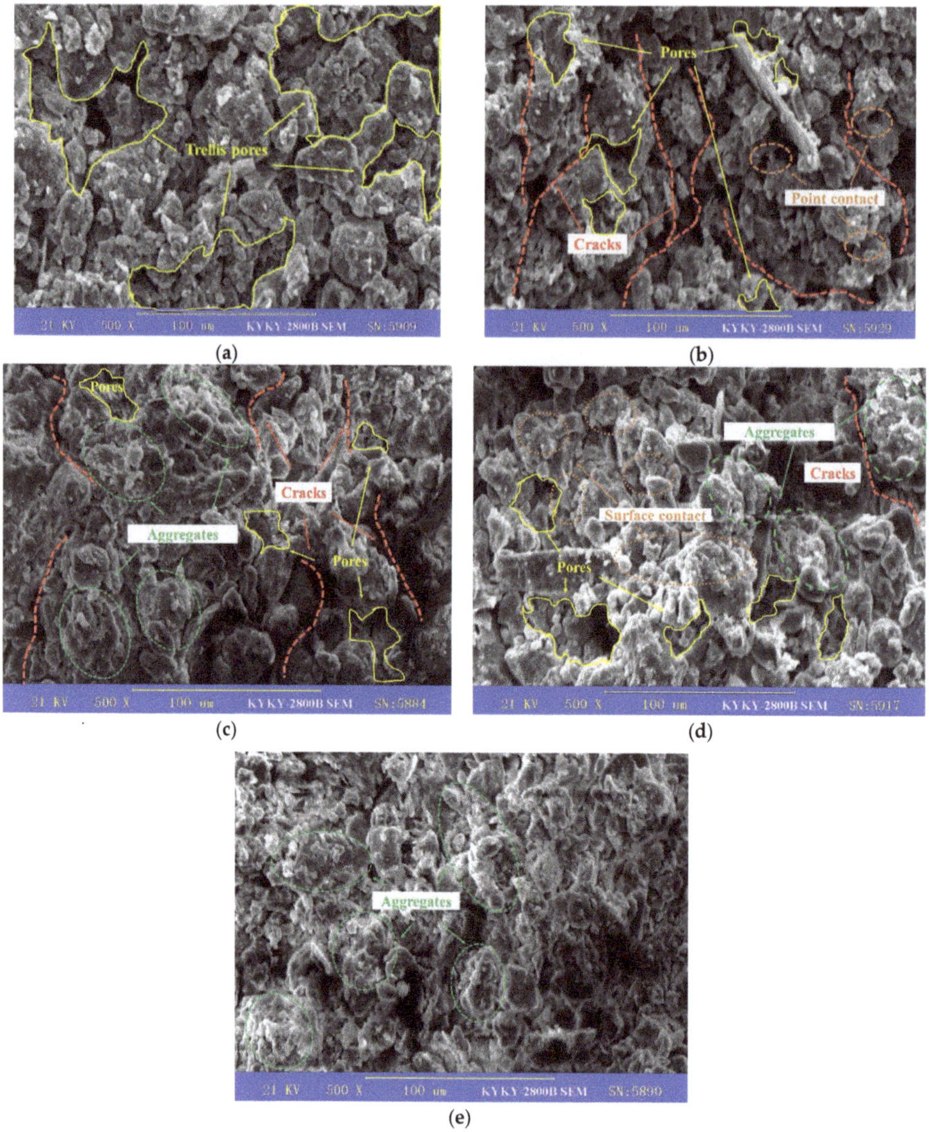

Figure 15. Microstructure characteristics of the saturated loess under different test conditions: (**a**) the original saturated loess with OCR = 1, (**b**) the aftershock loess with OCR = 1, (**c**) the aftershock loess with OCR = 2, (**d**) the aftershock loess with OCR = 3, and (**e**) the aftershock loess with OCR = 4.

This study investigates the characteristics of nonlinear dynamic properties of the post-cyclic saturated loess and analyzes the influence of pre-shock and reconsolidation on the dynamic properties of the saturated loess. Based on the test results and analysis, suggestions to aid subsequent research are as follows: (1) Seismic liquefaction is not only a kind of seismic geological hazard in the loess region but also a significant trigger.

For loess landslide and mudflow. Thus, studies on liquefaction behaviors of the post-cyclic saturated loess will be helpful in understanding post-earthquake landslides. (2) In recent years, geological hazards, including landslides and debris, flow frequently occur in strong earthquake-affected areas in the Loess Plateau. The variation of the moisture of the loess resulting from heavy rainfall and irrigation, and its coupling effect with strong earthquakes is the main cause of the disasters. Hence, the dynamic properties of the post-cyclic saturated loess under the coupling effect of aftershock and rainfall also deserve our consideration. (3) Based on the mercury injection experiment, CT scanning, and 3D modeling technology, a 3D model of the structural characteristics of the post-cyclic saturated loess may help us discover new findings and develop a better understanding of the mechanisms of variation of dynamic properties under strong earthquakes.

5. Conclusions

This study proposes the effect of pre-cyclic and reconsolidation on the nonlinear dynamic characteristics of the saturated loess and illustrates the possible mechanism for the variation of the nonlinear dynamic characteristics during the strong earthquake and post-earthquake reconsolidation via microstructure analysis of the loess in different scenarios. Through simulating the physical processes of saturated loess under the strong earthquake and reconsolidation in the laboratory and conducting a series of dynamic triaxial tests, the impaction of pre-shock and reconsolidation action on the nonlinear dynamic parameters of the saturated loess was investigated. The following conclusions are drawn:

(1) Strong earthquakes may lead to a significant attenuation of the kinetic stiffness of the saturated loess. The variation of the $G_d/G_{dmax} - \gamma_d$ curves before and after pre-seismic is more obvious in the large shear strain stage. With an increase in the OCR, the E_d of the saturated loess after pre-shock increased continuously. When the OCR reached two, the trend of $G_d/G_{dmax} - \gamma_d$ curves of the pre-shock saturated loess was similar to that of the original saturated loess.

(2) The D of the saturated loess tended to grow more obviously with the increase in ε_d after the action of a strong earthquake. The stability under the dynamic load tended to weaken. The influence of pre-shock on the $D - \gamma_d$ curves was mainly manifested in the large shear strain stage. The action of reconsolidation can lead to a significant decrease of D at the same strain level, and the growth of the curve tends to flatten with an increase in the OCR. This effect is expressed in the whole shear strain range.

(3) Compared with the original saturated loess, the G_d of the post-earthquake loess was significantly reduced and the D was obviously increased. The loess was more likely to be damaged when strong aftershocks occurred. Furthermore, the impaction of the reconsolidation on the nonlinear dynamic parameters of the post-earthquake loess is remarkable. The kinetic stiffness of the saturated loess is significantly enhanced after experiencing the over-consolidation, and its damage will be obviously weakened under strong aftershocks.

(4) Strong earthquake caused the weakening of the connection between particles of the soil. In addition, the trellis pores were consecutive and the structure was rearranged. The changes in structure led to the enhancement of dynamic nonlinear of the saturated loess. However, the compaction effect of reconsolidation increased the interparticle friction and healed some microfractures, which led to the increased stiffness of the saturated loess.

Author Contributions: Conceptualization, Q.W. and W.M.; methodology, Q.W.; tests, Y.W. and W.M.; analysis, Q.W. and D.T.; investigation, Q.W. and D.T.; data processing, Q.W. and Y.W.; writing—original draft preparation, Q.W.; writing—revised draft preparation, Q.W. and W.M.; project administration, Q.W.; funding acquisition, Q.W. All authors have read and agreed to the published version of the manuscript.

Funding: This research was conducted with the support of the Scientific Research Fund of Institute of Engineering Mechanics, China Earthquake Administration (Grant Nos. 2020D17, 2020EEEVL0409), the Open Fund of State Key Laboratory of Frozen Soil Engineering (Grant No. SKLFSE202005), the National Natural Science Foundation of China (Grant No. 51778590), and the Fund of Creative and Innovative Talents of Gansu Province (Grant No. 2022LQGR46).

Institutional Review Board Statement: Not applicable.

Informed Consent Statement: Not applicable.

Data Availability Statement: Not applicable.

Acknowledgments: The authors appreciate the support of the Gansu Earthquake Agency, and acknowledge Xiumei Zhong, Zhongnan Gao, Jun Wang, Jiemin Wang, Jinlian Ma, Master student Yi Wang, Xiaowei Xu, Qinyao Sun, and Yuxin Liang for help in field investigation, sampling, tests, and preparation of the revised manuscript, as well as to Mohamed Elmujtaba for his helpful comments in language editing during the revision process of this work. The authors also appreciate the editors and reviewers for their detailed and valuable comments and suggestions.

Conflicts of Interest: The authors declare no conflict of interest.

References

1. Wang, L.M. *Loess Dynamics*; Seismological Press: Beijing, China, 2003. (In Chinese)
2. Derbyshire, E.; Meng, X.M.; Dijkstra, T.A. *Landslides in the Thick Loess Terrain of North–West China*; Wiley: London, UK, 2000.
3. Wang, Q.; Ma, J.L.; Ma, H.P.; Wang, J.; Wang, L.M.; Gao, Z.N. Dynamic shear modulus and damping ratio of saturated loess. *Chin. J. Rock Mech. Eng.* **2019**, *38*, 1919–1927. (In Chinese) [CrossRef]
4. Jeferson, I.F.; Evstatiev, D.; Karastanev, D.; Mavlyanova, N.G.; Smalley, I.J. Engineering geology of loess and loess−like deposits: A commentary on the Russian literature. *Eng. Geol.* **2003**, *68*, 333–351. [CrossRef]
5. Salih, N.; Préat, A.; Gerdes, A.; Konhauser, K.; Proust, J.-N. Tracking the Origin and Evolution of Diagenetic Fluids of Upper Jurassic Carbonate Rocks in the Zagros Thrust Fold Belt, NE-Iraq. *Water* **2021**, *13*, 3284. [CrossRef]
6. Gao, G.R. Loess microstructure classifcation and loess collapsible. *Sci. China* **1980**, *12*, 203–204.
7. Ma, F.; Yang, J.; Bai, X. Water sensitivity and microstructure of compacted loess. *Transp. Geotech.* **2017**, *11*, 41–56. [CrossRef]
8. Feng, J.G.; Zhang, H.; Yang, P. Analysis on b−value Abnormality of the Minxian−Zhangxian, Gansu M MS 6.6 Earthquake. *Earthq. Res. China* **2016**, *30*, 186–192.
9. Ma, H.P.; Zhang, H.; Li, M.J.; Wu, S.Y.; Wang, P.T.; Wang, Q.; Zhao, J. Characteristics of the present crustal deformation in Tibetan Plateau and its relationship with strong earthquakes. *Open Geosci.*. 2022. *In press*. [CrossRef]
10. Zhang, D.X.; Wang, G.H. Study of the 1920 Haiyuan Earthquake−Induced Landslides in Loess (China). *Eng. Geol.* **2007**, *94*, 76–88. [CrossRef]
11. Wang, G.H.; Zhang, D.X.; Furuya, G.; Yang, J. Pore−pressure generation and fuidization in a loess landslide triggered by the 1920 Haiyuan earthquake, China: A case study. *Eng. Geol.* **2014**, *174*, 36–45. [CrossRef]
12. Zhong, X.; Xu, X.; Chen, W.; Liang, Y.; Sun, Q. Characteristics of loess landslides triggered by the 1927 Mw8.0 earthquake that occurred in Gulang County, Gansu Province, China. *Front. Environ. Sci.* **2022**, *10*, 973262. Available online: https://www.frontiersin.org/articles/10.3389/fenvs.2022.973262 (accessed on 9 September 2022). [CrossRef]
13. Tang, R.; Wu, J.; Ye, M.; Liu, W. Impact of Economic Development Levels and Disaster Types on the Short-Term Macroeconomic Consequences of Natural Hazard-Induced Disasters in China. *Int. J. Disaster Risk Sci.* **2019**, *10*, 371–385. [CrossRef]
14. Wei, Y.-N.; Fan, W.; Yu, B.; Deng, L.-S.; Wei, T. Characterization and evolution of three-dimensional microstructure of Malan loess. *Catena* **2020**, *192*, 104585. [CrossRef]
15. Wang, G.; Wei, J. Microstructure evolution of granular soils in cyclic mobility and post−liquefaction process. *Granul. Matter* **2016**, *18*, 51–64. [CrossRef]
16. Wang, J.D.; Li, P.; Ma, Y.; Vanapalli, S.K. Evolution of pore−size distribution of intact loess and remolded loess due to consolidation. *J. Soils Sediments* **2019**, *19*, 1226–1238. [CrossRef]
17. Wang, S.Y.; Luna, R.; Onyejekwe, S. Effect of initial consolidation condition on post−liquefaction undrained monotonic shear behavior of Mississippi river valley silt. *J. Geotech. Geoenviron. Eng.* **2016**, *142*, 04015075. [CrossRef]
18. Wen, B.-P.; Yan, Y.-J. Influence of structure on shear characteristics of the unsaturated loess in Lanzhou, China. *Eng. Geol.* **2014**, *168*, 46–58. [CrossRef]
19. Ma, J.L.; Zhong, X.M.; Wang, Q. Experimental Study on Shear Characteristics of Compacted Loess based on Micro−structure. *J. Disaster Prev. Mitig. Eng.* **2021**, *41*, 350–356. (In Chinese) [CrossRef]
20. Peng, J.; Wang, S.; Wang, Q.; Zhuang, J.; Huang, W.; Zhu, X.; Leng, Y.; Ma, P. Distribution and genetic types of loess landslides in China. *J. Asian Earth Sci.* **2018**, *170*, 329–350. [CrossRef]
21. Zhang, M.S.; Liu, J. Controlling factors of loess landslides in western China. *Environ. Earth Sci.* **2009**, *59*, 1671–1680. [CrossRef]
22. Pei, X.; Zhang, X.; Guo, B.; Wang, G.; Zhang, F. Experimental case study of seismically induced loess liquefaction and landslide. *Eng. Geol.* **2017**, *223*, 23–30. [CrossRef]
23. Xu, L.; Coop, M.R.; Zhang, M.; Wang, G. The mechanics of a saturated silty loess and implications for landslides. *Eng. Geol.* **2018**, *236*, 29–42. [CrossRef]
24. Rasouli, H.; Fatahi, B.; Nimbalkar, S. Re−liquefaction resistance of lightly cemented sands. *Can. Geotech. J.* **2021**, *59*, 2085–2101. [CrossRef]
25. Liu, W.; Wang, Q.; Lin, G.; Chen, W. Effect of pre-dynamic loading on dynamic liquefaction of undisturbed loess. *Bull. Earthq. Eng.* **2020**, *18*, 5779–5806. [CrossRef]

26. Wang, S.; Luna, R.; Onyejekwe, S. Postliquefaction behavior of low-plasticity silt at various degrees of reconsolidation. *Soil Dyn. Earthq. Eng.* **2015**, *75*, 259–264. [CrossRef]
27. Wang, S.; Luna, R.; Zhao, H. Cyclic and post-cyclic shear behavior of low-plasticity silt with varying clay content. *Soil Dyn. Earthq. Eng.* **2015**, *75*, 112–120. [CrossRef]
28. Wang, S.Y.; Luna, R. Compressibility characteristics of low–plasticity silt before and after liquefaction. *J. Mater. Civ. Eng.* **2014**, *26*, 040140141–1–6. [CrossRef]
29. Karakan, E.; Tanrinian, N.; Sezer, A. Cyclic undrained behavior and post liquefaction settlement of a nonplastic silt. *Soil Dyn. Earthq. Eng.* **2019**, *120*, 214–227. [CrossRef]
30. Liu, W.; Chen, W.; Wang, Q.; Wang, J.; Lin, G. Effect of pre-dynamic loading on static liquefaction of undisturbed loess. *Soil Dyn. Earthq. Eng.* **2019**, *130*, 105915. [CrossRef]
31. Ishihara, K.; Harada, K.; Lee, W.; Chan, C.; Safiullah, A. Post-liquefaction settlement analyses based on the volume change characteristics of undisturbed and reconstituted samples. *Soils Found.* **2016**, *56*, 533–546. [CrossRef]
32. Lade, P.; Yamamuro, J. Effects of nonplastic fines on static liquefaction of sands. *Canadian Geotech. J.* **1997**, *34*, 918–928. [CrossRef]
33. Song, B.; Tsinaris, A.; Anastasiadis, A.; Pitilakis, K.; Chen, W. Small-strain stiffness and damping of Lanzhou loess. *Soil Dyn. Earthq. Eng.* **2017**, *95*, 96–105. [CrossRef]
34. Sheander, N.; Raj, S.; John, A. Characteristics of nonlinear response of deep saturated soil deposits. *Bull. Seismol. Soc. Am.* **1997**, *87*, 342–355.
35. Wang, J.; Wang, Q.; Wang, J.M. Experimental study on the dynamic characteristics of post−earthquake loess. *Hydrogeol. Eng. Geol.* **2010**, *37*, 63–67. (In Chinese)
36. Wang, H.J.; Sun, P.; Liu, E.L.; Li, R.J. Dynamic properties of Tianshui saturated remolded loess: A laboratory study. *Eng. Geol.* **2020**, *272*, 105570. [CrossRef]
37. Suits, L.D.; Sheahan, T.C.; Elkady, T.Y.; Houston, S.L.; Houston, W.N. Static and Dynamic Behavior of Hydro-Collapsible Soils. *Geotech. Test. J.* **2011**, *34*, 103576. [CrossRef]
38. Zhang, Z.L.; Wang, T.; Wu, S.R.; Tang, H.M.; Xin, P.; Liang, C.Y. Dynamics stress–strain behavior of Tianshui soils. *Landslides* **2017**, *14*, 323–335. [CrossRef]
39. Araei, A.; Ghodrati, A. Predictive models for normalized shear modulus and damping ratio of modeled rockfill materials. *Acta Geodyn. Et Geomater.* **2017**, *14*, 27–40. [CrossRef]
40. Xu, Y.R.; Du, P.; Li, W.Q.; Zhang, W.H.; Tian, Q.J.; Xiong, R.W.; Wang, L. A case study on AD 1718 Tongwei M7. 5 earthquake triggered landslides application of landslide database triggered by historical strong earthquakes on the Loess Plateau. *Chin. J. Geophys. Chin. Ed.* **2020**, *63*, 1235–1248. (In Chinese) [CrossRef]
41. Duan, Y.M. Basic features of buried deep fault structure from Tongwei to Wudu. *ACTA Geol. Gausu* **2002**, *11*, 23–28. (In Chinese)
42. *GB 50123-2019*; Standard for Geotechnical Testing Method. China Metrology Publishing House: Beijing, China, 2019. (In Chinese)
43. *D4318−05;* Standard Test Methods for Liquid Limit, Plastic Limit, and Plasticity Index of Soils. ASTM: West Conshohocken, PA, USA, 2005.
44. *D854−06;* Standard Test Methods for Specific Gravity of Soil Solids by Water Pycnometer. ASTM: West Conshohocken, PA, USA, 2006.
45. *D422−63 (Reapproved 2007);* Standard Test Method for Particle−Size Analysis of Soils. ASTM: West Conshohocken, PA, USA, 2007.
46. *D3999−2011;* Standard Test Methods for the Determination of the Modulus and Damping Properties of Soils Using the Cyclic Triaxial Apparatus. ASTM: West Conshohocken, PA, USA, 2011.
47. Seed, H.B.; Idriss, I.M. Simplifed procedure for evaluating soil liquefaction potential. *J. Soil Mech. Found. Div.* **1971**, *97*, 1249–1273. [CrossRef]
48. *GB 18306−2015;* Seismic Ground Motion Parameters Zonation Map of China. China Standard Press: Beijing, China, 2015. (In Chinese)
49. *GB 50011−2010;* Code for Seismic Design of Buildings. China Building Industry Press: Beijing, China, 2010. (In Chinese)
50. Cheng, G.Y. Study on Correlation between Shear−Wave Velocity and Liquefaction Resistance of Saturated Sand. Ph.D. Thesis, Tianjin University, Tianjin, China, 2003. (In Chinese).
51. Li, W.; Lang, L.; Wang, D.; Wu, Y.; Li, F. Investigation on the dynamic shear modulus and damping ratio of steel slag sand mixtures. *Constr. Build. Mater.* **2018**, *162*, 170–180. [CrossRef]
52. Hardin, B.O.; Drnevich, V.P. Shear Modulus and Damping in Soils: Design Equations and Curves. *J. Soil Mech. Found. Div.* **1972**, *98*, 667–692. [CrossRef]
53. Jiang, M.; Zhang, F.; Hu, H.; Cui, Y.; Peng, J. Structural characterization of natural loess and remolded loess under triaxial tests. *Eng. Geol.* **2014**, *181*, 249–260. [CrossRef]
54. Huie, C.; Yaling, J.; Cencen, N.; Leng, G.J.; Tian, G.L. Dynamic characteristics of saturated loess under diferent confning pressures: A microscopic analysis. *Bull. Eng. Geol. Environ.* **2019**, *78*, 931–944. [CrossRef]

applied sciences

MDPI

Article

Effects of Cement Treatment on Mechanical Properties and Microstructure of a Granite Residual Soil

Xinxin Dong [1], Xiaohua Bao [1,*], Hongzhi Cui [1], Changjie Xu [2] and Xiangsheng Chen [1]

1 MOE Key Laboratory for Resilient Infrastructures of Coastal Cities, College of Civil and Transportation Engineering, Shenzhen University, Shenzhen 518060, China
2 School of Civil Engineering and Architecture, East China Jiao Tong University, Nanchang 330013, China
* Correspondence: bxh@szu.edu.cn

Abstract: A proper treatment of granite residual soil (GRS) in geotechnical practices requires both macro and microscopic evaluations. In this study, uniaxial and oedometric compression tests were conducted to investigate the mechanical properties of the saturated untreated and cement-treated GRS. Meanwhile, XRD, SEM, and MIP tests were conducted to identify the presence and types of C–S–H and the changes in the pore structure after cement treatment. The effects of cement treatment on the uniaxial compressive strength, secant modulus, compressibility, and vertical yielding pressure were revealed and the mechanisms of the soil structure to be modified through cement treatment were clarified based on the test results. A threshold volumetric cement content of 2–3% was determined based on the mechanical properties and microstructural characteristics of the saturated cement-treated GRS. Cement contents below this threshold would produce inadequate cementation between the soil particles. In contrast, cement contents above this threshold are considered inefficient because the transformation of the soil structure from single-porosity to dual-porosity increases the total porosity and retards the strength and stiffness gains.

Keywords: granite residual soil; cement treatment; mechanical property; microstructure

1. Introduction

Granite residual soil (GRS), one of the most commonly encountered soils in the tropics and subtropics, is often gap-graded and characterized by bimodal grain-size distribution curves from clay to gravel. For example, the GRS widely distributed in Shenzhen, China, is usually dominated by clay and coarse-sand particles whereas the fine- and medium-sand particles are found in very small amounts [1]. Due to its poorly graded nature, the GRS is difficult to compact similar to a clayey soil and disintegrate quickly after being soaked under free conditions similar to a sandy soil. Considering the distinct heterogeneity in its soil structure, GRS is treated as special soil in geotechnical engineering that needs to be dealt with carefully.

Cement treatment can improve the geotechnical performance of soil materials with small dosages of cement and allows for an immediate strength gain compared to lime and fly-ash treatment [2]. Therefore, it has been widely used in geotechnical practices such as embankment construction and ground improvement [3–5]. The mechanisms of cement treatment to improve soil properties include the occupation of large pores and a binding of the soil particles by the hydration products (i.e., C–(A)–S–H gel) of the cement component to form a denser and/or cemented soil structure [2,6]. The efficiency of cement treatment depends on the soil and cement types, cement content, and the treating technology [7–9]. In general, to achieve the same engineering indices, higher cement contents are required for clayey soils than for sandy soils. For a given geotechnical context and soil type, a proper cement dosage is important to meet the engineering requirements.

The effectiveness of cement treatment is judged primarily by the physicochemical and mechanical properties of the soil–cement mixtures. A review of past investigations showed that cement contents higher than 10% (on a dry-weight basis) are usually required for clayey soils to achieve satisfying engineering indices; in this case, higher cement contents produced lower consistency limits and a higher strength and stiffness of the treated soils [3,10,11]. For sandy soils, cement contents lower than 10% (on a dry-weight basis) are often adopted; in this case, higher cement contents yielded lower optimum water contents and maximum dry densities but higher strengths and stiffnesses of the treated soils [12–14]. As GRS is usually classified as silty sand or clayey sand (which is between sand and clay), the dosages for cement treatment and the physicomechanical properties of the treated soil are expected to vary with the mineralogy and grain compositions of the original soil. Nevertheless, the cement-treated GRS exhibits a higher compaction efficiency and durability than the original soil [15,16], and cement contents of around 5% (on a dry-weight basis) are preferred or suggested in the context of highway and high-speed railway embankment construction [17,18].

Apart from the mechanical characteristics, attention was paid to the alterations in the mineralogy and microstructure of soils following cement treatment. These efforts were mainly performed based on X-ray diffraction (XRD), scanning electron microscope (SEM), and mercury-intrusion porosimetry (MIP) tests. The presence of C–(A)–S–H can be identified from XRD patterns, Raman spectra, and thermogravimetry (TG) curves, and its morphology can be observed from SEM images. The C–(A)–S–H was observed to be fibrous at a very low cement content [19,20]. As the cement content increased, the C–(A)–S–H fibers became shorter and eventually reticular [21–23]. In addition to the cementation from C–(A)–S–H, changes in the pore structure (especially the pore size distribution, PSD) affect the mechanical properties of the treated soils and a few investigations were performed based on MIP tests. For sandy soils with single-porosity, an overall reduction in the pore size was revealed as the water–cement ratio increased [24,25]. For clayey soils with dual-porosity, a reduction in the intra-aggregate pores and an increase in the inter-aggregate pores were revealed as the cement content increased [23,26]. It can be inferred that changes in the pore structure with the cement content depend on the clay content of the origin soil. In the case of high clay content, an increase in the cement content may not produce an increase in the strength and stiffness of the treated soil because of the increase in the large pores [23]. Hence, a proper treatment of soils requires both macro- and micro-scopic evaluations.

This study investigates the mechanical properties and microstructure of a GRS treated with Portland fly-ash cement. The effects of cement treatment on the strength and stiffness of the treated soil are discussed based on the results of uniaxial and oedometric compression tests. Prediction formulas for the uniaxial compressive strength and secant modulus incorporating the porosity after curing, cement content, and curing time are also established. The presence and types of C–S–H and the changes in the pore structure following cement treatment are identified using XRD, SEM, and MIP tests. The effects of the pore structure change on the mechanical properties of the treated soil are discussed, and an appropriate cement dosage for treating the GRS is suggested based on mechanical and microstructural characteristics of the treated soil.

2. Experimental Procedures

2.1. Materials

The GRS used in this study is obtained from Shenzhen, Guangdong Province, China. The bulk GRS is air-dried and pulverized carefully using a rubber hammer to avoid breakage of the sand particles. The grain size distribution (GSD) of the air-dried GRS is measured by dry-sieving (after washing) in combination with the Hydrometer method [27]. The overall GSD curve of the GRS is shown in Figure 1; it is observed to be a gap-graded soil dominated by 40% fine particles ($d < 0.075$ mm) and 25% coarse particles ($d > 2$ mm) on a dry-weight basis. Sand particles coarser than 5 mm account for no more than 1%; for convenience, they are eliminated from the original soil in the following test procedures. The specific gravity of the fractions finer than 5 mm is measured using the Pycnometric method [28]. The consistency limits of the fractions finer than 0.5 mm are determined using a liquid-plastic limit combined tester [29]. A bimodal compaction curve is revealed by the standard Proctor test [30] (see

Figure 2); this may be due to the GRS's gap-graded nature. The optimum water content and maximum dry density are identified at the second peak of the compaction curve. The basic physical properties of the GRS are summarized in Table 1. According to the Unified Soil Classification System [31], the GRS is classified as silty sand.

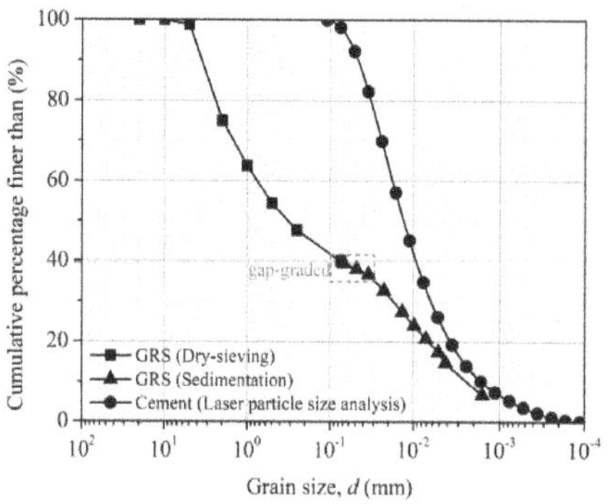

Figure 1. Grain-size distribution curves of GRS and Portland fly-ash cement.

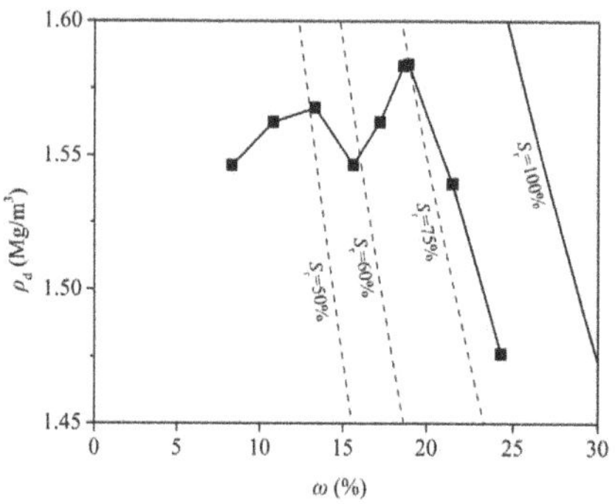

Figure 2. Compaction curve of GRS from standard Proctor test.

Table 1. Basic physical properties of granite residual soil.

G_s (-)	ω_L (%)	ω_P (%)	I_P (-)	ω_{opt} (%)	ρ_{dmax} (Mg/m^3)
2.64	53.5	30.7	22.8	18.8	1.58

Note: G_s = specific gravity; ω_L = liquid limit; ω_P = plastic limit; I_P = plastic index; ω_{opt} = optimum water content; and ρ_{dmax} = maximum dry density.

The XRD profile of the GRS with a grain size finer than 0.5 mm is shown in Figure 3, and the basic minerals are identified as Kaolinite (90%) and Quartz (10%) on a dry-weight basis. A SEM image of the GRS with a grain size finer than 0.5 mm is shown in Figure 4. The clay lamella of Kaolinite and its stacks (i.e., quasicrystal) are vivid in the SEM image, and the irregular quartz is wrapped in the clay lamellae. Despite the high clay content, clay aggregates (i.e., stacks of quasicrystals) are not recognized from SEM image, indicating a single-porosity structure of GRS compared to the dual-porosity structure of clays [32,33].

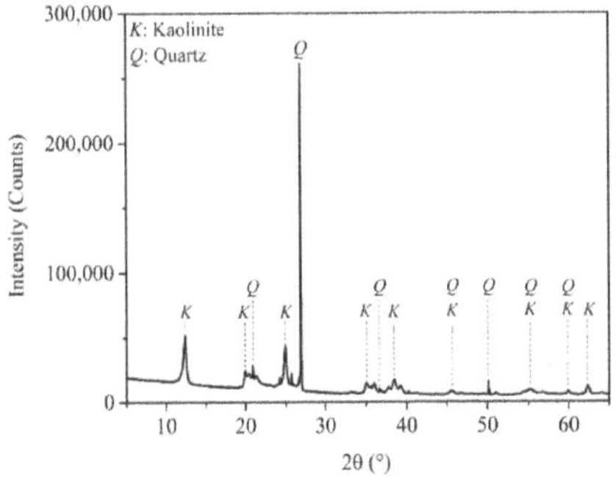

Figure 3. The XRD profile of GRS finer than 0.5 mm and identification of basic mineral compositions.

Figure 4. A SEM image of GRS finer than 0.5 mm and at a dry density of 1.44 Mg/m^3.

The Portland fly-ash cement used in this study is of a commercial type (PF32.5 [34]) containing 25% of active fly-ash on a dry-weight basis; this cement type is preferred in the context of embankment construction due to its economy and low hydration heat. The grain-size distribution of the powder cement is obtained by laser particle-size analysis with dry dispersion and the result is added to Figure 1. The specific gravity of the powder cement is estimated to be 3.10 via the Pycnometric method [28] with Kerosene (instead of water) used as the dispersion medium. The XRD profile of the oven-dried powder cement is shown in Figure 5, and the basic mineral compositions are identified as Tricalcium silicate (C_3S, 71%), Quartz (12%), Gypsum (8%), Calcite (6%), and Dicalcium silicate (C_2S, 2%) on a dry-weight basis. The XRD results do not reveal the presence of Tricalcium aluminate (C_3A) and Tetra-calcium aluminoferrite (C_4AF), possibly because of their low fractions.

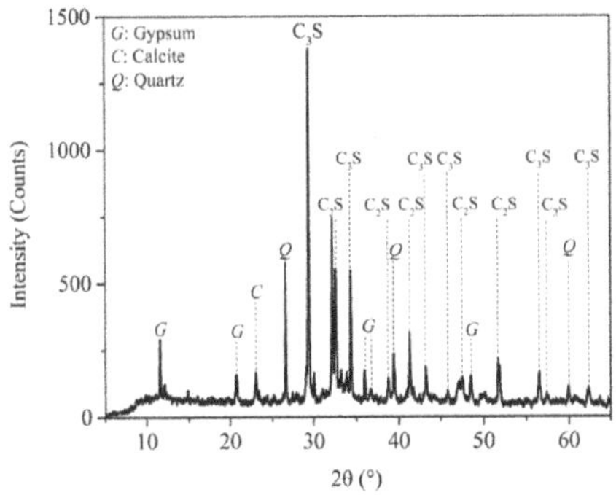

Figure 5. The XRD profile of Portland fly-ash cement and the identification of its basic mineral compositions.

2.2. Sample Preparation

To begin, the air-dried powder GRS is adjusted to a water content of 18.8% (i.e., ω_{opt}) and sealed in a plastic bag. This maturation lasts for 24 h to achieve a better homogenization of the moist condition. Following this, the powder cement is mixed with the maturated GRS and the specimen is compacted in three layers of equal thickness in a stainless steel mold or sample ring. The top surface of the lower layer is scarified prior to the compaction of the next layer to allow for better interlocking between the adjacent layers. The whole procedure is done in 30 min to avoid a hardening of the soil–cement mixture. Finally, the specimens are sealed with preservative film and placed into a curing chamber at a temperature of 20 °C and a relative humidity of 95%.

In all test types, the portion of GRS in a specimen is controlled to an identical dry density of 1.58 Mg/m^3 (i.e., ρ_{dmax}). The masses of the cement and maturated GRS required for preparing a specimen are calculated based on the dimensions of the specimen, target dry density of GRS (1.58 Mg/m^3), specific gravity of the cement, and cement content. The volumetric cement content (the ratio of the volume of cement particles to the volume of all solid particles, C_{vi}) is scheduled as 0% (untreated), 1%, 2%, 3%, 4%, and 5%, corresponding to a mass cement content (the ratio of the mass of cement particles to mass of all solid particles, C_{mi}) of 0% (untreated), 1.9%, 3.9%, 5.7%, 7.6%, and 9.4%, respectively. This yields an overall dry density of the specimens ranging from 1.580 Mg/m^3 (untreated) to 1.656 Mg/m^3 (C_{vi} = 5%). Detailed information on the as-compacted specimens, such as the initial moist condition, initial porosity, and equivalent grain density (ratio of the mass of all solids to volume of all solids), is listed in Table 2.

Table 2. Initial states of as-compacted untreated and cement-treated specimens.

C_{vi} (%)	C_{mi} (%)	ω_0 (%)	ρ_{d0} (Mg/m^3)	e_0 (-)	η_0 (-)	ρ_s (-)	S_{r0} (%)
0	0	18.8	1.580	0.671	0.402	2.640	74.0
1	1.94	18.4	1.595	0.660	0.398	2.648	74.0
2	3.85	18.1	1.610	0.649	0.393	2.655	74.0
3	5.72	17.7	1.626	0.638	0.389	2.663	74.0
4	7.56	17.4	1.641	0.627	0.385	2.670	74.0
5	9.36	17.0	1.656	0.617	0.381	2.677	74.0

Note: C_{vi} = volume cement content; C_{mi} = mass cement content; ω_0 = initial water content; ρ_{d0} = initial dry density; e_0 = inital void ratio; η_0 = initial porosity; ρ_s =equivalent grain density; and S_{r0} = initial degree of saturation.

2.3. Uniaxial Compression Tests

Uniaxial compression tests are conducted to determine the uniaxial compressive strength (UCS) and secant modulus (E_{50}) of the untreated and cement-treated GRSs following water saturation. The specimens for the uniaxial compression test are 50 mm in diameter and 100 mm in height. Four series of specimens with volumetric cement contents of 0% (untreated), 1%, 2%, 3%, 4%, and 5% in each series are cured to 1 d, 7 d, 14 d, and 28 d, respectively. A duplicate specimen is prepared for each cement content and curing time. Prior to uniaxial compression, the cured specimen is vacuum-saturated under confined conditions to ensure an identical saturation state (i.e., full saturation) and the surfaces of the saturated specimens are coated with a thin film of Vaseline to avoid air-drying. An electronic universal testing machine is utilized for the uniaxial compression tests. A loading speed of 0.1 mm/min is employed, as was suggested by ISRM [35]. Each test is ended after a peak state is achieved or when the axial strain exceeds 20%.

2.4. Oedometric Compression Tests

Oedometric compression tests are conducted to determine the compressibility and vertical yielding pressure of the untreated and cement-treated GRSs following water saturation. The specimens for the oedometric compression tests are 61.8 mm in diameter and 20 mm in height. Two series of specimens with volumetric cement contents of 0% (untreated), 1%, 2%, 3%, 4%, and 5% in each series are cured to 7 d and 28 d, respectively. Prior to the oedometer compression, the cured specimen is vacuum-saturated under confined conditions. A conventional oedometer with a loading capacity of 1600 kPa is employed. A loading–unloading–reloading cycle up to a vertical pressure of 1200 kPa is accomplished for each cement content and curing time. The loading cell is filled with water to maintain saturation of the specimen, and each loading and unloading stage lasts for 24 h. The reading deformation is calibrated based on the systematic deformation of the oedometer.

2.5. Microscopic Tests

XRD, SEM, and MIP tests are conducted to investigate the mineral compositions, soil fabric, and pore size distribution of the untreated and cement-treated GRSs following water saturation, respectively. The specimens for preparing the samples used in the microscopic tests are 61.8 mm in diameter and 20 mm in height. A series of specimens with volumetric cement contents of 0% (untreated), 1%, 2%, 3%, 4%, and 5% are cured to 28 d and subsequently vacuum-saturated under confined conditions. For each cement content, cube samples with a side length of approximately 2 cm are taken from the saturated specimen and immediately freeze-dried using liquid nitrogen and a vacuum-freezing dryer. The freeze-dried bulk samples are directly used for the SEM and MIP tests, and one of the freeze-dried bulk samples is ground into powder for use in the XRD tests. The equipment used for the XRD tests is the Bruker D8 Advance. XRD patterns in a range of 2θ = 5–80° are obtained using a Cu-$K\alpha$ (λ = 1.515 Å) X-ray tube and a continuous scan rate of 2°/min. SEM tests are conducted using a FEI Quanta 250; SEM images are captured from fresh fractures sprayed with gold. MIP tests are carried out using the Micromeritics AutoPore IV 9510 and each sample has accomplished an intrusion–extrusion cycle. The arrangements for the mechanical and microscopic tests are summarized in Table 3.

Table 3. Arrangements for mechanical and microscopic tests.

Test Types	C_{vi} (%)	t (d)
UCT	0, 1, 2, 3, 4, 5	1, 7, 14, 28
OCT	0, 1, 2, 3, 4, 5	7, 28
XRD, SEM, MIP	0, 1, 2, 3, 4, 5	28

Note: UCT= uniaxial compression test; OCT = oedometric compression test; C_{vi} = volume cement content; and t = curing time. All the cured specimens were saturated under confined conditions.

3. Results and Analysis

3.1. UCS and Secant Modulus

One set of the uniaxial compression curves of the saturated untreated and cement-treated GRSs is shown in Figure 6. For the untreated specimen, the stress–strain relationship is of strain-hardening type (see Figure 6a). In this case, the axial stress at an axial strain of 15% is adopted as the UCS of the specimen. For the treated specimens cured to 1–28 d, the stress–strain relationship is of strain-softening type (see Figure 6b–e). In this case, the UCS of the specimen is identified at the peak state. The relationship between the UCS and the cement content at four curing times is shown in Figure 7a. As the volumetric cement content increased from 0% to 3%, UCS increased linearly from 78 kPa to 413 kPa, 763 kPa, 1306 kPa, and 1277 kPa at curing times of 1 d, 7 d, 14 d, and 28 d, respectively. As the volumetric cement content exceeded 3%, the increase of the UCS with the cement content diminished. The relationship between the UCS and the curing time at five cement contents is shown in Figure 7b. For a given cement content, UCS increased linearly in the first 14 d and the increase of UCS with the curing time was then diminished (most notably in the case of C_{vi} = 1–3%). In detail, the UCS of the specimens cured to 1 d, 7 d, and 14 d reached 31–53%, 59–71%, and 79–107%, respectively, of those cured to 28 d, and these ratios decreased as the cement content increased. In the case of $C_{vi} \geq 4\%$, the UCS developed successively within 28 d; in the case of $C_{vi} \leq 3\%$, the development of the UCS ended at an early age of 14 d.

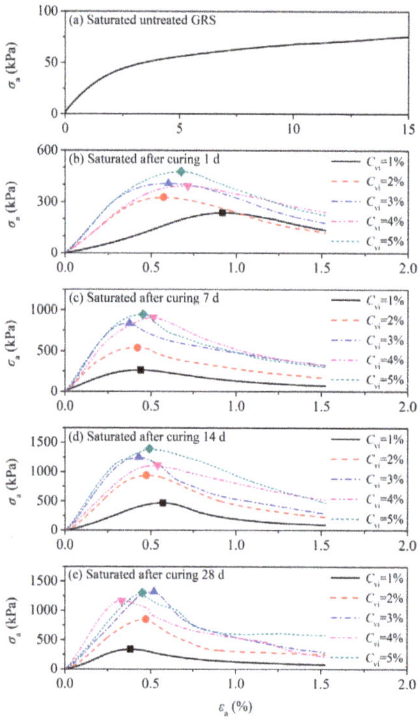

Figure 6. Stress–strain curves of saturated untreated and cement-treated GRSs at various cement contents and curing times.

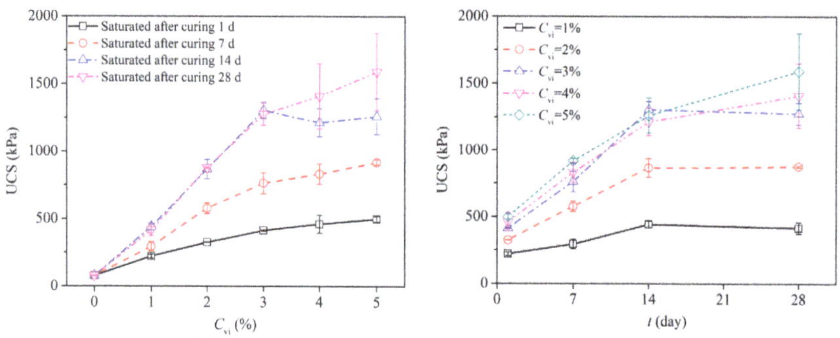

Figure 7. Uniaxial compressive strength (UCS) of saturated untreated and cement-treated GRSs at various cement contents and curing times. (**a**) Changes in UCS with cement content at various curing times; (**b**) Changes in UCS with curing time at various cement contents.

In all the cases, the uniaxial compression curves are nonlinear before the peak state (see Figure 6). In this study, the secant modulus is determined to describe the stiffness of the specimen. The dependence of the secant modulus on the cement content and curing time is shown in Figure 8. The experimental trends are similar to that of the UCS. As the volumetric cement content increased from 0% to 3%, the secant modulus increased almost linearly from 2.8 MPa to 116 MPa, 240 MPa, 319 MPa, and 292 MPa at a curing time of 1 d, 7 d, 14 d, and 28 d, respectively (see Figure 8a). As the volumetric cement content exceeded 3%, the increase of the secant modulus with the cement content diminished. For a given cement content, the secant modulus of the specimens increased rapidly in the first 7–14 days, after which the increase of the secant modulus with the curing time diminished (see Figure 8b). Specifically, the secant modulus of the specimen cured to 1 d, 7 d, and 14 d reached 28–43%, 67–109%, and 84–109%, respectively, of those cured to 28 d. A similar conclusion can be drawn that the secant modulus of the specimens developed successively in 28 d in the case of $C_{vi} \geq 4\%$ and may have ended in 7–14 d in the case of $C_{vi} \leq 3\%$.

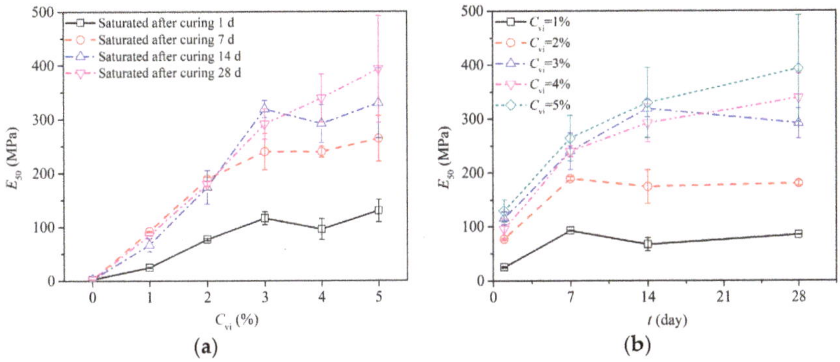

Figure 8. Secant modulus (E_{50}) of saturated untreated and cement-treated GRSs at various cement contents and curing times. (**a**) Changes in E_{50} with cement content at various curing times; (**b**) Changes in E_{50} with curing time at various cement contents.

3.2. Compressibility and Vertical Yielding Pressure

Oedometric compression curves of the saturated untreated and cement-treated GRSs cured to 7 d and 28 d are shown in Figures 9 and 10, respectively. The $e-\log\sigma_v$ relationship is nonlinear in all the cases, indicating initial over-consolidation states of the specimens due to compaction. The untreated specimen entered the normal consolidation state as the vertical pressure exceeded 400 kPa (see Figure 9). However, the states of the treated specimen cannot be confirmed due to the moderate stress level (see Figure 10). Nevertheless, the compression index (C_c) of the specimens was determined via linear fitting of the $e-\log\sigma_v$ relationship in the range of σ_v > 400 kPa (for C_{vi} = 0–4%) or σ_v > 800 kPa (for C_{vi} = 5%). The unloading–reloading cycle yielded a hysteresis loop of the $e-\log\sigma_v$ relationship; its diagonal slope (absolute value) was accepted as the rebound index (C_r) of the specimen. Given the subjectivity of Casagrande method [36], an alternate approach was employed to determine the vertical yielding pressure (σ_y) of the specimen [37]. As all the compression curves are horizontal in the case of σ_v < 12.5–25 kPa, the vertical pressure at which the intrinsic compression line (ICL) intersects the initial void ratio was deemed to be the vertical yielding pressure of the specimen (see Figures 9 and 10).

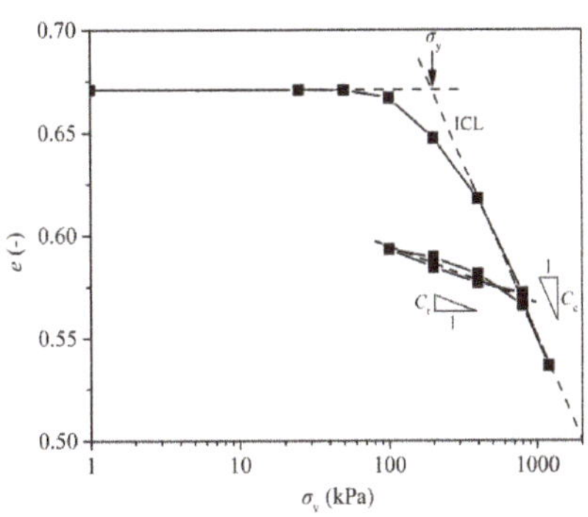

Figure 9. $e-\log\sigma_v$ relationship of saturated untreated GRS.

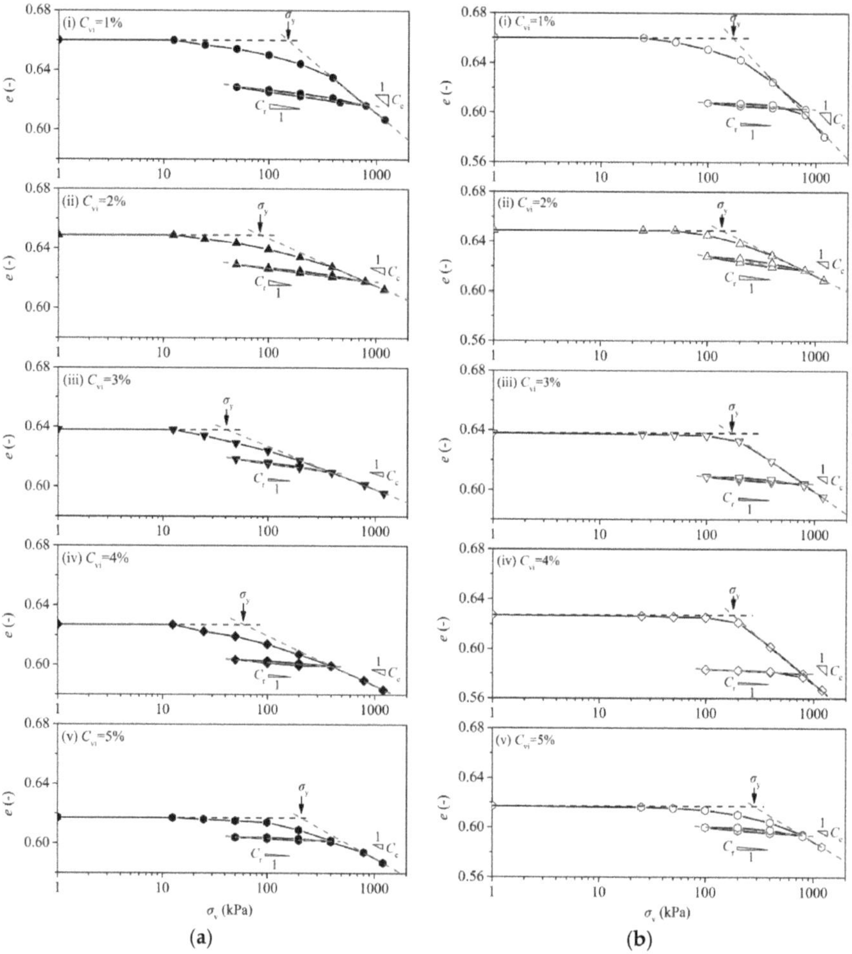

Figure 10. $e-\log\sigma_V$ relationship of saturated cement-treated GRS. (**a**) Curing time = 7 d; (**b**) Curing time = 28 d.

The dependence of the compression index, rebound index, and vertical yielding pressure on the cement content and curing time is shown in Figure 11. It is observed that, for a given cement content, the specimen cured to 28 d has a slightly larger compression index compared with the case of 7 d (see Figure 11a). As the volumetric cement content increased to 2%, the compression index of the specimens cured to 7 d and 28 d dropped sharply from 0.17 to 0.03 and 0.04, respectively. As the volumetric cement content exceeded 2%, the variation of the compression index with the cement content became less significant. The rebound index of the treated specimens is much smaller than that of the untreated one (0.003–0.01 versus 0.028) and changes in the rebound index with the cement content and curing time are less significant (see Figure 11a). For a given cement content, the vertical yielding pressure of the specimens cured to 28 d is much higher than that of those cured to 7 d (see Figure 11b). For a given curing time, the vertical yielding pressure of the specimens decreased successively and was followed by a continuous increase as the volumetric cement content increased. The volumetric cement content at which the tendency reversed is 2–3%.

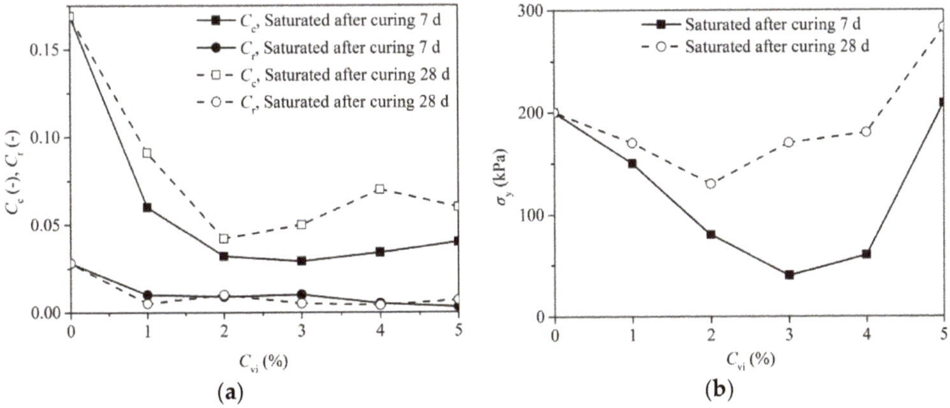

Figure 11. Compressibility and vertical yielding pressure of saturated untreated and cement-treated GRSs depending on cement content and curing time. (**a**) Compressibility; (**b**) Vertical yielding pressure.

3.3. Mineralogy and Microstructure

Cementitious products include the C–(A)–S–H and CH (calcium hydroxide). The C–A–S–H and C–A–H are unidentifiable due to the negligible C_3A in the cement used in this study. XRD patterns of the saturated untreated and cement-treated GRSs cured to 28 d are shown in Figure 12. The crystalline minerals include the Kaolinite and Quartz from the original soil and the non-hydratable Calcite from the cement (see Figure 12a). For the treated specimens cured to 28 d, C–S–H was identified at 2θ = 29.0°, 29.2°, 29.5°, and 32.1° (see Figure 12b) [38–40].

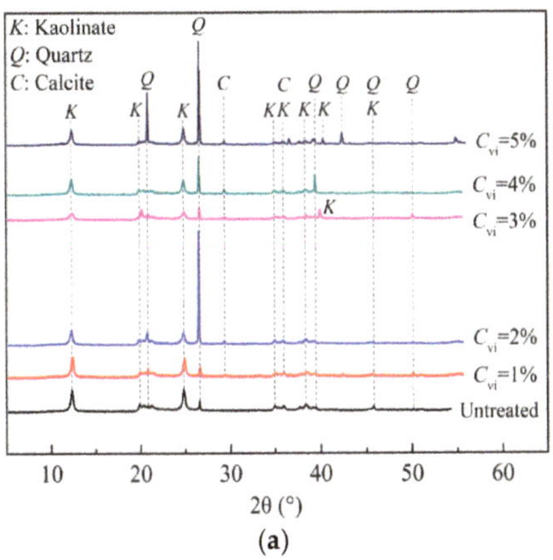

Figure 12. *Cont.*

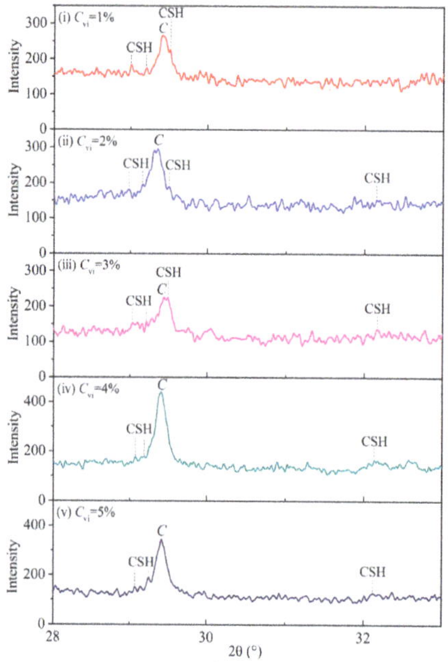

(b)

Figure 12. XRD profiles of saturated untreated and cement-treated GRSs cured to 28 d. (**a**) Identification of crystalline minerals; (**b**) Identification of C–S–H.

SEM images of the saturated untreated and cement-treated GRSs are shown in Figure 13. The untreated specimen consists of clay lamella and quasicrystals (see Figure 13a). By convention, the pores within the quasicrystals are deemed to be micro-pores, and those between the quasicrystals are referred to as macro-pores. For the treated specimens, their microscopic morphology and the types of C–S–H [41] were observed to depend on the cement content. The C–S–H is fibrous (C–S–H of type I) in the case of C_{vi} = 1–2% (see Figure 13b,c); the long C–S–H fibers were observed to fill the macro-pores (see Figure 13b), whereas the short ones attached to the surfaces of the clay lamellae (see Figure 13c). The C–S–H became reticular (C–S–H of type II) in the case of C_{vi} = 3–5%; the quasicrystals were assembled by the C–S–H and aggregates (stacks of quasicrystals) of larger size and more rounded geometry were observed to form (see Figure 13d–f). CH of a hexagonal tabular shape can be observed in the inter-aggregate pores as the volumetric cement content increased to 5% (see Figure 13f).

The changes in the pore structure due to cement treatment can be assessed based on the MIP results. The pore size distribution and cumulative curves of the saturated untreated and cement-treated GRSs cured to 28 d are shown in Figure 14. The PSD curves were derived from the differential intrusion volume of the mercury (per unit mass, V_{in}) multiplied by the equivalent grain density (i.e., $de_{in}/d\log D = dV_{in}/d\log D \times \rho_s$) (see Figure 14a). The sudden increase of intrusion at a pore size of 5.4 μm is an artifact due to the transfer of the samples from the low-pressure chamber to the high-pressure chamber during MIP tests. The untreated specimen is of single-porosity, characterized by a single peak at a pore diameter of about 0.4 μm in the PSD curve; this single pore family represents both the pores within and between the quasicrystals. The treated specimens are of dual-porosity with two peaks in the PSD curve representing two distinguishable pore families; the first pore family represents the pores within the aggregates, whereas the second represents the pores between the aggregates. The first peak was observed to decrease with the cement content in the case of $C_{vi} \leq 2\%$, and this tendency reversed as the cement content further increased. The pore diameter corresponding to the first peak decreased successively as the cement content increased. The second peak is not distinct in the case of $C_{vi} \leq 2\%$ and becomes notable as the cement content further increases. Contrary to the first peak, the second peak and its corresponding pore diameter increased sequentially as the cement content increased.

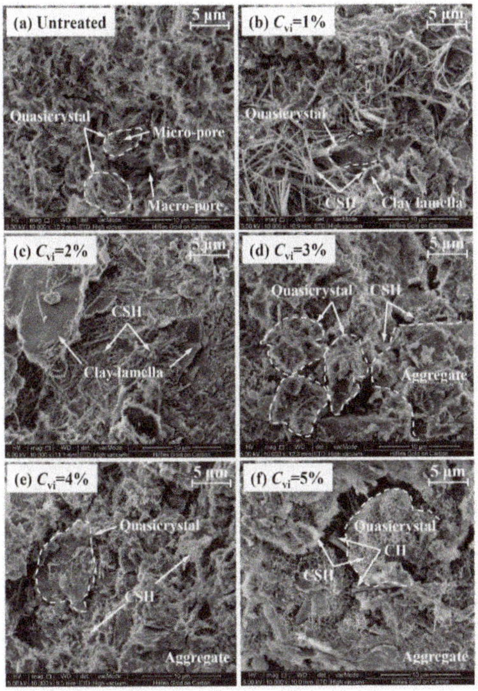

Figure 13. SEM images of saturated untreated and cement-treated GRSs cured to 28 d.

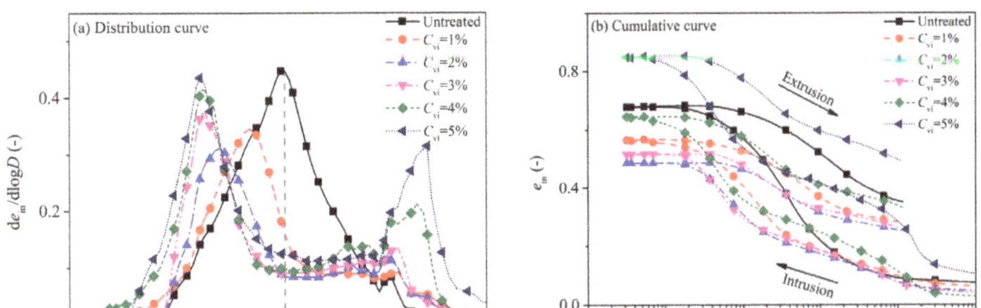

Figure 14. MIP results of saturated untreated and cement-treated GRSs cured to 28 d.

The cumulative curves came from the cumulated intrusion volume of the mercury in an intrusion–extrusion cycle multiplied by the equivalent grain density (i.e., $e_{in} = V_{in} \times \rho_s$) (see Figure 14b). It was assumed that the mercury was retained in the macro-pores after the intrusion pressure was removed [32,33]. Therefore, the macro-void ratio (ratio of the volume of macro-pores to the volume of all solid particles, e_M) equals the intruded void ratio after an intrusion–extrusion cycle. In addition, the maximum e_{in} was identified and adopted as the intruded void ratio of the specimen (e_{in}). The micro-void ratio (ratio of the volume of micro-pores to the volume of all solid particles, e_m) is the difference between the void ratio (e) of the specimen and the macro-void ratio (i.e., $e_m = e - e_M$). As the void ratio of the specimen may change during curing, the exact micro-void ratio cannot be determined based on the initial void ratio (e_0). Figure 15 illustrates the delimitation of the macro- and micro-pores for the saturated untreated specimens as an example. The e_{in} and e_M of the specimens were determined based on this criterion, and the dependences of e_0, e_{in}, and e_M on the cement content are shown in Figure 16. The e_{in} and e_M were observed to decrease with the cement content in the range of $C_{vi} \leq 2\%$, and these tendencies reversed as the cement content further increased. For the untreated specimen, the e_{in} equals the e_0, indicating no changes in the void ratio during confined wetting; this allows the e_M and e_m of the untreated specimen to be determined (giving e_M = 0.35 and e_m = 0.33). For the treated specimens, the differences between the e_{in} and e_0 are the combined effects of the formation of C–S–H and aggregates. The detailed mechanisms are discussed and the void ratios of the cured specimens are estimated in Section 4.3.

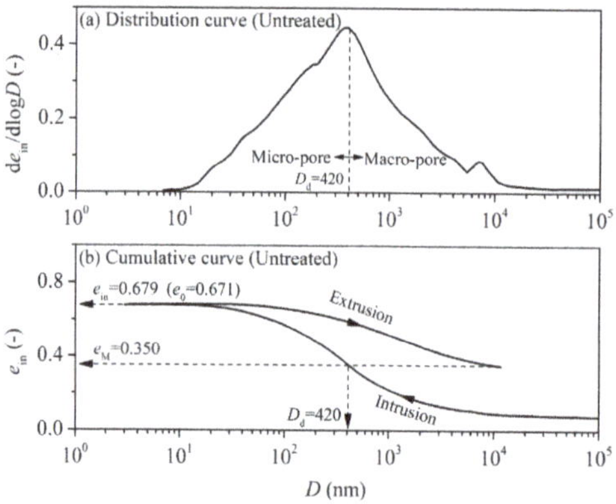

Figure 15. Delimitation of macro- and micro-pores in saturated untreated GRS based on results of an intrusion–extrusion cycle.

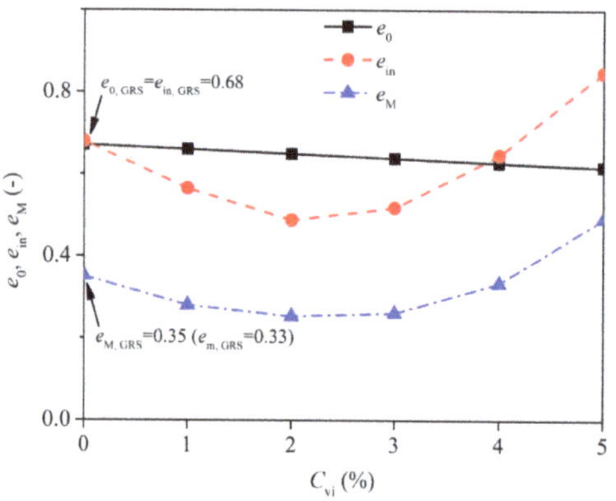

Figure 16. Initial void ratio, intruded void ratio, and macro-void ratio of saturated untreated and cement-treated GRSs cured to 28 d.

4. Discussion

4.1. Prediction of UCS and Secant Modulus

Various methods have been proposed to predict the mechanical properties of cement-treated soils, among which the unconfined compressive strength, triaxial shear strength, and tensile strength were well-linked to an adjusted porosity/cement ratio, $\eta/(C_{vi})^{\xi}$ [42–44].

The exponent ξ was found to be a positive value smaller than unity, meaning that, for a given soil type, various strengths of the treated soil cannot be predicted based on those of the untreated one (C_{vi} = 0). According to the literature review, the strength and stiffness of soils increases with the increasing cement content and the decreasing porosity (and the decreasing $\eta/(C_{vi})^{\xi}$ as a result). Hence, in this study, an alternative approach using a parameter $\eta(C_{vi})^{\xi}$ ($\xi > 0$) was employed to merge the gap between the treated and untreated states, allowing for a prediction of the strength and stiffness of the treated soil based on its untreated state. The term $\eta(C_{vi})^{\xi}$ can describe the net effects of the porosity (η) and cement content (C_{vi}), and the exponent ξ adjusts the weights of the two contradictory factors. Figure 17a presents the UCS–$\eta(C_{vi})^{\xi}$ relationship of the saturated untreated and treated specimens in the case of ξ = 1. For a given curing time, the UCS of the specimens exhibited a diminished increase with the increasing ηC_{vi} and an exponential correlation can be established; these exponential tendencies do not change in the case of $\xi > 1$ but become irregular in the case of $\xi < 1$. Therefore, the ηC_{vi} (i.e., ξ = 1) was adopted for simplicity and an exponential function was formulated to fit the testing data at each curing time:

$$\mathrm{UCS}(\eta, C_{vi}) = \mathrm{UCS}(\eta, C_{vi} = 0) + \alpha_{UCS}[1 - \exp(-\eta C_{vi}/\beta_{UCS})] \quad (1)$$

where UCS(η, C_{vi} = 0) is the UCS of the untreated specimen and depends on the porosity. The α_{UCS} and β_{UCS} describe the diminished increase of the UCS with the ηC_{vi}; they depend on the curing time and take into account the effects of curing on the porosity.

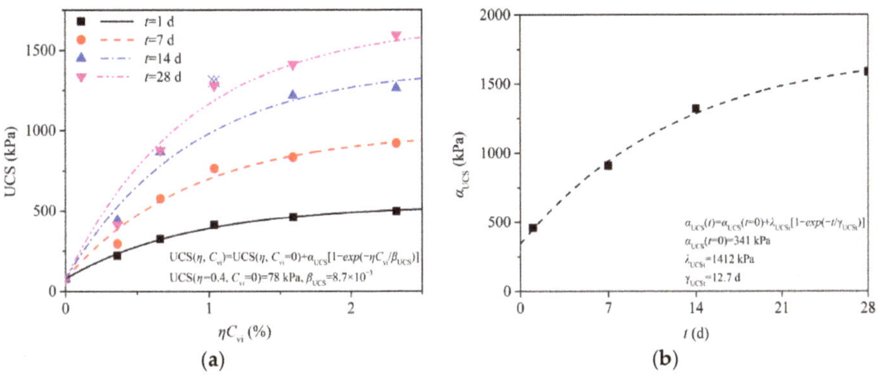

Figure 17. Modeling of the UCS of saturated cement-treated GRS (Hollow symbol with cross indicates data excluded in the fitting). (**a**) Fitting of UCS at four curing times; (**b**) α_{UCS} dependent on curing time.

The UCS(η = 0.4, C_{vi} = 0) was fixed at 78 kPa for the untreated specimen based on the test results and a direct fitting of the four data sets gave $\beta_{UCS} = (8.7 \pm 0.5) \times 10^{-3}$. Given its small variation, β_{UCS} was fixed at 8.7×10^{-3}, independent of the curing time. The adjusted fitting is presented in Figure 17a, and the dependence of α_{UCS} on the curing time is shown in Figure 17b. The α_{UCS} also presents a diminished increase with the curing time and an exponential function can be formulated:

$$\alpha_{UCS}(t) = \alpha_{UCS}(t = 0) + \lambda_{UCSt}[1 - \exp(-t/\gamma_{UCSt})] \quad (2)$$

where, $\alpha_{UCS}(t = 0)$ is the α_{UCS} for the non-cured specimen. The λ_{UCSt} and γ_{UCSt} describe the diminished increase of α_{UCS} with the curing time. The $\alpha_{UCS}(t = 0)$, λ_{UCSt}, and γ_{UCSt} were fitted to be 341 kPa, 1412 kPa, and 12.7 d, respectively (see Figure 17b).

Given the same experimental trends as that of UCS, the secant modulus of the saturated treated specimens was predicted based on the E_{50}–ηC_{vi} relationship in Figure 18a:

$$E_{50}(\eta, C_{vi}) = E_{50}(\eta, C_{vi} = 0) + \alpha_{E50}[1 - \exp(-\eta C_{vi}/\beta_{E50})]. \quad (3)$$

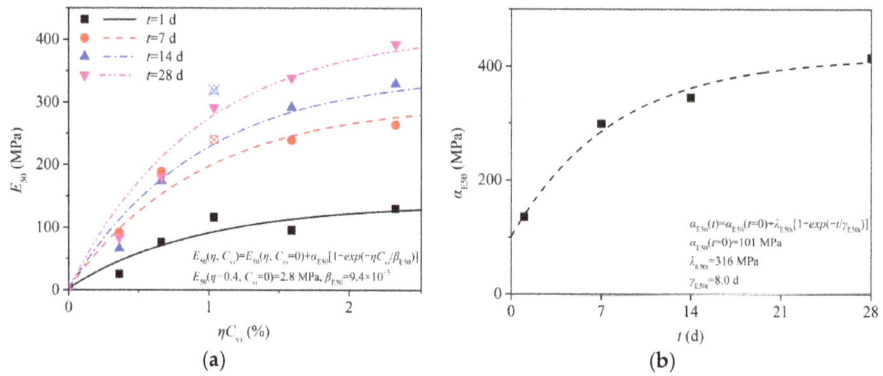

Figure 18. Modeling of E_{50} of saturated cement-treated GRS (Hollow symbol with cross indicates data excluded in the fitting). (**a**) Fitting of E_{50} at four curing times; (**b**) α_{E50} dependant on curing time.

Similarly, $E_{50}(\eta = 0.4, C_{vi} = 0)$ was fixed at 2.8 MPa for the saturated untreated specimen, and a direct fitting of the four data sets gave $\beta_{E50} = (9.4 \pm 1.8) \times 10^{-3}$. Despite its slightly larger variation than that of the β_{UCS}, β_{E50} was fixed at 9.4×10^{-3} for the four curing times. The adjusted fitting is plotted in Figure 18a, and the dependence of α_{E50} on the curing time is shown in Figure 18b. The α_{E50} demonstrates a diminished increase with the curing time and an exponential function was established:

$$\alpha_{E50}(t) = \alpha_{E50}(t = 0) + \lambda_{E50t}[1 - \exp(-t/\gamma_{E50t})] \quad (4)$$

where $\alpha_{E50}(t = 0)$ is the α_{E50} for the non-cured specimen. The λ_{E50t} and γ_{E50t} describe the diminished increase of α_{E50} with the curing time. The $\alpha_{E50}(t = 0)$, λ_{E50t}, and γ_{E50t} were fitted to be 101 MPa, 316 MPa, and 8.0 d, respectively (see Figure 18b).

It is noted that the β_{UCS} and β_{E50} are of comparable values. Hence, an approximate value of $\beta = 0.009$ can be adopted for the β_{UCS} and β_{E50}. An integrated formula can be established to predict the strength and stiffness of the saturated cement-treated GRS:

$$\Phi(\eta, C_{vi}) = \Phi(\eta, C_{vi} = 0) + \alpha(t)[1 - \exp(-\eta C_{vi}/\beta)] \quad (5)$$

$$\alpha(t) = \alpha(t = 0) + \lambda_t[1 - \exp(-t/\gamma_t)] \quad (6)$$

where $\Phi(\eta, C_{vi} = 0)$ is the strength and stiffness of the saturated untreated specimen. The parameter α describes the increase of the strength and stiffness with the ηC_{vi}, and β adjusts its diminished trends. However, α still depends on the curing time and needs to be determined via Equations (2) and (4).

4.2. *Effect of Cement Treatment on Oedometric Compression Behaviour*

The cementation induced by C–S–H would deteriorate to varying degrees during the loading of a cured specimen. The complete destruction of the cementation at high stress was supposed to yield a consistent intrinsic compression line (ICL) or paralleled ICLs for weakly cemented soils [45–47], however, this may not be the case for strongly cemented soils. Figure 19 shows the integrated plots of the $e-\log\sigma_v$ relationship during successive loading of the saturated untreated and cement-treated specimens. The ICL of the saturated untreated specimen was presented as a reference, and the oedometric compression curves of the saturated treated specimens were extrapolated to judge the potential convergence or parallelism of ICLs at high stress. The oedometric compression curve of the untreated specimen intersects those of the treated specimens due to its highest initial void ratio and compressibility. This is also the case at $C_{vi} = 1\%$ due to its secondary initial void ratio and much higher compressibility when compared to the cases of $C_{vi} = 2–5\%$ (see Figure 11a). In the case of C_{vi} = 1%, the ICL of the specimen cured to 28 d is likely to parallel with that of the untreated specimen at high stress. In the case of C_{vi} = 2–5%, the ICLs of the specimens cured to 7 d and 28 d tend to parallel each other at high stress despite the initial void ratio decreasing with the cement content. This resulted in an overall increase of the vertical yielding pressure with the cement content in the range of C_{vi} = 2–5% (see Figure 11c).

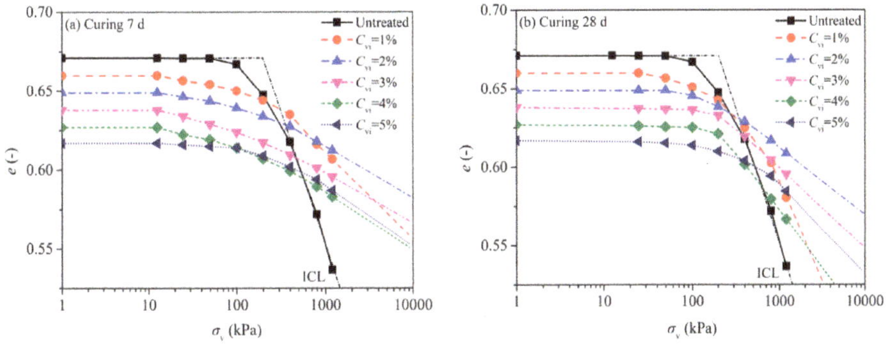

Figure 19. Integrated plot of the $e-\log\sigma_v$ relationship of saturated untreated and cement-treated GRSs. (**a**) Curing time = 7 d; (**b**) Curing time = 28 d.

The deterioration of the soil cementation can also be judged via the evolution of the compression index upon compression. The compression index (C_c) at a vertical pressure of σ_v is calculated by:

$$C_c(\sigma_{vi}) = -(e_{i+1} - e_i)/(\log\sigma_{v,i+1} - \log\sigma_{v,i}) \quad (7)$$

where e_i is the void ratio of the specimen after being loaded to $\sigma_{v,i}$.

The evolution of the compression index during successive loading of the specimens is shown in Figure 20. In all the cases, a diminishing increase of the compression index with the vertical pressure was revealed, indicating a gradual approach of the specimens into the normal consolidation state; this is a consequence of the breakage of the interparticle cementation and rearrangement of the discrete soil particles upon compression. In addition, for the specimens cured to 7 d, the difference in the compression index is not distinct in the case of $C_{vi} \geq 2\%$ compared to the case of $C_{vi} \leq 1\%$; this may be because the interparticle cementation was not fully established after curing to 7 d and the differences in the compressibility depend mainly on the presence and types of C–S–H (see Figure 13). The interparticle cementation may have been fully established after the specimens were cured to 28 d and, in this case, the compression index at a given vertical pressure depends largely on the cement content. Nevertheless, the compression index tends to converge at high vertical pressures in the case of $C_{vi} \geq 2\%$.

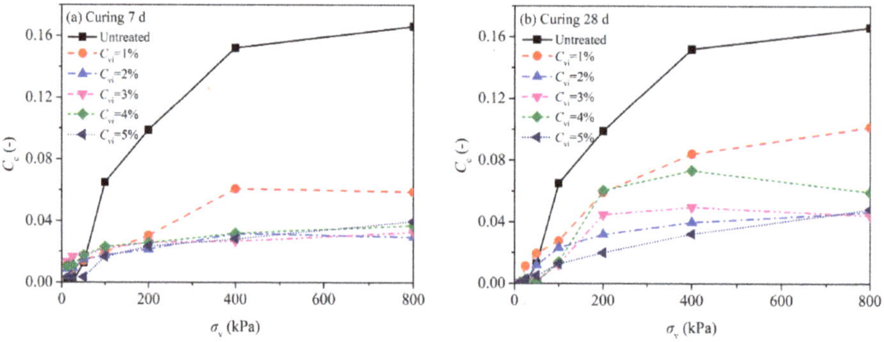

Figure 20. Compression index of saturated untreated and cement-treated GRSs depending on vertical pressure. (**a**) Curing time = 7 d; (**b**) Curing time = 28 d.

4.3. Effect of Cement Treatment on Pore Structure

To further quantify the effects of cement treatment on pore structure, a detailed analysis must be performed based on the MIP results. The volumetric constitutions of the as-compacted untreated and cement-treated specimens are illustrated in Figure 21. For a given cement content, various void ratios of the as-compacted specimen are defined as:

$$e_{\mathrm{Mi0}} = V_{\mathrm{vMi0}} / (V_{\mathrm{si0}} + V_{\mathrm{s0,GRS}}) \tag{8}$$

$$e_{\mathrm{mi0}} = V_{\mathrm{vmi0}} / (V_{\mathrm{si0}} + V_{\mathrm{s0,GRS}}) \tag{9}$$

$$e_{\mathrm{i0}} = e_{\mathrm{Mi0}} + e_{\mathrm{mi0}} \tag{10}$$

where V_{vMi0} and V_{vmi0} are the volumes of the macro- and micro-pores in an as-compacted specimen, respectively. V_{si0} and $V_{\mathrm{s0,GRS}}$ are the volumes of the solid particles constituting the cement and GRS components in an as-compacted specimen, respectively.

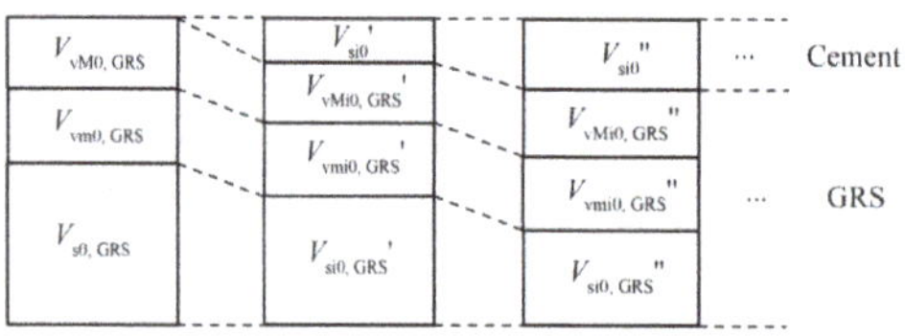

Figure 21. Illustration of volumetric constitutions of an as-compacted specimen with different cement contents.

Additionally, the volumetric cement content was previously defined as:

$$C_{\mathrm{vi}} = V_{\mathrm{si0}} / (V_{\mathrm{si0}} + V_{\mathrm{s0,GRS}}). \tag{11}$$

For the untreated specimen, the macro- and micro-void ratios that followed confined wetting were determined based on the MIP results (see Figure 15). Given the non-swelling nature of Kaolinite, it is likely that the pore structure of the GRS altered slightly upon confined wetting; this is evidenced by MIP results that the intruded void ratio of the saturated untreated specimen is close to its initial void ratio (see Figure 16). Hence, the macro- and micro-void ratios of the as-compacted untreated specimen are adopted as:

$$e_{\mathrm{M0,GRS}} = V_{\mathrm{vM0,GRS}} / V_{\mathrm{s0,GRS}} = 0.35 \tag{12}$$

$$e_{\mathrm{m0,GRS}} = V_{\mathrm{vm0,GRS}} / V_{\mathrm{s0,GRS}} = 0.32 \tag{13}$$

where $V_{vM0,GRS}$ and $V_{vm0,GRS}$ are the volumes of the macro- and micro-pores in an as-compacted untreated specimen, respectively.

Moreover, the GRS components in all the as-compacted specimens were scheduled to an identical dry density of 1.58 Mg/m^3 (which gives $e_{0,GRS}$ = 0.67); therefore, the following correlations can be established according to Figure 21:

$$V_{vMi0,GRS}/V_{si0,GRS} = V_{vM0,GRS}/V_{s0,GRS} = e_{M0,GRS} \tag{14}$$

$$V_{vmi0,GRS}/V_{si0,GRS} = V_{vm0,GRS}/V_{s0,GRS} = e_{m0,GRS} \tag{15}$$

where $V_{vMi0,GRS}$ and $V_{vmi0,GRS}$ are the volumes of the macro- and micro-pores of the GRS component in an as-compacted treated specimen, respectively. The $V_{si0,GRS}$ is the volume of the solid particles of the GRS component in an as-compacted treated specimen.

Finally, the macro- and micro-void ratios of the as-compacted treated specimens are derived as:

$$e_{Mi0} = V_{vMi0}/(V_{si0} + V_{si0,GRS}) = (1 - C_{vi})e_{M0,GRS} \tag{16}$$

$$e_{mi0} = V_{vmi0}/(V_{si0} + V_{si0,GRS}) = (1 - C_{vi})e_{m0,GRS} \tag{17}$$

During the curing of a treated specimen, the hydration of its cement component modifies the pore structure in three possible ways: (i) the formation of C–S–H and the micro-pores within the C–S–H; (ii) the intrusion of C–S–H into the macro- and micro-pores between the quasicrystals and clay lamellae (see Figure 13b,c); and (iii) a wrapping of the quasicrystals by the C–S–H to form aggregates and the emergence of the macro-pores (between the aggregates) and micro-pores (within the aggregates) (see Figure 13d–f). The C–S–H was regarded as a porous medium with nano-scale pores of 0.5–2.5 nm [48]. The minimum pore size that the MIP test can detect is 3 nm; hence, the formation of C–S–H produced non-detectable pores (sorted into the micro-pores) within the C–S–H. In addition, it was concluded that the porosity of C–S–H ($\eta_{CSH} = V_{v,CSH}/V_{CSH}$, where V_{CSH} and $V_{v,CSH}$ are the volumes of C–S–H and the pores within the C–S–H, respectively) in Portland cement paste had a fixed value of 0.28 [48]. Given the small cement dosage, changes in the volume of the solid particles (particularly the cement particles) upon hydration are ignored, and the void ratio of the non-detectable micro-pores in a cured specimen is estimated by:

$$e_{m,CSH} = V_{v,CSH}/(V_{si0} + V_{s0,GRS}) = \eta_{CSH}/(1 - \eta_{CSH})C_{vi}. \tag{18}$$

As almost all the pores in the GRS component can be detected via MIP tests (as was revealed in Figure 15), the first mechanism is responsible for the difference between the intruded void ratio and (total) void ratio of the treated specimens after curing. Hence, the void ratios and micro-void ratios of the saturated treated specimens can be calculated by:

$$e = e_{in} + e_{m,CSH} \tag{19}$$

$$e_m = e - e_M. \tag{20}$$

The macro- and micro-void ratios and (total) void ratios of the as-compacted and saturated specimens cured to 28 d are presented in Figure 22a. In the case of $C_{vi} \leq 2\%$, the macro- and micro-void ratios decreased successively as the cement content increased; this may be because the occupation of the macro- and micro-pores by the C–S–H (i.e., the second mechanism) surpassed the formation of the micro-pores within the C–S–H (i.e., the first mechanism) and the emergence of the aggregates (i.e., the third mechanism) is invalid in this case (as was revealed in Figure 13b,c). As the volumetric cement content increased from 2% to 4%, aggregates gradually formed (see Figure 13d,e), and the macro- and micro-void ratios started to increase; as long as the second mechanism dominated, the macro- and micro-void ratios would be lower than their initial values. As the volumetric cement content exceeded 4%, the dual-porosity characteristic of the cured specimens became remarkable (see Figure 13f). In this case, the third mechanism is believed to dominate and the macro- and micro-void ratios of the cured specimens exceed their initial values.

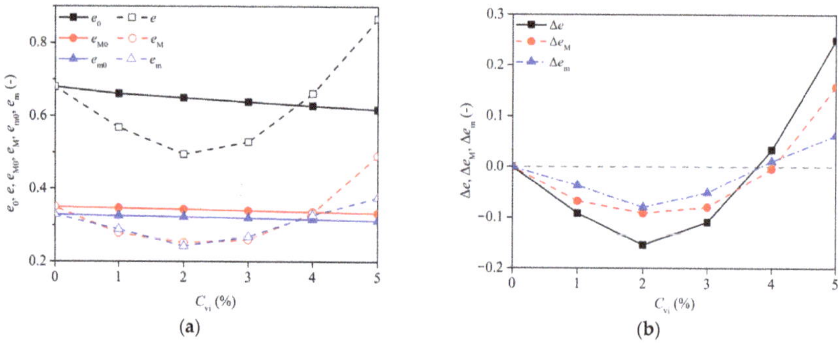

Figure 22. Macro-, micro-, and total void ratios of as-compacted and cured cement-treated GRS depending on cement content (Hollow symbol indicates a curing time of 28 d). (**a**) Initial state and after curing to 28 d; (**b**) Difference between initial and cured states.

The changes in the macro-, micro-, and total void ratios after the curing and confined wetting of the treated specimens are shown in Figure 22b. As the cement content increases, the macro-, micro-, and total void ratios of the specimens decreased to 73–76% of their initial values at C_{vi} = 2% and recovered to their initial states at C_{vi} = 4%. The macro-, micro-, and total void ratios of the specimens would exceed their initial values in the case of C_{vi} > 4%. Hence, in a microscopic view, volumetric cement contents of 2–3% are suggested for treating the GRS due to producing the densest pore structure after curing.

5. Conclusions

Mechanical and microscopic tests were conducted on the GRS treated with Portland fly-ash cement, and the effects of cement treatment on the uniaxial and oedometer compression behaviors and the soil structure were analyzed. The following conclusions were drawn based on the findings of this study:

(1) The UCS and secant modulus increased linearly with the volumetric cement content in the range of $C_{vi} \leq 3\%$ and with the curing time in $t \leq$ 7–14 d. As the volumetric cement content exceeded 3% or the curing time exceeded 14 d, the increase of the UCS and secant modulus with the cement content and curing time diminished. A unified formula using ηC_{vi} as the key parameter can estimate the UCS and secant modulus of the cement-treated specimens based on the untreated states.

(2) Compressibility decreased markedly as the volumetric cement content increased to 2% and then varied slightly as the cement content further increased. Vertical yielding pressure decreased with the cement content in the range of C_{vi} < 2–3%, followed by a continuous increase as the cement content further increased. The diminished increase of the compressibility with the vertical pressure indicated the breakage of the interparticle cementation and rearrangement of the soil particles upon compression.

(3) The C–S–H is fibrous (C–S–H of Type I) in the case of C_{vi} = 1–2% and becomes reticular (C–S–H of Type II) in the case of C_{vi} = 3–5%. The soil structure remained of single-porosity in the case of C_{vi} < 2% and a dual-porosity structure formed as the volumetric cement content exceeded 3%.

(4) The mechanisms of soil structure to be modified through cement treatment were supposed in three aspects: (i) the formation of C–S–H and non-detective micro-pores within C–S–H; (ii) the intrusion of C–S–H into macro- and micro-pores; and (iii) the emergence of aggregates along with inter- and intra-aggregate pores. The macro- and micro-void ratios decreased with the cement content in the case of $C_{vi} \leq 2\%$, indicating that the role of the second aspect surpassed that of the first aspect (the third aspect is invalid) in this case.

The macro- and micro-void ratios started to increase due to the third aspect in the case of $C_{vi} \geq 3\%$ and exceeded their initial values as the third aspect began to dominate in the case of $C_{vi} > 4\%$. However, the coupling effects of the second and third aspects on the macro- and micro-void ratios need to be quantified in future investigations.

(5) A volumetric cement content of 2–3% is suggested for treating the GRS (a silty sand) based on mechanical and microstructural characteristics of the saturated treated specimens. Volumetric cement contents lower than 2% provide inadequate interparticle cementation, whereas volumetric cement contents higher than 3% are inefficient as the inter-aggregate pores emerge along with the aggregates. In view of economic efficiency and environmental protection, combined treatment of cement and green binders is needed to reduce the cement content to its lower level and obtain more favorable physicomechanical properties in a given geotechnical context.

Author Contributions: Writing—original draft: X.D.; Writing—review & editing: X.B.; Validation: H.C.; Supervision: C.X.; Conceptualization: X.C. All authors have read and agreed to the published version of the manuscript.

Funding: This research was supported by the National Natural Science Foundation of China (Grant Nos. 52022060, 42202298), Postdoctoral Science Foundation of China (Grant No. 2021M702234), and Joint Fund of National Natural Science Foundation of China-Railway Corporation for basic research of high-speed railway (Grant No. U1934208).

Institutional Review Board Statement: Not applicable.

Informed Consent Statement: Not applicable.

Data Availability Statement: Not applicable.

Conflicts of Interest: The authors declare no conflict of interest.

References

1. Zhao, Y.; Sun, X.; Wen, T.; Chen, R.; Huang, L. Micro-structural evolution of granite residual soil under external loading based on X-ray micro-computed tomography. *KSCE J. Civ. Eng.* **2021**, *25*, 2836–2846. [CrossRef]
2. Firoozi, A.A.; Olgun, C.G.; Firoozi, A.A.; Baghini, M.S. Fundamentals of soil stabilization. *INT J. Geomech.* **2017**, *8*, 26. [CrossRef]
3. Smaida, A.; Mekerta, B.; Gueddouda, M.K. Physico-mechanical stabilization of a high swelling clay. *Constr. Build. Mater.* **2021**, *289*, 123197. [CrossRef]
4. Feng, R.; Wu, L.; Liu, D.; Wang, Y.; Peng, B. Lime- and cement-treated sandy lean clay for highway subgrade in China. *J. Mater. Civil Eng.* **2020**, *32*, 4019335. [CrossRef]
5. Khalid, U.; Liao, C.C.; Ye, G.; Yadav, S.K. Sustainable improvement of soft marine clay using low cement content: A multi-scale experimental investigation. *Constr. Build. Mater.* **2018**, *191*, 469–400. [CrossRef]
6. Wu, J.; Liu, L.; Deng, Y.; Zhang, G.; Zhou, A.; Wang, Q. Distinguishing the effects of cementation versus density on the mechanical behavior of cement-based stabilized clays. *Constr. Build. Mater.* **2021**, *271*, 121571. [CrossRef]
7. Croft, J.B. The influence of soil mineralogical composition on cement stabilization. *Geotechnique* **1967**, *17*, 119–135. [CrossRef]
8. Williamson, S.; Cortes, D.D. Dimensional analysis of soil–cement mixture performance. *Geotech. Lett.* **2014**, *4*, 33–38. [CrossRef]
9. Sasanian, S.; Newson, T.A. Basic parameters governing the behaviour of cement-treated clays. *Soils Found.* **2014**, *54*, 209–224. [CrossRef]
10. Ho, T.; Chen, W.; Yin, J.; Wu, P.; Tsang, D.C.W. Stress-strain behaviour of cement-stabilized Hong Kong marine deposits. *Constr. Build. Mater.* **2021**, *274*, 122103. [CrossRef]
11. Yao, K.; Xiao, H.; Chen, D.; Liu, Y. A direct assessment for the stiffness development of artificially cemented clay. *Geotechnique* **2019**, *69*, 741–747. [CrossRef]
12. Nazari, Z.; Tabarsa, A.; Latifi, N. Effect of compaction delay on the strength and consolidation properties of cement-stabilized subgrade soil. *Transp. Geotech.* **2021**, *27*, 100495. [CrossRef]
13. Wei, X.; Ku, T. New design chart for geotechnical ground improvement: Characterizing cement-stabilized sand. *Acta Geotech.* **2020**, *15*, 999–1011. [CrossRef]
14. Choo, H.; Nam, H.; Lee, W. A practical method for estimating maximum shear modulus of cemented sands using unconfined compressive strength. *J. Appl. Geophys.* **2017**, *147*, 102–108. [CrossRef]
15. Ling, J.; Lin, S.; Qian, J.; Zhang, J.; Han, B.; Liu, M. Continuous compaction control technology for granite residual subgrade compaction. *J. Mater. Civil Eng.* **2018**, *30*, 04018316. [CrossRef]
16. Zhou, Y.H.; Wang, Y.H.; Qing, Q.X.; He, Q. Experimental study of long-term stability of improved granitic residual soil subgrade for high-speed railway. *Rock Soil Mech.* **2011**, *32*, 596–602. (In Chinese)

17. Yi, W.; Wang, Y. An experimental study on engineering characteristics of improved weathered granite soil for a high-speed railway subgrade in south China. *J. Test. Eval.* **2017**, *45*, 20140358. [CrossRef]
18. Zhou, Y.H.; Wang, Y.H.; Qing, Q.X.; He, Q. Experimental study of appropriateness of improved granitic residual soil for high-speed railway subgrade. *Chin. J. Rock Mech. Eng.* **2011**, *30*, 625–634. (In Chinese)
19. Chamling, P.K.; Biswal, D.R.; Sahoo, U.C. Effect of moulding water content on strength characteristics of a cement-stabilized granular lateritic soil. *Innov. Infrastruct. Solut.* **2021**, *6*, 82. [CrossRef]
20. MolaAbasi, H.; Naderi Semsani, S.; Saberian, M.; Khajeh, A.; Li, J.; Harandi, M. Evaluation of the long-term performance of stabilized sandy soil using binary mixtures: A micro- and macro-level approach. *J. Clean. Prod.* **2020**, *267*, 122209. [CrossRef]
21. He, X.; Chen, Y.; Tan, X.; Wang, S.; Liu, L. Determining the water content and void ratio of cement-treated dredged soil from the hydration degree of cement. *Eng. Geol.* **2020**, *279*, 105892. [CrossRef]
22. Pu, S.; Zhu, Z.; Song, W.; Wan, Y.; Wang, H.; Song, S.; Zhang, J. Mechanical and microscopic properties of cement stabilized silt. *KSCE J. Civ. Eng.* **2020**, *24*, 2333–2344. [CrossRef]
23. Horpibulsuk, S.; Rachan, R.; Chinkulkijniwat, A.; Raksachon, Y.; Suddeepong, A. Analysis of strength development in cement-stabilized silty clay from microstructural considerations. *Constr. Build. Mater.* **2010**, *24*, 2011–2021. [CrossRef]
24. Cardoso, R.; Ribeiro, D.; Néri, R. Bonding effect on the evolution with curing time of compressive and tensile strength of sand-cement mixtures. *Soils Found.* **2017**, *57*, 655–668. [CrossRef]
25. Cardoso, R. Porosity and tortuosity influence on geophysical properties of an artificially cemented sand. *Eng. Geol.* **2016**, *211*, 198–207. [CrossRef]
26. Ural, N. Effects of additives on the microstructure of clay. *Road Mater. Pavement.* **2015**, *17*, 104–119. [CrossRef]
27. *ASTM D422-63(2007)e2*; Standard Test Method for Particle-Size Analysis of Soils. ASTM International: West Conchhocken, PA, USA, 2007.
28. *ASTM D854-14*; Standard Test Methods for Specific Gravity of Soil Solids by Water Pycnometer. ASTM International: West Conchhocken, PA, USA, 2014.
29. GB/T 50123-2019*Standard for Geotechnical Testing Method*; Ministry of Housing and Urban-Rural Development of the People's Republic of China: Beijing, China, 2019.
30. *ASTM D698-12(2021)*; Standard Test Methods for Laboratory Compaction Characteristics of Soil Using Standard Effort. ASTM International: West Conchhocken, PA, USA, 2021.
31. *ASTM D2487-17e1*; Standard Practice for Classification of Soils for Engineering Purposes (Unified Soil Classification System). ASTM International: West Conchhocken, PA, USA, 2017.
32. Alonso, E.E.; Pinyol, N.M.; Gens, A. Compacted soil behaviour: Initial state, structure and constitutive modelling. *Geotechnique* **2013**, *63*, 463–478. [CrossRef]
33. Romero, E. A microstructural insight into compacted clayey soils and their hydraulic properties. *Eng. Geol.* **2013**, *165*, 3–19. [CrossRef]
34. GB 175-2007*Common Portland Cement*; China National Standardization Management Committee: Beijing, China, 2007.
35. Ulusay, R.; Hudson, J.A. (Eds.) *The Complete ISRM Suggested Methods for Rock Characterization, Testing and Monitoring: 1974–2006*; International Society for Rock Mechanics: Ankara, Turkey, 2007.
36. Casagrande, A. The determination of pre-consolidation load and its practical significance. In Proceedings of the 1st International Conference on Soil Mechanics and Foundation Engineering, Cambridge, MA, USA, 22–26 June 1936.
37. Cui, Y.J.; Delage, P. Yielding and plastic behaviour of an unsaturated compacted silt. *Geotechnique* **1996**, *46*, 291–311. [CrossRef]
38. Mutaz, E.; Dafalla, M.A. Chemical analysis and X-ray diffraction assessment of stabilized expansive soils. *Bull. Eng. Geol. Environ.* **2014**, *73*, 1063–1072. [CrossRef]
39. Lemaire, K.; Deneele, D.; Bonnet, S.; Legret, M. Effects of lime and cement treatment on the physicochemical, microstructural and mechanical characteristics of a plastic silt. *Eng. Geol.* **2013**, *166*, 255–261. [CrossRef]
40. Gates, W.P.; Bouazza, A. Bentonite transformations in strongly alkaline solutions. *Geotext. Geomembr.* **2010**, *28*, 219–225. [CrossRef]
41. Diamond, S. *Hydraulic Cement Pastes: Their Structure and Properties*; Cement and Concrete Association: Slough, UK, 1976.
42. Consoli, N.C.; Cruz, R.C.; Floss, M.F.; Festugato, L. Parameters controlling tensile and compressive strength of artificially cemented sand. *J. Geotech. Geoenviron.* **2010**, *136*, 759–763. [CrossRef]
43. Consoli, N.C.; Dalla Rosa, A.; Corte, M.B.; Lopes, L.D.S.; Consoli, B.S. Porosity-cement ratio controlling strength of artificially cemented clays. *J. Mater. Civil Eng.* **2011**, *23*, 1249–1254. [CrossRef]
44. Consoli, N.C.; Foppa, D. Porosity/cement ratio controlling initial bulk modulus and incremental yield stress of an artificially cemented soil cured under stress. *Geotech Lett.* **2014**, *4*, 22–26. [CrossRef]
45. Horpibulsuk, S.; Rachan, R.; Suddeepong, A.; Liu, M.D.; Du, Y.J. Compressibility of lightweight cemented clays. *Eng. Geol.* **2013**, *159*, 59–66. [CrossRef]
46. Rios, S.; Viana Da Fonseca, A.; Baudet, B.A. Effect of the porosity/cement ratio on the compression of cemented soil. *J. Geotech. Geoenviron.* **2012**, *138*, 1422–1426. [CrossRef]
47. Kasama, K.; Zen, K.; Iwataki, K. Undrained shear strength of cement-treated soils. *Soils Found.* **2006**, *46*, 221–232. [CrossRef]
48. Powers, T.C. Physical properties of cement paste. In Proceedings of the Fourth International Symposium on the Chemistry of Cement, Washington, DC, USA, 2–7 October 1960; pp. 577–609.

applied sciences

MDPI

Article

Mechanical Properties and Engineering Applications of Special Soils—Dynamic Shear Modulus and Damping of MICP-Treated Calcareous Sand at Low Strains

Xinlei Zhang [1], Jun Guo [1], Yumin Chen [2,*], Yi Han [2], Ruibo Yi [1], Hongmei Gao [1], Lu Liu [1], Hanlong Liu [3] and Zhifu Shen [1]

1 Research Center of Urban Underground Space, Nanjing Tech University, Nanjing 211816, China
2 Key Laboratory of Ministry of Education for Geomechanics and Embankment Engineering, College of Civil and Transportation Engineering, Hohai University, Nanjing 210098, China
3 Key Laboratory of New Technology for Construction of Cities in Mountain Area, College of Civil Engineering, Chongqing University, Chongqing 400045, China
* Correspondence: ymch@hhu.edu.cn

Abstract: Calcareous sand deposits are widespread along the shoreline in tropical and subtropical regions. Microbially induced calcite precipitation (MICP) treatment is a new method for improving the soil's stiffness and strength. The small-strain shear modulus and damping ratio of MICP-treated calcareous sand, two critical parameters for predicting the dynamic behavior of soil, are little known. This study conducts a series of resonant column tests to investigate the dynamic characteristics of MICP-treated calcareous sand, emphasizing the influence of treatment duration and confining stress on the stiffness and damping characteristics. It analyzes the relationship between the initial dynamic shear modulus and unconfined compressive strength. In addition, empirical relationships between the reference shear strain and treatment duration or confining stress are provided. The results show that the normalized shear modulus G/G_0 of MICP-cemented calcareous sand has a higher strain sensitivity than that of untreated sand, and the Hardin–Drnevich model can describe its attenuation pattern. The effective confining stress σ_c affects the degradation characteristics of the dynamic shear modulus of MICP-treated calcareous sand with a low cementation level; however, its impact decreases as the treatment duration increases. There is a linear relationship between the reference shear strain and confining stress. While the relationship between the reference shear stain and treatment duration is a power law.

Keywords: calcareous sand; microbially induced calcite precipitation; dynamic shear modulus; resonant column test; reference shear strain

1. Introduction

Calcareous sand is widespread along the coastline of the South China Sea, the Red Sea, Australia, the Gulf of Mexico, and other tropical or subtropical regions [1], making it an essential component of construction materials in coastal structures. Usually, calcareous sand consists of the reworked shell fragments and skeletal debris of marine organisms [2,3]. Calcareous sand has unique features, such as irregularly shaped particles, a porous interior, a polygonal angle, and crushable grain [4]. As a result, the mechanical behavior of calcareous sand can be considerably different from siliceous sands due to their diverse mineral resources and particle characteristics [5,6].

Several conventional geotechnical applications, including grouting, densification and deep mixing, have been explored to enhance siliceous sand's bearing capacity and liquefaction resistance [7]. The development of new technologies and materials, such as microbially induced calcite precipitation treatment [8–10] and nano-modification [11–13], have attracted great attention due to their low cost, low pollution, and harmless impact on the environment. However, when the MICP is performed in clay soil, the ground improvement cost is increased due to clay soil's low permeability [14]. The production of carbonate precipitation is affected by environmental factors, such as calcium concentration, bacterial concentration, pH, and temperature [15]. MICP treatment is suitable for treating artificial calcareous sand islands far from the inland and with a poor construction environment. Microbially induced calcite precipitation (MICP) is a modern technology that utilizes biological metabolic processes to improve soil properties [16–18]. The fundamental chemical reactions governing the MICP process can be defined as follows [8,19]:

$$CO(NH_2)_2 + 2H_2O \rightarrow 2NH_4^{+} + CO_3^{2-}$$

$$Ca^{2+} + CO_3^{2-} \rightarrow CaCO_3 \downarrow$$

Soil improvement methods include environmental friendliness, a far transmission distance, low disturbance, and controllability of the process [10]. MICP has been applied to solve practical engineering problems, such as increasing soil strength and stiffness [19,20], reducing soil permeability [21], repairing cracks in the structure [22], and controlling erosion [23]. MICP treatment has recently been demonstrated to improve calcareous sand's static strength and liquefaction resistance due to the calcite crystal precipitation between the sand particles, which strengthens the soil structure [1,16,24]. The composition, shape, and pore structure of calcareous sand particles are more suitable for calcium carbonate precipitation than siliceous sand [1,25,26].

However, studies on the dynamic properties of MICP-treated calcareous sand, such as the small-strain dynamic shear modulus and damping ratio, are rare. The dynamic shear modulus and damping ratio are essential for analyzing soil's dynamic response during dynamic loading conditions, such as earthquakes, traffic, and wave loads [27]. Over the past few decades, the dynamic properties of naturally occurring calcareous sand have been extensively investigated in the laboratory [28,29]. For example, the small-strain shear modulus and damping ratio of calcareous sand under an isotropic or anisotropic stress consolidated state have been studied [30–33]. The effects of confining stress, pore ratio, relative density, grading and particle size, and particle breakage on the dynamic properties have also been discussed [5,29–35]. Morsy et al. [34] discovered that the effect of the void ratio on shear modulus is more pronounced at low void ratios, and the maximum shear modulus increases as the void ratio decreases. Javdanian and Jafarian [5] conducted resonant column and cyclic triaxial tests on two marine calcareous sand samples and reported that as the effective confining pressure grows, the shear stiffness ratio rises while the damping ratio decreases. Hassanlourad et al. [35] examined the shear behavior of calcareous sand using drained and undrained triaxial tests, revealing the effect of particle breakage on specimen response. Khalil et al. [36] performed a series of bender element and cyclic triaxial tests to study the low-strain and high-strain characteristics of calcareous sand and terrigenous sand.

Although numerous researchers have studied the dynamic shear modulus and damping ratio at low strains on natural and cement-stabilized soil [30,34,36–41], considerably less information is known about the dynamic response of marine soil, such as calcareous sand at different MICP-treatment levels. Understanding how the shear modulus and damping ratio of MICP-treated calcareous sand varies with shear strain and other factors for engineering applications is necessary. Many studies have shown that the dynamic properties of cemented soils are affected by the cementation level [20,42] and confining stress [43–47]. Accordingly, the main objective of this investigation is to evaluate the dynamic behavior of MICP-treatment calcareous sands at a low strain amplitude, focusing on the effect of confining stress and treatment time on the precipitated calcite contents. A series of resonant column tests are conducted to determine calcareous sand's dynamic shear modulus and damping ratio at a low MICP-treated level. Initially, this paper describes the test materials and implementation methods. The effect of MICP treatment on the dynamic response of saturated calcareous sand at a low shear strain, including the responses of the dynamic shear modulus, degradation characteristics of a normalized dynamic shear modulus, and the damping ratio, is then discussed. In addition, the correlations of the initial dynamic shear modulus with an unconfined compression strength, confining stress, and treatment time are addressed. Finally, the relationship between the reference shear strain and the treatment duration or confining stress are proposed. The study's results provide a quantification of dynamic parameters of MICP-treated calcareous sand and relevant support for the development of constitutive models.

2. Materials and Methods

2.1. Material Properties

The calcareous sand used in the current study was collected from an island in the South China Sea, and its particle size distribution curve is depicted in Figure 1. The calcium carbonate content of this sand exceeds 90.25%, as determined by the chemical componential analysis; its microscope (SEM) images are shown in Figure 2, and a summary of the sand's parameters tested with the ASTM standard is listed in Table 1. Calcareous sands have a coefficient of uniformity of C_u equal to 3.55 and a coefficient of curvature of C_c equal to 0.97, and they are classified as poorly graded sand based on the ASTM D 2487 standard testing procedures.

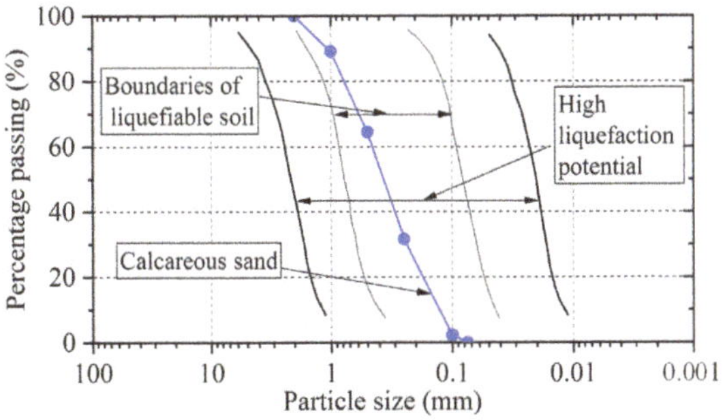

Figure 1. Particle size distribution curve of calcareous sand used in this study (The liquefiable boundaries are taken from Tsuchida [48]).

Figure 2. Scanning electron microscope (SEM) images of the calcareous sand used in this study.

Table 1. Index properties for the calcareous sand used in this study.

Sand	D_{50} (mm)	D_{10} (mm)	C_u	C_c	G_s	e_{min}	e_{max}
Calcareous sand	0.38	0.13	3.55	0.97	2.73	1.02	1.44

2.2. MICP Treatment Procedure

Sporosarcina pasteurii (American Type Culture Collection, ATCC11859) is an aerobic uratolytic bacterium that uses urea as a nutrient and produces CO_3^{2-} as the critical part of

$CaCO_3$ precipitation. The bacterial fluid was cultured in a medium solution (with a ratio of 1:10), whose composition is presented in Table 2. The bacteria were enlarged in an incubator shaker at 30 °C at a speed of 210 rpm for approximately 24–32 h. The harvested bacteria's optical density (OD600) was about 0.71, and the urease activity was approximately 1.1 mM urea hydrolyzed per min. The cementation solution (CS) used in the experiments was a mixture of 0.5 mol/L urea and 0.5 mol/L $CaCl_2$. The volume of the cementation solution utilized for each sample in the tests was about seven times the sample pore volume, i.e., 720 mL.

Table 2. Biological media.

Reagent	Concentration
Yeast extract	20 g/L
NH_4Cl	10 g/L
$MnCl_2 \cdot H_2O$	12 mg/L
$NiCl_2 \cdot 6H_2O$	24 mg/L
Distilled water	1000 g/L
1 mol/L NaOH	pH value of approximately 9

The calcareous sand samples were prepared using a series of specially designed devices to achieve batches of intact samples, as illustrated in Figure 3. The specimen's preparation device consisted of two semi-cylindrical PVC molds that could be quickly disassembled, two pieces of filter fabric to prevent the fine sand particles from being flushed out during the treatment process, a small aspirating hole on the mold to connect with the vacuum pump, and two PVC pedestals. The PVC split mold had an internal diameter and height of 5 cm and 10 cm, respectively. A thin rubber membrane was attached to the inner walls of the columniform mold. A vacuum pump was connected to the aspirating hole on the mold during the sample preparation process to draw a negative pressure between the rubber membrane and the inner wall of the mold without leaving any gaps to produce standard diameter samples.

Samples of 50 mm diameter and 100 mm high were prepared with dry tamping in the split mold. The sample preparation process was divided into several steps. First, to obtain a more uniform specimen, about 345 g of dry calcareous sand was spread into the thin rubber membrane in five layers. Each layer should be filled evenly, and the last layer's surface should be as smooth as possible; because the calcareous sand grain shows crushability under a shocking load, a rubber hammer was used to slightly tamp the mold outside to achieve the designed height. The relative density of the reconstituted specimens was controlled at about 40%. Second, once the sand sample was produced, 110 mL of the bacterial liquid was injected into the sand specimen utilizing a peristaltic pump at a rate of 180 mL/h. The bacterial solution was inoculated in the soil pore space for at least 6 h to allow the bacteria to diffuse and adhere to the soil particles. Third, after the detention of the bacterial fluid, the cementation solution was pumped into the soil sample at a rate of 180 mL/h and was allowed to flow out freely by the pipe on the top cover. All specimens were flushed with distilled water equal to four times the volume of the voids at the end of the treatment to purge them of the bacterial solution and residual cementation solution and to saturate the sample.

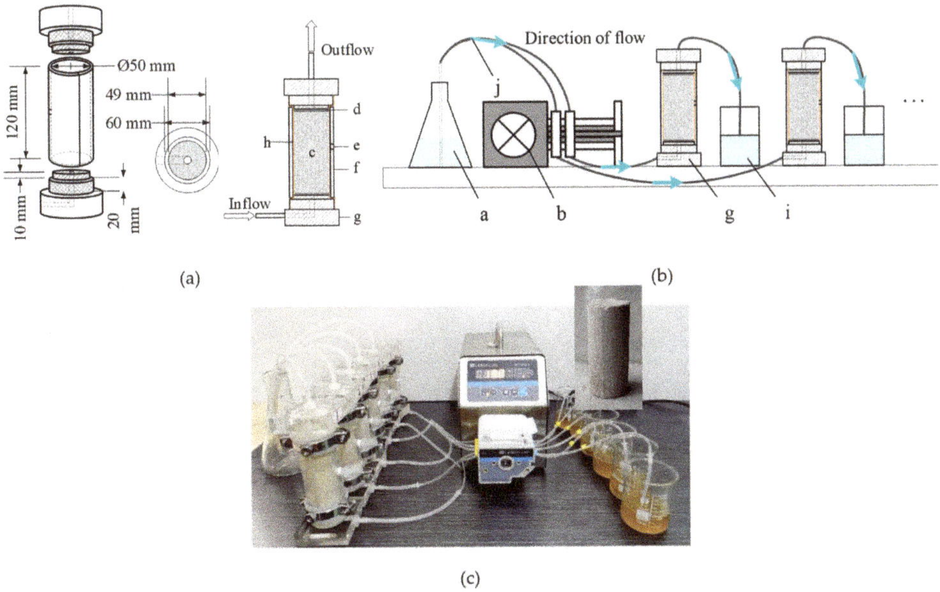

Figure 3. Schematics of MICP treatment device for (**a**) schematic diagram of mold for MICP-treated cylindrical specimens, (**b**) schematic of MICP-treated calcareous sand, and (**c**) sketch of MICP treatment procedure. (a. Bacterial liquid/cementation solution; b. peristaltic pump; c. sand sample; d. filter fabric; e. aspirating hole; f. PVC split mold; g. PVC pedestal; h. rubber membrane; i. effluent; j. plastic tube).

2.3. Equipment and Testing Procedure

The experiments related to the MICP-treated calcareous sand were conducted using the resonant column apparatus TSH-100 developed by GCTS, as shown in Figure 4a. The shear modulus can be calculated by combining the frequency–response curve data, specimen size, and physical characteristics of the resonant column test apparatus. The configuration of the resonant column apparatus is bottom-fixed and top-free, and it can be modified according to the size of prepared soil specimens. The maximum pressure of the air-filled cell is 1 MPa, and there is an internal high-resolution LVDT for measuring the deformation of specimens. The specimen and vibration generator were enclosed together within the pressure chamber. The specimen's resonant frequency was determined by changing the excitation frequency and force applied to the bottom of the specimens. The shear modulus G was calculated using the theory of elastic wave propagation as follows:

$$G = \rho V_S^2 \tag{1}$$

$$V_S = \frac{2\pi f_n l}{\beta} \tag{2}$$

where V_S is shear wave velocity, ρ is density, and f_n is the natural frequency of the system. β is a function of the mass polar moment of inertia of both specimen I and the mass attached at the free end of the specimen I_0, as determined by:

$$\beta \tan(\beta) = I/I_0 \tag{3}$$

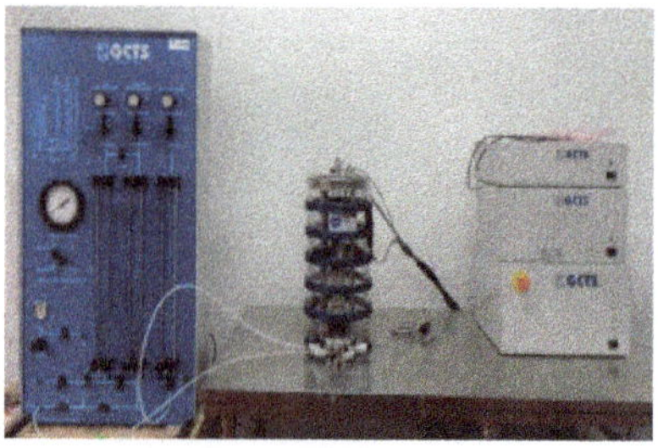

(a)

(b)

Figure 4. This study's experimental equipment consists of (**a**) a resonant column device and (**b**) an unconfined compression test.

In this study, the damping ratio is determined from the frequency–response curve, which can be calculated using:

$$D = \left(\frac{f_2 - f_1}{2f_n}\right) \times 100 \tag{4}$$

where f_1 and f_2 are the frequencies at $A = A_{max}/\sqrt{2}$, in which A is the acceleration of resonant frequency, and A_{max} is the maximum acceleration.

The test procedure can be divided into four steps. First, the prepared sample was installed on the base, and the top of the sample was connected to a vibration generator. Second, the back-pressure saturated method was applied to ensure that the specimens were saturated and the back pressure was 400 kPa. In this study, a B-value of more than 0.97 was reached for all the specimens. The effective confining stress was then increased to the target value and maintained. When the average strain rate of the sample was less than $1 \times 10^{-3}\%/\text{min}$, the consolidation step was completed. Finally, the resonant frequency was determined by changing the excitation frequency and force applied to the bottom of the specimens, and the values of the dynamic shear strain, shear wave velocity, and dynamic shear modulus were recorded.

A series of unconfined compressive tests were performed to analyze the relationship between the static strength and the dynamic properties of the MICP-treated calcareous sand. The unconfined compressive strength of the MICP-treated samples was measured using a YSH-2 type of unconfined compression apparatus (Figure 4b). The upper and lower surfaces of the samples were leveled and placed on the unconfined compressive apparatus. The unconfined compression strength of the soil samples was measured with a loading speed of 1 mm/min.

The samples were oven-dried at a temperature of 60° for a duration of 72 h after unconfined compressive and resonant column tests to determine the content of $CaCO_3$ that can be computed utilizing the following formula:

$$C_C(\%) = \frac{m_0 - m}{m} = \frac{\Delta m}{m} \times 100 \quad (5)$$

where C_C is the content of $CaCO_3$ in the specimen, m and m_0 are the oven-dried mass of the dry specimen before and after MICP treatment, and Δm is the mass of $CaCO_3$.

2.4. Experimental Program

This study conducted 18 groups of resonance column tests, including 15 groups of treated conditions and 3 groups of untreated conditions for comparison to the treated samples. The effects of treatment times (e.g., 6 h, 12 h, 18 h, 24 h, and 48 h) and the effective confining stress (e.g., 25 kPa, 50 kPa, and 100 kPa) were considered. In addition, five groups of unconfined compressive tests were carried out to establish the relationships between the initial dynamic shear modulus and the unconfined compressive strength. Table 3 displays the details of the samples and the contents of $CaCO_3$.

Table 3. Summary of the conducted experiments and $CaCO_3$ generation.

Group	Case	Test	Treatment Time (Hours)	Effective Confining Stress σ_c (kPa)	D_r (%)	m (g)	e_0	Δm (g)	C_C (%)
Group UN	UN1	RCT	/	25	41.3	236.4	1.27	/	/
	UN2	RCT	/	50	39.2	235.5	1.28	/	/
	UN3	RCT	/	100	40.5	236.0	1.27	/	/
Group M-A	M11	RCT	6	25	40.1	235.9	1.27	5.28	2.24
	M12	RCT	6	50	38.6	235.2	1.28	5.02	2.13
	M13	RCT	6	100	42.3	236.8	1.26	4.98	2.10
	M21	RCT	12	25	39.5	235.6	1.27	8.56	3.63
	M22	RCT	12	50	40.9	236.2	1.27	8.67	3.67
	M23	RCT	12	100	41.1	236.3	1.27	8.21	3.47
	M31	RCT	18	25	39.5	235.6	1.27	11.42	4.85
	M32	RCT	18	50	39.9	235.8	1.27	12.61	5.35
	M33	RCT	18	100	40.2	235.9	1.27	11.89	5.04
	M41	RCT	24	25	41.8	236.6	1.26	17.67	7.47
	M42	RCT	24	50	39.8	235.7	1.27	16.93	7.18
	M43	RCT	24	100	40.8	236.2	1.27	17.24	7.30
	M51	RCT	48	25	40.2	235.9	1.27	25.24	10.70
	M52	RCT	48	50	37.9	234.9	1.28	26.96	11.48
	M53	RCT	48	100	39.4	235.5	1.27	26.68	11.33
Group M-B	M1	UCT	6	/	42.1	236.7	1.26	5.98	2.53
	M2	UCT	12	/	40.8	236.2	1.27	8.65	3.66
	M3	UCT	18	/	42.5	236.9	1.26	13.54	5.72
	M4	UCT	24	/	41.1	236.3	1.27	17.65	7.47
	M5	UCT	48	/	39.5	235.6	1.27	27.16	11.53

Note: UN means an untreated condition; M means MICP-treated condition; RCT means resonant column test; UCT means unconfined compressive test.

3. Experimental Results

3.1. Dynamic Shear Modulus of MICP-Treated Calcareous Sands

Significant nonlinearity in soil stress–strain behavior is presented by two parameters of the shear modulus G and damping ratio D, which change with shear strain amplitude. Figure 5a illustrates the variation in G with the shear strain γ ranging from 10^{-4} to 4×10^{-2}%. The shear modulus of the clean sands used in this investigation oscillated from

58 MPa to 20 MPa, 63 MPa to 23 MPa, and 95 MPa to 50 MPa at confining stress of 25 kPa, 50 kPa, and 100 kPa, respectively.

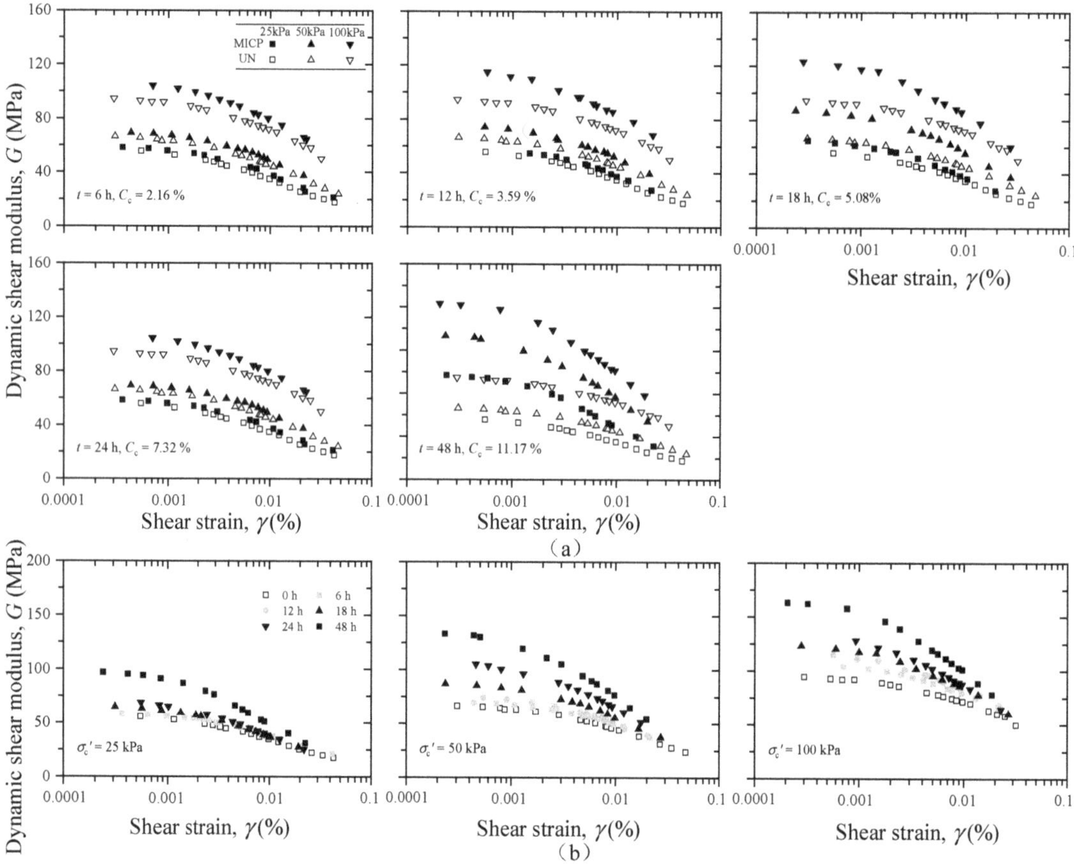

Figure 5. Dynamic shear modulus attenuation curves of calcareous sand and MICP-treated calcareous sand under (**a**) different treatment times and (**b**) different confining stress.

The dynamic shear modulus of the soil after treatment was not significantly increased at a low confining stress of 25 kPa when the treatment duration was less than 18 h. Because the amount of precipitated calcium carbonate in specimens with a low treated level is limited, and the stiffness of these cementations is low, a higher $CaCO_3$ content is required for the loose and porous calcareous sands to form an effective bonding network. The treated sands have a higher shear modulus than the untreated sands at a small shear strain under the same loading conditions.

As the shear strain increases, the shear modulus of the MICP-treated sand gradually decreases and eventually approaches that of the untreated calcareous sand at approximately 0.03% shear strain. This observation indicates that when the shear strain exceeds 0.03%, the beneficial effect of calcite bonding between sand particles induced by MICP treatment disappears. Simatupang et al. [43] discovered comparable outcomes in their study of cemented sand. Under conditions of lower confining stress and a shorter treatment duration, the shear strain level at which the shear modulus of the MICP-treated sand decreases to the untreated value is generally lower.

Figure 5b depicts the dynamic shear modulus for various confining stresses. As the confining stress increases, the differences in the behavior of the shear modulus between treated and untreated sand become more pronounced. This is because the confining stress impacts the amount of effective contact between sand particles and/or calcium carbonate precipitation, thereby altering the dynamic shear modulus.

3.2. Initial Shear Modulus, G_0

The initial shear modulus G_0 is defined as the shear modulus at a small strain amplitude of 10^{-4}%.

Figure 6a illustrates the effect of the confining stress on the initial shear modulus. The relationship between the initial shear modulus and the confining stress can be satisfactorily expressed by a power function of Equation (6) proposed by Hardin et al. [49]:

$$G_0 = \alpha \cdot \sigma_c^{\beta} \tag{6}$$

where G_0 is the initial shear modulus, α and β are the fitting parameters related to cement level, void ratio, and other factors, and σ_c is the confining stress. The *R-square* is greater than 0.96 in all cases. As the treatment time increased from 12 h to 48 h, α rose from 10.9 to 31.8, and β decreased from 0.51 to 0.36.

Figure 6b reveals the relationship between the initial shear modulus and the treatment duration. It indicates that the increasing treatment time contributed to the increase in G_0, which can be attributed to the calcite cementation gradually precipitating with the increasing time. The shear modulus of the untreated sands at confining stress of 25 kPa, 50 kPa, and 100 kPa was 58 MPa, 68 MPa, and 95 MPa, respectively, and it was about 98 MPa, 133 MPa, and 164 MPa after being treated for 48 h.

When the MICP treatment time was less than 18 h, the difference between the dynamic shear modulus of the treated samples under a low confining stress of 25 kPa and untreated samples was not obvious. Under a confining stress of 100 kPa, the dynamic shear modulus of the treated samples was much higher than that of the untreated ones. This is because the calcareous sands used in the tests are loose and porous, making it challenging for calcium carbonate precipitated in the pores to form effective contacts in cases with short treatment time. The neighboring particles become clustered under a higher confining pressure, and the calcium carbonate particles precipitated in the pores, which do not contribute to the shear strength, have the potential to become effective contact points, thereby improving the initial dynamic shear modulus of the samples. Whiffin et al. [8] reported that a lower content of $CaCO_3$ (C_{CaCO_3} = 3.5%, which is consistent with the content of $CaCO_3$ in the specimen treated for 12 h in this study) had no significant effect on the strength or stiffness properties of sand compared to untreated sand.

Researchers have stated that a reliable correlation can be established between the G_0 and q_{ucs} for the cemented soil [49]. This study uses the average peak stress of two replicate specimens to represent the unconfined compressive strength. Figure 6c depicts the variation in G_0 with q_{ucs} for bio-treated samples under different confining stresses. The G_0 increases with the q_{ucs}, and the relationship between them can be defined by a power function presented by Consoli et al. [50]:

$$G_0 = m \times q^{n}{}_{ucs} \tag{7}$$

where G_0 is described in MPa, q_{ucs} is described in kPa, and constants m and n are the fitting parameters in the formula. The *R-square* exceeds 0.98 in all cases. The G_0 of the treated sands rises nonlinearly with an increase in q_{ucs}, which is related to cementation between particles.

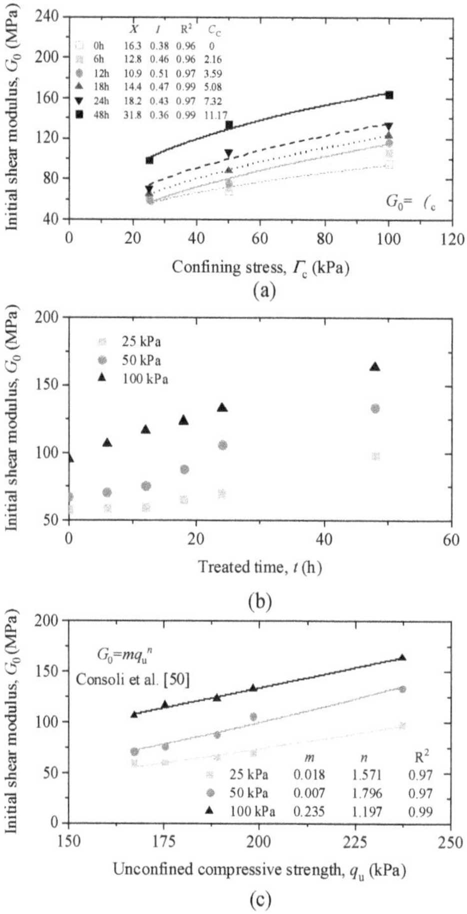

Figure 6. Initial shear modulus versus (a) confining stress, (b) treatment time, and (c) unconfined compressive strength.

3.3. Degradation Characteristics of Normalized Dynamic Shear Modulus

The degradation characteristics of the strain-dependent normalized dynamic shear modulus (G/G_0) are essential for predicting and characterizing the non-linear dynamic response of the soil. Figure 7 exhibits the normalized shear modulus degradation curves of the MICP-treated sands at low strains. As shown in Figure 7a, the degradation of G/G_0 with γ in the higher confining stress is much slower for the same treatment time.

After MICP treatment, the G/G_0 of the calcareous sandy sample decreases quickly with increasing strain due to the gradual deterioration of the bonding between sand particles. This indicates that the non-linear dynamic characteristic of MICP-treated calcareous sand is more pronounced, which has also been reported for other types of treated soil [5,35,38,40]. In addition, the strain sensitivity of samples with longer treatment times is more evidence indicating that the brittleness of MICP-treated calcareous sands grows during the treatment process.

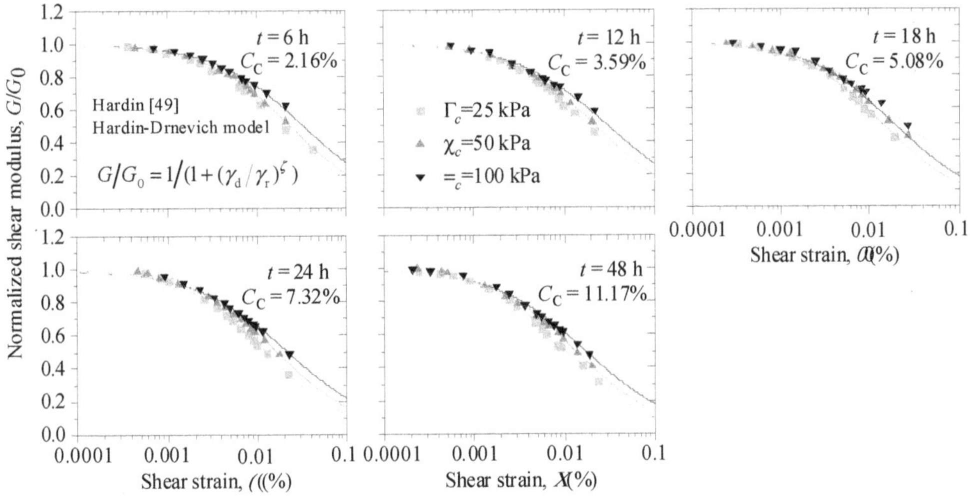

(a)

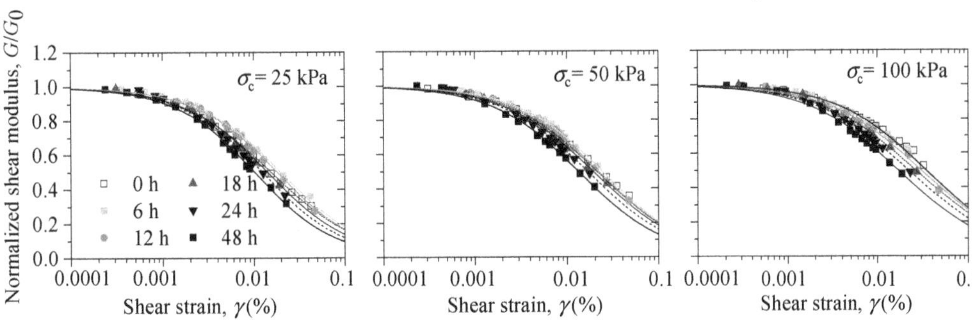

(b)

Figure 7. Effect of (**a**) confined pressure and (**b**) treatment time on normalized shear modulus.

The Hardin–Drnevich model [49] (Equation (8)) has been widely applied to describe the nonlinear relationship between the G/G_0 and γ. It allows the normalized shear modulus reduction to be expressed by two parameters:

$$G/G_0 = 1/(1 + (\gamma/\gamma_r)^{\zeta}) \tag{8}$$

where γ_r is the reference shear strain, defined as the shear strain corresponding to G/G_0 = 0.5, and ζ is the curvature coefficient.

All the experiments in this study reveal that the G/G_0-lnγ curves of MICP-treated calcareous sands are well-fitted by the Hardin–Drnevich model, where *R-square* is greater than 0.98 in all cases. Table 4 lists the parameters of the Hardin–Drnevich model. The value of ζ ranges from 0.83 to 0.99 and decreases as the confining stress improves. Generally, the reference shear strain is less when the confining stress and treatment duration are reduced.

Table 4. Summary of experiments conducted.

Treatment Time, t (h)	Confining Stress, σ_c								
	25 kPa			50 kPa			100 kPa		
	ζ	γ_r/%	R^2	ζ	γ_r/%	R^2	ζ	γ_r/%	R^2
6	0.93	0.01989	0.997	0.93	0.02389	0.997	0.88	0.03362	0.991
12	0.95	0.01844	0.997	0.92	0.02186	0.996	0.84	0.02990	0.984
18	0.94	0.01423	0.996	0.90	0.01893	0.998	0.83	0.02590	0.998
24	0.97	0.01264	0.992	0.92	0.01660	0.990	0.84	0.02180	0.988
48	0.99	0.01029	0.997	0.93	0.01317	0.995	0.87	0.01679	0.993

Figure 8 depicts a best-fit hyperbolic curve for all the data points (G/G_0-lnγ/γ_r) when the shear strain γ is divided by the reference shear strain γ_r. The fitting parameter ζ is 0.9. The *R-square* value of 0.99 indicates that the model can be utilized to describe the behavior of stiffness attenuation in both treated and untreated calcareous sand. Hence, only one parameter, γ_r, can adequately represent the degradation characteristics of the normalized dynamic shear modulus.

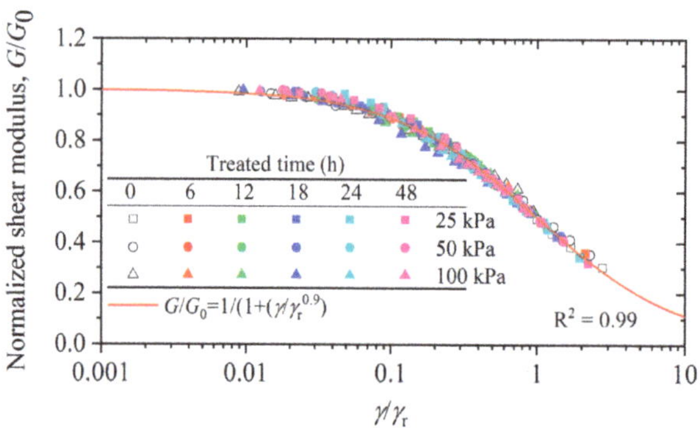

Figure 8. Relationship between normalized dynamic shear modulus and lnγ/γ_r.

3.4. Correlation between γ_r and t, σ_c

In this study, the relationship between γ_r and the treatment time or confining stress waws examined. Figure 9a illustrates the correlation between the reference strain and the treatment time. The reference shear strain can be described as a power function of processing time:

$$\gamma_r = 1/(a + b \cdot t^c) \tag{9}$$

where t is the treatment time, and a, b, and c are the fitting parameters. The fitting parameters a and b decrease from 3503.8 to 2396.9 and 383.9 to 106.2, respectively, and c increases from 0.74 to 0.92 as the confining stress increases from 25 kPa to 100 kPa.

Figure 9b indicates the correlation between the reference shear strain and the confining stress. The γ_r increases as σ_c rises, and the difference in γ_r between various treatment times grows with the increase in σ_c. This also demonstrates that the confining stress still significantly influences the shear modulus attenuation curve of the sand sample with a lower cementation level. In addition, as shown in Figure 9b, the difference of γ_r under various confining stresses decreases gradually as the treatment time rises, indicating that the effect of σ_c on the attenuation of the shear modulus drops during the treatment process. The relationship between γ_r and σ_c is linear and can be described by Equation (10), and the fitting *R-square* for all cases is greater than 0.99:

$$\gamma_r = \delta \cdot \sigma_c + \phi \tag{10}$$

where δ and ϕ are the fitting parameters. Both δ and ϕ drop gradually as the treatment time increases, resulting in a decrease in the slope and intercept of γ_r–σ_c.

It should be noted the empirical relationships presented in this study are only applicable to MICP-treated calcareous sand. The applicability of these empirical formulas to pure calcareous sand needs further study and verification.

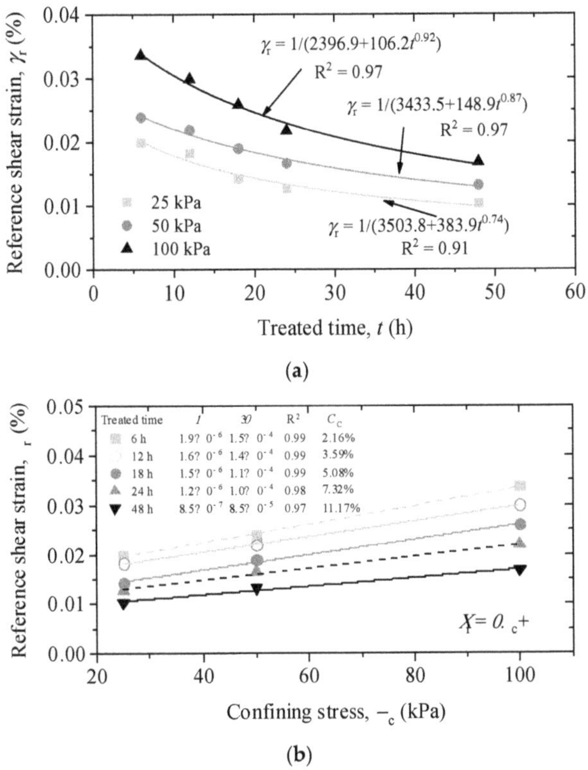

Figure 9. Variation in the reference shear strain with (**a**) treatment time and (**b**) effective confining stress.

3.5. Dynamic Shear Modulus of MICP-Treated Calcareous Sands

Figure 10 depicts the variation in the damping ratio with strain amplitude (D-logγ relationship). The confining stress significantly impacts the damping ratio in the treated sample, with a lower value at a higher confining stress due to an increase in interparticle contacts. Hence, less energy will be dissipated for the wave to propagate.

Figure 11 indicates that the damping ratio rises slightly when the shear strain amplitudes are less than 1×10^{-3}% and increases significantly once the shear strain amplitude is greater than 1×10^{-3}%. It should be noted that the damping ratio initially grows with the treatment time (6 h–24 h) and then decreases gradually after around 24 h. Similar outcomes have been discovered in the studies of treated soils by a few researchers [50–54]. Delfosse-Ribay et al. [55] evaluated the damping ratio and shear modulus of sand samples grouted with three different cementitious materials and found that the samples with a greater dynamic shear modulus also had higher damping ratios. Saxena et al. [45] investigated the dynamic moduli and damping ratios of cemented sands at low strains using resonant column test. They reported that the damping ratio of cemented sands with a low cementation level increases with cement content and reaches its maximum value at a cement content of about 5% to 8% (defined as threshold cement content), then decreases with further cement content. In this study, the threshold calcium carbonate content was around 7% after 24 h of treatment.

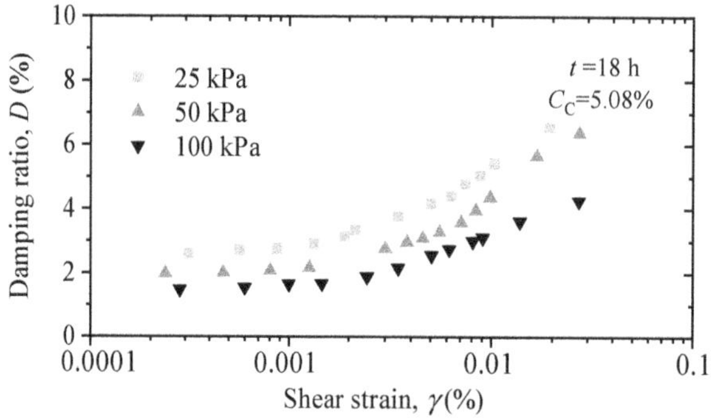

Figure 10. Damping ratio of samples treated for 18 h under different confining pressure.

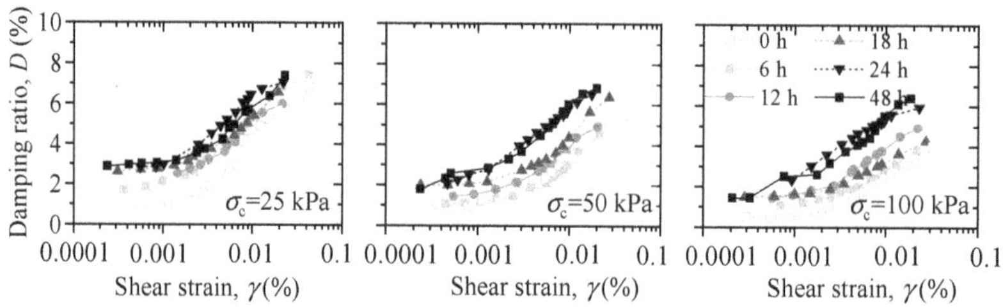

Figure 11. Comparison of damping ratio curves of calcareous sandy specimens with different treatment times.

4. Discussion

The composition, shape, and pore structure of calcareous sand particles are more suitable for calcium carbonate precipitation than siliceous sand [1]. However, the crystals developed mainly on the surface of the sand particles to coat the particle surfaces (in grain coating) due to the porosity and irregularly shaped particles of calcareous sand, especially in the lightly treated specimen. Identical results were observed by Xiao [1] and van Paassen [28].

San Pablo [56] discovered that various grains of sand subjected to similar treatments showed varying degrees of V_s improvement despite similar calcite contents. Identical findings were reported by Cheng [57] and Chu [58]. In reality, three distinct spatial distributions of $CaCO_3$ precipitate exist in silica sand during MICP: (a) contact cementing, (b) grain coating, and (c) matrix supporting [10,59]. The increase in the modulus and strength of the MICP-treated specimen is primarily contributed by the interparticle cementation (contact cementing). Accordingly, in this paper, although the calcium carbonate content ranged from 2% to 11%, the rise in q_u and G_0 was limited compared to that of silica sand, because most of the $CaCO_3$ precipitated in calcareous sand samples was wrapped on the surface of the particles during MICP treatment.

The influence of σ_c was removed by normalizing the G_0, as shown in Figure 12. The normalized G_0 rose linearly with the increase in C_C. In future experiments, it would be beneficial to better understand the relationship between strength/stiffness and calcium carbonate content for calcareous sand with special composite and particle characteristics.

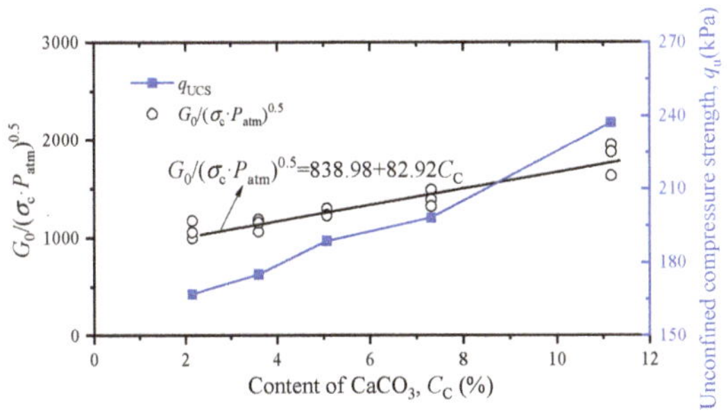

Figure 12. Relationships between normalized G_0 and C_C, and q_u and C_C.

5. Conclusions

This study conducted a series of resonant column tests and unconfined compressive tests to evaluate the static strength and dynamic characteristics of MICP-treated calcareous sand. It investigated the effect of the treatment time and confining stress on the shear modulus degradation and damping ratio. The relationship between the unconfined compressive strength and the initial shear modulus was also analyzed. The correlation between the relative shear strain, treatment time, and confining stress was discussed. The conclusions can be stated as follows:

1. The initial dynamic shear modulus and the unconfined compressive strength rose significantly after MICP treatment, indicating that MICP treatment improves the dynamic properties and strength of calcareous sand. The treated calcareous sand sample showed a higher strain sensitivity (strain-dependent) than that of untreated sands, and the strain sensitivity increased with the increase in treatment duration, demonstrating that the brittleness of the MICP-treated calcareous sands grew during treatment. The Hardin–Drnevich model can describe the G/G_0 attenuation law of MICP-treated calcareous sand. The shear modulus of the MICP-treated sand eventually approached that of the untreated calcareous sand when the shear strain was approximately 0.03%.
2. The difference between the shear modulus of the treated sand and the at of untreated sand became more evident as the confining stress increased because of the reduced pore volume of samples and the increase in the contact between the particles. The MICP-cemented calcareous sand improved significantly in the initial dynamic shear modulus G_0. The relationship between G_0 and the confining pressure σ_c, as well as between G_0 and the unconfined compressive strength q_u, can be described by a power function.
3. The variation in the reference shear strain γ_r under various confining pressures dropped gradually with increasing treatment time, indicating that the influence of the confining pressure on the attenuation curve of the dynamic shear modulus of the MICP-treated calcareous sand decreased during the treatment process. The power function $\gamma_r = 1/(a + b \cdot t^c)$ and the linear function $\gamma_r = \alpha \cdot \sigma_c + \beta$ can describe the relationship between γ_r, t, and σ_c. The confining stress substantially impacted the response of the damping ratio in the treated sample. The threshold calcium carbonate content was about 7%.

However, the relationship between γ_r and treatment duration or confining stress is provided in this manuscript. Nevertheless, the experimental data analyzed in this study are limited, and further efforts are needed to demonstrate the empirical formulas and their applicability to other types of liquefiable soils and treatment conditions. In addition, the empirical formulations between γ_r and other physical properties such as relative density should be conducted in further work.

Author Contributions: Conceptualization, X.Z. and Y.C.; Methodology, X.Z. and H.G.; Validation, J.G. and R.Y.; Formal analysis, Y.C. and Y.H.; Investigation, J.G.; Resources, H.G. and H.L.; Data curation, J.G. and R.Y.; Writing—original draft, X.Z. and Z.S.; Writing—review & editing, L.L.; Visualization, J.G. All authors have read and agreed to the published version of the manuscript.

Funding: This research was funded by the National Natural Science Foundation of China, grant number 52179101, 52108324, and 52008207; the Natural Science Foundation of Jiangsu for Young Researchers, grant number BK20190667; the Natural Science Research Project of Colleges and Universities in Jiangsu Province of China, grant number 19KJB560015; and the Postgraduate Research and Practice Innovation Program of Jiangsu Province, grant number KYCX22_1354.

Institutional Review Board Statement: Not applicable.

Informed Consent Statement: Not applicable.

Data Availability Statement: The data are contained within the article.

Conflicts of Interest: The authors declare no conflict of interest.

References

1. Xiao, P.; Liu, H.L.; Xiao, Y.; Stuedlein, A.W.; Evans, T.M. Liquefaction resistance of bio-cemented calcareous sand. *Soil Dyn. Earthq. Eng.* **2018**, *107*, 9–19. [CrossRef]
2. Giretti, D.; Fioravante, V.; Been, K.; Dickenson, S.; Mesri, G.; Kane, T. Mechanical properties of a carbonate sand from a dredged hydraulic fill. *Geotechnique* **2020**, *70*, 937–942. [CrossRef]
3. Sasanakul, I.; Gassman, S.; Ruttithivaphanich, P.; Dejphumee, S. Characterization of shear wave velocity profiles for South Carolina Coastal Plain. *AIMS Geosci.* **2019**, *5*, 303–324. [CrossRef]
4. Sharma, S.S.; Ismail, M.A. Monotonic and Cyclic Behavior of Two Calcareous Soils of Different Origins. *J Geotech. Geoenviron. Eng.* **2006**, *132*, 1581–1591. [CrossRef]
5. Javdanian, H.; Jafarian, Y. Dynamic shear stiffness and damping ratio of marine calcareous and siliceous sands. *Geo Mar. Lett.* **2018**, *38*, 315–322. [CrossRef]
6. Shahnazari, H.; Rezvani, R. Effective parameters for the particle breakage of calcareous sands: An experimental study. *Eng. Geol.* **2013**, *159*, 98–105. [CrossRef]
7. Qin, Z.G.; Yuan, X.M.; Cao, Z.Z.; M, H.G. Applicability and quality evaluation of foundation treatment method for backfilled coral sand site. *J. Nat. Disasters* **2021**, *1*, 78–88. (In Chinese)
8. Whiffin, V.S.; van Paassen, L.; Harkes, M.P. Microbial Carbonate Precipitation as a Soil Improvement Technique. *Geomicrobiol. J.* **2007**, *24*, 417–423. [CrossRef]
9. Ivanov, V.; Chu, J. Applications of microorganisms to geotechnical engineering for bioclogging and biocementation of soil in situ. *Rev. Environ. Sci. Bio. Technol.* **2008**, *7*, 139–153. [CrossRef]
10. DeJong, J.T.; Mortensen, B.M.; Martinez, B.C.; Nelson, D.C. Bio-mediated soil improvement. *Ecol. Eng.* **2010**, *36*, 197–210. [CrossRef]
11. Lang, L.; Chen, B.; Duan, H. Modification of nanoparticles for the strength enhancing of cement-stabilized dredged sludge. *J. Rock Mech. Geotech. Eng.* **2021**, *13*, 694–704. [CrossRef]
12. Lang, L.; Chen, B. Strength properties of cement-stabilized dredged sludge incorporating nano-SiO2 and straw fiber. *Int. J. Geomech.* **2021**, *21*, 04021119. [CrossRef]
13. Lang, L.; Chen, B. Stabilization of Dredged Sediment Using Activated Binary Cement Incorporating Nanoparticles. *J. Mater. Civ. Eng.* **2022**, *34*, 04021381. [CrossRef]
14. Tiwari, N.; Satyam, N.; Sharma, M. Micro-mechanical performance evaluation of expansive soil biotreated with indigenous bacteria using MICP method. *Sci. Rep.* **2021**, *11*, 10324. [CrossRef] [PubMed]
15. Rajasekar, A.; Moy, C.K.S.; Wilkinson, S. MICP and advances towards eco-friendly and economical applications[C]//IOP Conference Series: Earth and Environmental Science. *IOP Publ.* **2017**, *78*, 012016.
16. Xiao, Y.; Zhang, Z.; Stuedlein, A.W.; Evans, T.M. Liquefaction Modeling for Biocemented Calcareous Sand. *J. Geotech. Geoenviron. Eng.* **2021**, *147*, 04021149. [CrossRef]
17. Xiao, Y.; Wang, Y.; Wang, S.; Evans, T.M.; Stuedlein, A.W.; Chu, J.; Zhao, C.; Wu, H.; Liu, H. Homogeneity and mechanical behaviors of sands improved by a temperature-controlled one-phase MICP method. *Acta Geotech.* **2021**, *16*, 1417–1427. [CrossRef]
18. Cheng, L.; Shahin, M.A.; Mujah, D. Influence of Key Environmental Conditions on Microbially Induced Cementation for Soil Stabilization. *J. Geotech. Geoenviron. Eng.* **2017**, *143*, 1–11. [CrossRef]
19. DeJong, J.T.; Fritzges, M.B.; Nüsslein, K. Microbially Induced Cementation to Control Sand Response to Undrained Shear. *J. Geotech. Geoenviron. Eng.* **2006**, *132*, 1381–1392. [CrossRef]
20. Feng, K.; Montoya, B.M. Influence of confinement and cementation level on the behavior of microbial-induced calcite precipitated Sands under monotonic drained loading. *J. Geotech. Geoenviron. Eng.* **2015**, *2*, 04015057. [CrossRef]

21. Minto, J.M.; Lunn, R.J.; El Mountassir, G. Development of a Reactive Transport Model for Field-Scale Simulation of Microbially Induced Carbonate Precipitation. *Water Resour. Res.* **2019**, *55*, 7229–7245. [CrossRef]
22. Jonkers, H.M.; Thijssen, A.; Muyzer, G.; Copuroglu, O.; Schlangen, E. Application of bacteria as self-healing agent for the development of sustainable concrete. *Ecol. Eng.* **2010**, *36*, 230–235. [CrossRef]
23. Wang, Z.; Zhang, N.; Jin, Y.; Li, Q.; Xu, J. Application of microbially induced calcium carbonate precipitation (MICP) in sand embankments for scouring/erosion control. *Mar. Georesources Geotechnol.* **2020**, *39*, 1459–1471. [CrossRef]
24. Zhang, X.; Chen, Y.; Liu, H.; Zhang, Z.; Ding, X. Performance evaluation of a MICP-treated calcareous sandy foundation using shake table tests. *Soil Dyn. Earthq. Eng.* **2019**, *129*, 105959. [CrossRef]
25. Cheng, L.; Shahin, M.; Cord-Ruwisch, R. Bio-cementation of sandy soil using microbially induced carbonate precipitation for marine environments. *Geotechnique* **2014**, *64*, 1010–1013. [CrossRef]
26. Lin, H.; Suleiman, M.T.; Brown, D.G.; Kavazanjian, E. Mechanical behavior of sands treated by microbially induced carbonate precipitation. *J. Geotech. Geoenviron. Eng.* **2016**, *142*, 04015066. [CrossRef]
27. Paassen, L.A.V. *Biogrout, Ground Improvement by Microbial Induced Carbonate Precipitation;* Delft University of Technology The Nethelands: Delft, The Netherlands, 2009.
28. Fioravante, V.; Giretti, D.; Jamiolkowski, M. Small strain stiffness of carbonate Kenya Sand. *Eng. Geol.* **2013**, *161*, 65–80. [CrossRef]
29. Giang, P.H.H.; Van Impe, P.O.; Van Impe, W.F.; Menge, P.; Haegeman, W. Small-strain shear modulus of calcareous sand and its dependence on particle characteristics and gradation. *Soil Dyn. Earthq. Eng.* **2017**, *100*, 371–379. [CrossRef]
30. Jafarian, Y.; Javdanian, H.; Haddad, A. Dynamic properties of calcareous and siliceous sands under isotropic and anisotropic stress conditions. *Soils Found.* **2018**, *58*, 172–184. [CrossRef]
31. He, H.; Li, W.; Senetakis, K. Small strain dynamic behavior of two types of carbonate sands. *Soils Found.* **2019**, *59*, 571–585. [CrossRef]
32. Shi, J.; Haegeman, W.; Cnudde, V. Anisotropic small-strain stiffness of calcareous sand affected by sample preparation, particle characteristic and gradation. *Géotechnique* **2021**, *71*, 305–319. [CrossRef]
33. Shi, J.; Haegeman, W.; Xu, T. Effect of non-plastic fines on the anisotropic small strain stiffness of a calcareous sand. *Soil Dyn. Earthq. Eng.* **2020**, *139*, 106381. [CrossRef]
34. Morsy, A.M.; Salem, M.A.; Elmamlouk, H.H. Evaluation of dynamic properties of calcareous sands in Egypt at small and medium shear strain ranges. *Soil Dyn. Earthq. Eng.* **2018**, *116*, 692–708. [CrossRef]
35. Hassanlourad, M.; Salehzadeh, H.; Shahnazari, H. Dilation and particle breakage effects on the shear strength of calcareous sands based on energy aspects. *Int. J. Civ. Eng.* **2008**, *6*, 108–119.
36. Khalil, A.; Khan, Z.H.; Attom, M.; El Emam, M.; Fattah, K. Dynamic Properties of Calcareous Sands from Urban Areas of Abu Dhabi. *Appl. Sci.* **2022**, *12*, 3325. [CrossRef]
37. Liang, K.; He, Y.; Chen, G.X. Experimental study of dynamic shear modulus and damping ratio characteristics of coral sand from Nansha Islands. *Rock Soil Mech.* **2020**, *1*, 23–38. (In Chinese)
38. Jafarian, Y.; Javdanian, H. Small-strain dynamic properties of siliceous-carbonate sand under stress anisotropy. *Soil Dyn. Earthq. Eng.* **2020**, *131*, 106045. [CrossRef]
39. Shi, J.; Xiao, Y.; Hu, J.; Wu, H.; Liu, H.; Haegeman, W. Small-strain shear modulus of calcareous sand under anisotropic consolidation. *Can. Geotech. J.* **2022**, *59*, 878–888. [CrossRef]
40. Lang, L.; Li, F.; Chen, B. Small-strain dynamic properties of silty clay stabilized by cement and fly ash. *Constr. Build. Mater.* **2019**, *237*, 117646. [CrossRef]
41. Subramaniam, P.; Banerjee, S.; Ku, T. Shear Modulus and Damping Ratio Model for Cement Treated Clay. *Int. J. Géoméch.* **2019**, *19*, 06019010. [CrossRef]
42. Feng, K.; Montoya, B.M. Drained Shear Strength of MICP Sand at Varying Cementation Levels. In Proceedings of the IFCEE 2015, San Antonio, TX, USA, 17–21 March 2015; pp. 2242–2251.
43. Simatupang, M.; Okamura, M.; Hayashi, K.; Yasuhara, H. Small-strain shear modulus and liquefaction resistance of sand with carbonate precipitation. *Soil Dyn. Earthq. Eng.* **2018**, *115*, 710–718. [CrossRef]
44. Acar, Y.B.; El-Tahir, E.T.A. Low strain dynamic properties of artificially cemented sand. *J. Geotech. Eng.* **1986**, *112*, 1001–1015. [CrossRef]
45. Saxena, S.K.; Avramidis, A.S.; Reddy, K. Dynamic moduli and damping ratios for cemented sands at low strains. *Can. Geotech. J.* **1988**, *25*, 353–368. [CrossRef]
46. Lin, B.; Zhang, F.; Feng, D.; Tang, K.; Feng, X. Dynamic shear modulus and damping ratio of thawed saturated clay under long-term cyclic loading. *Cold Reg. Sci. Technol.* **2018**, *145*, 93–105. [CrossRef]
47. Gao, H.; Xia, S.; Chen, F.; Stuedlein, A.W.; Li, X.; Wang, Z.; Shen, Z.; Chen, X. Dynamic shear modulus and damping of cemented and uncemented lightweight expanded clay aggregate (LECA) at low strains. *Soil Dyn. Earthq. Eng.* **2021**, *142*, 106555. [CrossRef]
48. Tsuchida, H. Prediction and remedial measures against liquefaction of sandy soil. In *Annual Seminar of Port and Harbour Research Institute;* International Association of Ports and Harbors: Tokyo, Japan, 1970; Volume 3, pp. 1–33. (In Japanese)
49. Hardin, B.O.; Drnevich, V.P. Shear Modulus and Damping in Soils: Design Equations and Curves. *J. Soil Mech. Found. Div.* **1972**, *98*, 667–692. [CrossRef]
50. Consoli, N.C.; da Fonseca, A.V.; Cruz, R.C.; Heineck, K.S. Fundamental Parameters for the Stiffness and Strength Control of Artificially Cemented Sand. *J. Geotech. Geoenviron. Eng.* **2009**, *135*, 1347–1353. [CrossRef]

51. Porcino, D.; Ghionna, V.N.; Granata, R.; Marcianò, V. Laboratory determination of mechanical and hydraulic properties of chemically grouted sands. *Geomech. Geoengin.* **2016**, *11*, 164–175. [CrossRef]
52. Liu, Q.F.; Zhuang, H.Y.; Wu, Q.; Zhao, K.; Chen, G.X. Experimental study on dynamic modulus and damping ratio of rubber-sand mixtures over a wide strain range. *J. Earthq. Tsunami* **2022**, *16*, 2140006. [CrossRef]
53. Hardin, B.O.; Drnevich, V.P. Shear Modulus and Damping in Soils: Measurement and Parameter Effects (Terzaghi Leture). *J. Soil Mech. Found. Div.* **1972**, *98*, 603–624. [CrossRef]
54. Wang, Z.Y.; Mei, G.X.; Yu, X.B. Dynamic Shear Modulus and Damping Ratio of Waste Granular Rubber and Cement Soil Mixtures. *Adv. Mater. Res.* **2011**, *243-249*, 2091–2094. [CrossRef]
55. Delfosse-ribay, E.; Djeran-maigre, I.; Cabrillac, R.; Gouvenot, D. Shear modulus and damping ratio of grouted sand. *Soil Dyn. Earthq. Eng.* **2004**, *24*, 461–471. [CrossRef]
56. San, P.A.; Lee, M.; Graddy, C.; Kolbus, C.M.; Khan, M.; Zamani, A. Meter-scale biocementation experiments to advance process control and reduce impacts: Examining spatial control, ammonium by-product removal, and chemical reductions. *J. Geotech. Geoenviron. Eng.* **2020**, *146*, 1–14.
57. Cheng, L.; Shahin, M.A.; Cord-Ruwisch, R. Soil stabilization by microbial-induced calcite precipitation (MICP). In Proceedings of the Investigation Into Some Physical and Environmental Aspects, 7th International Congress on Environmental Geotechnics: ICEG 2014. Barton: Engineers Australia, Melbourne, Australia, 10–14 November 2014.
58. Chu, J.; Ivanov, V.; Naeimi, M. Optimization of calcium-based bioclogging and biocementation of sand. *Acta Geotech.* **2014**, *9*, 277–285. [CrossRef]
59. Choi, S.G.; Chang, I.; Lee, M.; Lee, J.H.; Han, J.T.; Kwon, T.H. Review on geotechnical engineering properties of sands treated by microbially induced calcium carbonate precipitation (MICP) and biopolymers. *Constr. Build. Mater.* **2020**, *246*, 118415. [CrossRef]

applied sciences

MDPI

Article

The Shrink–Swell Process of the Granite Residual Soil with Different Weathering Degree in a Gully System in Southern China

Honghu Liu [1,2,*], Jing Liu [2], Xianwei Zhang [3] and Xinyu Liu [3]

1 Changjiang River Scientific Research Institute, Changjiang Water Resource Commission, Wuhan 430010, China
2 State Key Laboratory of Soil Erosion and Dryland Farming on the Loess Plateau, Institute of Soil and Water Conservation, Chinese Academy of Sciences and Ministry of Water Resources, Xinong Rd. 26, Yangling 712100, China
3 State Key Laboratory of Geomechanics and Geotechnical Engineering, Institute of Rock and Soil Mechanics, Chinese Academy of Sciences, Wuhan 430071, China
* Correspondence: liuhh@mail.crsri.cn

Abstract: The soil shrink–swell phenomenon produces crack networks and slope instability. However, few studies have involved the continuous shrink–swell process of granite residual soils. The objective of the study is to explore the shrink–swell process of weathered granite soils and its effects on gully development in southern China. The bulk density, soil water content (SWC), shrink–swell ratio (SSR), clay mineral content, and mechanical composition, etc., of soil samples from five soil layers (at depths of 0.3 m, 3.0 m, 7.0 m, 12.0 m, and 16.0 m) along a profile in Yudu County was analyzed. After quantifying the soil properties at different soil depths, we analyzed these data statistically in an effort to identify strong parametric relationships. The results indicated that some properties such as bulk density and shear stress increased with soil depth, while other soil properties, such as plasticity index and liquid limit, were inversely related to depth. Soil cohesion, the angle of internal friction, and shear stress were closely related to the SWC. Every 1% decrease in the SWC resulted in a shear stress reduction of 6.62 kPa. The SSR values exhibited significant variation between the three dry–wet cycles and were closely related to the bulk density values of our kaolin and montmorillonite samples. As an environmental factor, the SWC can trigger changes in internal soil properties such as shear stress and the SSR. Using these data and observations made during our field survey, it can be proposed that continuous shrink–swell variation in deep granite-weathering crust can result in crack formation and gully erosion. It can be inferred that crack development velocity and gully retreat rate may be affected by the soil's shrink–swell process. Consequently, this information provides insight to understanding the mechanism of gully development in southern China.

Keywords: water content; clay minerals; dry–wet cycles; shrinking; swelling

1. Introduction

Gully erosion is defined as an erosion process whereby runoff water accumulates and often recurs in narrow channels and, over short periods, moves the soil from this narrow area to considerable depths [1]. Gully erosion closely relates to rainfall amount and rainfall erosivity [2,3] because rainfall affects soil water content (SWC), and results in a change in soil shrinkage and swelling [4]. Continuous shrink–swell (SS) processes generate cracks in soil [5–7]. These make gullies develop as soil cracks extend. More importantly, the soil shrink–swell extent determines the velocity of soil cracking and collapse [8,9]. Therefore, it is very necessary to study the soil shrink–swell process.

There are numerous reports expounding the soil shrinkage process and its effects on soil properties. Chertkov [6] modelled the shrinkage curves of soil clay pastes. Peng et al. [10] quantified 2D soil cracks on the soil surface using digital image processing. Krisdani et al. [11]

analyzed the relationship between shrinkage rate and SWC. Kalkan [12] reported the effects of silica fumes on swelling pressure and the potential of mixed clayey soil. Chertkov [13] constructed a physical model of the soil swelling curve and shrinkage curve. Stoltz et al. [14] analyzed the swelling and shrinkage of lime-treated expansive clayey soil. Fernandes et al. [15] measured the swelling and shrinkage of clay soil at various depths and explored the effects of SWC and temperature on shrinkage and swelling. Leong and Wijaya [16] developed a universal soil shrinkage curve equation. Zolfaghari et al. [17] reported the relationships of soil shrinkage between intrinsic soil properties and environmental variables for calcareous soils. Houben et al. [18] measured the swelling and shrinkage of rocks using a 1 mm sample cube with an environmental scanning electron microscope (ESEM) in a 3D dilatometer. Fang et al. [19] proved that paddy soil's bulk density and aggregate stability increased after being dried and decreased upon saturation, while soil shrinkage capacity was the opposite. Mishra et al. [20] reviewed soil shrinkage characteristic curves and estimated the relationship between void ratio and gravimetric moisture content. Tang et al. [7] measured the 3D crack parameters of compacted clayey soil in a different water content condition using X-ray-computed tomography. Zenero et al. [21] found the overall weakness of shrinkage and high soil shrinkage curves along the forest and pasture top sequences in Amazonia. Zúñiga et al. [22] thought the soil shrinkage had a high dependency on soil organic carbon. Generally, most of these studies focused on the shrinkage curve of clayey soil, but few studies were conducted on the shrinkage of granite residual soil and fewer studies focused on the swelling process of soil.

Until 2005, there were 239.1 thousand gullies in 362 counties of seven provinces of China [23]. Greater than 80% of these gullies were distributed in the granite's residual soil [24]. Liu et al. [3] proved the retreat rate of a small- and medium-sized gully over 7.5 years. Chen et al. [5] found that 27 units were still collapsing or collapsed, consisting of 12 very active ones, while only 3 units were fully stabilized, among 30 restored Benggang units investigated after 2–10 years of restoration in the Fujian province. Effects of water content on soils derived from granite were very significant [25], which resulted in numerous gullies developing. Thus, a series of laboratory investigations were performed to examine the soil property changes of granite residual soil due to the drying cycles. For example, Lin [26] measured the expansion and contraction characteristics of surface soil and found that soil expansion rates decreased with the increase in initial water contents. Zhang et al. [27] measured that the shear stress of surface soil decreased with SWC. These studies focused on surface soil properties and shrinkage characteristics. Duan et al. [28] proved that the red soil layer had a higher water retention capacity and shear strength than the sandy soil layer. Huang et al. [29] examined that the average values of the maximum linear shrinkage in the laterite, transition, and sandy layers were 1.50%, 2.09%, and 1.74%, respectively. Soil shrinkage rate was positively correlated with clay and Fe_2O_3 content and negatively correlated with sand content. Liu et al. [30] thought the weakening of cementation triggered the breakup of soil aggregates and led to soil disintegration and the occurrence of a gully collapse. However, little attention was paid to the shrink–swell process of the deep granite residual soil layers in the continuous dry–wet processes.

The objective of the study is to analyze the shrink–swell process of weathered granite soils and reveal its factors and impact on gully development. Thus, it is very helpful to understand crack formation, crack propagation, and gully wall failure processes developed in granite residual soil in southern China.

2. Materials and Methods

2.1. The Test Site

The studied soil profile is located in Zuoma, a small catchment with about 0.72 km^2, Yudu County, Jiangxi province, South China (115.40° N, 25.91° E) (Figure 1a, Liu et al. [3]. This region has a typical warm and humid subtropical climate with an average annual precipitation of 1508 mm and a mean annual temperature of 19.7 °C. The main soil type is Orthic Acrisol, with a soil depth of about 1.0 m, according to the Soil Survey Staff

(Chen et al. [31]). The parent material of the soil is a weathered granite regolith located at depths of 20–60 m. The soil has a pH value of 5–6. Each year, there are two distinct rice crop periods: one from April to July and another from July to November. Land use in the catchment is a mixture of field crops (rice), tree crops, and forests.

Figure 1. The location of Zuoma small catchment (**a**), the sampled gully (**b**), sampling points (**c**), and exhibition of crack (**d**), pipe (**e**), and slide (**f**) along the gully wall in Yudu county, Jiangxi province.

2.2. Soil Sampling

Five soil layers were ensured along the soil profile based on the characteristics of granite residual soil. The depths at which the samples were collected were 0.3 m, 3.0 m, 7.0 m, 12.0 m, and 16.0 m, respectively (Figure 1b,c). Two types of soil samples collected from each soil layer were then transported to the laboratory, including 9 undisturbed soil samples using cutting rings (5.46 cm in diameter and 5 cm in height), and 1 mixed soil sample weighting ~1 kg.

2.3. Soil Analysis

2.3.1. Experimental Analysis of Undistributed Soil

The undistributed soil samples were used to determine bulk density (BD), shear stress (τ), plasticity index (wL_p), SSR, etc.

(1) Soil bulk density was determined by the cutting ring method [32].

(2) The stress increased frictional force between soil particles and enhanced resistance to shear failure. Under constant stress, the shear stress of soil is linearly related to the normal stress of the section, which can be described by the Mohr–Coulomb formula [33]:

$$\tau = C_q + \sigma \tan \psi_q \quad (1)$$

where τ is the shear stress (kPa); C_q is the cohesion (kPa); σ is the normal stress on the failure surface (kPa); ψ_q is the angle of internal friction (°); and $\tan \psi_q$ represents the friction coefficient; and φ are determined by soil properties, which are defined as soil shear stress parameters.

(3) The liquid and plastic limits of every soil sample were measured by using a liquid–plastic limit tester. The plasticity index was determined according to the liquid and plastic limits (Equation (2)).

$$wL_p = IP_L - w_p \quad (2)$$

where wL_p is the plasticity index, IP_L is the liquid limit (%), w_p is the plastic limit (%).

(4) For every soil sample, soil cylinders in stainless steel sample retainers were submitted to three water absorption and desiccation processes in 20 to 30 °C. The soil was weighted with a table balance every 30 min in the water absorption process and every 3 days in the desiccation process to determine the SWC. At the same time, the height of the soil sample was measured using a caliper. This study hypothesizes that a zero shrink–swell ratio appears when the SWC is minimal. The shrink–swell (SSR) was computed according to Equation (3).

$$SSR = \frac{h_i - h_0}{h_0} \times 100\% \quad (3)$$

where SSR is the shrink–swell ratio (%), h_i is the height of soil at i time (0.1 mm), h_0 is the height of soil at the start (0.1 mm).

2.3.2. Experimental Analysis of Distributed Soil

The mixed soil samples were air-dried and sieved to remove stone friction (>2 mm diameter), and these mixed soil samples were used to analyze particle composition, mineral components and clay mineral components. The mechanical composition was determined using the sieving and pipette method [34]. Soil oxide compositions (SiO_2, Fe_2O_3) were measured using the lithium carbonate–boric acid melting method [32]. The type and relative contents of primary minerals were determined using an X-ray diffractometer (XRD) [35].

2.4. Statistical Analysis

The figures of soil properties and SSRs were drawn by using originPro 9.0. The differences in all soil properties of different soil layers were compared. Pearson's correlation analysis was used to determine the correlation coefficients between soil parameters and the shrink–swell ratio. Stepwise regression analysis was employed to analyze the relationship between maximum and average soil shrink–swell ratio and other soil parameters. All tests were performed using the statistical software MATLAB 2010.

3. Results

3.1. Change of Soil Properties with Depth

Table 1 lists the basic properties of granite residual soil at different depths. BD increased with soil depth, which ranged from 1.44 g cm^{-3} to 1.79 g cm^{-3}. The coarse particle content (0.25–2 mm) increased from 12.9% to 24.3% with depth, while the clay particle content (<0.002 mm) decreased from 27.0% to 11.2%, indicating that the soils at the surface are more weathered than those closer to the parent material. Overall, with increased depth, the quartz content decreased, and the clay content increased overall. Despite this prevailing trend, there is an outlier occurrence of illite in Layer 3. Plasticity indexes wL_{17} and wL_{10}, the plastic limit (w_p), and liquid limits IP_{17} and IP_{10} all generally decreased with depth, with the exception of the outlier values observed in Layer 3.

Figure 2a–c shows the changes in the soil cohesion (C_q), angle of internal friction (ψ_q), and shear stress (τ) for the different soil layers with variable water contents. All three indicators exhibited strong negative correlations with the SWC and significant variations among the different soil layers. For example, every 1% decrease in the water content resulted in a shear stress reduction of 6.62 kPa. These parameters were positively correlated with depth, because the SWC was higher at the surface.

Table 1. Basic properties of soil samples.

Layer No.	Depth (m)	Bulk Density (g/cm^3)	Particle Composition (mm, %)				Mineral Component (%) (Total 100%)			Clay Mineral Component (Total 100%)			wL_{17}	wL_{10}	w_p	IP_{17}	IP_{10}
			>2	2–0.25	0.25–0.002	<0.002	Quartz	Pyrite	Clay Mineral	Kaolinite	Illite	Montmorillonite	%				
1	0.3	1.44	1.1	12.9	59.0	27.0	96.2	0.51	3.29	76.57	-	23.43	34.2	28.6	16.6	17.6	12.0
2	3.0	1.67	8.0	15.0	59.9	17.1	94.58	0.55	4.87	89.84	5.76	4.39	30.5	25.6	15.1	15.4	10.5
3	7.0	1.71	3.8	16.7	67.6	11.9	93.71	-	6.29	77.7	20.08	2.22	26.2	22.3	13.8	12.4	8.5
4	12.0	1.73	5.6	19.8	59.3	15.3	94.27	-	5.73	94.17	3.29	2.54	29.8	24.9	14.5	15.3	10.4
5	16.0	1.79	12.5	24.3	52.0	11.2	89.42	-	10.58	91.54	3.28	5.18	25.7	21.9	13.6	12.1	8.3

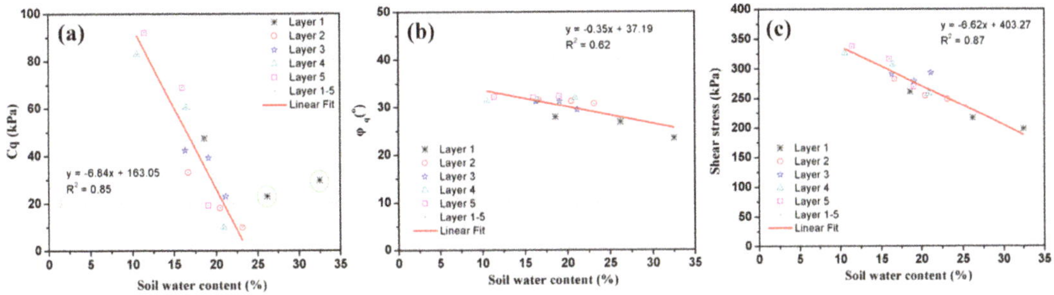

Figure 2. Variations of C_q (**a**), ψ_q (**b**), and shear stress (**c**) with soil water content at five soil layers. The points in the green circle are not included in the linear fit.

3.2. Variation in the Shrink–Swell Ratio during Three Dry–Wet Cycles

Figure 3 shows the SSR variation in samples from the five soil layers during three dry–wet cycles. The SSR values exhibited period variations during the three dry–wet cycles; the shrinking process was relatively slow, while the swelling process was relatively fast. There were significant differences in maximum SSR values among samples from the five soil layers, while their minimum values were relatively similar. For the maximum SSR values, the layer ranking order was Layer 1 (7.53%) > Layer 3 (6.06%) > Layer 5 (5.92%) > Layer 2 (4.81%) > Layer 4 (3.66%). The magnitude of the maximum and minimum SSR values decayed with each cycle.

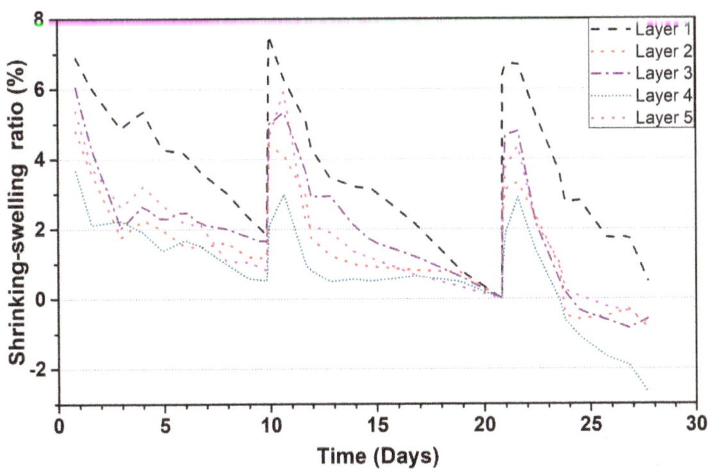

Figure 3. SSR of five layers during 3 dry–wet cycles.

3.3. Correlation between the Shrink–Swell Ratio and Soil Parameters

We computed the maximum and average SSR values, as well as the average values for twelve other soil parameters; the correlation matrix for these parameters is shown in Table 2. Some parameters exhibited significant correlations with others, including the maximum SSR value with kaolin content ($p < 0.05$), the average SSR value with montmorillonite content ($p < 0.01$), and the average SSR value with BD, the average SWC, and kaolin content ($p < 0.05$). Because the average SSR had a strong relationship with the maximum SSR ($p < 0.01$), it is a good proxy for the maximum SSR value of granite soils in southern China.

Table 2. Correlation coefficient of shrink–swell ratio and the other 12 soil indicators.

Parameter	Maximum SSR	Average SSR	BD	Average SWC	>2 mm	<0.002 mm	Quartz	Clay Mineral	Kaolinite	Montmorillonite	wL_{17}	wL_{10}	wp	IP_{17}	IP_{10}
Maximum SSR	1														
Average SSR	0.96 **	1.00													
BD	−0.64	−0.83 *	1.00												
Average SWC	0.63	0.82 *	−1.00 **	1.00											
>2 mm	−0.37	−0.54	0.77	−0.73	1.00										
<0.002 mm	0.50	0.72	−0.97 **	0.97 **	−0.67	1.00									
Quartz	0.11	0.35	−0.77	0.76	−0.88 *	0.76	1.00								
Clay mineral	−0.14	−0.38	0.80 *	−0.79	0.85 *	−0.79	−1.00 **	1.00							
Kaolinite	−0.82 *	−0.83 *	0.68	−0.64	0.74	−0.47	−0.47	0.47	1.00						
Montmorillonite	0.76	0.90 **	−0.92 **	0.92 **	−0.55	0.91 **	0.49	−0.52	−0.58	1.00					
wL_{17}	0.25	0.50	−0.88 *	0.90 **	−0.64	0.96 **	0.82 *	−0.85 *	−0.28	0.78	1.00				
wL_{10}	0.28	0.53	−0.90 **	0.91 **	−0.64	0.97 **	0.82 *	−0.85 *	−0.30	0.79	1.00 **	1.00			
wp	0.42	0.66	−0.95 **	0.96 **	−0.65	0.99 **	0.79	−0.83 *	−0.42	0.86 *	0.98 **	0.99 **	1.00		
IP_{17}	0.15	0.42	−0.84 *	0.85 *	−0.62	0.93 **	0.83 *	−0.85 *	−0.20	0.72	0.99 **	0.99 **	0.95 **	1.00	
IP_{10}	0.16	0.43	−0.84 *	0.86	−0.62	0.94 **	0.83 *	−0.85 *	−0.21	0.73	1.00 **	0.99 **	0.96 **	1.00 **	1.00

Note: ** $p < 0.05$, * $p < 0.01$.

3.4. Controlling Factor of the Shrink–Swell Ratio

The parameters that had the strongest correlation with the average SSR value, as well as the strength of the correlation between BD and the average SWC, are shown in Figure 4. The average SSR exhibited a strong positive correlation with the average SWC, because it was inversely related to BD, and BD was positively related to the average SWC. The changing trend between kaolin and montmorillonite was also opposite. Out of all of these parameters, the relationship between the average SSR value and the montmorillonite content had the strongest correlation and the smallest confidence interval; based on these statistical analyses, we assume that the montmorillonite content exerts the strongest influence over the soil SSR value.

While the soil BD, montmorillonite content, and kaolin content values were relatively consistent between the different soil layers, the SWC changed quickly during rainfall events and slowly during the drying process. Figure 5 shows how the SWC affects the SSR values. The SSR values exhibited strong variation between the five soil layers, with the SSR value being much higher in Layer 1 than it was in the other four layers. While the SSR values varied between the five layers, they all followed similar trends with respect to the SWC.

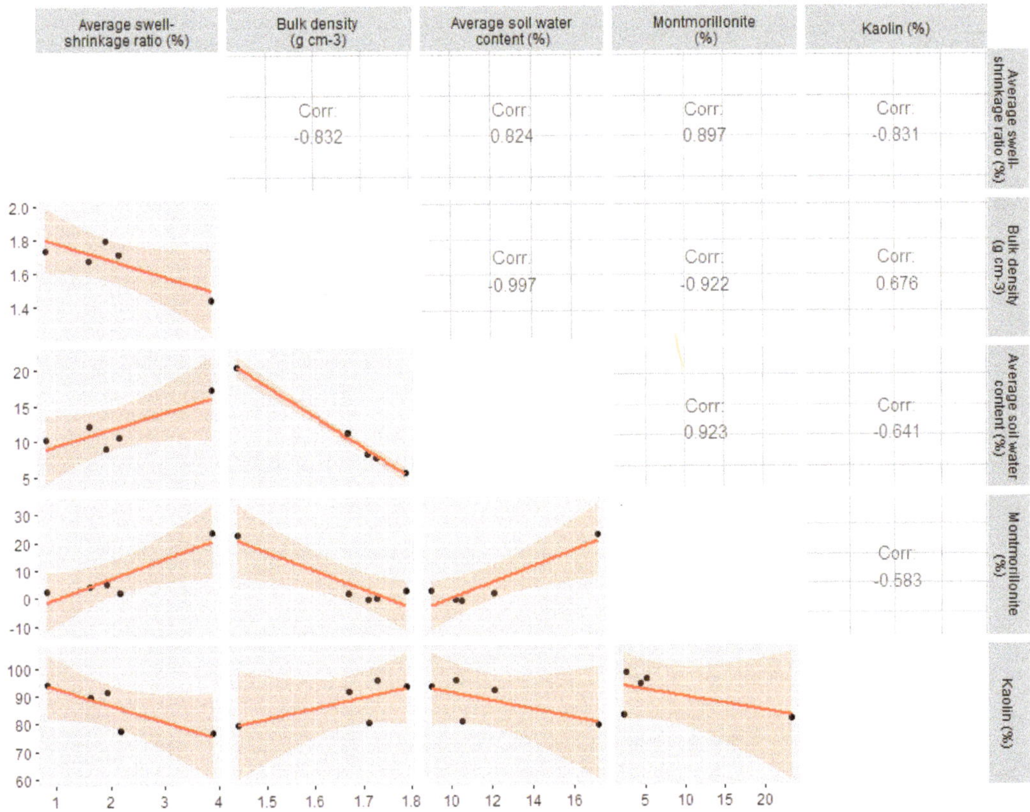

Figure 4. The correlation matrix among five good correlation indicators.

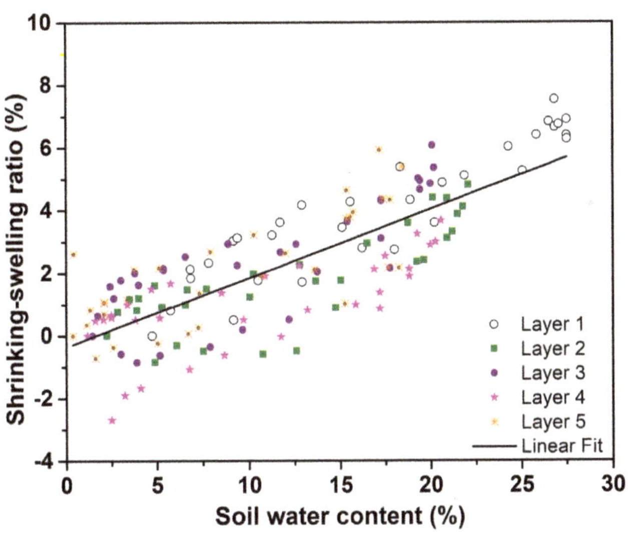

Figure 5. The relationship between shrink–swell ratio and soil water content.

4. Discussion

4.1. *Explanation on the Change of Shrink–Swell Ratio of Granite Residual Soil*

Our attention is inconsistent with other studies, which pay more attention to the shrinkage process (e.g., Leong and Wijaya [16]; Mishra et al. [20]). Our results cast a new light on a continuous quick-swelling and slow-shrinking processes of granite residual soil. This was also found by Lin [26], in which the swelling of surface soil was stable at about 0.5 h, and its shrinkage at more than 36 h. Another paper found higher swelling and lower shrinkage in granite residual soil. To reduce these errors, the SSR range of five soil layers is displayed in Figure 6. It shows that: (1) the SSR range except for Layer 2 reduces with the increase in soil depth; (2) the SSR difference ranging between 5.6% and 7.5% is less than the expansive soil (Stoltz et al. [14]) or clayey soil (Kalkan [12]).

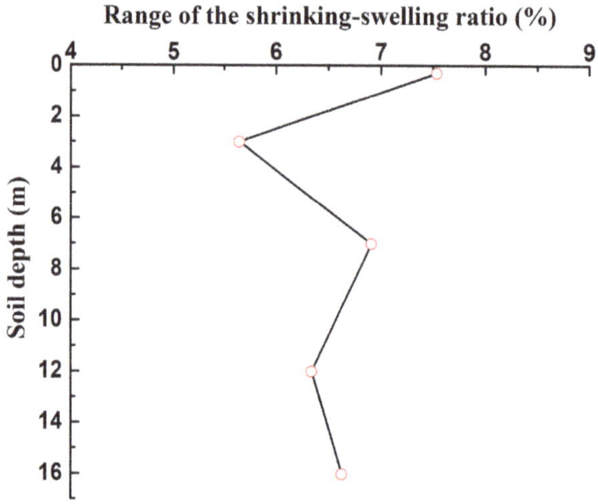

Figure 6. Range of shrink–swell ratio of the studied weathering profile.

The relatively low SSR is caused by the low quantity of clay minerals (e.g., illite, montmorillonite). Liang et al. [36] thought that soil moisture easily reached saturation and exceeded the plastic limit of the soil during rainfall, the soil swelled, and the shear thus stress rapidly reduced. When the water content increased to 35%, the soil's shear stress decreased by 60% [36]. The relationship between shear stress and shear displacement at the different normal stress in Figure 7 shows that once water content was increased, shear stress became smaller and shear displacement became larger. In addition, shear stress increased with depth at the same shear displacement.

4.2. *Effects of Shrink–Swell of Granite Residual Soil on Gully Development*

When the rainfall intensity is greater than 15–20 mm h^{-1}, it can have a significant impact on weathered granite soils [37]. During rainfall events, the shrink–swell process results in crack development (Figure 1d–f). Cracks develop along the gully wall (Figure 1d) when rain infiltrates the soil, reduces the soil shear stress, and increases the overriding soil weight [27]. With enough rain, cracks can widen into pipes (Figure 1e) or result in slides (Figure 1f). These slides typically move very quickly, but obstacles or other obstructions can reduce the slide velocity (Figure 1f). Once the energy of rain and runoff exceeds the resistance of obstacles, the gully develops further. Liu et al. [3] proved that two gully retreat rates were 0.46 and 1.10 m yr^{-1}, respectively, which resulted from mass movement triggered by gravity and soil saturation. This illustrates that the velocity of crack development and gully evolution can also be affected by soil saturation. As a result, analyzing crack formation and propagation can provide us with valuable information about the gully development process. Consequently, it is very necessary to study the shrink–swell process and crack formation and propagation for understanding the mechanism of gully erosion.

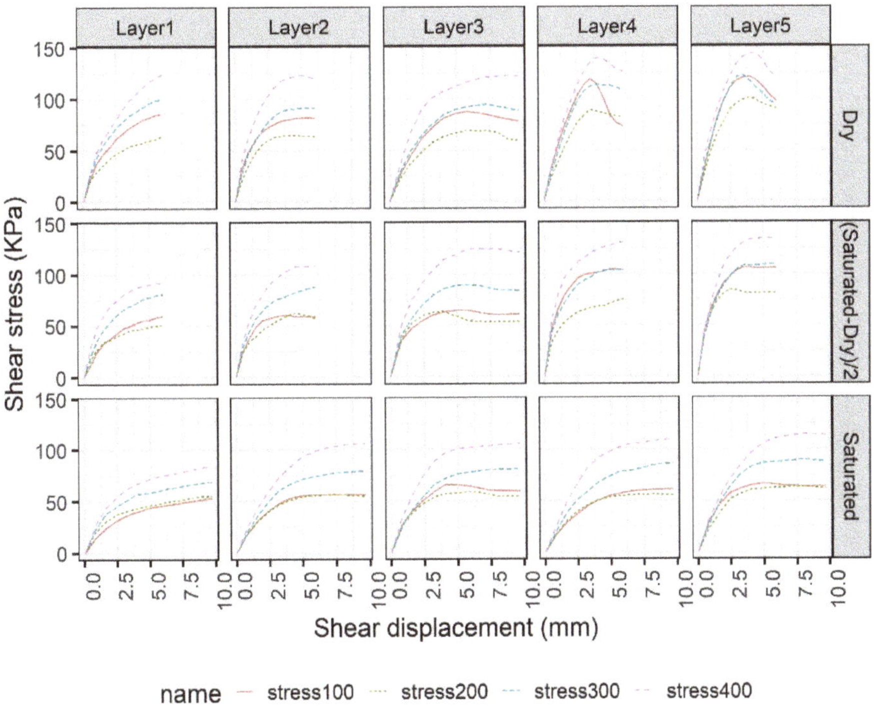

Figure 7. Shear stress–shear displacement relationship of five soil layers at the different saturated conditions.

5. Conclusions

This study measured the shrink–swell behavior of granite residual soil deposited between 0.3 m to 16 m subjected to three dry–wet cycles. Some effects of soil properties and environmental factors in southern China were analyzed. The following conclusions were obtained: (1) Soil cohesion, angle of internal friction, and shear stress had a strong linear relationship with water content. (2) SSR was reduced with the increase in soil depth. BD, water content, kaolinite, and montmorillonite had a very close correlation with SSR. (3) The SWC was the main factor affecting shear stress and the SSR of weathered granite soils. As the water content increased, shear stress decreased, and shear displacement increased. The shrink–swell process promotes crack formation, crack propagation, and gully development in weathered granite soils. We recommend that crack formation and propagation, and spatial distribution of cracks in weathered granite soils of southern China be further explored in future studies in order to better understand the process and distribution characteristics of gully erosion in southern China.

Author Contributions: Data curation, J.L.; Writing—original draft, H.L.; Writing—review & editing, X.Z. and X.L. All authors have read and agreed to the published version of the manuscript.

Funding: This research was financially supported by Key R & D projects of Hubei Province (2021BAA186), National Natural Science Foundation of China (Grant No. 413101297).

Institutional Review Board Statement: Not applicable.

Informed Consent Statement: Not applicable.

Data Availability Statement: Not applicable.

Acknowledgments: We thank Georg Hörmann, Liang Liu and Wenting Wang for revising the manuscript.

Conflicts of Interest: The authors declare no conflict of interest.

References

1. Poesen, J.; Nachtergaele, J.; Verstraeten, G.; Valentin, C. Gully erosion and environmental change: Importance and research needs. *Catena* **2003**, *50*, 91–133. [CrossRef]
2. Bouchnak, H.; Felfoul, M.S.; Boussema, M.R.; Snane, M.H. Slope and rainfall effects on the volume of sediment yield by gully erosion in the Souar lithologic formation (Tunisia). *Catena* **2009**, *78*, 170–177. [CrossRef]
3. Liu, H.; Qian, F.; Ding, W.; Gómez, J.A. Using 3D scanner to study gully evolution and its hydrological analysis in the deep weathering of southern China. *Catena* **2019**, *183*, 104218. [CrossRef]
4. Chen, J.-L.; Zhou, M.; Lin, J.-S.; Jiang, F.-S.; Huang, B.-F.; Xu, T.-T.; Wang, M.-K.; Ge, H.-L.; Huang, Y.-H. Comparison of soil physicochemical properties and mineralogical compositions between noncollapsible soils and collapsed gullies. *Geoderma* **2018**, *317*, 56–66. [CrossRef]
5. Chen, P.; Liu, X.; Yu, S.; Xu, J.; Hong, B.; Ma, J.; Ding, J.; Chen, Y.; Chen, Y.; Lu, C. Stability assessment of the restored Benggang units in a weathered granite crust region of South China. *Ecol. Eng.* **2022**, *182*, 106709. [CrossRef]
6. Chertkov, V. Modelling the shrinkage curve of soil clay pastes. *Geoderma* **2003**, *112*, 71–95. [CrossRef]
7. Tang, C.-S.; Zhu, C.; Leng, T.; Shi, B.; Cheng, Q.; Zeng, H. Three-dimensional characterization of desiccation cracking behavior of compacted clayey soil using X-ray computed tomography. *Eng. Geol.* **2019**, *255*, 1–10. [CrossRef]
8. Chen, N.; Zhou, W.; Yang, C.; Hu, G.; Gao, Y.; Han, D. The processes and mechanism of failure and debris flow initiation for gravel soil with different clay content. *Geomorphology* **2010**, *121*, 222–230. [CrossRef]
9. Vanmaercke, M.; Poesen, J.; Van Mele, B.; Demuzere, M.; Bruynseels, A.; Golosov, V.; Bezerra, J.F.R.; Bolysov, S.; Dvinskih, A.; Frankl, A.; et al. How fast do gully headcuts retreat? *Earth-Sci. Rev.* **2016**, *154*, 336–355. [CrossRef]
10. Peng, X.; Horn, R.; Peth, S.; Smucker, A. Quantification of soil shrinkage in 2D by digital image processing of soil surface. *Soil Tillage Res.* **2006**, *91*, 173–180. [CrossRef]
11. Krisdani, H.; Rahardjo, H.; Leong, E. Effects of different drying rates on shrinkage characteristics of a residual soil and soil mixtures. *Eng. Geol.* **2008**, *102*, 31–37. [CrossRef]
12. Kalkan, E. Impact of wetting–drying cycles on swelling behavior of clayey soils modified by silica fume. *Appl. Clay Sci.* **2011**, *52*, 345–352. [CrossRef]
13. Chertkov, V. Physical modeling of the soil swelling curve vs. the shrinkage curve. *Adv. Water Resour.* **2012**, *44*, 66–84. [CrossRef]
14. Stoltz, G.; Cuisinier, O.; Masrouri, F. Multi-scale analysis of the swelling and shrinkage of a lime-treated expansive clayey soil. *Appl. Clay Sci.* **2012**, *61*, 44–51. [CrossRef]
15. Fernandes, M.; Denis, A.; Fabre, R.; Lataste, J.-F.; Chrétien, M. In situ study of the shrinkage-swelling of a clay soil over several cycles of drought-rewetting. *Eng. Geol.* **2015**, *192*, 63–75. [CrossRef]
16. Leong, E.; Wijaya, M. Universal soil shrinkage curve equation. *Geoderma* **2015**, *237–238*, 78–87. [CrossRef]
17. Zolfaghari, Z.; Mosaddeghi, M.; Ayoubi, S. Relationships of soil shrinkage parameters and indices with intrinsic soil properties and environmental variables in calcareous soils. *Geoderma* **2016**, *277*, 23–34. [CrossRef]
18. Houben, M.; Barnhoorn, A.; Peach, C.; Drury, M. Potential permeability enhancement in Early Jurassic shales due to their swelling and shrinkage behavior. *Int. J. Coal Geol.* **2018**, *196*, 115–125. [CrossRef]
19. Fang, H.; Zhang, Z.; Li, D.; Liu, K.; Zhang, K.; Zhang, W.; Peng, X.; Zhou, H. Temporal dynamics of paddy soil structure as affected by different fertilization strategies investigated with soil shrinkage curve. *Soil Tillage Res.* **2019**, *187*, 102–109. [CrossRef]
20. Mishra, P.N.; Scheuermann, A.; Bore, T.; Li, L. Salinity effects on soil shrinkage characteristic curves of fine-grained geomaterials. *J. Rock Mech. Geotech. Eng.* **2019**, *11*, 181–191. [CrossRef]
21. Zenero, M.D.O.; Grimaldi, M.; Cooper, M. Variability in soil shrinkage along forest and pasture toposequences in Amazonia. *Geoderma* **2019**, *338*, 291–301. [CrossRef]
22. Zúñiga, F.; Horn, R.; Rostek, J.; Peth, S.; Uteau, D.; Dörner, J. Anisotropy of intensity–capacity parameters on Aquands with contrasting swelling–shrinkage cycles. *Soil Tillage Res.* **2019**, *193*, 101–113. [CrossRef]
23. Feng, M.H.; Liao, C.Y.; Li, S.X.; Lu, S.L. Investigation on status of hill collapsing and soil erosion in southern China. *Yangtze River* **2009**, *40*, 66–68, 75. (In Chinese)
24. Li, S.X.; Gui, H.Z.; Ding, S.W. Features of special layout of hill collapse in South China. *J. Huazhong Agric. Univ.* **2013**, *32*, 83–86. (In Chinese)
25. Zhang, X.M.; Ding, S.W.; Cai, C.F.; Liu, J.B. Mechanism of effects of wetting–drying on nonuniform settlement and caved wall collapse in slope disintegration erosion area. *Rock Soil Mech.* **2013**, *32* (Suppl. S2), 299–305. (In Chinese)
26. Lin, J.L. Study on the expansion and contraction characteristics of red soil layer in Benggang of granite area. *J. Soil Water Conserv.* **2019**, *33*, 87–92. (In Chinese)
27. Zhang, Y.; Zhong, X.; Lin, J.; Zhao, D.; Jiang, F.; Wang, M.-K.; Ge, H.; Huang, Y. Effects of fractal dimension and water content on the shear strength of red soil in the hilly granitic region of southern China. *Geomorphology* **2020**, *351*, 106956. [CrossRef]

28. Duan, X.; Deng, Y.; Tao, Y.; He, Y.; Lin, L.; Chen, J. The soil configuration on granite residuals affects Benggang erosion by altering the soil water regime on the slope. *Int. Soil Water Conserv. Res.* **2021**, *9*, 419–432. [CrossRef]
29. Huang, W.-X.; Deng, Y.-S.; Cai, C.-F.; Jiang, D.-H. Effects of soil shrinkage in permanent gullies formation: The case of Benggang erosion in the granite area of southern China. *J. Mt. Sci.* **2021**, *18*, 2328–2344. [CrossRef]
30. Liu, X.; Zhang, X.; Kong, L.; Wang, G.; Liu, H. Formation mechanism of collapsing gully in southern China and the relationship with granite residual soil: A geotechnical perspective. *Catena* **2022**, *210*, 105890. [CrossRef]
31. Chen, Z.C.; Gong, Z.T.; Zhang, G.L.; Zhao, W.J. Correlation of soil taxa between Chinese soil genetic classification and Chinese soil taxonomy on various scales. *Soil* **2004**, *36*, 584–595.
32. Institute of Soil Science, the Chinese Academy of Science (ISSCAS). *Soil Chemical and Physical Analysis*; Shanghai Science and Technology Press: Shanghai, China, 1981.
33. Lin, H.; Hua, Y.S.; Yong, R.; Lei, D.X.; Xu, W.Z.; Du, S.G. Strength parameters of rock considering area and stress correction during shearing. *Geotech Geol Eng* **2020**, *38*, 961–970. [CrossRef]
34. Gee, G.W.; Bauder, J.W. Particle size analysis. In *Methods of Soil Analysis, Part 1*; Agronomy Monographs 9; Klute, A., Ed.; American Society of Agronomy: Madison, WI, USA, 1986.
35. Li, X.Y. *Soil Chemistry and Experimental Guidance*; China Agricultural Press: Beijing, China, 1997.
36. Liang, Y.; Ning, D.H.; Pan, X.Z.; Li, D.C.; Zhang, B. The characteristics and control of the collapsing gullies in the red soil area of South China. *SWCC* **2009**, *1*, 31–34. (In Chinese)
37. Zhang, J.W.; Yao, J.Y. *Studies on Slopeland in the Southern China*; Science Press: Beijing, China, 1994. (In Chinese)

applied sciences

MDPI

Article

Triaxial Experimental Study of Zinc Contaminated Red Clay under Different Temperature Conditions

Jianqiang Wang [1,2], Yu Song [1,2,*], Shuaishuai Dong [1,2], Song Ding [1,2], Yukun Geng [1,2] and Xiaotong Gao [1,2]

1 College of Civil and Architecture Engineering, Guilin University of Technology, Guilin 541004, China
2 Guangxi Key Laboratory of Geomechanics and Geotechnical Engineering, Guilin 541004, China
* Correspondence: songyu119@glut.edu.cn

Abstract: Temperature is one of the important factors affecting the mechanical properties of geotechnical soils, and its role in engineering construction in China cannot be underestimated. In order to study the effects of temperature and zinc contamination concentration on the mechanical properties of Guilin local red clay, a temperature-controlled triaxial shear test was conducted on Guilin red clay under three variables of temperature, zinc contamination concentration and surrounding pressure. The test findings revealed that there are significant differences in the effects of temperature, zinc contamination concentration and surrounding pressure on the mechanical properties of Guilin red clay. The stress–strain curves of the red clay at various temperatures, contamination concentrations and envelope pressures are of the strain-hardening type, and the deformation modulus showed a tendency to increase rapidly with increasing strain, then decrease rapidly, and finally, decrease slowly. With the increase of temperature, the cohesion of Zn-contaminated red clay increases, while the angle of internal friction increases and then decreases, both of which increase the shear strength of red clay. As the concentration of Zn contamination grows, the shear strength of the red clay increases, while the internal friction angle increases and then decreases, and the shear strength of the soil increases and then decreases. The shear strength of the Zn-contaminated red clay improved as the surrounding pressure increased.

Keywords: temperature; stress–strain curves; modulus of deformation; shear strength; shear strength index

1. Introduction

Red clay soil is a regional unique soil [1,2] that is relatively widespread in the southern region of China. In comparison to other soils, it has relatively obvious engineering properties, such as high pore ratio, high dispersion, low compressibility, obvious water loss and shrinkage, and soft and hard strata at the top and bottom of the strata [3–10]. These engineering properties have attracted extensive attention and research from many scholars. Guangxi Zhuang Autonomous Region, commonly known as "the hometown of non-ferrous metals", possesses a wealth of mineral resources, such as lead, zinc, tin, and sulfur iron ore, among which zinc and lead have the largest reserves. In the past, the local government's model for economic development was a rough economy, and there were some industrial "three wastes" improperly treated in the process of producing and utilizing local mineral resources, which led to the deterioration of the ecosystem surrounding the mining area. For example, large amounts of heavy metal pollutants in high concentrations are discharged into the natural environment, resulting in severe soil and water pollution [11,12]. Excessive heavy metal pollutants in certain soils lead to a decrease in water quality in the associated watersheds, and some relevant indicators are exceeded, affecting the surrounding ecological environment posing a threat [13–15]. Li and Wang et al. [16,17] found that zinc contamination in the investigation and study of lead-zinc and zinc-manganese mining areas in Guangxi poses a threat to the natural environmental safety of the adjacent areas. The improper discharge of heavy metals from mines such as these, resulting in their infiltrating and migrating into the geotechnical body, causes soil contamination that poses many difficulties for engineering construction.

Since the beginning of the new century, governments and enterprises have paid increasing attention to environmental protection issues; consequently, countries have increased their investment in research and treatment of contaminated soil and water bodies to cope with the deteriorating environmental problems. In recent decades, domestic and international scholars have conducted a great deal of research and treatment on soils and contaminated soils under a variety of conditions. Some scholars have studied the variation pattern of shear strength and index of red clay under conventional physical independent variables [18], including dry density, water content, fracture degree, and other variables. Some scholars have studied the evolution pattern of mechanical strength of red clay under different engineering conditions [19–21], and explored the microstructural changes that accompany the mechanical evolution process under different engineering conditions [20]. In terms of soil improvement, researchers have improved the engineering properties of the original soil by incorporating various amendments into the soil [1,3,10,22–24], such as increasing the strength of the soil, altering the distribution of soil pores and volume, etc. In their investigation of contaminated soils, researchers have also achieved some research results. Currently, some researchers are concentrating on the properties of non-heavy metal contaminated soils [25,26], including shear strength and microstructure. As for the research on soils contaminated with heavy metals, the main focus is on the study and management of the properties of contaminated soils. Li J. M. et al. investigated the effects and mechanisms of zinc and chromium ions on the shear strength and structure of red clay [2,27,28]; Chen et al. [29] revealed the intrinsic link between shear strength and resistivity of copper ion-contaminated red clay through experimental studies; Yang Z. P. et al. [30] investigated the variation law of unconfined compressive strength of lead-contaminated soil after freeze-thawing. Chen et al. [31] (2019) explored the variation law of mineral composition and the microstructure trend in copper-contaminated red clay. Hazreek et al. [32] used the ERI technique to study the variation law of electrical resistance in heavy metal contaminated fields. All of these studies are about the physical and mechanical properties of various metal-contaminated soil types. In the treatment of soils polluted with heavy metals, some researchers have attempted to solidify heavy metal contaminated soils with various curing agents before studying their mechanical properties, mineral composition, and the evolution of the microstructure [4,33–35]. Dong et al. [36] examined the cementation mechanism of cadmium-contaminated clay cement by analyzing the properties such as the pore structure of the cement contaminants. In the study of soils under temperature conditions, Chang et al. [37] investigated the effects of drying temperature, moisture content, and dry density on the strength properties and fracture development of unsaturated red clay soils. In summary, the engineering conditions involved so far in the study of heavy metal contaminated soils primarily consist of dry and wet cycles, freeze–thaw cycles, etc., which all involve the influence of ambient temperature on contaminated soils, but none of them combine a single temperature with heavy metal contaminated soils to study their mechanical properties as well as microstructural changes, and the corresponding mechanisms.

As a significant factor in nature, temperature has a crucial impact on the mechanical properties of red clay. There are few results combining temperature and heavy metal contamination to study red clay, and the available literature is rather limited, so it is vital for us to research the mechanical properties of heavy metal contaminated red clay under temperature conditions.

In this paper, the mechanical property parameters of red clay soils of varying temperature, zinc contamination concentration, and surrounding pressure were tested by temperature-controlled triaxial tests under strictly controlled test temperature conditions, respectively. Based on the analysis of the mechanical parameters of the soils, the evolution of their mechanical properties is summarized, and the evolution mechanism of the Zn-contaminated soils under the temperature conditions is investigated. The research results of this paper improve the information of mechanical strength evolution of red clay soils in Guilin region under seasonal temperature variation. These findings provide reliable basic research for the management of heavy metal contaminated soils in Guilin and the subtropical region of southern China.

2. Materials and Methods

2.1. Test Soil

The test soil sample was taken from a building pit in Lingui District, Guilin City, Guangxi Zhuang Autonomous Region, China. Within five meters of the surface was a Quaternary residual layer (Q3el) that is reddish brown. It contains kaolinite, illite, and montmorillonite as its primary clay minerals. It also contains quartz, acerulite, and a small amount of water. Its basic physical indices were obtained by the basic physical properties test in the room, as shown in Table 1.

Table 1. Basic physical indexes of red clay in Lingui area of Guilin.

Optimum Moisture Content/ω	Maximum Dry Density	Specific Gravity G_S	Compression Modulus /MPa	Liquid Limit (W_L/%)	Plastic Limit (W_P/%)	Plasticity Index I_P
30.00	1.515	2.73	10.15	57.95	31.22	26.73

2.2. Pollution Source of the Test Soil Sample

The contaminant utilized in the test was analytically pure zinc nitrate hexahydrate $Zn(NO_3)_2{\cdot}6H_2O$ produced by Xilong Technology Co., Ltd. (Shantou, China). It is colorless and transparent, crystallized easily, soluble in water, with a melting point of 36 °C, and a density of 2.065. Its technical conditions are shown in Table 2.

Table 2. Technical conditions of zinc nitrate hexahydrate ($Zn(NO_3)_2{\cdot}6H_2O$) (analytical purity) (unit: %).

[$Zn(NO_3)_2{\cdot}6H_2O$] Content	pH Value (50g/L, 25 °C)	Clarity Test	Water Insoluble Matter	Chloride (Cl)	Sulfate (SO_4)	Iron (Fe)	Lead (Pb)	Ammonium Sulfide Does Not Precipitate
≥99.0	≥3.5	qualified	0.005	0.001	0.002	0.0003	0.005	0.10

2.3. Test Preparation

The red clay retrieved from a building pit in Lingui District was air-dried, removed from debris, crushed, and sieved through a 2 mm soil sieve to be utilized as a backup soil for the test. The amount of distilled water required for soil preparation was calculated by the water content of the original red clay, the target value of dry density of the contaminated red clay (1.4 g/cm^3), the target water content (30%), and other parameters; then, the amount of distilled water required for soil preparation was calculated according to the doping of zinc ions of 0%, 0.05%, 0.1%, 0.2%, and 0.5% (the doping of zinc ions is the mass ratio of zinc ions to dry soil, and the doping amount in this test was determined by Zhou Xing et al. in the contaminated red clay). The required $Zn(NO_3)_2{\cdot}6H_2O$ mass was calculated based on the desired Zn ion concentration [38,39] and the actual experimental results. Next, the proper amount of $Zn(NO_3)_2{\cdot}6H_2O$ solution was prepared according to the mass ratio needed for sample preparation (i.e., the mass ratio of $Zn(NO_3)_2{\cdot}6H_2O$ to distilled water), and finally, the $Zn(NO_3)_2$ solution was sprayed onto the red clay soil and stirred well. After mixing the soil sample, the zinc-contaminated wet soil was sealed in a constant temperature and humidity chamber for 24 h to make it evenly moist and fully react between the substances, then an appropriate amount of soil was taken for remeasurement of moisture content. When the desired moisture content of the test soil sample and the actual moisture content error are acceptable, the hydrostatic method is employed to generate an 80 mm × 39.1 mm triaxial specimen. When specimen preparation is completed, the prepared specimen must be promptly inserted into the triaxial specimen saturator. Then, it is placed in a vacuum saturation cylinder for at least 2 h, after which it is soaked in distilled water for 12 h to make the specimen completely saturated.

2.4. Test Method

The temperature-controlled SLB-1 type stress–strain controlled chamber triaxial shear penetration tester produced by Nanjing Soil Instrument Factory Co., Ltd. (Nanjing, China) was utilized for this test, as seen in Figure 1. Triaxial saturated specimens with various Zn ion concentrations were loaded onto the test bench of the triaxial apparatus at temperatures of 8.3 °C, 19.4 °C, 28.5 °C, and 40 °C (the temperature variables were selected from the local winter, year-round, and summer average temperatures in Guilin and a temperature from the local high temperature history records), the surrounding pressure of 100 kPa, 200 kPa, 300 kPa, and 400 kPa, and the shear rate of the solidification undrained shear test was carried out under the condition of 0.05 mm/min, and the test was stopped when the axial strain was 20%.

Figure 1. Temperature-controlled SLB-1 type stress–strain controlled triaxial shear penetration tester.

3. Results and Discussion

3.1. Changes in Stress–Strain Curves of Red Clay with Different Concentrations of Zinc Contamination under Different Temperature Conditions

3.1.1. Effect of Temperature on Stress–Strain Curves of Zinc-Contaminated Red Clay

In order to study the deformation characteristics of soil environment temperature on zinc-contaminated red clay, four different temperature red clay stress–strain curves were selected under the conditions of zinc pollution concentration of 0% and 0.5% and confining pressure of 300 kPa and 400 kPa, as shown in Figure 2.

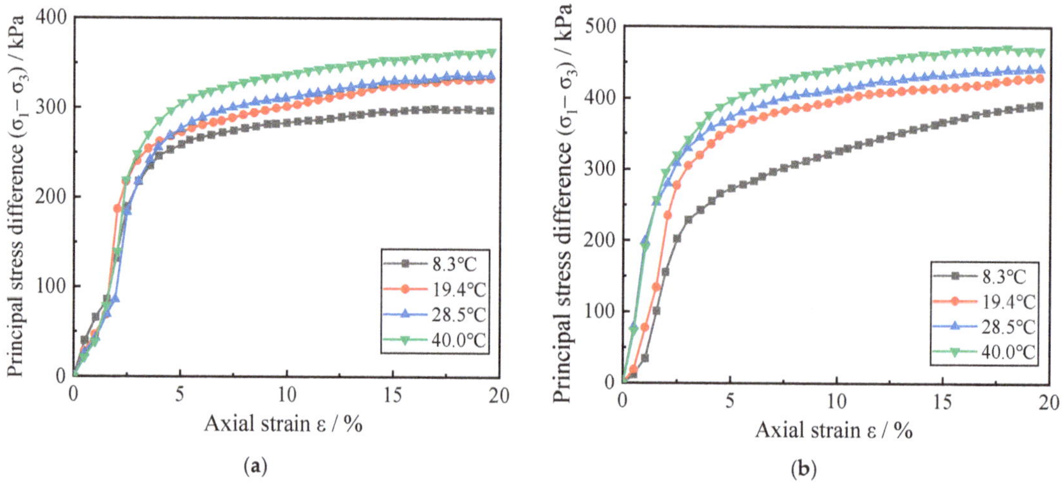

Figure 2. Stress–strain curves of red clay under different temperature conditions. (**a**) The zinc ion concentration is 0% and the confining pressure is 300 kPa; (**b**) The zinc ion concentration is 0.5% and the confining pressure is 400 kPa.

According to the classification of mechanical properties of the soil stress–strain relationship, there are usually three types of strain hardening type, strain softening type, and ideal elastic-plastic type. As can be judged from the change of curves in Figure 2, the stress–strain relationship of red clay at different temperatures is of the strain-hardening type. As axial strain increases, the corresponding principal stress difference demonstrates a linear increase followed by a gradual increase. Initially, the deformation modulus of the red clay increases sharply; then, it decreases rapidly and then slowly with the increase of the axial strain. In Figure 2a, the elastic limits appear at ε = 2.5%, 2%, 2.5%, and 2.5% for the soils at each temperature condition with the increase of ambient temperature; and the elastic limits are 189.7 kPa, 187.1 kPa, 184.1 kPa, and 219.5 kPa, respectively. In Figure 2b, with the increase of ambient temperature, the elastic limits appear at ε = 2%, 2%, 1%, and 1% for the soils at each temperature condition, and the elastic limits are 155.2 kPa, 235.4 kPa, 253.4 kPa, and 257.3 kPa, respectively. This indicates that the ε corresponding to the elastic limit of uncontaminated red clay is basically constant with increasing temperature, while the ε corresponding to the elastic limit of zinc-contaminated red clay is gradually decreasing. Meanwhile, the elastic limit of uncontaminated red clay showed an overall decreasing trend with increasing temperature, while the elastic limit of zinc-contaminated red clay showed an overall increasing trend with increasing temperature. In Figure 2a, it can be seen that when ε = 15%, the stress in the uncontaminated red clay increases from 295.3 kPa to 353.8 kPa as the temperature increases from 8.3 °C to 40 °C, and the soil stresses at each temperature increase by 10.13%, 12.06%, and 19.81%, respectively; while in Figure 2b, the stress of red clay with 0.5% Zn ion concentration increased from 366.8 kPa to 463.7 kPa, and the soil stresses at each temperature increased by 13.25%, 17.91%, and 26.42%, respectively. It indicates that the principal stress difference corresponding to the same axial strain location basically increases with increasing temperature when the zinc ion concentration and the surrounding pressure in the red clay are the same. This suggests that temperature can greatly influence the mechanical strength of Zn-contaminated red clay. The reason is that as the ambient temperature of the soil rises, the red clay soil in the consolidation stage converts more bonded water into free water and discharges it from the soil due to thermal expansion and contraction. The contact between soil particles is closer, which enhances the ability of the soil to resist external shear damage. Meanwhile, the red clay soil in the undrained shear stage will make the movement of residual water more active due to thermal expansion and contraction. The moisture is more easily dispersed uniformly inside the soil body, and bears part of the external damage force through the super-pore water pressure.

3.1.2. Effect of Contamination Concentration on the Stress–Strain Relationship of Red Clay under Temperature Conditions

In order to study the deformation characteristics of red clay under the temperature conditions of pollution concentration, the stress–strain curves of red clay with five kinds of zinc pollution concentration under the ambient temperature of 8.3 °C, 40 °C, and the surrounding pressure are 300 kPa are selected, as shown in Figure 3.

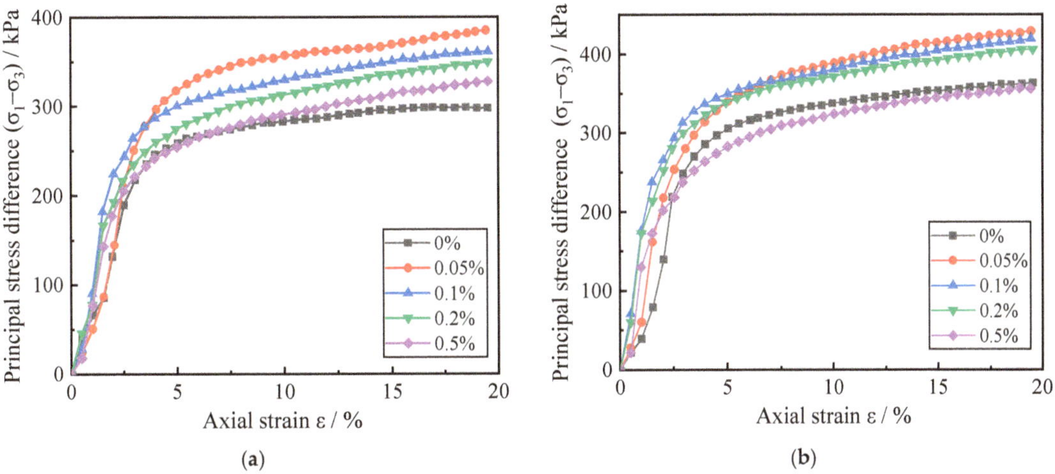

Figure 3. Stress–strain curve of red clay with Zn pollution concentration. (**a**) The temperature is 8.3 °C and the confining pressure is 300 kPa; (**b**) The temperature is 40 °C and the confining pressure is 300 kPa.

Based on the type of soil stress–strain relationship and the change of stress–strain curve in Figure 3, it is possible to conclude that the stress–strain curve of red clay with varying contamination concentrations at the same ambient temperature exhibits strain hardening. During the shearing process, the primary stress difference exhibited a linear increase, followed by a slow increase, as the axial strain increase, while the deformation modulus also showed a trend of rapid increase and then a slow decrease. Comparing the curves in Figure 3, it was discovered that when the ambient temperature and the surrounding pressure of the soil were the same, the principal stress difference corresponding to the same axial strain of the red clay with Zn ion concentration from 0 to 0.05% exhibited an increasing trend, while the principal stress difference corresponding to the same axial strain of the red clay with Zn ion concentration from 0.05% to 0.5% exhibited a decreasing trend, and the degree of decrease was more obvious. In Figure 3a, it can be seen that when ε is 15%, the stress in the red clay at 8.3 °C increased from 295.3 kPa to 369.7 kPa, then decreased to 314.7 kPa as the concentration of zinc ions in the soil increased from 0 to 0.5%, and the stress in the soil at each concentration increased by 25.19%, 18.90%, 13.85%, and 10.66%, respectively. In Figure 3b, with the increase of zinc ion concentration in soil, the stress of red clay at 8.3 °C first increases from 353.8 kPa to 414.6 kPa and then decreases to 345.3 kPa, and the soil stress at each concentration increases by 17.18%, 14.47%, 10.96%, and −2.45% respectively. It indicates that the amount of Zn ions in red clay can influence its macroscopic strength via altering the red clay deformation. The reason is that when small amounts of zinc ions are present in red clay, crystalline water is present in the zinc salt produced by the reaction of zinc ions with soil particles. The crystalline water causes the surrounding soil particles to have a fitting effect, which in turn causes the local soil particles to aggregate. The denseness of the soil increases, making the soil more resistant to external destructive forces. When large amounts of zinc ions are present in the red clay, the zinc ions react with a large number of soil particles to produce more zinc salts. The bound water in zinc salts can attract to more soil particles and form larger agglomerates of soil particles, which destroys the overall structure of the soil body. At the same time, the reaction of zinc ions with soil particles destroys the cementing material between soil particles and also reduces the attraction between soil particles. Therefore, the macroscopic manifestation is that the ability of the soil body to resist external destructive forces is reduced.

3.1.3. Effect of Enclosure Pressure on Stress–Strain Curves of Zn-Contaminated Red Clay at Different Temperature Conditions

In order to study the deformation characteristics of Zn-contaminated red clay under different temperature conditions, the stress–strain curves of red clay with ambient temperatures of 8.3 °C and 40 °C and Zn-contaminated concentrations of 0.05% and 0.5% at different enclosure pressures were selected, as shown in Figure 4.

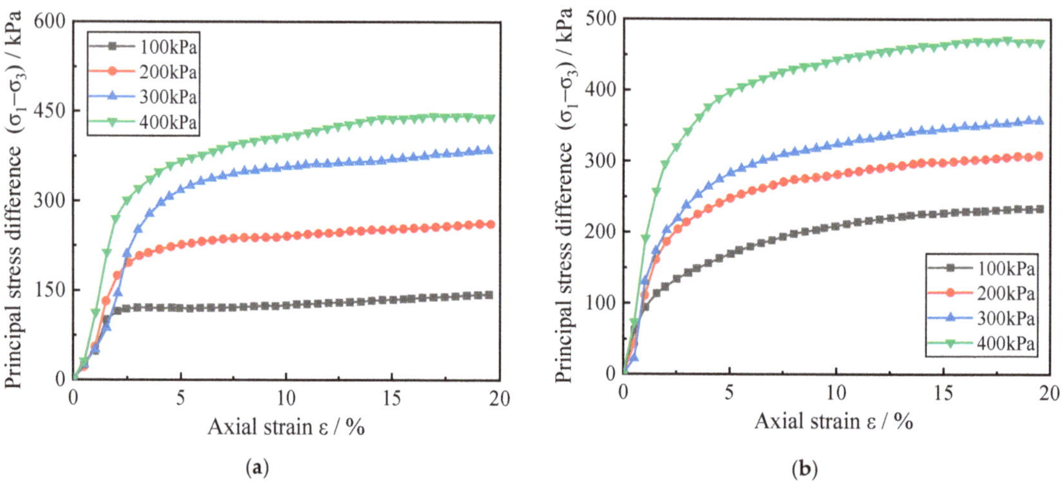

Figure 4. Stress–strain curves of zinc-contaminated red clay under different confining pressures. (**a**) The temperature is 8.3 °C, and the zinc ion concentration is 0.05%; (**b**) The temperature is 40 °C, and the zinc ion concentration is 0.5%.

In addition to temperature and contamination concentration, the circumferential pressure to which the red clay is subjected also has an effect on its stress–strain relationship. From Figure 4, it can be judged that the stress–strain relationship of the Zn-contaminated red clay under different circumferential pressures also belongs to the strain-hardening type when the test temperature and the Zn contamination concentration of the red clay are the same. In Figure 4a, when ε is 15%, the stress of the red clay with 0.05% zinc ion concentration increases from 134.8 kPa to 437.3 kPa as the increase of the enclosing pressure from 100 kPa to 400 kPa. The stress of the soil at each enclosing pressure increases by 86.87%, 174.26%, and 224.41%, respectively. In Figure 4b, the stress of red clay with 0.5% Zn ion concentration increased from 227.5 kPa to 463.7 kPa with the increase of the enclosing pressure. The stress of the soil under each enclosing pressure increased by 13.25%, 17.91%, and 26.42%, respectively. In addition, the deformation modulus of the red clay increased linearly during the initial stage of shear; at a certain critical point, it began to fall rapidly and subsequently decreased gradually. At the same time, the principal stress difference of the red clay grows as the surrounding pressure becomes larger, and the phase of linear increase of the deformation modulus is extended. It indicates that the surrounding pressure can have a significant effect on the mechanical strength of zinc-contaminated red clay. This is due to the fact as the surrounding pressure applied to the soil increases, more water is removed from the red clay specimen during the consolidation stage, and it is known from the principle of effective stress of soil that with the discharge of pore water, the pore pressure decreases and the effective stress borne by the soil skeleton increases. In the undrained shear stage, the water in the soil is not completely discharged. The residual water in the soil can bear part of the shear stress with the increase of the surrounding pressure, so that the soil can resist the external damage ability.

3.2. Changes in Shear Strength of Zn-Contaminated Red Clay under Different Temperature Conditions

3.2.1. Effect of Temperature on the Shear Strength of Zinc-Contaminated Red Clay

In order to better study the effect of temperature on the shear strength of Zn-contaminated red clay, the shear strength variation of red clay with contamination concentration of 0% and 0.5% under different temperature conditions were selected in this section. Figure 5 illustrates the shear strength variation curves of Zn-contaminated red clay at various ambient temperatures.

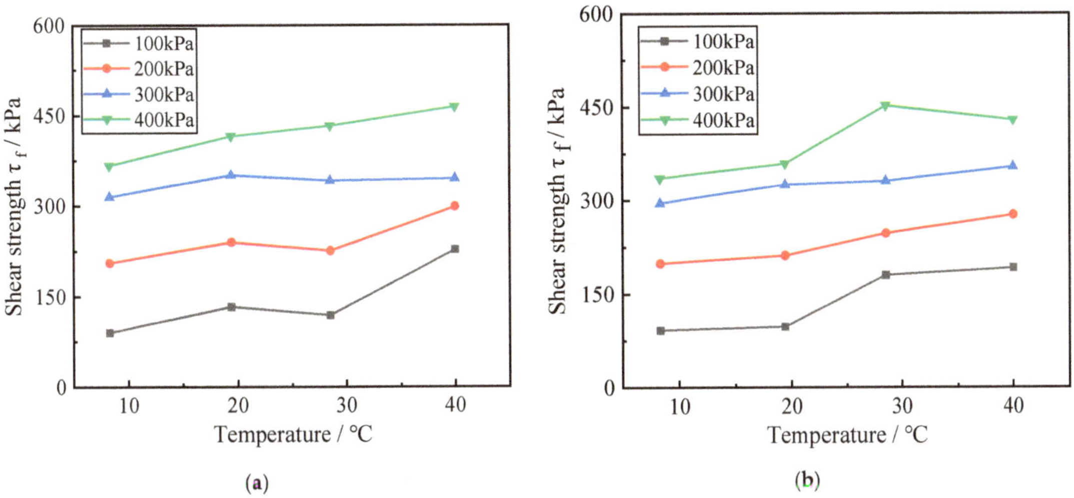

Figure 5. Influence of temperature on the shear strength of red clay. (**a**) The zinc ion concentration is 0%; (**b**) The zinc ion concentration is 0.5%.

It can be seen in Figure 5 that the effect of temperature on the shear strength of red clay is relatively obvious, that with an increase in temperature, the shear strength of red clay under different contamination concentrations exhibits an upward trend, and that the relationship between the two is essentially linear. In addition, when the red clay is subjected to the same zinc contamination concentration is at the same temperature, its shear strength increases as the surrounding pressure rises, demonstrating a clear positive correlation.

This is because, when the ambient temperature of the red clay soil increases, the molecular movement inside the soil becomes more active, and the particles and pore water expand in volume due to the effect of thermal expansion and contraction. In the consolidation stage, due to the effect of the surrounding pressure, the expansion of both causes the pore water within the soil to leave the soil, the volume of the soil decreases, and the distance between the particles shrinks, bringing a specific soil particle into contact with more soil particles and increasing the effective stress of the soil. Therefore, the stress required to produce axial strain in red clay during shear likewise increases as ambient temperature rises, and the macroscopic expression is that the shear strength of zinc-contaminated red clay increases with the increase of ambient temperature.

3.2.2. Effect of Contamination Concentration on the Shear Strength of Zinc-Contaminated Red Clay

In order to better explore the influence of contamination concentration on the shear strength of Zn-contaminated red clay, the variation of the shear strength of Zn-contaminated red clay at temperatures of 8.3 °C and 40 °C, respectively, was selected. Figure 6 illustrates the shear strength variation curves of Zn-contaminated red clay with various contamination concentrations at the same ambient temperature conditions.

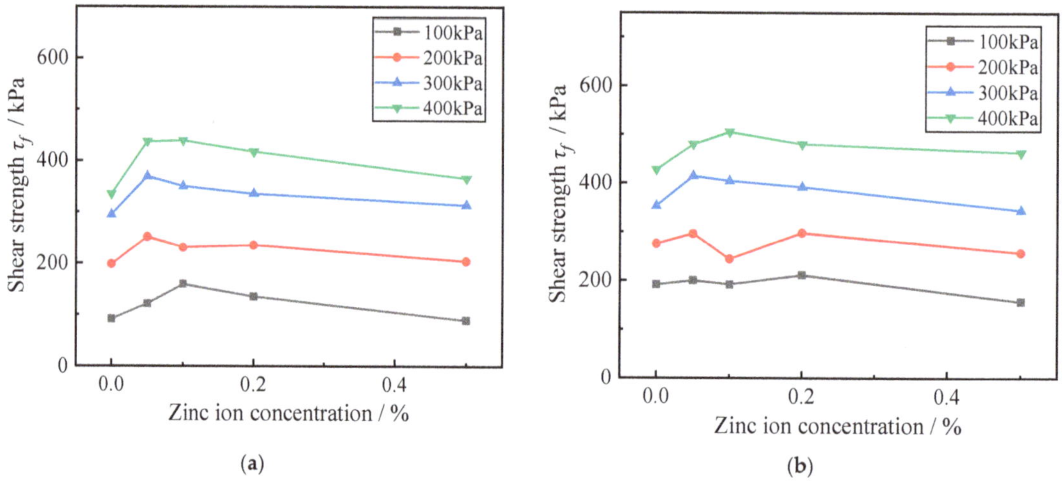

Figure 6. Influence curve of contamination concentration on shear strength of zinc-contaminated red clay. (**a**) The temperature is 8.3 °C; (**b**) The temperature is 40 °C.

From Figure 6a, it can be seen that when the ambient temperature is 8.3 °C, the shear strength of red clay with different zinc contamination concentration changes very obviously; from Figure 6b, it can be seen that when the ambient temperature is 40 °C, the shear strength of red clay with different zinc contamination concentration changes more obviously; however, the changes of zinc ion concentration and soil shear strength in red clay under the two temperature conditions are not significantly different. At the same temperature, the shear strength of the red clay increased with the increase of zinc contamination concentration, then declined gradually. Comparing Figure 6a,b, we discovered that the shear strength of red clay first increased at the zinc ion content in red clay grew from zero to zero; after that, the shear strength of red clay basically showed a significant decreasing trend as the zinc ion content continued to increase. It demonstrates that the Zn ion content in the soil has a significant effect on the strength of the red clay, and a small amount of Zn ion can improve the shear strength of the red clay, whereas a large amount of Zn ion can significantly reduce the shear strength of the red clay, resulting in the deterioration of the engineering properties of the red clay and bringing adverse geological hazards to the construction. In addition, there is a positive association between the shear strength and the surrounding pressure of red clay soils under the same temperature settings and with the same concentration of zinc contamination also shows a positive correlation. That is, the greater the surrounding pressure of the red clay, the greater the shear strength of the red clay itself. This is because in the consolidation stage of the specimen, with the increase of the surrounding pressure, more pore water in the soil is eliminated; according to the principle of effective stress of soil, it is known that with the elimination of pore water in the soil, the pore pressure in the soil decreases, and the contact of soil particles becomes more compact, and the pressure borne by the soil skeleton is enhanced, leading to the enhancement of the effective stress of the soil, which is macroscopically expressed as the increase of the shear strength of the soil.

3.3. Changes of Shear Strength Index of Zinc-Contaminated Red Clay under Different Temperature Conditions

3.3.1. Effect of Temperature on the Shear Strength Index of Zinc-Contaminated Red Clay

According to the data-processing method of the triaxial compression test in the Standard for Geotechnical Test Methods (GB/T 50123-2019), the values of the damage points in the test data are obtained, calculated, and processed in accordance with the relevant formulas, the total stress circle of damage under different circumferential pressure is drawn, and the common tangent is drawn, and finally, the values of cohesion and internal friction angle under each condition are obtained.

Figure 7 illustrates the variation curves of the shear strength index of Zn-contaminated red clay at various temperatures. In Figure 7, it can be seen that when the concentration of zinc contamination in the red clay remains constant, the cohesive force exhibits an overall increasing trend with the increase in temperature, whereas the internal friction angle exhibits a first increasing and then decreasing trend with the increase in temperature. Due to the thermal consolidation, as the ambient temperature rises, the volume of particles and pore water within the soil expands, the bound water between soil particles is converted into free water, the hydration film surrounding soil particles gradually decreases, the diffusion layer becomes thinner, and the hydrogel connection force between soil particles increases, resulting in a strengthening of the cohesion of red clay. Due to the confining effect of the surrounding pressure on the soil body during the heat consolidation stage, more pore water is discharged from the soil body, resulting in a decrease in pore pressure and an increase in pressure to be borne by the soil body, prompting the volume reduction of the soil body, the enhancement of the occlusion and friction between the soil particles, and a rise in the increase of the internal friction angle between the soil particles; after that, as the pore water continues to be discharged, the soil body deforms seriously, leading to the destruction of the occlusion point between the soil particles. Therefore, the internal friction angle of the red clay maintains the trend of first increasing and then decreasing with the increase of temperature.

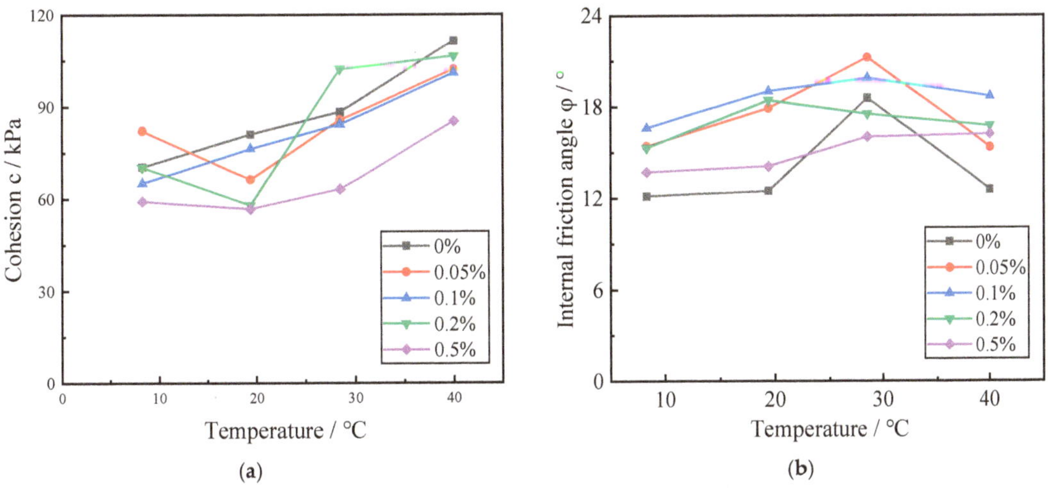

Figure 7. Curve of influence of temperature on the shear strength index of red clay. (**a**) The graph of the change of cohesion force with temperature; (**b**) The graph of internal friction angle with temperature.

3.3.2. Effect of Pollution Concentration on the Shear Strength Index of Red Clay under the Same Temperature Condition

Figure 8 depicts the variation curves of shear strength index of red clay with different Zn contamination concentrations at the same temperature condition. From Figure 8, it can be shown that when the red clay is at the same ambient temperature, its cohesive force basically shows a decreasing trend with the increase of zinc contamination concentration, while its internal friction angle shows a trend of first increasing and then decreasing with the increase of zinc contamination concentration in the soil.

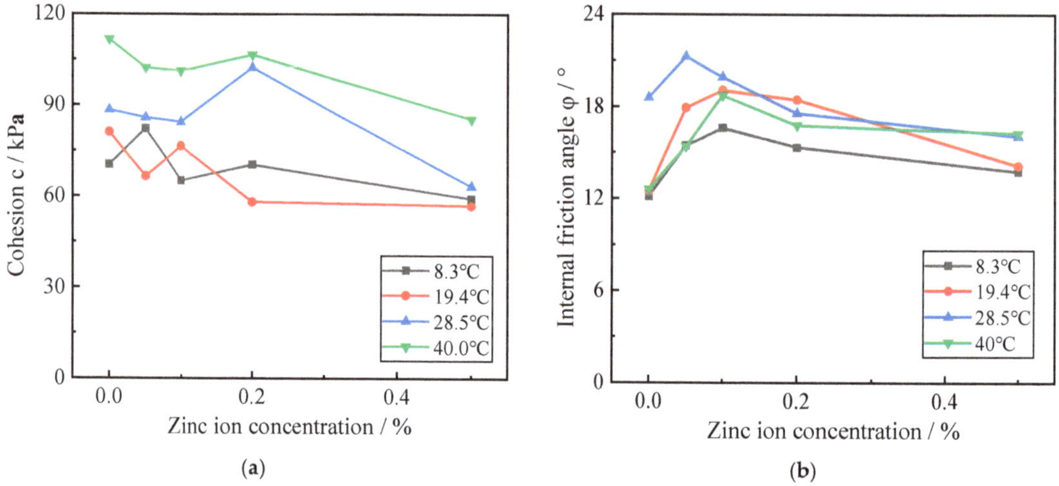

Figure 8. Influence curve of zinc contamination concentration on shear strength index of red clay. (**a**) The variation of cohesion with zinc ion concentration; (**b**) The variation of internal friction angle with zinc ion concentration.

This is because, when the ambient temperature of the red clay remains constant, with the increase of Zn contamination concentration in the soil, more Zn ions react with the cemented material within the soil and destroy its structure, reducing the attraction between soil particles, and the macroscopic expression of the cohesion of the red clay gradually decreases with the increase in Zn contamination concentration. When zinc ions appear in the red clay, zinc ions react with the surrounding soil particles and produce zinc salts; after the crystalline water in zinc salts enters the pores of soil particles, it causes a kind of fitting effect on the surrounding soil particles, which then causes the present soil particles in the local area within the soil to gather, and the occlusal force and slip friction between them to increase, which is expressed as an increase in the internal friction angle of the soil body; when the content of zinc ions in the red clay continues to increase when the content of zinc ions in the red clay continues to increase, the number of zinc salts generated within the soil increases significantly, the number of agglomerates formed by the agglomerates of soil particles increases, the larger the volume of the agglomerates, the more the structure of the contaminated red clay is destroyed, followed by the formation of small particles of soil, an increase in the number of pores around them increases, and the friction between the soil particles around the pores decreases, and the macroscopic expression is that as the concentration of zinc contamination in the red clay increases, the angle of internal friction first increases and then decreases.

4. Conclusions

In this paper, red clay soil from a foundation pit in Guilin City, Guangxi Zhuang Autonomous Region was used as the test soil, and the triaxial shear test of the red clay soil was conducted with temperature, zinc ion concentration, and surrounding pressure as variables to analyze the effects of these three variables on the deformation and strength of the local zinc-contaminated red clay soil in Guilin City. The following conclusions were obtained.

(1) The stress–strain relationship of red clay is strain-hardened at varying temperatures, different Zn ion concentrations, and different surrounding pressures, depending on which variable is altered. In the stress–strain relationship, the deformation modulus of the red clay tends to grow rapidly with an increase of axial strain before decreasing slowly.

(2) Temperature can change the strength of red clay. Under the same zinc contamination concentration and circumferential pressure, the red clay increases with the temperature, which boosts the thermal consolidation and changes the cohesion and internal friction angle of the red clay, resulting in a continuous increase in shear strength.

(3) The concentration of zinc contamination can also affect the shear strength of red clay. Under the same temperature and surrounding pressure, the shear strength of red clay tends to increase and subsequently drop as the concentration of zinc contamination rises.

(4) The surrounding pressure is also related to its shear strength. As the volume of red clay decreases in response to an increase of the surrounding pressure, the effective stress of the soil increases, hence increasing its shear strength.

Author Contributions: Conceptualization, J.W. and Y.S.; methodology, J.W.; validation, Y.G., S.D. (Shuaishuai Dong) and S.D. (Song Ding); investigation, X.G.; writing—original draft preparation, J.W.; writing—review and editing, Y.S. and J.W.; supervision, X.G. and S.D. (Shuaishuai Dong); project administration, Y.S. All authors have read and agreed to the published version of the manuscript.

Funding: (1) Guangxi Innovation-Driven Development Special Project "Research and Demonstration of Key Technologies for Water Resource Utilization and Synergistic Development of Water Ecological Industry in Typical Karst Wetlands in the Lijiang River Basin" (Guike AA20161004-1); (2) National key research and development program subject "R&D and experimental demonstration of key technologies for water resource regulation in Karst wetlands in the Lijiang River Basin" (2019YFC0507502); (3) The National Natural Science Foundation of China Project "Research on the Collapse Mechanism of Karst Water-soil Coupling in Guilin under Extreme Climate Conditions" (41967037).

Institutional Review Board Statement: Not applicable.

Informed Consent Statement: Not applicable.

Data Availability Statement: Not applicable.

Conflicts of Interest: The authors declare no conflict of interest.

References

1. Chen, L.J.; Chen, X.J.; Wang, H.; Huang, X.; Song, Y. Mechanical properties and microstructure of lime-treated red clay. *KSCE J. Civ. Eng.* **2021**, *25*, 70–77. [CrossRef]
2. Song, Y.; Wang, J.Q.; Chen, X.J.; Yu, S.Z.; Ban, R.L.; Yang, X.; Zhang, X.C.; Han, Y. Study the effects of dry-wet cycles and cadmium pollution on the mechanical properties and microstructure of red clay. *Environ. Pollut.* **2022**, *302*, 119037. [CrossRef] [PubMed]
3. Chen, L.J.; Chen, X.J.; Yang, X.; Bi, P.Y.; Ding, X.; Huang, X.; Wang, H. Effect of calcium carbonate on the mechanical properties and microstructure of red clay. *Adv. Mate. Sci. Eng.* **2020**, *2020*, 5298186. [CrossRef]
4. Song, Y.; Li, H.; Gan, X.H.; Zhang, M.Z.; Xue, M.M.; Li, J. Comparative experimental study on calcium oxide, calcium hydroxide, and calcium carbonate solidified zinc-contaminated red clay. *Geofluids* **2022**, *2022*, 8428982. [CrossRef]
5. James, K.M.; Kenichi, S. *Fundamentals of Soil Behavior*; John Wiley & Sons Inc.: Hoboken, NJ, USA, 2005.
6. Chen, Y.F.; Liu, Z.K. Influence of different amount of red mud on shear strength of red clay. *Yangtz. Riv.* **2017**, *48* (Suppl. S1), 264–267.
7. Lu, H.B.; Ji, C.S.; Xie, C.; Zeng, Z.T.; Wang, Y.; Xu, Y.S. Effect of acids on physical and mechanical properties of Liuzhou lateritic soil. *J. Guangxi Univ. Nat. Sci. Ed.* **2017**, *42*, 1465–1471.
8. Zhang, Y.Z.; Zuo, S.Y.; Ji, Y.X. Experimental Study on undisturbed and reconstituted mechanical properties of red clay. *J. Undergr. Space Eng.* **2017**, *13*, 1477–1482.

9. Chen, J.Y.; Liu, Z.K. Strength change mechanism of undisturbed red clay and remolded red clay. *J. Henan Univ. Sci. Technol. Nat. Sci. Ed.* **2019**, *27*, 53–59.
10. Li, J.M.; Chen, X.J.; Yang, Y.; Huang, X.; Song, Y. Experiment of modified red clay by graphite nanoparticles. *Henan Univ. Sci. Technol. Nat. Sci. Ed.* **2019**, *40*, 69–73+80.
11. Mukiza, E.; Zhang, L.L.; Liu, X.M.; Zhang, N. Utilization of red mud in road base and subgrade materials: A review. *Resour. Conserv. Recycl.* **2019**, *141*, 187–199. [CrossRef]
12. Zou, W.S.; Feng, X.Z.; Wang, R.H.; Wei, W.F.; Luo, S.Y.; Zheng, R.J.; Yang, D.Z.; Mi, H.W.; Chen, H. High-efficiency core-shell magnetic heavy-metal absorbents derived from spent $LiFePO_4$ battery. *J. Hazard. Mater.* **2021**, *402*, 123583. [CrossRef] [PubMed]
13. Chen, S.B.; Wang, M.; Li, S.S.; Zhao, Z.Q.; E, W.D. Overview on current criteria for heavy metals and its hint for the revision of soil environmental quality standards in China. *J. Integr. Agr.* **2018**, *17*, 765–774. [CrossRef]
14. Mishra, S.; Bharagava, R.N.; More, N.; Yadav, A.; Zainith, S.; Mani, S.; Chowdhary, P. Heavy metal contamination: An alarming threat to environment and human health. In *Environmental Biotechnology: For Sustainable Future*; Springer: Singapore, 2019; pp. 103–125.
15. Shi, T.R.; Ma, J.; Wu, X.; Ju, T.N.; Lin, X.L.; Zhang, Y.Y.; Li, X.H.; Gong, Y.W.; Hou, H.; Zhao, L.; et al. Inventories of heavy metal inputs and outputs to and from agricultural soils: A review. *Ecotox. Environ. Safe.* **2018**, *164*, 118–124. [CrossRef] [PubMed]
16. Li, J.C.; Yin, R.Z.; Luo, Y.P.; Lu, Y.L.; Zhang, L.D. Assessment of heavy metal contamination of soils in Daxin manganese mine Guangxi. *Environ. Sci. Technol.* **2010**, *33*, 183–190.
17. Wang, Y.H.; Chen, X.J.; Zhao, Y.L.; Qi, S.H. Heavy metal pollution in soils and plant accumulation in a restored lead zinc mineland in Guangxi, South China. *J. China Univ. Min. Technol.* **2007**, *36*, 487–493.
18. Chen, H.B.; Chen, X.J.; Qi, Y.L.; Huang, X.; Yu, S.Z.; Xiao, G.Y. Effects of dry density and moisture content on shear strength parameters of remolded red clay soil. *J. Eng. Geol.* **2019**, *27*, 1035–1040.
19. Gong, X.P.; Tang, C.S.; Shi, B.; Wang, H.S.; Leng, T.; Tan, Y.Z.; Deng, Y.F. Evolution of soil microstructure during drying and wetting. *J. Eng. Geol.* **2019**, *27*, 775–793.
20. Xie, Y.H.; Zhang, B.H.; Liu, B.C.; Zeng, Z.T.; Zhang, Y.; Zheng, Y. Shrinkage cracking and strength deterioration of red clay under cyclic drying and wetting. *Alex. Eng. J.* **2022**, *61*, 2574–2588. [CrossRef]
21. Zeng, L.; Yu, H.C.; Gao, Q.F.; Liu, J.; Liu, Z.H. Evolution of Tensile Properties of Compacted Red Clay under Wet and Dry Cycles. *KSCE J. Civ. Eng.* **2022**, *26*, 606–618. [CrossRef]
22. Chen, X.J.; Chen, L.J.; Song, Y.; Wang, H.; Zhou, X.; Li, J.M. Experimental study on improvement of red clay with lime. *J. Eng. Geol.* **2017**, *25*, 230–234.
23. Chen, X.J.; Chen, L.J.; Song, Y.; Chen, H.B.; Wang, H. Experimental study on strength characteristics of red clay under different particle size of calcium carbonate. In *International Symposium on Energy Geotechnics*; Springer: Cham, Switzerland, 2018; pp. 236–242.
24. Qi, Y.L.; Bai, M.Z.; Zhou, H.; Shi, H.; Li, P.X.; He, B.H. Study on the mechanical properties of red clay under drying-wetting cycles. *Adv. Mater. Sci. Eng.* **2021**, *2021*, 8665167. [CrossRef]
25. Chen, Y.; Wang, Q.; Yu, M.Y.; Wu, Z.H.; Gao, B.; Dong, J. Study on shear strength and microstructure characteristics of alkali contaminated red clay. *J. Eng. Geol.* **2018**, *26*, 1300–1310.
26. Liu, K.; Ye, W.J.; Gao, H.J.; Dong, Q. Evolutionary law and microscopic mechanism of shear strength of acid-alkaline contaminated loess. *Geotechnics* **2022**, *43* (Suppl. S1), 1–12.
27. Li, J.M.; Tang, S.B.; Chen, X.J. Analysis of the Mechanical properties and mechanism of zinc ion-contaminated red clay. *Adv. Mater. Sci. Eng.* **2021**, *2021*, 6649691. [CrossRef]
28. Song, Y.; Zhang, M.Z.; Li, H.; Yang, X.; Chen, X.J.; Li, J. Effect of chromium ion on the strength characteristics and damage law of red clay. *Geofluids* **2022**, *2022*, 8451476. [CrossRef]
29. Chen, Y.C.; Huang, X.; Chen, X.J.; Zhang, X.C.; Ban, R.L. Relationship between shear strength and electrical resistivity of cu^{2+} contaminated red clay. *J. Yangtze Riv. Sci. Res. Inst.* **2021**, *38*, 121–127+136.
30. Yang, Z.P.; Li, D.H.; Deng, R.F.; Tian, X.; Ren, S.P. An experimental study for influence of freeze thaw cycles on strength and pore characteristics of solidified lead contaminated soil. *J. Eng. Geol.* **2019**, *27*, 539–549.
31. Chen, X.J.; Chen, Y.C.; Song, Y.; Li, J.M.; Yu, S.Z.; Chen, L.J. Analysis on variation of physical-mechanical properties of red clay contaminated by Cu^{2+}. *J. Eng. Geol.* **2019**, *27*, 1010–1018.
32. Hazreek, Z.A.M.; Azhar, A.T.S.; Rosli, S. Field miniature study on heavy metal detection using electrical resistivity imaging (ERI). *Internat. J. Civ. Eng. Technol.* **2018**, *5*, 284–292.
33. Fan, H.L.; Shen, X.D.; Zhou, H.L.; Liu, X.H. Mechanical property and strength prediction of cement-solidified/stabilized zinc-contaminated red clay. *Bullet. Chin. Ceram. Soci.* **2016**, *12*, 3964–3971.
34. Zhang, C.H.; Sheng, X.D.; Fan, H.L.; Yuan, Q. Solidification rate and micro pore characteristics of cement stabilized zinc contaminated red clay. *Environ. Eng.* **2019**, *37*, 172–176.
35. Song, Y.; Gan, X.H.; Li, H.; Zhang, M.Z.; Xue, M.M.; Gao, X.T.; Li, J. Curing effect and resistivity evolution of zinc-contaminated red clay cured by phosphate-based binder. *Geofluids* **2022**, *2022*, 8404620. [CrossRef]
36. Dong, Y.Q.; Lu, H.J.; Li, J.X. Ion release and micro structure of cd-polluted clay solidified by cement. *Chin. J. Environ. Eng.* **2015**, *9*, 4578–4584.

37. Chang, L.C.; Wang, H.Y.; Xue, K.X. Study on thechanges of cracks and shear strength in unsaturated red clay under different drying temperature. *J. Eng. Geol.* **2021**, 1–14.
38. Li, J.M.; Chen, X.J.; Huang, X.; Cheng, Y.; Yu, S. Test on mechanical properties and microstructure of zinc-contaminated red clay. *J. Henan Univ. Sci. Technol. Nat. Sci. Ed.* **2020**, *41*, 99–104.
39. Chen, X.J.; Zhou, X.; Song, Y.; Huang, Y.Y. Experimental study on the characteristics of zn^{2+} pollution on the strength damage of red clay under leaching conditions. *J. Eng. Geol.* **2017**, *25*, 1264–1269.

applied sciences

MDPI

Article

Study on the Mechanical Strength of Calcium Oxide-Calcium Phosphate Cured Heavy Metal Zinc Contaminated Red Clay Soil

Yu Song [1,2], Song Ding [1,2], Yukun Geng [1,2], Shuaishuai Dong [1,2] and Hongbin Chen [1,*]

[1] College of Civil and Architecture Engineering, Guilin University of Technology, Guilin 541004, China
[2] Guangxi Key Laboratory of Geomechanics and Geotechnical Engineering, Guilin 541004, China
* Correspondence: tcchb0325@163.com

Abstract: With the continuous improvement of the construction of the ecological economic system in the new era, the problem of heavy metal pollution has become an important issue in urban construction. In this paper, Zn^{2+} contaminated red clay is used as the research object, and calcium superphosphate and calcium oxide are used as curing agents to conduct the simultaneous test of unconfined compressive strength-resistivity. The mechanical properties and resistivity of the contaminated soil under different test conditions were analyzed to investigate their effects on the cured red clay. The results showed that different contamination concentrations showed different weakening effects on the unconfined compressive strength of red clay soils, and the unconfined compressive strength of cured soils increased significantly. The age of maintenance affects the unconfined compressive strength of cured soil, and the growth of unconfined compressive strength is most obvious in the age of 0–7 day. After that, it tends to be stable with the growth of age. The deformation modulus of contaminated soil before and after curing was reduced to different degrees. Before and after curing, the resistivity of contaminated soil decreased with the increase of ion concentration, and the resistivity of cured soil increased with the increase of curing agent incorporation rate under the same contamination concentration. The research results can enrich the soil treatment problem of heavy metal contaminated sites and provide theoretical support for the application of this type of curing agent in the field engineering of Zn^{2+} contaminated red clay soil.

Keywords: heavy metal pollution; resistivity; compressive strength; age of maintenance; deformation modulus

1. Introduction

Red clay, as a class of special clay, is mainly spread in the two provinces, Yunnan and Guizhou, with characteristics such as high porosity ratio, high water content, high liquid limit, high strength and low compressibility, etc. [1,2]. As a large resource province, Guangxi is rich in mineral resources with sufficient reserves, which improves the economic benefits for the society and at the same time raises the problem of land pollution. With the rapid development of the mining industry in the last decade or so, tailings disposal and mineral processing have caused heavy metal pollution in the soil of mining areas; a large amount of industrial wastewater discharge has caused the waters and soils around mining areas to be polluted. Since heavy metals have the characteristics of long-term retention and poor degradability in the soil, heavy metal substances in the soil have a lasting impact on the surrounding environment, often triggering geotechnical construction disasters such as roadbed destabilization and uneven foundation settlement [3]. In this paper, we explore the effect of zinc ion pollution on the mechanical properties of Guilin red clay soil from the aspect of soil mechanical properties.

The research on heavy metal contaminated soils is nowadays receiving more and more attention from scholars at home and abroad. Voglar G E et al. studied in depth the changes of compressive properties of soils before and after acid and alkali contamination with various chemical concentrations using artificially prepared acid and alkali contaminated soils, and the compressibility coefficient of acid and alkali contaminated soils increased and the rebound index became larger due to the increase of contaminant content [4]. Li Jiaming et al. found that the greater the doping of zinc ions, the more the pores of red clay gradually increased, the fractal dimension became larger, and the porosity also became larger, while the shear strength decreased, and the cohesion and internal friction angle also became smaller, leading to the instability of the structure [5]. Li Chi et al. applied MICP technology to remediate Pb-Zn contaminated soils and revealed the remediation mechanism of MICP technology and porous adsorbent materials for curing contaminated soils [6]. Dai Chengyuan et al. discovered the compressive properties of cured heavy metal Zn-ion contaminated red clay under different conditions of temperature, Zn ion concentration, and cement admixture, and thus derived an empirical equation reflecting the variation of the compression coefficient on the soil under three different conditions [7]. Guo Mingshuai et al. provided a theoretical basis for the proportioning of new curing agents by studying the comparison between new curing agents and cement curing agents [8]. Saeek et al. considered the type and concentration of heavy metals and the type and concentration of curing agents as important factors affecting the mechanical strength of contaminated soils [9]. Li Jiaming et al. studied the effects of different concentrations of Zn ions on the mechanical properties, material composition and microstructure of red clay, and the results showed that the shear strength of the soil decreases continuously with the increase of Zn ion concentration in red clay, and the content of iron oxide and quartz is decreasing [10]. Song Yu et al. investigated the effects of dry and wet cycling and cadmium ions on the microstructure of Guilin red clay, and the results showed that although both of them destroyed the structure of red clay, they had opposite effects, with dry and wet cycling making the structure of red clay more compact, while cadmium ions destroyed the cementing material of red clay and made the soil structure looser [11]. Ioannou Z et al. studied the adsorption of Zn and desorption of Zn, Mg and Fe in different systems of Clinoptilolite-$Fe(NO_3)_3$ under different conditions. It was shown that the adsorption of Zn started to increase with increasing concentration and pH; with decreasing KCI, the adsorption of Zn increased. The percentage of Zn adsorption decreases with increasing initial concentration, and Zn, Mg and Fe desorption increase with decreasing pH and electrolytes [12]. Dimirkou A et al. mainly studied the adsorption and desorption of cadmium and arsenic by Gobi stone and limonite, and the results showed that the adsorption and desorption of cadmium by Gobi stone and limonite increased with the increase of pH value and the decrease of electrolyte concentration [13].Therefore, it is of great significance to promote the development of urban construction in China by studying new environmentally friendly curing agents and further developing the curing and stabilization effect, mechanism, and process application related to the disposal of high content-heavy metal compounds contaminated soil by using them.

Based on the above, this paper investigates the effectiveness of the composite curing agent on Zn in contaminated soil by using the composite curing agent calcium superphosphate-calcium oxide as the treatment raw material. The effectiveness of this curing agent on Zn^{2+} curing is further investigated based on the effect and evolution pattern of the variation of the curing agent's admixture, unconfined compressive strength and resistivity on the mechanical strength of Guilin red clay soil. To study the use of this type of curing agent in the construction of sites in red clay-covered areas and the possibility of assessing the coagulation effect by resistivity method, to provide technical support for the remediation of heavy metal contaminated soils.

2. Experimental Materials and Methods

The red clay used in this test came from a construction site in Yanshan District, Guilin City, Guangxi Zhuang Autonomous Region, about 3–5 m below ground and was brownish-red in color. The red clay was dried outdoors in a sunny place, then crushed with a wooden mallet, and finally passed through a 2 mm standard sieve for geotechnical tests and stored under sealed conditions. The compaction curve of the soil sample was measured according to the indoor compaction test method (ASTM D698-07) as shown in Figure 1. The optimum moisture content of the soil sample was 30.78% and the maximum dry density was 1.54 g/cm^{-3}. The basic physical properties of the red clay for the test were obtained through basic geotechnical tests as shown in Table 1.

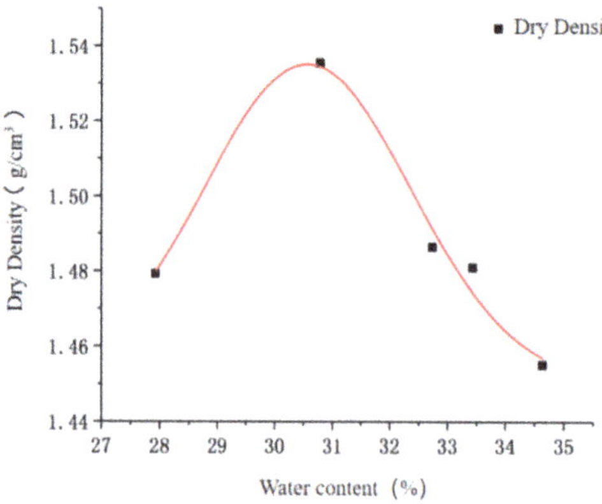

Figure 1. Combat curve.

Table 1. Basic physical properties of red clay.

Liquid Limit (%)	Plastic Limit (%)	Plasticity Index	Optimum Moisture Content (%)	Maximum Dry Density (g/cm^{-3})	Specific Gravity
55.8	31.9	23.9	30.78	1.536	2.74

The heavy metal ion studied in this paper is a zinc ion, and the zinc nitrate solution prepared by deionized water and zinc nitrate hexahydrate ($Zn(NO_3)_2{\cdot}6H_2O$) crystals is selected. Due to its fast dissolution rate in water and the small influence of nitrate ions on the test process, zinc ions are better retained in red clay. The soil nature and heavy metal content in the Guangxi region are different, and the Zn^{2+} contamination concentration can be set as 0%, 0.05%, 0.1%, 0.2% and 0.3% respectively through the report of land quality geochemical evaluation in Guilin. The test used calcium superphosphate ($Ca(H_2PO_4)_2$-H_2O) + quicklime (CaO) as the curing agent, selected the mass ratio of the two as 3:1, and also set the curing agent incorporation rate in the test as 0%, 3%, 6%, 12%, i.e., the curing agent to the dry soil mass. In this paper, calcium superphosphate produced by Shanghai Chemical Reagent Co., Ltd., Shanghai, China is used, which is characterized by colorless and transparent, easily soluble in water, and has oxidizing and certain corrosiveness; quicklime from Xilong Science Co., Ltd., Guangdong, China.

Zn^{2+} contaminated soil is prepared according to the target moisture content and the determined pollution concentration. Three sets of parallel samples were taken from the prepared contaminated soil for water content testing, and the average of the water content of the three sets of parallel samples was taken. Then the remaining contaminated soil samples were put into the standard sample maker with a diameter of 39.1 mm and a height of 80 mm four times by static pressure method, and the soil samples were compacted by oil pressure jack, and the surface layer of the soil samples was thrown with a scraper after each compaction to make the upper and lower soil layers form a better whole. The soil column was then demolded with a hydraulic jack and placed in a tray after demolding with cling film.

The unrestricted compressive strength is one of the indicators of the mechanical properties of polluted soil. The sideless limit test adopts the UTM5305 electronic universal testing machine, and sets the axial strain rate of the universal machine to 1 mm/min. The resistivity meter adopts a VC4091C type LCR precision digital bridge with an effective range of 0.0001 Ω to 99.999 MΩ. The test uses low frequency, AC, for convenience, its voltage frequency refers to the 50 Hz used by most scholars in China. A pressure of 2–5 kPa should be applied to both ends of the electrode during the test to ensure good contact between the specimen and the copper sheet. The two are now combined for an unlimited-side compressive strength-resistivity synchronization test. Keep the room temperature at 25 °C during the test. A layer of conductive paste is applied to both ends of the specimen to increase the contact surface between the soil electrode plates and reduce the contact resistance. The back of the copper sheet is pasted with insulating glue to avoid conductivity with the compression surface of the electronic universal testing machine. Place the primary copper sheet on the compression platform surface and place the specimen on the copper sheet. The other pole copper sheet is then placed at the upper end of the specimen, and then the pressure plate is lowered in close contact with the upper copper sheet and a force of 2–5 kPa is applied. Finally, the resistivity meter and the electronic universal testing machine are turned on synchronously for testing.

3. Mechanical Properties Analysis of Zinc Ion Contaminated Red Clay

3.1. Lateral Limitless Compressive Strength of Contaminated Soil

The unconfined compressive strength is an important evaluation index of the treatment effect of curing/stabilization technology. To investigate the relationship between the contamination concentration of soil samples and their unconfined strength, I carried out a unilateral limit compressive test of polluted soil at different concentrations (0%, 0.05%, 0.1%, 0.2% and 0.3%), and the data can be derived from the test as shown in Table 2.

Table 2. Lateral limitless compressive strength of contaminated soil (kPa).

Pollution Concentration	0%	0.05%	0.1%	0.2%	0.3%
Compressive strength without lateral limit	257.63	215.06	121.92	126.74	145.77

As shown in Table 2, Zn ion contamination affects the strength of the red clay, resulting in a decrease in the lateral limit compressive strength of the red clay, which decreases to 16.5%, 52.6%, 50.5%, and 43.4% with increasing contamination concentrations, respectively. Different weakening effects occur at each concentration, and at increasing concentrations of heavy metals, the strength of contaminated soils decreases continuously compared to plain soils.

The relationship between the contamination concentration and the lateral-free strength of the red clay can be obtained by analyzing the experimental data as shown in Figure 2.

From Figure 2, it can be seen that the overall trend of lateral limit compressive strength of contaminated soil decreases and then increases at different concentrations. It can be seen that when the contamination concentration is lower than 0.1%, the lateral limit compressive strength of the soil gradually decreases while the contamination concentration increases; when it exceeds 0.1%, the strength slowly increases with the increase of the contamination concentration, but both are lower than the strength of the plain red clay. This is because after the soil is contaminated with heavy metal Zn^{2+}, various substances inside the soil will react with Zn^{2+}, which will reduce the cementation inside the soil and make the cohesion inside the soil decrease; thus changing the soil structure and leading to the decrease of soil stability and unconfined compressive strength.

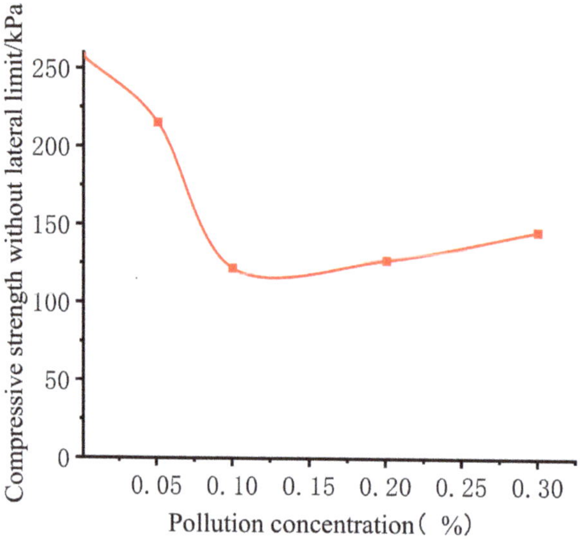

Figure 2. Relationship between different pollution concentrations and unconfined compressive strength.

3.2. *Effect of Contamination Concentration on Unconfined Compressive Strength*

To study the effect of different concentrations on the cured soil, it is necessary to control the amount of different curing agents incorporated according to the controlled variable method. The changes in contamination concentration and unconfined compressive strength of the contaminated soil before and after curing were obtained by analyzing the experimental data as shown in Figure 3.

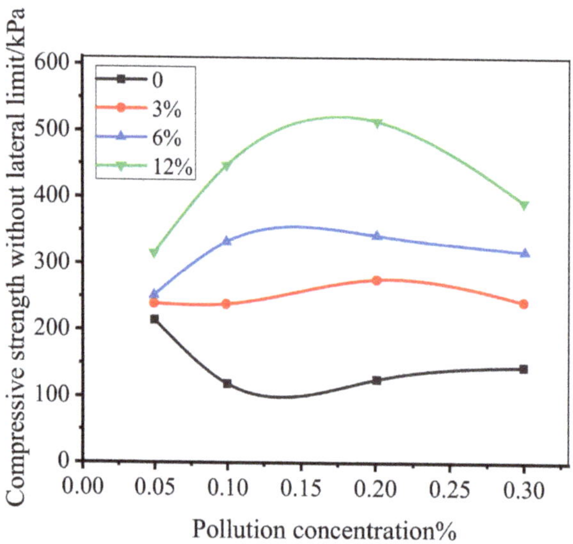

Figure 3. Changes in contamination concentration and unconfined compressive strength of contaminated soil before and after curing.

From Figure 3, it can be seen that the change law of unconfined compressive strength of soil under different contamination concentrations is related to the amount of curing agent.

When no curing agent is added, the overall trend of the lateral limit compressive strength of contaminated soil decreases first and then increases, and the overall decreases, all smaller than the compressive strength of plain soil; when the curing agent is added, the overall trend of the lateral limit compressive strength of contaminated soil increases first and then decreases, and the greater the amount of curing agent, the more obvious the change. At 0.05% contamination concentration, the compressive strength of the three doses of curing soil is 1.12, 1.17, 1.47 times the compressive strength of contaminated soil, while the high concentration of 0.3% contamination, the compressive strength of the three doses of curing soil is 1.66, 2.18, 2.69 times of the compressive strength of contaminated soil, when the amount of curing agent is different, with the change in the concentration of contamination the higher the 201 dose, the higher the lateral compressive strength. The higher the concentration of 202 contamination, the greater the unconfined compressive strength.

3.3. Effect of Curing Agent Incorporation Rate on Unconfined Compressive Strength

From the preliminary tests, it can be seen that the amount of curing agent admixture affects the compressive strength of the contaminated cursed soil. From the data obtained from the tests, it was found that the unconfined compressive strength of the cured soil further increased due to the increase in the rate of curing agent incorporation at the same contamination concentration, and the corresponding difference in the curing effect was more obvious due to the difference in the size of the contaminant concentration.

The data of unconfined compressive strength and curing agent incorporation rate obtained from the test are discussed in terms of low contamination concentration of 0.05% and high contamination concentration of 0.3%. As shown in Figure 4.

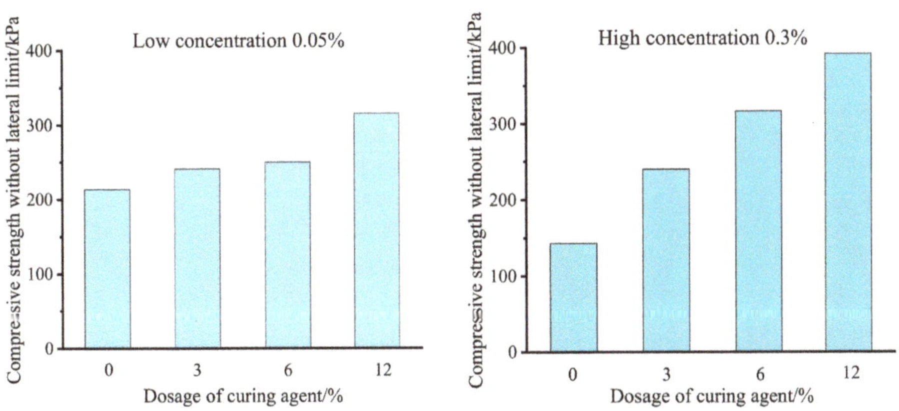

Figure 4. Compressive strength of cured soils with high and low contamination concentrations versus the amount of curing agent.

From Figure 4, it can be seen that the compressive strength of the cured soil increases at different pollution concentrations when the amount of curing agent is increased. Comparing the compressive strength of cured soil at high and low pollution concentrations, it can be obtained that the increase is particularly prominent at high pollution concentrations. The main factor is that after the addition of the curing agent to the soil, some of the quicklime and calcium superphosphate in the curing agent can generate hydroxyapatite in the soil pore water by acid-base reaction, which can effectively cure the Zn^{2+} contaminated soil [14]. The active oxides in the surface layer of clay particles undergo volcanic ash reaction with $Ca(OH)_2$ and produce a large amount of gelling material to cement the soil particles to form a whole. At the same time, $Ca(OH)_2$ and $CaCO_3$ produced during the hardening and carbonization of lime can fill the pores of soil particles and make the internal structure of cured soil more compact, which in turn improves the stability of cured soil and increases the compressive strength of cured soil, and when the pore water of soil meets lime reaction to form a strongly alkaline environment, it speeds up the curing reaction process.

3.4. Effect of Maintenance Age on Unconfined Compressive Strength

According to the initial test data, the unconfined compressive strength test of soil samples under different conditions (curing agent dosing, maintenance age) was carried out, and the sample with 0.3% contamination concentration was selected as an example, and the results showed that the unconfined compressive strength of cured soil was increasing under different maintenance age conditions, among which, the unconfined compressive strength of soil samples in the age of 0–7 day increased particularly, and in the same age The effect of the amount of curing agent on the compressive strength of the soil samples was particularly prominent when compared with the soil samples of the same age. The compressive strength of cured soil with high admixture is significantly higher than the compressive strength of cured soil with low admixture.

The variation pattern of the unconfined compressive strength of the cured soil with the age of curing is shown in Figure 5.

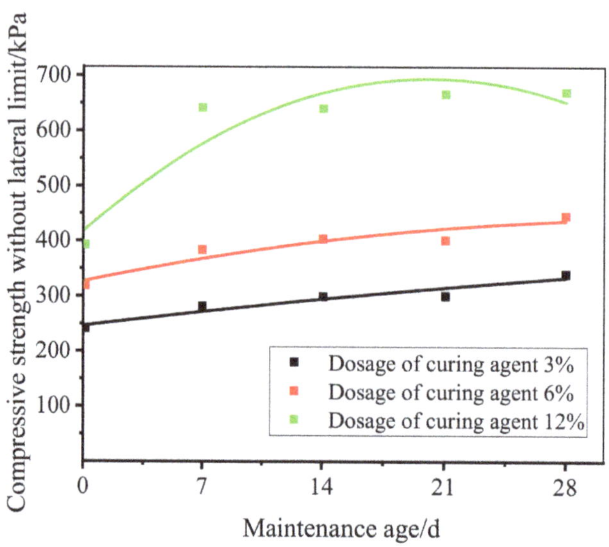

Figure 5. Relationship between unconfined compressive strength of soil samples at different ages.

From Figure 5, it is known that the variation of the unconfined compressive strength of cured soil is related to its curing age, and the longer the curing age, the greater its unconfined compressive strength. The data obtained were fitted and the curve for the 3% curing agent incorporation rate was obtained as follows: $y = 3.0485x + 248.54$, $R^2 = 0.9345$; the curve for the 6% curing agent incorporation rate was fitted as follows: $y = 3.9867x + 334.19$, $R^2 = 0.8846$; the curve for the 12% curing agent incorporation rate was fitted as follows: $y = -0.6772x^2 + 27.245x + 419.61$, $R^2 = 0.8846$; the curve for the 12% curing agent incorporation rate was fitted as follows: $y = -0.6772x^2 + 27.245x+ 419.61$, $R^2 = 0.8862$.

3.5. Deformation Modulus of Consolidated Soil

The soil deformation modulus E50 can be obtained from the stress-strain curve. To study the soil deformation modulus in-depth, the deformation modulus of cured soil was compared for five different curing agent admixtures and five different contamination concentrations as shown in Table 3 below.

Table 3. Deformation modulus of cured soil at different contamination concentrations E50.

Curing Agent Incorporation Rate	Modulus of Deformation of Contaminated Soil at Different Concentrations E50 (kPa)				
	0%	0.05%	0.1%	0.2%	0.3%
0%	11,236.8	7646.93	4101.53	2235.26	4380.19
3%	/	11,705.08	5033.19	10,762.38	9551.88
6%	/	10,904.57	6352.83	3511.53	14,046.1
12%	/	22,076	5652.93	2200	9539.25

As can be seen in Table 3, the E50 before and after curing showed different levels of reduction compared to the plain soil. Before curing, when the contamination concentration of the soil increases, its E50 decreases significantly. With the addition of curing agent admixture it can be found that the deformation modulus of the contaminated soil is bounded by the contamination concentration of 0.1%, and the E50 increases with the increase of curing agent admixture when the contamination concentration is 0.05%, i.e.

By measuring the relationship between the unconfined compressive strength and E50 at different contamination concentrations, the data were compiled and analyzed, and the relationship between the soil E50 and the unconfined compressive strength was plotted in Figure 6. the ratio of the two was (10–48) qu. when the unconfined compressive strength increased, the E50 of the cured soil also increased, and it was in a positive proportion. The Technical Specification for Construction Foundation Treatment (JGJ79-2012) specifies that the deformation modulus of cement soil is taken as (100–120) qu [15]. Liu Songyu suggested that the E50 of the cured soil used for highway foundations should be taken as 126 qu [16]. Xu Zhijun et al. found that the E50 of three different soils in Japan is recommended to be taken as 150 qu [17]. Cai Guanghua et al. studied MgO carbonated soils by unconfined compressive strength tests and found that their E50 was (25–250) qu [18]. According to the studies of other scholars, it can be seen that the E50 measured in this paper is smaller, which is because when the soil sample in this paper is subjected to external forces, it is easy to produce larger deformation than the E50 is also smaller.

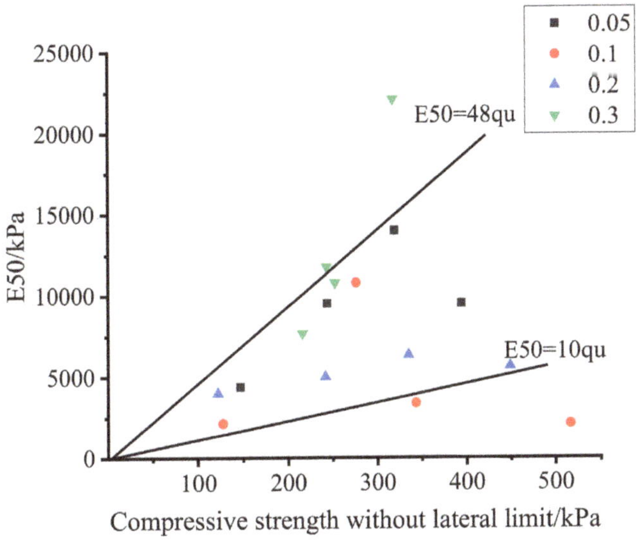

Figure 6. Relationship between modulus of deformation and unconfined compressive strength.

4. The Resistivity of Solidified Soil

4.1. Initial Resistivity of Solidified Soil

To measure the relationship between resistivity and contamination concentration, the control variable method was used to obtain their evolution laws under different condition factors (contamination concentration, curing agent incorporation rate). The resistivity both decreases with the increase of heavy metal ion concentration, the initial resistivity of the soil is related to the heavy metal ion content, when the red clay is contaminated, the ion concentration increases, then the resistivity decreases; at the same contamination concentration, the resistivity decreases with the increase of curing agent admixture, this is because when the curing reaction occurs, the soil pores are filled, when the soil pores become less, making the soil structure more dense, so resistivity decreases [19]. The strength of the cured soil is affected by the curing reaction and the pore structure, when some of the crystals generated by the reaction from a network structure inside the soil, form the skeleton of the cured soil, resulting in a stable soil structure, increased strength and decreased resistivity. To study the effect of different curing agent doping on the resistivity of red clay, the variation of initial resistivity with contamination concentration under different curing agent doping conditions was plotted in Figure 7.

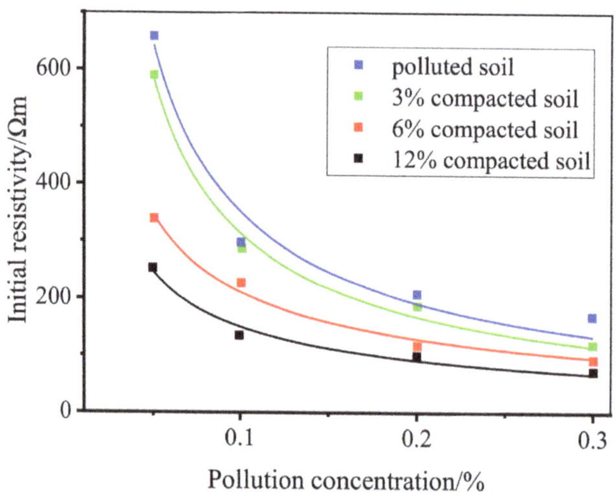

Figure 7. Relationship between different pollution concentrations and initial resistivity.

It can be seen from the above figure that when the heavy metal content of the soil increases, the overall trend of the initial resistivity of the soil is decreasing, and the obtained data are fitted. The fitting curves of different curing agent incorporation rates are obtained as follows: at 0% admixture, y = 2.0645x, $R^{-0.7432}$ = 0.9524; at 3% admixture, y = 0.9652x, $R^{-0.8382}$ = 0.9884; at 6% admixture, y = 1.166x, $R^{-0.7512}$ = 0.9767; at 12% admixture, y = 1.8724x, $R^{-0.6362}$ = 0.9743.

4.2. Damage Resistivity of Solidified Soil

To study the effect of zinc contamination concentration on the red clay destruction resistivity before and after curing, the variation of red clay destruction resistivity with zinc contamination concentration before and after curing was plotted, as shown in Figure 8.

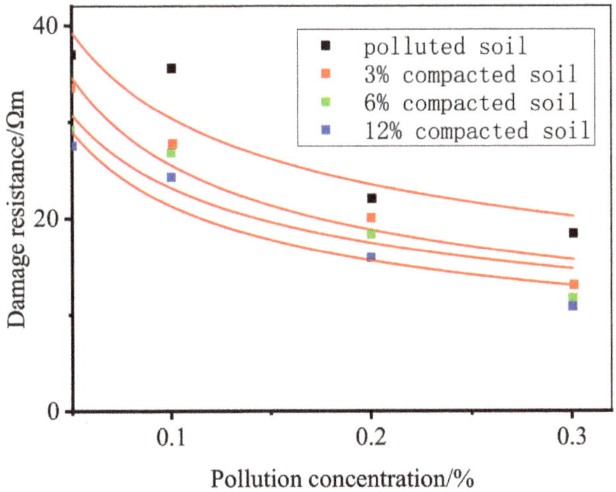

Figure 8. Damage resistivity-pollution concentration relationship curve.

It can be seen from Figure 8 that when the content of zinc ions in the soil increases, its damage resistivity is decreasing, which is because when zinc ions enter the soil, they start to react with the cementing material in the soil and destroy the local structure of the soil, but they will combine with the pore water to form new agglomerates and fill the soil pores, making the soil compact, resulting in the reduction of the pores of the red clay soil and its damage resistivity.

The data obtained were fitted to the following curves for different curing agent incorporation rates: $y = 1.7504x$, $R^2 = 0.9019$ at 0%; $y = 0.8001x - 0.501$, $R^2 = 0.9324$ at 3%; $y = 0.8599x - 0.478$, $R^2 = 0.8773$ at 6%; $y = 0.5872x - 0.521$, $R^2 = 0.9108$ at 12%.

5. Conclusions

This article analyzes the fixes of calcium phosphate and calcium oxide curing agents to repair zinc polluting red clay in different pollution concentrations, different curing agents, and unlimited pressure resistance, deformation modulus and resistivity of different maintenance ages. Get the following conclusions.

(1) The heavy metal Zn^{2+} pollution of red clay will lead to the decrease of the unconfined compressive strength of red clay. With the increase of pollution concentration, the unconfined compressive strength decreased to 16.5%, 52.6%, 50.5% and 43.4%, respectively. The unconfined compressive strength of red clay first decreased and then increased slowly with the increase of pollution concentration. In addition, the weakening effect will be different at each concentration, and the curing effect will also be different.

(2) The unconfined compressive strength of the cured soil increased under different maintenance age conditions, especially in the age of 0–7 day. After that, the unconfined compressive strength increased with the age and then gradually stabilized. In addition, the effect of the amount of curing agent on the compressive strength of the soil samples was particularly prominent when comparing the soil samples at the same age. The compressive strength of cured soil with high dosing is significantly higher than the compressive strength of cured soil with low dosing.

(3) Both before and after curing E50 showed different levels of reduction, before curing, when the contamination concentration of the soil increased, its E50 significantly reduced when the contaminated soil was cured, making the compressive strength of the soil increased, deformation modulus increased, the ratio of deformation modulus to compressive strength is (10–48) qu, when the unconfined compressive strength

increased, the E50 of the cured soil also increased, and a positive proportional relationship.

(4) Before and after the curing of contaminated soil, the resistivity of contaminated soil decreases with the increase of heavy metal ion concentration, and the resistivity of cured soil increases with the increase of curing agent incorporation rate at the same contamination concentration.

Author Contributions: Conceptualization, Y.S., H.C. and S.D. (Song Ding); methodology, S.D. (Song Ding); validation, Y.G., S.D. (Shuaishuai Dong) and H.C.; investigation, S.D. (Song Ding); writing—original draft preparation, S.D. (Song Ding); writing—review and editing, Y.S.; supervision, Y.G. and S.D. (Shuaishuai Dong); project administration, Y.S. All authors have read and agreed to the published version of the manuscript.

Funding: (1) Guangxi Innovation-Driven Development Special Project "Research and Demonstration of Key Technologies for Water Resource Utilization and Synergistic Development of Water Ecological Industry in Typical Karst Wetlands in the Lijiang River Basin" (Guike AA20161004-1); (2) National key research and development program subject "R&D and experimental demonstration of key technologies for water resource regulation in Karst wetlands in the Lijiang River Basin" (2019YFC0507502); (3) The National Natural Science Foundation of China Project "Research on the Collapse Mechanism of Karst Water-soil Coupling in Guilin under Extreme Climate Conditions" (41967037).

Institutional Review Board Statement: Not applicable.

Informed Consent Statement: Not applicable.

Data Availability Statement: Not applicable.

Conflicts of Interest: The authors declare no conflict of interest.

References

1. Lv, H.B.; Ji, C.S.; Xie, C.; Zeng, Z.T.; Wang, Y.; Xu, Y.S. Effect of strong acid on physical and mechanical properties of Liuzhou red clay. *J. Guangxi Univ. Nat. Sci. Ed.* **2017**, *42*, 1465–1471. [CrossRef]
2. Zhang, Z.Y.; Zuo, S.Y.; Ji, Y.X. Experimental study on mechanical properties of red clay in in-situ and remodeled states. *J. Undergr. Space Eng.* **2017**, *13*, 1477–1482.
3. Chen, J.Y.; Liu, Z.K. Mechanisms of strength change of in situ and remodeled red clay. *J. Henan Univ. Sci. Technol. Nat. Sci. Ed.* **2019**, *40*, 53–59+65+7–8. [CrossRef]
4. Voglar, G.E.; Leštan, D. Solidification/stabilization of metals contaminated industrial soil from former Zn smelter in Celje, Slovenia, using cement as a hydraulic binder. *J. Hazard. Mater.* **2010**, *178*, 926–933. [CrossRef] [PubMed]
5. Li, J.M.; Chen, X.J.; Yang, Y.; Huang, X.; Song, Y. Experiment on the improvement of red clay by nano-graphite powder. *J. Henan Univ. Sci. Technol. Nat. Sci. Ed.* **2019**, *40*, 69–80. [CrossRef]
6. Chi, L.; Lei, T.; Caihuan, D.; Yongfeng, Z.; Yanxing, W. Experimental study of MICP technology combined with porous silica adsorbent material for the remediation of Zn-Pb composite contaminated soil solidification/stabilization. *Geotechnics* **2022**, *43*, 307–316. [CrossRef]
7. Dai, C.Y.; Feng, Z.H.; Fan, C.W. Analysis of factors influencing the compression characteristics of cement-cured zinc-contaminated soil. *J. Water Resour. Water Transp. Eng.* **2020**, 89–93.
8. Guo, M.S.; Pan, H.; Wang, F. Strength and leaching characteristics of red mud-based curing/stabilization of lead-zinc-cadmium contaminated soil. *J. Rock Mech. Eng.* **2021**, *40*, 2968–2975. [CrossRef]
9. Saeed, K.A.; Kassim, K.A.; Nur, H.; Yunus, N.Z.M. Strength of lime-cement stabilized tropical lateritic clay contaminated by heavy metals. *KSCE J. Civ. Eng.* **2015**, *19*, 887–892. [CrossRef]
10. Li, J.M.; Tang, S.B.; Chen, X.J. Analysis of the mechanical properties and mechanism of zinc ion-contaminated red clay. *Adv. Mater. Sci. Eng.* **2021**, *2021*, 1–12. [CrossRef]
11. Song, Y.; Wang, J.Q.; Chen, X.J.; Yu, S.-Z.; Ban, R.-L.; Yang, X.; Zhang, X.; Han, Y. Study the effects of dry-wet cycles and cadmium pollution on the mechanical properties and microstructure of red clay. *Environ. Pollut. Barking Essex 1987* **2022**, *302*, 119037. [CrossRef] [PubMed]
12. Ioannou, Z.; Dimirkou, A.; Golia, E.E.; Ioannou, A. Sorption of zinc by clinoptilolite-Fe $(NO_3)_3$ systems. *Commun. Soil Sci. Plant Anal.* **2009**, *40*, 240–258. [CrossRef]
13. Dimirkou, A.; Ioannou, Z.; Golia, E.E.; Danalatos, N.; Misios, I.K. Sorption of cadmium and arsenic by goethite and clinoptilolite. *Commun. Soil Sci. Plant Anal.* **2009**, *40*, 259–272. [CrossRef]
14. Feng, Y.S.; Wang, S.; Zhou, L.; Xia, W.Y.; Ge, Y.X.; Zhong, D.X.; Du, Y.J. Stability study of cured stabilized heavy metal contaminated clay under natural exposure and indoor maintenance conditions. *J. Geotech. Eng.* **2021**, *43*, 154–157.

15. *JGJ79-2012*; Technical Specification for Building Foundation Treatment. China Architecture and Building Press: Beijing, China, 2021.
16. Liu, S.Y. *Highway Foundation Treatment*; Southeast University Press: Nanjing, China, 2009.
17. Xu, Z.J.; Cao, M.B. *Cement Soil Mixing Method for Treating Foundations*; Machinery Industry Press: Beijing, China, 2004.
18. Cai, G.H.; Liu, S.Y.; Cao, J.J. Effect of initial water content on the strength and resistivity of MgO carbonized chalk. *Chin. J. Highw.* **2017**, *30*, 18–26. [CrossRef]
19. Wang, X.H.; Ma, J.B.; Wang, Y.N.; He, B.L. Reinforcement of calcium phosphate cement with phosphorylated chitin. *Chin. J. Polym. Sci.* **2002**, *20*, 325–332.

applied sciences

MDPI

Article

Effect of Lead Ion Contamination on the Microstructure of Guilin Red Clay

Yu Song [1,2], Shuaishuai Dong [1,2], Song Ding [1,2], Yukun Geng [1,2], Yaoyi Huang [1,2] and Hui Li [1,2,*]

1 College of Civil and Architectural Engineering, Guilin University of Technology, Guilin 541004, China
2 Guangxi Key Laboratory of Geomechanics and Geotechnical Engineering, Guilin 541004, China
* Correspondence: 1020200597@glut.edu.cn

Abstract: The heavy metal contamination of red clay in Guilin is a serious problem. Lead ions pollute red clay and have a series of effects, which affect the macroscopic properties of red clay. However, fundamentally, the effects occur because the internal microstructure of red clay is eroded by Pb^{2+}, which results in the change in the macroscopic properties of red clay. Therefore, we adopted a mercury injection experiment and used electronic microscope Pb^{2+} to explore the microscopic mechanism through which red clay is internally influenced. From the mercury injection experiment, we found that an increase in the concentration of Pb^{2+} increased soil pore diameter and volume, and that a higher heavy metal content of Pb^{2+} had a greater effect on red clay cementation. Using scanning electron microscopy, we found that when the micro-image magnification was 500 and 20,000 times, the inside of the red clay pore increased with the increase in the concentration of Pb^{2+}, showing that the heavy metal within the microstructure damaged the red clay. The above two experiments showed that heavy metal ions increase the intergranular fractures of red clay, and the thickness of the double layer reduces, which results in the weakening of the interaction force between particles.

Keywords: lead ion contamination; mercury injection experiment; scanning electron microscope experiment; microscopic mechanism

1. Introduction

Red clay has a high porosity ratio, water content, and liquid limit; at the same time, it has high strength and low bearing capacity for the compressibility of special clay. It is mainly distributed in two broad areas in Guangzhou, Guangxi Zhuang Autonomous Region, Guizhou, Yunnan, and other places. Its unique engineering properties have also attracted wide attention from experts, scholars, and researchers. In 2012, Lv used X-ray diffraction, total chemical element analysis, and other methods to analyze the minerals of local red clay in Guilin, and research results showed that kaolinite, trilbite, and quartz were the main mineral components of red clay [1]. In 2015, Dong Wei explored the effect of different dispersants on the particle size analysis test results of red clay and found the unique aggregate cementation structure produced by the joint attraction of particles and free iron oxide [2].

The Guangxi Zhuang Autonomous Region is known as the "hometown of nonferrous metals", with rich nonferrous metal resources, such as lead, zinc, antimony, mercury, silver, and other heavy metals. However, due to unadvanced mining technology and improper personnel management, the environment inside and outside the mines are polluted; the soil is especially polluted with heavy metal ions. For example, Wang Yinghui investigated the waste soil of the Saiding lead–zinc mine in Guangxi and found that the lead content in the soil was high [3]. Li Jincheng conducted an investigation on the Daxin zinc mining area in Guangxi, and the results showed that the contents of heavy metals in the waste area significantly changed, and the main pollution factor, in addition to Zn, was Pb [4]. Chen Xuejun added lignin to red clay in 2021, which improved the particle structure of soil [5].

In 2020, Li Jiaming found that the influence of heavy metals on the mineral composition of red clay changed, which reduced the viscosity of red clay and damaged the structure of red clay particles [6].

In 1983, Tan Luorong made a developmental summary of the previous soil microstructure [7]. In 1994, Liao Yiling and Yu Peihou reported the intergranular bonding and van der Waals forces of red clay and constructed a generalization model of the microstructure of red clay [8]. Zhou Xunhua found in 2004 that the microstructure change in red clay was caused by the change in the cementation of free iron oxide [9]. In 2014, Lu Haijun's SEM results showed that heavy metal Pb^{2+} promoted the increase in soil pores, and the pore size also increased with the increase in pollution concentration [10]. Jiang's research in 2015 showed that under the influence of lead ions, the cohesion degree of clay minerals in the soil and the thickness of the double electric layer significantly changed, thus changing the microstructure of the soil [11]. Zhang Tingting's mercury injection test in 2016 showed that the pore structure of soil polluted by nonferrous heavy metals was reduced, the agglomeration of soil particles was obvious, the soil bond strength was improved, and the aggregation of soil particles was increased [12]. According to the mercury injection test results, Lv Haibo in 2020 compared the pore size distribution curves of soil samples and found that the partial pores with constant pore size represented the internal pores of the aggregates [13]. Jin Pan, in 2020, obtained the pore distribution through mercury injection experiments, indicating that dry density has a good water retention behavior for red clay within a wide suction range [14].

To explore the effects of heavy metal pollution on the microstructure of the red clay in guilin, we studied lead ion pollution, for example, under different lead ion concentrations and action times, prepared lead ions in indoor artificial pollution, reshaped guilin red clay, and performed electron microscope tests. The results of the mercury injection experiment showed that lead ion pollution reshapes the influence of the microscopic mechanism of guilin red clay.

2. Experimental Method

The source of the red clay was Yanshan District, Guilin City; its physical properties are shown in Table 1.

Table 1. Basic physical properties of red clay.

Natural Moisture Content (%)	Soil Particle Relative Density	Liquid Limit (%)	Plastic Limit (%)	Maximum Dry Densityg·cm^{-3}	Optimum Moisture Content (%)
27.3	2.69	62	47	1.55	30

The specimens for this test were prepared in accordance with the Geotechnical Test Procedure (GB/T50123–1999). The soil samples were air-dried and milled through a 2 mm sieve to keep the soil in reserve. The indoor test simulated the heavy metal lead ion contaminated red clay soil, and the concentration of lead ion contaminated soil was taken as 0%, 0.1%, 0.5%, and 1.0%, which we mixed into the red clay soil by spraying. The moisture content was measured after 24 h of sealing with sufficient mixing and controlled at 30 $\pm$ 1%. The reaction time of the solution with the soil was 5 days. After the mercury injection test, the sample was cut into small soil strips about 2 cm^3 in size. Before the test, the small soil strips were cut into 1 cm^3 soil blocks. In this process, the fresh section of the sample had to be kept from being worn. A conductive glue was affixed to the surface of the metal base of the sample, and then a small soil block was fixed on the conductive glue with tweezers, keeping the fresh section upward. Then, the sample was evacuated, and a layer of gold film was plated on the surface of the sample with a gold spray instrument to prevent the image from being unclear when the charge was accumulated and discharged.

Mercury injection test using the POREMASTER 33GT automatic mercury injection meter from the United States of America produced by Konta Instruments. The POREMASTER

33GT instrument's measured pressure was in the range of 140 KPa~231 MPa. This type of instrument was used to determine the pore diameter range of 0.007 μm~1000 μm. The sample was prepared using the Geotechnical Test Procedures (GB/T50123–1999), adding Pb^{2+} to the sample at a passivation time of 5 days. The experiment was set for a total of 4 groups, each group consisting of 2 parallel experimental specimens for a total of 8 test specimens. The 8 specimens were vacuum-saturated, and the central soil body was extracted; the size of the long strip was 1 × 1 × 2 cm. The specimens were then cooled by liquid nitrogen for 1 h, dried for 24 h, and taken out for mercury compression.

The SEM test equipment used in this study was an S-4800-type field emission scanning electron microscope jointly produced by Oxford Company and Japanese High-tech Company. Its basic parameters were as follows: the maximum accelerating voltage was 30 kV, the maximum magnification was up to 800,000 times, and the resolution was extremely high. The line resolution and point resolution were 0.14 nm and 0.19 nm, respectively. Four soil samples were available in total. This test involved magnifications of 500×, 5000×, and 20,000×.

3. Experimental Results and Analysis

3.1. Mercury Pressure Test and Analysis of Results

The basic principle of the mercury pressure test involves assuming a cylindrical pore of radius r. Mercury is pressed into the pore by pressure, and when equilibrium is reached, the pressure $P \cdot \pi r^2$ acting in the direction normal to the mercury contact ring cross-section is equivalently reversed with the component $2\pi r^2 \cdot \cos\alpha$ of the tension on the same cross-section normal to that surface, i.e., we have:

$$P \cdot \pi r^2 = -2\pi R^2 - \alpha \cos \alpha \tag{1}$$

The above equation yields:

$$p = -\frac{2\sigma \cos \alpha}{r} \tag{2}$$

where p is the required pressure; r is the pore radius; σ is the surface tension coefficient of mercury; α is the infiltration angle of mercury on the material.

The mercury pressure test is one of the most commonly used methods for conducting tests, such as for studying soil microstructure. The mercury intake versus pore distribution curve and cumulative mercury intake curve can be obtained, and microstructure pore distribution can be visualized through images. Because of metallic mercury's inability to enter closed pores, part of the fine pores can be easily overlooked. The results of the mercury pressure test on red clay contaminated with different concentrations of Pb^{2+} in soil samples contaminated for five days are plotted in Figures 1 and 2.

As can be seen in Figure 1a, the pore distribution of the red clay soil that was not contaminated with Pb^{2+} consisted of two peaks: the first peak area occurred at 0.01 μm~0.1 μm, and the second peak area occurred at 4 μm~20 μm. The pores in the 0.01 μm~0.1 μm region did not considerably change as the Pb^{2+} concentration in the soil rose, whereas the pores in the 4 μm~20 μm region dramatically changed as the Pb concentration in the soil rose. The magnified image in the range of 0.1 μm to 100 μm is a cut-off of Figure 1b from Figure 1a. The number of pores within the range of 0.2 μm~4 μm decreased as the heavy metal (Pb^{2+}) concentration inside the red clay grew, whereas the number of pores within the range of 4 μm~20 μm accordingly increased. This demonstrated that the heavy metal Pb^{2+} had a significant impact on the microstructure of the red clay, and that the complex physical and chemical properties of red clay cause the cementation of the soil to be eroded between the particles, and increased agglomeration and particle clusters, resulting in a smaller and gradually disappearing cementation structure between the particles. Therefore, the original small pores gradually enlarged into larger pores, and there was a significant increase in pores in the region of 4~20 μm, which also indicated that the microstructure was limited by the influence of heavy metal ions, and heavy metal ions were incapable of eroding the finer pore structure.

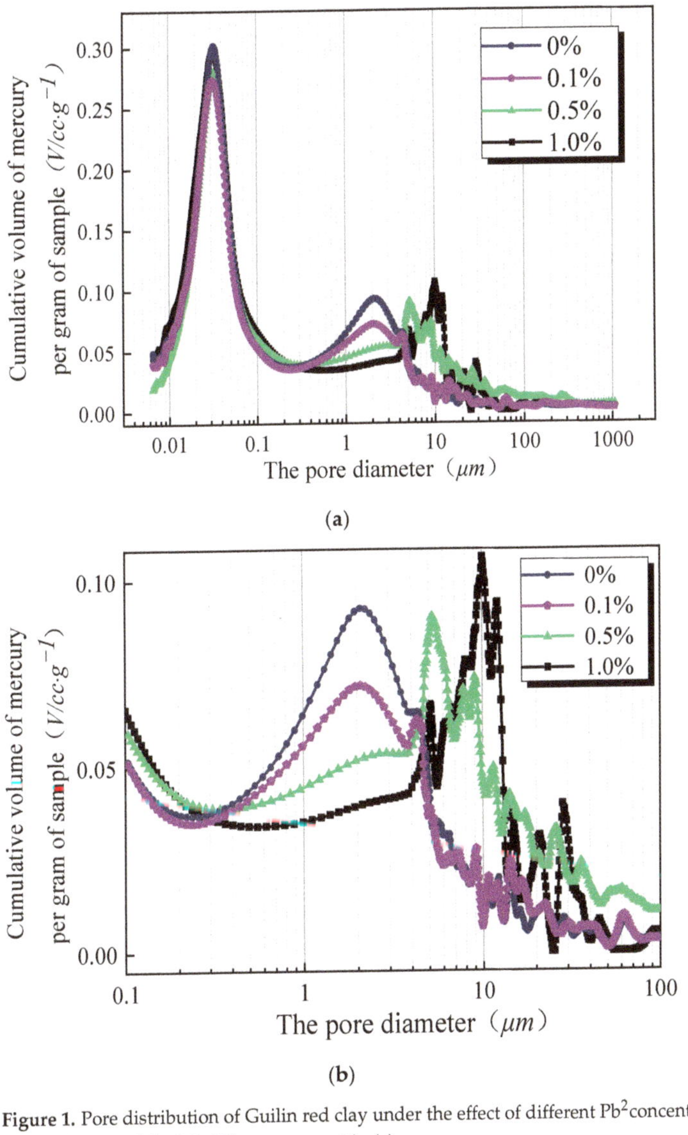

Figure 1. Pore distribution of Guilin red clay under the effect of different Pb2concentrations. (**b**) is an enlarged view of the 0.1–100 μm segment in (**a**).

Figure 2 depicts the relationship between the pore size of red clay and the accumulated mercury input volume for various concentrations of Pb^{2+}, which was mercury pressed into a long strip of the soil sample with external pressure. The accumulated mercury input volume corresponds to the unit horizontal coordinate in the figure, which is expressed as the incremental magnitude. With the increase of mercury injection pressure, the content of metal mercury in soil increases, which is consistent with this phenomenon. As shown in the figure, the initial stage occurred when the pore space (100~1000 μm) was large, and the volume of mercury in the soil contaminated with different concentrations of heavy metals was the same; during the rising stage, the increase in pressure in the pore space distribution range of 0.01~100 μm as accompanied by a large amount of mercury leaching into the red clay contaminated by Pb^{2+}. In the stable stage, the level of mercury pressure gradually reached its maximum and tended to stabilize. The metallic mercury injected into the red clay contaminated with various concentrations of heavy metals eventually stabilized and approached the nanoscale pores (range < 0.01 μm). Additionally, due to the increase in heavy metal (Pb^{2+}) concentration in the red clay, the accumulated mercury input in the red clay that was not contaminated with Pb^{2+} was less than that in the red clay contaminated with Pb^{2+}. This finding coincides with that depicted in Figure 1, where the pore space decdreased in the range of 0.2–4 μm and grew in the range of 4–20 μm. We confirmed that various concentrations of the heavy metal Pb^{2+} cause different internal pore damage in Guilin red clay. The principle is that Pb^{2+} erodes a portion of the cemented material within the red clay, causing the original small pores to enlarge. The result is an increase in the total pore volume of the soil.

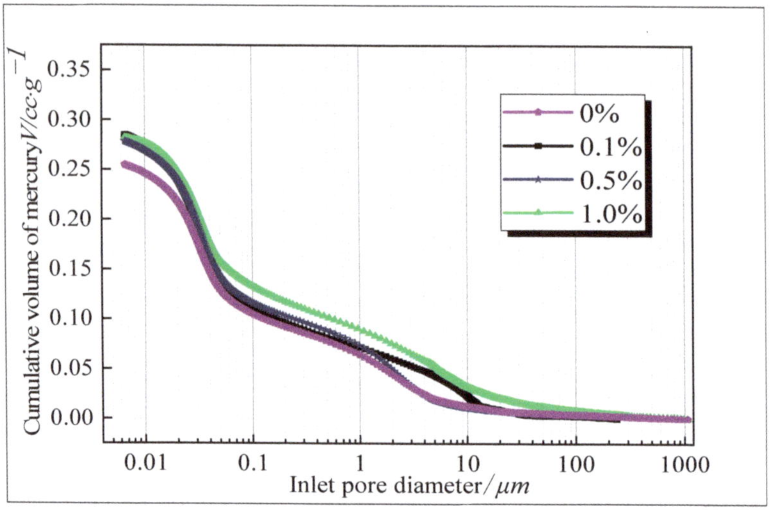

Figure 2. Pore size–cumulative mercury input relationship curve of Guilin red clay under the effect of different concentrations of heavy metal ions.

3.2. *Scanning Electron Microscopy Test and Analysis of Results*

Secanning lectron microscopy (SEM) uses a fine beam of electron probe, irradiates the surface of a sample, and accepts the reflected secondary electrons, which are transformed through a series of processing into a picture of the microstructure of the sample. It is a common method used for studying microstructure, and the SEM process is much easier to perform due to the ease of the sample preparation. SEM has a higher resolution and has been widely used in major disciplines. For the mercury pressure test, we used the same specimens, and liquid nitrogen was used to freeze-dry the test soil. Samples were directly taken around the mercury pressure soil samples. The more flat and uniform small pieces of soil were used for the electron microscope scanning test. The sample is illustrated in Figure 3.

Figure 3 depicts the microstructure map of Pb^{2+}-contaminated soil under the action of different concentrations for 5 days using electron microscopy, when the soil was magnified to 500 times. Soil is a complex three-phase soil (solid–liquid–gas); the solids in the soil are different; the size of different substances and their different nature result in the conductivity of the soil also varying, so it was necessary to spray high-conductivity substances on the soil. The surface of the soil contained some potholes and inequalities. These factors affected the final imaging effect of the microscopic result map. As seen in Figure 3, the corresponding black areas were produced in the corners of the image, which were also caused by the aforementioned factors. Even though it was not possible to binarize the image of the electron microscope at 500 times magnification, in general, the phenomenon of fissures in Guilin red clay increased with the increase in heavy metal Pb^{2+} concentration when the image was magnified to 500 times, which was caused by the erosion of the cementing material in the soil caused by the entry of Pb^{2+} into the interior of the soil. This is consistent with the results of the previous mercury pressure test described in Equation (1). When the soil sample was magnified 20,000 times with an electron microscope, it resulted in the original and binarized processed images depicted in Figure 4.

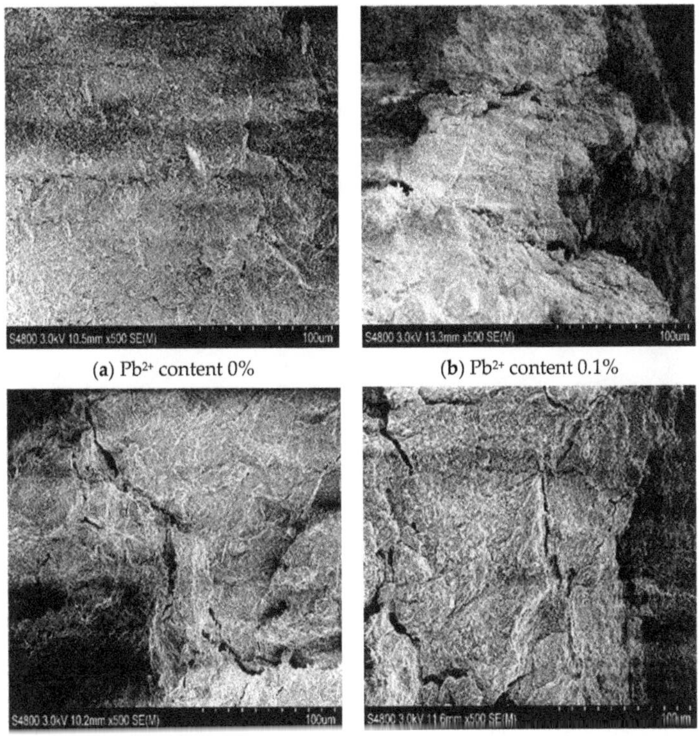

(**a**) Pb^{2+} content 0% (**b**) Pb^{2+} content 0.1%

(**c**) Pb^{2+} content 0.5% (**d**) Pb^{2+} content 1.0%

Figure 3. Microstructure of soil after 5 days of Pb^{2+} contaminations at different concentrations (500×).

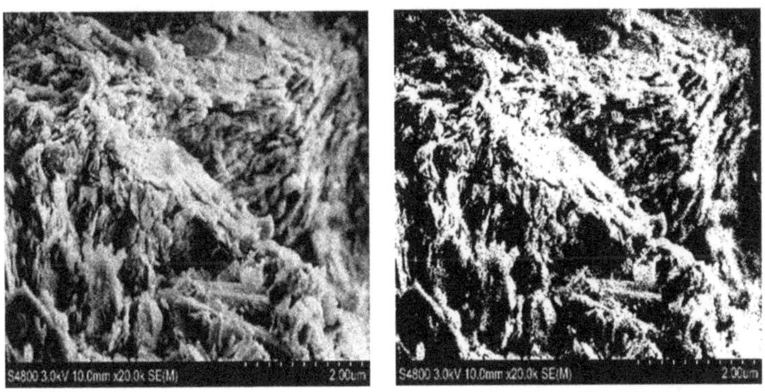

(**a**) Pb^{2+} content 0% (original figure) (**b**) Pb^{2+} content 0% (binarized)

Figure 4. *Cont.*

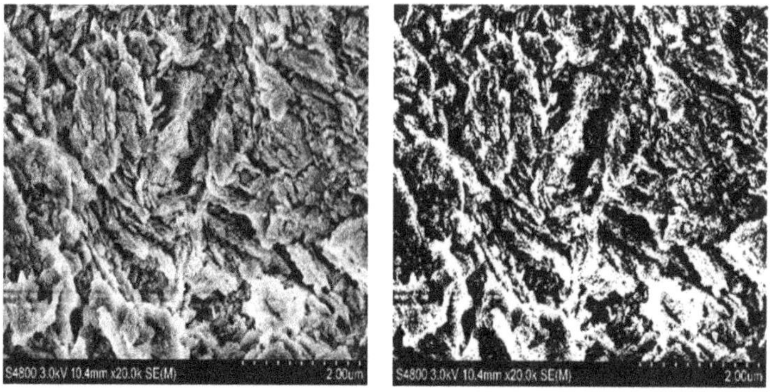

(**c**) Pb^{2+} content 0.1% (original figure) (**d**) Pb^{2+} content 0.1% (binarized)

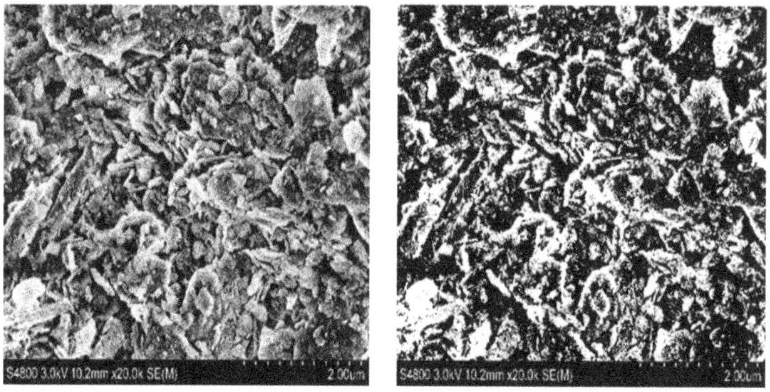

(**e**) Pb^{2+} content 0.5% (original figure) (**f**) Pb^{2+} content 0.5% (binarized)

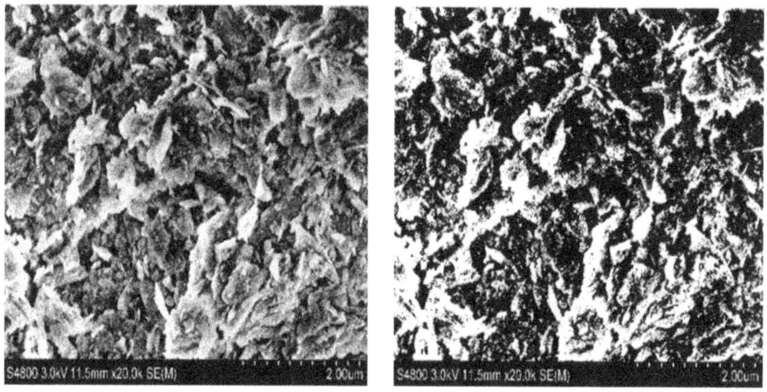

(**g**) Pb^{2+} content 1.0% (original figure) (**h**) Pb^{2+} content 1.0% (binarized)

Figure 4. Microstructure of soil after 5 days of contamination with different concentrations of lead ions (20,000×).

Figure 4 shows that when the electron microscope was magnified 20,000 times, the concentration of lead increased, the pores on the clay surface gradually developed and arranged in a point-to-point manner, and the soil particles were loosely arranged. The unit presented in the figure is at the particle level, which represents the spacing between clay particles and the tightness of the arrangement. The microstructure of the remodeled red clay soil was relatively complex. Because the main mineral components of Guilin red clay are kaolinite, trilbite, goethite, and quartz, the bulk, strip, and scale shapes could be seen in the electron microscope scanning. Using MATLAB to select different gray levels of the original image in Figure 4a,c,e,g, binarization processing for the image Figure 4b,d,f,h, and two functions Figure 4b,d,f,h, we concluded that the porosity was the black and white pixels in the binary images. The number calculated according to the proportion of its porosity is shown in Table 2.

Table 2. Porosity of Pb^{2+}-contaminated Guilin red clay.

Heavy Metal Pollution Concentration	0%	0.1%	0.5%	1.0%
Porosity	56.82%	59.62%	62.97%	63.73%

As seen in Table 2, with the increase in heavy metal ion concentration, the binary porosity data of Guilin remolded red clay at 20,000 times magnification obtained by SEM represented the spacing relationship between the clay minerals. According to the theory of electric double layers, the electric double layer is formed by a fixed layer and a diffusion layer [15]. When heavy metals enter the soil, the chemical composition of the soil internally changes, the ionic balance is broken, the negative charge on the surface of the soil particles of heavy metal cations increases, the negative charge on the surface of the soil particles of heavy metal cations increases,, and the thickness of the electric double layer diffusion is affected. and the thickness of the electric double layer diffusion is affected. The attraction between soil particles weakens; that is, the distance between clay minerals increases with the increase in heavy metal ion concentration, proving that Guilin red clay is reshaped when heavy metal ions enter it. Due to the internal changes, the original ion balance inside the body of the soil breaks, decreasing the particle zeta potential. The thickness of the electric double layer diffusion and the attraction between soil particles decreased, and the distance between the soil particles increased. The result is similar to the role of lubrication under the same effect of normal stress. The friction between the particles weakened.

(1) The results of the mercury pressure test indicated that the heavy metal (Pb^{2+}) has a significant effect on destroying the microstructure of the soil. In addition, as the concentration of Pb^{2+} increases, the pore space continuously expands, and the small pore space gradually becomes large pore space, thereby increasing the volume of the soil unit. The smaller the pore structure, the weaker the erosion effect.
(2) The results of the electron microscope scanning test further illustrated the results of the mercury piezoelectric test. Under the influence of Pb^{2+}, an increase in the concentration of Pb^{2+} reduces the thickness of the red clay's double electric layer, and the interparticle interaction force likewise becomes weaker as a result, producing a similar lubricating effect.
(3) The results of the above two preceding experiments indicate that the microscopic mechanism of Pb^{2+} contaminated Guilin red clay is as follows: on the one hand, when the heavy metal ions enter the soil body, the soil body generates fissures, which increases the pore volume and causes the structure to loosen; on the other hand, the thickness of the double electric layer on the surface of the soil particles decreases, which weakens the particle-particle interaction. In conclusion, the macroscopic mechanical properties of the Pb^{2+} contaminated Guilin red clay are affected by two aspects.

Author Contributions: Y.S., S.D. (Shuaishuai Dong), S.D. (Song Ding), Y.G., Y.H. and H.L. conceived the presented ideas and elaborated the method. Y.S., S.D. (Shuaishuai Dong) and Y.H. wrote the manuscript. S.D. (Shuaishuai Dong), S.D. (Song Ding) and Y.G. implemented the methods and performed the experiments. Y.H. and H.L. supervised the research. All authors have read and agreed to the published version of the manuscript.

Funding: 1. Guangxi Innovation-Driven Development Special Project "Research and Development and Experimental Demonstration of Key Technologies for the Synergistic Development of Water Resources Utilization and Water Ecological Industry in Typical Karst Wetlands of Lijiang River Basin" (Guik AA20161004-1). 2. National Key Research and Development Program Project "Research and Development and Experimental Demonstration of Key Technologies of Water Resources Regulation in Karst Wetland of Lijiang River Basin" (2019YFC0507502). 3. National Natural Science Foundation of China "Study on Collapse Mechanism of Karst Soil and water coupling in Guilin under extreme Climate Conditions" (41967037).

Institutional Review Board Statement: Not applicable.

Informed Consent Statement: Not applicable.

Data Availability Statement: Not applicable.

Acknowledgments: All individuals have agreed to confirm.

Conflicts of Interest: The authors declare no conflict of interest.

References

1. Haibo, L.V.; Zhaotian, Z.E.; Guoqiang, Y.I.; Yanlin, Z.H. Analysis of mineral composition of Guangxi red clay. *J. Eng. Geol.* **2012**, *20*, 651–656.
2. Dong, W.; Wang, B.; Yao, Y.; Tian, P. Effect of different dispersants on the analytical test of red clay particles. *Sci. Technol. Eng.* **2015**, *19*, 165–168+187.
3. Wang, Y.; Chen, X.; Zhao, Y.; Qi, S. Characteristics of heavy metal contamination and dominant plant accumulation in soils of lead and zinc mining areas. Zinc mining areas. *J. China Univ. Min. Technol.* **2007**, *36*, 487–493.
4. Li, J.; Yin, R.; Luo, Y.; Lu, Y.; Zhang, L. Evaluation of soil heavy metal pollution in the Daxin manganese mining area of Guangxi. *Environ. Sci. Technol.* **2010**, *33*, 183–190.
5. Chen, X.; Ding, X.; Song, Y. Effects of lignin on physical and mechanical properties of red clay. *Sci. Technol. Eng.* **2021**, *21*, 5922–5928.
6. Li, J.; Chen, X.; Huang, X.; Xu, K. Mechanical properties and microstructure test of zinc-contaminated red clay. *J. Henan Univ. Sci. Technol.* **2020**, *41*, 99–104.
7. Tan, L. Overview and development of research on microstructure of soils. *Geotech. Mech.* **1983**, *1*, 73–86.
8. Liao, Y.; Yu, P. Microstructure of red clay and its probabilistic model. *J. Eng. Geol.* **1994**, *1*, 27–37.
9. Zhou, X.; Liao, Y. Colloidal chemical characteristics of structural linkages between red clay particles. *J. Guizhou Univ. Technol.* **2004**, *1*, 26–29.
10. Lu, H.; Liao, Z.; Wang, Q.; Zhao, Y.; Chen, W. Microstructure and deformation strength properties of lead-contaminated clay. *J. Rock Mech. Eng.* **2014**, *33* (Suppl. S2), 4252–4257.
11. Jiang, S.L.; Qiang, X.; Ping, W. Effect of lead (II) on the mechanical behavior and microstructure development of a Chinese clay. *Appl. Clay Sci.* **2015**, *105–106*, 92–199.
12. Zhang, T.; Li, J.; Wang, P.; Huang, X.; Xue, Q. Experimental study on the mechanical properties and microscopic mechanism of lead-contaminated soil cured by magnesium phosphate cement. *Geotechnics* **2016**, *37* (Suppl. S2), 279–286. [CrossRef]
13. Lv, H.; Xue, X.; Jiang, Y.; Yin, G.; Zeng, Z. Microstructure analysis of Liuzhou red clay based on mercury injection test and nitrogen adsorption test. *Sci. Technol. Eng.* **2020**, *20*, 14619–14624.
14. Jin, P.; Zhen, W.; Chen, B.; Sun, D.A.; Gao, Y.; Xiong, Y. Effect of microstructure on water retention behavior of lateritic clay over a wide suction range. *Geomech. Eng.* **2021**, *25*, 417–428.
15. Song, Y.; Wang, J.Q.; Chen, X.J.; Yu, S.Z.; Ban, R.L.; Yang, X.; Zhang, X.; Han, Y. Study the effects of dry-wet cycles and cadmium pollution on the mechanical properties and microstructure of red clay. *Environ. Pollut.* **2022**, *302*, 119037. [CrossRef] [PubMed]

MDPI AG
Grosspeteranlage 5
4052 Basel
Switzerland
Tel.: +41 61 683 77 34

Applied Sciences Editorial Office
E-mail: applsci@mdpi.com
www.mdpi.com/journal/applsci